Fluctuations of Glaciers 1965 - 1970

This volume continues the earlier
work published under the title

FLUCTUATIONS OF GLACIERS 1959-1965

Paris, IAHS - Unesco, 1967

FLUCTUATIONS OF GLACIERS

1965 - 1970

A contribution to the
International Hydrological Decade

Compiled for the
Permanent Service on the Fluctuations
of Glaciers of the IUGG-FAGS/ICSU

by Peter Kasser
Laboratory of Hydraulics, Hydrology and
Glaciology at the Federal Institute of Technology
Zurich

International Commission on Snow and Ice of the
International Association of Hydrological Sciences
and
Unesco, Paris
1973

Published jointly by the

International Association of Hydrological Sciences
19 rue Eugène-Carrière, 75010 Paris

and the

United Nations Educational, Scientific
and Cultural Organization
7 Place de Fontenoy, 75700 Paris

Printed by Vontobel Druck AG,
Feldmeilen, Switzerland

ISBN-92-3-101045-X (Unesco)

PREFACE

The first session of the Co-ordinating Council of the International Hydrological Decade adopted as a project of the IHD a proposal for the measurement of glacier variations throughout the world. The Co-ordinating Council also included in this project a related proposal by the General Assembly of the International Union of Geodesy and Geophysics held in August, 1960, in Helsinki, and designated the International Commission on Snow and Ice of the International Association of Hydrological Sciences to act as the technical secretariat for the project.

The study of recent glacier fluctuations constituting the present publication was entrusted by the IAHS to Mr. Peter Kasser at the Federal Institute of Technology, Zurich, Director of the Permanent Service on the Fluctuations of Glaciers, a member of the Federation of Astronomical and Geophysical Sciences (FAGS) of ICSU. Such a Service ensures greater continuity in these essential observations and permits their extension to all regions of the world which possess large glacial masses, in addition to the Arctic and Antarctic which are already the object of international campaigns and studies.

The study is, at the same time, a continuation, on an extended basis, of the international reports on the variations of glaciers for the periods 1894-95 to 1958-59 and 1959-60 to 1964-65, with addenda from earlier years. It assembles data from several hundred glaciers in 15 countries for the period 1965 - 1970 with addenda from earlier years.

In view of the importance the study will have in the implementation of the IHD programme of snow and ice investigations, it was decided that it should be published jointly by IAHS and Unesco. The opinions expressed in it are, of course, those of the author and do not necessarily reflect the views of Unesco.

The completion of the present study constitutes another outstanding example of co-operation between the IAHS, Unesco and the National Committees of the International Hydrological Decade and it is hoped it will contribute appreciably to the success of the Decade.

José A. da Costa
Secretary, Co-ordinating Council, IHD
Unesco

FOREWORD

This volume is a publication of the Permanent Service on the Fluctuations of Glaciers, which was established in 1967. The idea of a program of regular observations of glaciers had been suggested as early as 1773, and the International Commission on Glaciers was founded in 1894 with that objective. The successor to this commission, the International Commission on Snow and Ice (ICSI) of the International Association of Hydrological Sciences / International Union of Geodesy and Geophysics (IAHS/IUGG) began negotiations in the early 1960's which led to the creation of the Permanent Service. A major part of the success of this undertaking is due to Professor H. Hoinkes of the University of Innsbruck (President of ICSI 1964-67) and to the present Director, whose diligent and selfless work was essential to the formative stages; we also acknowledge with pleasure the understanding and cooperation of Unesco, IUGG, IAHS, the Swiss Glacier Commission, and many other interested people and organizations.

One of the three International Hydrological Decade snow and ice programs is Measurements of Glacier Variations on a World-Wide-Basis. Unesco and ICSI cooperated in the publication of a Technical Paper (No. 3, 1969) outlining this program, which dovetails closely with the work of the Permanent Service, a major difference being the limited time frame of the IHD (1965-1974). Results from the IHD program, Combined Water-, Ice -, and Heat Balances at Representative Glacier Basins also help the reports of the Permanent Service. The IHD program World Inventory of Perennial Snow and Ice Masses is providing the basic catalog of glacier sizes and volumes, against which these data on fluctuations can be referenced.

The Permanent Service on the Fluctuations of Glaciers is a member of the Federation of Astronomical and Geophysical Sciences (FAGS) of the International Council of Scientific Unions. It is operated by the Laboratory for Hydraulics, Hydrology and Glaciology of the Federal Institute of Technology, Zurich, under the Directorship of Prof. P. Kasser. A special grant from Unesco permitted the publication of a first volume in 1967 which served as a pilot study in preparation for the formal establishment of the Permanent Service. This second volume will be useful to all those who require data on variations of glacier termini, thickness or mass balance, and it should thus provide information of value to hydrology, meteorology, geophysics, and environmental studies in all lands blessed with high mountains or cool climates. We especially thank the National Correspondents who compiled the data - much of it unpublished - in their respective countries, and Prof. Kasser and his staff for putting it all together.

Directing Board of the Permanent Service
on the Fluctuations of Glaciers

International Commission on Snow and Ice (IAHS)

	Past Chairman	Chairman
Tacoma and Bristol, 1973	Mark F. Meier	John F. Nye

TABLE OF CONTENTS

LIST OF TABLES

LIST OF FIGURES

LIST OF ANNEXED MAPS

1 Glacier map of Southern Norway, scale 1 : 500 000, compiled from topographical maps revised by aid of air photographs mainly from the period 1958-68 by Tor Ziegler and Gunnar Østrem. Norges Vassdrags- og Elektrisitetsvesen, Oslo 1969.

2 Map of the Mattmark Glaciers, scale 1 : 10 000, contour intervals 10 metres, surveys 1956 and 1967, including precipitation, specific net balance values and surface velocities. Jointly published by Topographical survey of Switzerland, Berne, Laboratory of Hydraulics, Hydrology and Glaciology at the Federal Institute of Technology, Zurich, Water power plant Mattmark AG, Saas Grund and Swiss Glacier Commission, Zurich. Berne 1971.

3 Map of the Vernagtferner, scale 1 : 10 000, contour intervals 10 metres, survey 1969, including the locations of hydrometeorological, glaciological and topographical observations. Published by the Commission for Glaciology of the Bavarian Academy of Sciences, Munich, 1972.

4 Maps (three sheets) with former states of the Vernagtferner, scale 1 : 10 000, contour intervals 50 metres. Published by the Commission for Glaciology of the Bavarian Academy of Sciences, Munich, 1972.
Sheet 1: surveys 1889 and 1912
Sheet 2: surveys 1912 and 1938
Sheet 3: surveys 1938 and 1969.

5 Map of the Vernagtferner with bedrock topography and geological survey, scale 1 : 10 000, contour intervals 10 metres. Published by the Commission for Glaciology of the Bavarian Academy of Sciences, Munich, 1972.

EXPLANATIONS, REMARKS, COMMENTS AND REFERENCES

1 INTRODUCTION

This second volume of "Fluctuations of Glaciers 1965-1970" has been compiled by the Permanent Service on the Fluctuations of Glaciers (PSFG). The PSFG belongs to the International Union of Geodesy and Geophysics (IUGG) and is one of the eleven Permanent Services incorporated in the Federation of Astronomical and Geophysical Services (FAGS), an organization attached to the International Commission of Scientific Unions (ICSU). The Directing Board of the PSFG comprises the President, the three Vice-Presidents and the Secretary of the International Commission on Snow and Ice (ICSI), a commission of the International Association of Hydrological Sciences (IAHS) attached to the IUGG, and the Chairmen of the two ICSI Working Groups for the World Inventory of Perennial Ice and Snow Masses and the Combined Heat, Ice and Water Balances, and the Director of the PSFG.

The PSFG received the measurement figures partly from its National Correspondents, partly from collaborators who are not the official representatives of a country. The National Correspondents of the PSFG are:

Australia: Dr. U. Radok, Meteorology Department, University of Melbourne, Parkville 3052, Victoria, Australia.

Austria: Prof. Dr. Herfried Hoinkes, Institut für Meteorologie und Geophysik der Universität Innsbruck, Schöpfstrasse 41, A-6020 Innsbruck, Austria.

Canada: Dr. Olav Løken, Head of Glaciology Sub-Division, Inland Waters Branch, Department of Energy, Mines and Resources, 1170 Bonnie Crescent, Ottawa 5, Ontario, Canada.

Chile: Dr. Cedomir Marangunic, Departemento de Geologia, Casilla 13518, Correo, Santiago de Chile, Chile.

Finland: Dr. E. Palosuo, Merentutkimuslaitos, Tähtitornink 2, Helsinki 14, Finland.

German Federal Republic: Kommission für Glaziologie bei der Bayerischen Akademie der Wissenschaften, Marstallplatz 8, D-8 München 22, German Federal Republic (Secretary: Prof. Dr. Julius Büdel; Assistant: Oskar Reinwarth, Dipl. Met.).

Iceland: Dr. Sigurjón Rist, Hjardarhagi 60, Reykjavik, Iceland.

India: Shri S. Banerji, Director General, Geological Survey of India, 27 Chowringhee Road, Calcutta, India.

Italy: Prof. Dr. M. Vanni, Secretary of the Comitato Glaciologico Italiano, Piazza Adriano 17, Torino, Italy.

Japan: Dr. H. Higuchi, Disaster Prevention Research Institute, Kyoto University, Gokasho, Uji, Kyoto, Japan (for Japan and Japanese work in Nepal). Dr. A. Higashi, Applied Physics Department, Faculty of Engineering, Hokkaido University, Sapporo, Japan (for Japanese work in other regions).

Norway: Dr. Gunnar Østrem, Glaciology Section, Vassdrags- og Elektrisitetsvesen, Hydrologisk Avdeling, P.O. Box 5091, Majorstua, Oslo 3, Norway.

New Zealand: Dr. Peter Anderton, Ministry of Works, Glaciology Section, P.O. Box 1479, Christchurch, New Zealand.

Sweden: Dr. Valter Schytt, Naturgeografiska Institutionen, Drottninggatan 120, Stockholm Va, Sweden.

Turkey: Prof. Dr. Sirri Erinç, Istanbul Üniversitesi, Coğrafya Bölümü, Istanbul, Turkey.

United Kingdom: Prof. Dr. J. F. Nye, H.H. Wills Physics Laboratory, University of Bristol, Royal Fort, Tyndall Avenue, Bristol B 68 1 TL, United Kingdom.

United States of America: Dr. William O. Field, One West 81st Street, New York, New York 10024, United States of America.

Union of Soviet Socialist Republics: Dr. K. G. Makarevich, Alma Ata, Kalinina 69-A, Section of Geography of the Academy of Sciences of Kaz. Union of Soviet Socialist Republics.

As its title implies, Volume 2 contains primarily results of glacier fluctuations in 1965-1970. There are, however, a number of exceptions. Thus the published observations of the variations in the position of glacier fronts begin as a general rule in 1964, so as to connect up to Volume 1. For technical reasons the series of measurements from the Alpine countries Austria, Italy and Switzerland run only up to Autumn 1968. More recent results will be found in the publications cited in Chapter 3. New features in this volume are the results from the USA, USSR, West Irian, New Zealand, Heard Island and Deception Island, which were almost wholly absent from Volume 1. As far as possible, results for 1959-1965 have also been included for the glaciers in these regions. A few of the last series to be prepared extend into 1971.

The sequence of chapters and tables is indicated in the Table of Contents. The correspondents who supplied the data and the scientists who made the measurements are mentioned in the various chapters of text, where references to the literature are also given, together with particulars which could not be included in the table section and some supplementary comments.

In each chapter a brief introduction is followed by information arranged in the following order:

Northern Hemisphere		Southern Hemisphere
Canada	France	West Irian
USA	Germany	New Zealand
Iceland	Italy	Heard Island
Norway	Switzerland	Deception Island
Sweden	USSR	
Austria		

2 GENERAL INFORMATION ON THE OBSERVED GLACIERS

In accordance with Resolution No. 12 of the first session of the UNESCO-IHD Co-ordinating Council in 1965 (cf. "Fluctuations of Glaciers", Vol. 1, page 46), guidelines for the compilation of a "World Inventory of Perennial Ice and Snow Masses" were prepared by a Working Group of the ICSI/IAHS under the chairmanship of Prof. Dr. F. Müller. They were published jointly by the UNESCO and the IAHS in 1970 as No. 1 of the series "Technical Papers in Hydrology". On the basis of this publication, which contains an introduction followed by pilot studies for the south-west part of Axel Heiberg Island (C. S. L. Ommanney), for the glaciers in the Waputik Mountains of the Canadian Rockies (A. D. Stanley) and for the glaciers in the Eastern Himalayas (F. Müller), an inventory is at present being made in a number of other countries. A Status Report on the World Glacier Inventory, from which some of the information that follows is taken, was presented by F. Müller and C. S. L. Ommanney in the Symposium in Reading on World Water Balance (IAHS-UNESCO-WMO, 1970, IAHS Publication No. 94, 1970). This information was unfortunately not available during the preparation of this second volume of "Fluctuations". A reference will be made, however, to the published inventories known to the author under the various countries.

Chapter 2, "General Information on the Observed Glaciers", will contain all information which does not belong in the special chapters 3 - 7 or which relates to several of these chapters.

The following particulars apply to the various countries and regions.

Canada

The geographical co-ordinates are given in Tables 9.1.2, 9.2.2 and 9.2.3 for every glacier observed. The "Inventory of Canadian Glaciers" is being compiled under the direction of C. S. L. Ommanney, Head of the Glacier Inventory Section, Glaciology Subdivision, Inland Waters Branch, Department of Energy, Mines and Resources, Government of Canada, Ottawa.

USA

Fluctuations of Glaciers, 1964 - 1968: United States Report by William O. Field, National Correspondent, April 1970.
(cf. Tables 2.1, 2.2, 9.1.3, 9.2.4, 9.2.5, 9.3.1 and 9.5.1, Figures 2.1 and 2.2)

1. Introduction

The data on fluctuations of glaciers in the United States, or observed in Antarctica by United States scientists, during the period 1964 - 1968 are divided into four categories: change of terminus, change of thickness, mass balance, and hydrometeorological data. These are covered in Tables 9.1.3, 9.2.4, 9.2.5, 9.3.1 and 9.5.1. Selection of the glaciers listed in the four categories is not considered sufficiently systematic to permit a satisfactory statistical analysis of fluctuation trends.

2. Order of listing

The glaciers are grouped according to individual mountain ranges, and are listed proceeding from north to south and west to east. Locally the listing is clockwise around the head of each drainage basin. The numbers assigned each glacier are for this report only, and are not for identification in a national inventory. The distribution of the data is given in Table 2.1, and Figures 2.1 and 2.2.

3. Explanatory notes, including symbols for all tables

3.1. Geographical coordinates refer to a point near the present center of the ablation area.

3.2. Lengths of glaciers are measured along the flowline from the uppermost source to the terminus as shown on the latest topographic maps, with adjustments for subsequent changes in their termini. These figures are considered approximate and designed only to indicate the relative size of the various glaciers. Fractions of kilometers are omitted except for the very short glaciers.

3.3. Abbreviations for the names of organizations used in the tables, listed alphabetically, are as follows:

AGS	- American Geographical Society, New York, N.Y.
AINA	- Arctic Institute of North America, Montreal, Canada, and Washington, D.C.
IPS	- Institute of Polar Studies, The Ohio State University, Columbus, Ohio
MAZ	- Mazamas, Portland, Oregon
MC	- Muskingum College, New Concord, Ohio
MSU	- Montana State University, Missoula, Montana
NDGS	- North Dakota Geological Survey, Grand Forks, North Dakota
NPS	- National Park Service, Dept. of the Interior, Washington, D.C.
NPS (GB)	- Glacier Bay National Monument, National Park Service, Gustavus, Alaska
NSF	- National Science Foundation, Washington, D. C.
UC (INSTAAR)	- Institute of Arctic and Alpine Research, University of Colorado, Boulder, Colorado
UND	- Dept. of Geology, University of North Dakota, Grand Forks, North Dakota

USC&GS	- U.S. Coast & Geodetic Survey, Dept. of Commerce, Washington, D.C.
USFS	- U.S. Forest Service, District Office, Anchorage, Alaska
USGS (D)	- U.S. Geological Survey, Denver, Colorado
USGS (F)	- U.S. Geological Survey, Fairbanks, Alaska
USGS (S)	- U.S. Geological Survey, Water Resources Division, Sacramento, California
USGS (T)	- U.S. Geological Survey, Water Resources Division, Tacoma, Washington
USN	- U.S. Navy Southeast Alaska Aerial Survey, 1948
UW (AS)	- Dept. of Atmospheric Sciences, University of Washington, Seattle, Washington
UW (EE)	- Dept. of Electrical Engineering, University of Washington, Seattle, Washington
UW (G)	- Geophysics, University of Washington, Seattle, Washington

3.4. The sponsoring agencies maintain the files of original field data, unless otherwise indicated. The files at the AGS currently include those of World Data Center A: Glaciology.

4. The tables for glaciers in Alaska require some additional explanatory notes:

4.1. Only glaciers which presently terminate in Alaska have been listed. However, a few of these have all or parts of their sources in Canada.

4.2. Altitudes in Alaska are based on topographic maps with 100 ft. (30.5 m) contour intervals. In the case of termini which are between contours, the altitude range between the two contours is given. If considerable change has occurred since the survey on which the map is based, the year of the survey is given.

4.3. Much of the observational data in Alaska is derived from aerial photography, either as the primary source of information or to supplement terrestrial observations. These sources are indicated under "remarks". Sources are as follows:

USGS (T)	- vertical and aerial photography by Austin Post; 1964, 1965, 1966, 1967, 1968
AINA	- oblique photography by John Sater, 1964
USC&GS	- vertical photography, 1964
NPS (GB)	- oblique photography by Ranger Charles Janda of Glacier Bay National Monument, 1965, 1966

5. Acknowledgements

The National Correspondent for the United States is deeply indebted to Dr. Mark F. Meier and his colleagues at the Water Resources Division, U. S. Geological Survey, Tacoma,Washington, for compiling the mass balance and hydrometeorological data and for valuable suggestions in regard to the organization of the terminus and thickness change data. Responsibility for the presentation of the latter, however, rests with the National Correspondent.

Thanks are also due to the various investigators who have submitted data for this report. The aerial photographs taken each year by Austin Post, U. S. Geological Survey, Tacoma, have been a particularly valuable source of information.

April 1970.

The inventory of glaciers in the USA has started with the publication:

Inventory of Glaciers in the North Cascades, Washington, by Austin Post, Don Richardson, Wendell V. Tangborn and F. L. Rosselot - Geological Survey, Professional Paper 705 - A, United States Government Printing Office, Washington 1971.

Fig. 2. 1. USA - Index map of the observed glaciers

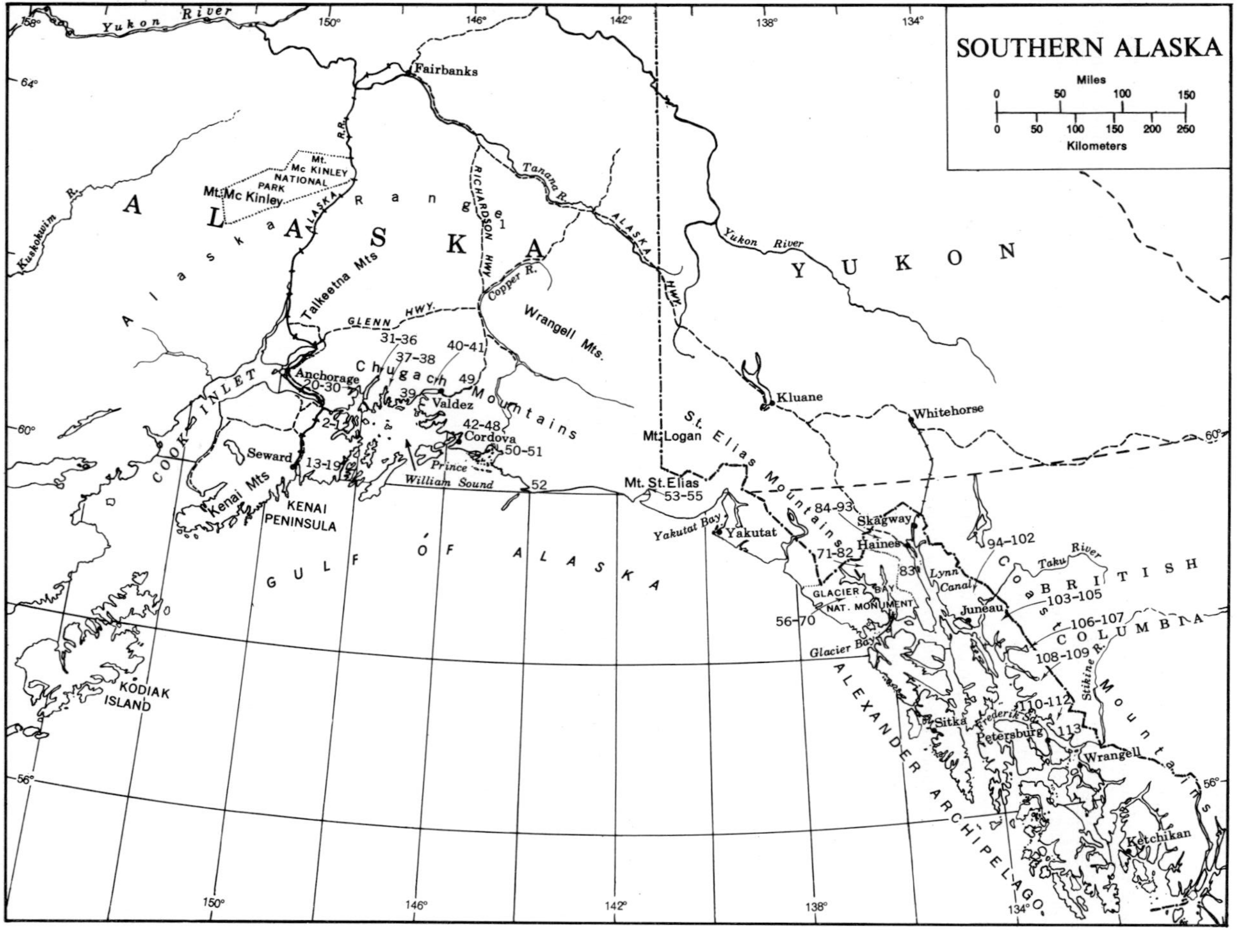

Fig. 2. 2. USA - Index map of the observed glaciers

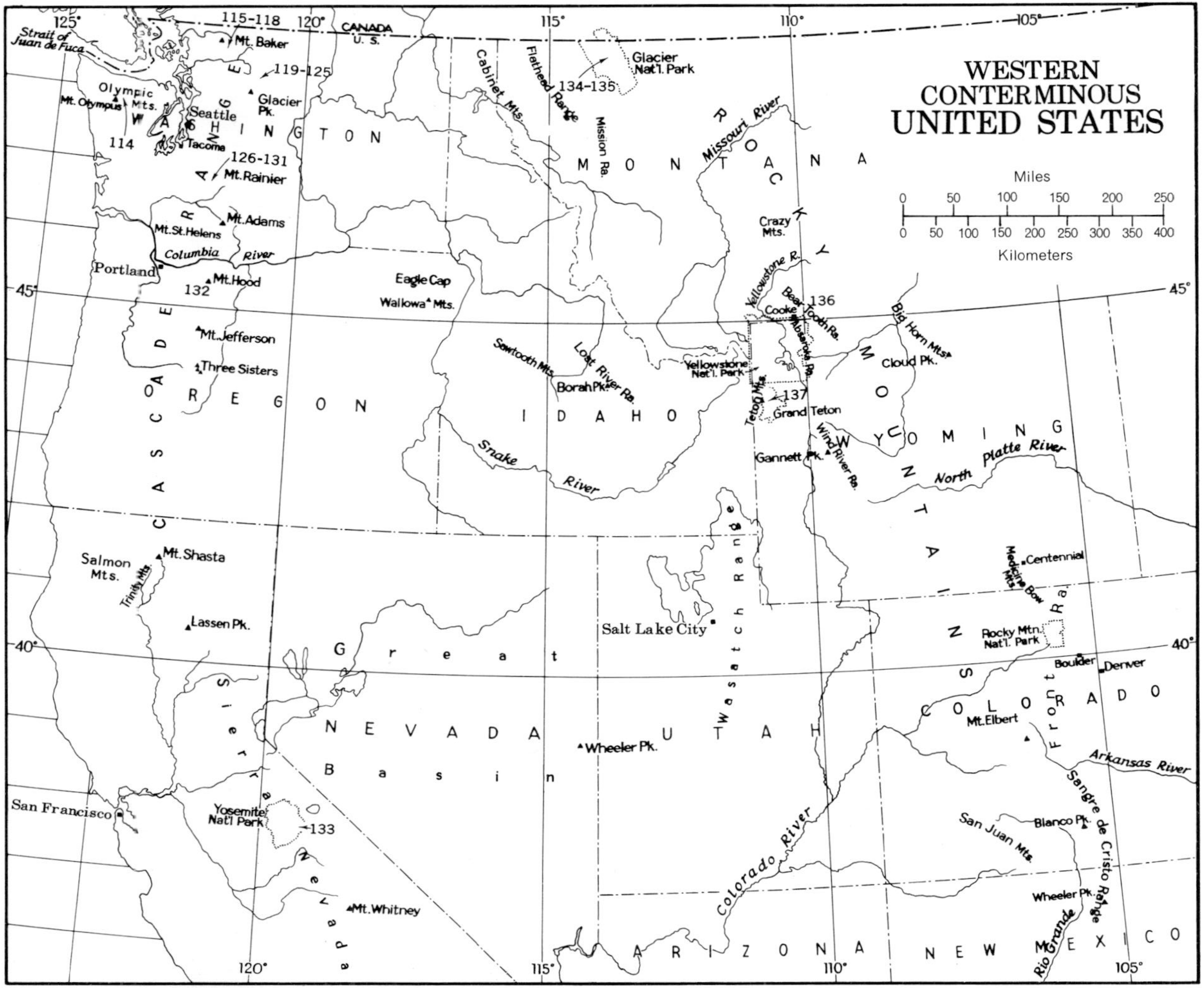

Table 2.1. USA - Distribution of Observations

State or Region	Mountain Range	Glacier numbers	Number of glaciers observed with respect to various parameters			
			Terminus change	Thickness change	Mass Balance	Hydrometeoro-logical Data
Alaska	Brooks Range	140			1	
	Alaska Range	1			1	
	Kenai Mts.	2 - 19	17		1	
	Chugach Mts.	20 - 51	32	2	1	
	Chugach/St. Elias Mts.	52 - 55	4			
	St. Elias Mts.	56 - 93	39			
	Coast Mts.	94 - 113	20			
Washington	Olympic Mts.	114			1	1
	Cascade Mts.	115 - 132	18	3	3	2
California	Sierra Nevada	133	1		1	1
Montana	Rocky Mts.	134 - 136	3	2	1	1
Wyoming	Rocky Mts.	137	1			
Colorado	Front Range	141			1	
Antarctica	Royal Society Range	138 - 139	2			
		Totals	137	7	11	5

Table 2.2 USA - Description of the observed glaciers.

Part 1 of 10

	No.	Name	Location	Lat.N	Long.W	Length km	Type of Glacier	Terminus: Type of Terminus	Terminus: Alt. m	Sponsoring Agency
ALASKA, KENAI MOUNTAINS	1	Gulkana	Alaska Range	63°18'	145°25'	6.7	branched valley glacier		1136	USGS
	2	Portage	Portage Pass	60°45'	148°48'	9	branched valley glacier	in lake, actively calving	30- 60	USFS
	3	Spencer	Placer River	60°40'	148°57'	19	outlet glacier	active ice on gently sloping terrain	30- 60	AGS
	4	Bartlett	Placer River	60°37'	148°57'	8	simple valley glacier	active ice on steep slope	200-220	AGS
	5	Trail	Trail Creek	60°33.5'	148°58'	10	branched valley glacier	active ice on gently sloping terrain	305-335	AGS
	6	Wolverine	Kenai Mts.	60°24'	148°55'	7.2	simple valley glacier		400	USGS (F)
	7	Taylor	Kings Bay	60°34.5'	148°37'	8	branched valley glacier	in shallow tidewater	0	AGS
	8	Tebenkof	Blackstone Bay	60°43'	148°29'	13	simple valley glacier	active ice on gently sloping terrain	0- 30	AGS
	9	Lawrence	Blackstone Bay	60°39.5'	148°36'	4.8	simple valley glacier	active ice on steep slope	0- 30	AGS
	10	Marquette	Blackstone Bay	60°38.5'	148°38'	3.8	simple valley glacier	active ice on steep slope	0- 30	AGS
	11	Beloit	Blackstone Bay	60°38'	148°41'	9	outlet glacier	in shallow tidewater , actively calving	0	AGS
	12	Blackstone	Blackstone Bay	60°39'	148°43'	11	outlet glacier	in shallow tidewater , actively calving	0	AGS
	13	Unnamed	Ice Bay	60°10'	148°26'	1.5	ice cap on ridge crest	active ice on steep slope	275-305	AGS
	14	Tiger	Ice Bay	60°10.5'	148°30'	11	branched valley glacier	in deep tidewater, actively calving	0	AGS
	15	Tigertail	Nassau Fiord	60°15'	148°24'	5	simple valley glacier	active ice on gently sloping terrain	0- 30	AGS
	16	Chenega	Nassau Fiord	60°17'	148°30'	22	outlet glacier	in deep tidewater, actively calving	0	AGS

Table 2.2. USA - Description of the observed glaciers.

Part 2 of 10

	No, Name	Location	Lat.N.	Long.W	Length km	Type of Glacier	Terminus: Type of Terminus	Terminus: Alt. m	Sponsoring Agency
KENAI MTS.	17 Nellie Juan	Port Nellie Juan	$60^{\circ}27'$	$148^{\circ}24'$	9	outlet glacier	in deep tidewater, actively calving	0	AGS
	18 Falling	Kings Bay	$60^{\circ}29'$	$148^{\circ}31'$	11	branched valley glacier	active ice on gently sloping terrain	0- 10	AGS
	19 Langdon	Kings Bay	$60^{\circ}25'$	$148^{\circ}38'$	6	branched valley glacier	in lake	30	AGS
ALASKA, CHUGACH MOUNTAINS	20 Toboggan	Harriman Fiord	$61^{\circ}01'$	$148^{\circ}17'$	38	branched valley glacier	active ice on gently sloping terrain	60	AGS
	21 Dirty	Harriman Fiord	$60^{\circ}57.5'$	$148^{\circ}25.5'$	3.4	branched valley glacier	active ice on gently sloping terrain	30- 60	AGS
	22 Harriman	Harriman Fiord	$60^{\circ}57'$	$148^{\circ}30'$	13	branched valley glacier	in shallow tidewater, negligible calving	0	AGS
	23 Roaring	Harriman Fiord	$60^{\circ}59.8'$	$148^{\circ}26.5'$	1.7	cirque glacier	active ice on steep slope	425 (1950)	AGS
	24 Cataract	Harriman Fiord	$61^{\circ}02'$	$148^{\circ}25'$	3.9	simple valley glacier	active ice on steep slope	300 (est)	AGS
	25 Surprise	Harriman Fiord	$61^{\circ}02'$	$148^{\circ}29'$	13	branched valley glacier	in deep tidewater, actively calving	0	AGS
	26 Baker	Harriman Fiord	$61^{\circ}04.5'$	$148^{\circ}22'$	3	cirque glacier	active ice on steep slope	275-305 (1957)	AGS
	27 Serpentine	Harriman Fiord	$61^{\circ}07'$	$148^{\circ}16'$	10	branched valley glacier	in shallow tidewater, negligible calving	0	AGS
	28 Cascade	Barry Arm	$61^{\circ}08'$	$148^{\circ}11'$	9	simple valley glacier	active ice on steep slope	0	AGS
	29 Barry	Barry Arm	$61^{\circ}10'$	$148^{\circ}05'$	24	simple valley glacier	in deep tidewater, negligible calving	0	AGS
	30 Coxe	Barry Arm	$61^{\circ}08'$	$148^{\circ}05'$	11	simple valley glacier	in shallow tidewater, negligible calving	0	AGS

Table 2.2. USA - Description of the observed glaciers.

Part 3 of 10

	No.	Name	Location	Lat.N	Long.W.	Length km	Type of Glacier	Terminus: Type of Terminus	Terminus: Alt. m	Sponsoring Agency
ALASKA, CHUGACH MOUNTAINS	31	Wellesley	College Fiord	61°12'	147°55'	6	simple valley glacier	in shallow tidewater	0	AGS
	32	Vassar	College Fiord	61°12.5'	147°52'	7	simple valley glacier	stagnant, heavy debris cover	0- 30	AGS
	33	Bryn Mawr	College Fiord	61°14'	147°49'	8	branched valley glacier	in shallow tidewater, actively calving	0	AGS
	34	Smith	College Fiord	61°16'	147°46.5'	10	branched valley glacier	in deep tidewater, actively calving	0	AGS
	35	Harvard	College Fiord	61°21'	147°35'	39	branched valley glacier	in deep tidewater, actively calving	0	AGS
	36	Yale	College Fiord	61°15'	147°35'	35	branched valley glacier	active ice on steep slope,in deep tidewater, actively calving	0	AGS
	37	Meares	Unakwik Inlet	61°11'	147°29'	25	branched valley glacier	in deep tidewater, actively calving	0	AGS
	38	Brilliant	Unakwik Inlet	61°07.5'	147°27'	3.7	simple valley glacier	active ice on steep slope	180-210	AGS
	39	Columbia	Columbia Bay	61°09'	147°05'	66	branched valley glacier	in deep tidewater, actively calving	0	AGS
	40	Shoup	Port Valdez	61°12'	146°32'	30	branched valley glacier	in shallow tidewater, negligible calving	0	AGS
	41	Valdez	Port Valdez	61°15'	146°10'	34	branched valley glacier	in lake	90-120	AGS
	42	Sheridan	Glacier River	60°36'	145°15'	24	branched valley glacier	in lake, negligible calving	40- 50	AGS
	43	Sherman	Glacier River	60°33.3'	145°08.8'	13	branched valley glacier	active ice on gently sloping terrain	112	IPS
	44	Saddlebag	Copper River (Delta)	60°31'	144°05.5'	8	simple valley glacier	in lake	83	AGS
	45	Childs	Copper River (Delta)	60°41'	144°55'	19	branched valley glacier	in river, actively calving	30- 40	AGS
	46	Allen	Copper River	60°48'	144°45'	31	branched valley glacier	active ice on gently sloping terrain	30- 60	AGS

Table 2.2. USA - Description of the observed glaciers.

Part 4 of 10

	No. Name	Location	Lat.N	Long.W	Length km	Type of Glacier	Terminus: Type of Terminus	Terminus: Alt. m	Sponsoring Agency
ALASKA, CHUGACH MOUNTAINS	47 Schwan	Tasnuna River	$60^{o}55'$	$145^{o}10'$	26	branched valley glacier	in lake, stagnant, heavy debris cover	90-120	AGS
	48 Woodworth	Tasnuna River	$60^{o}57'$	$145^{o}27'$	23	branched valley glacier	active ice on gently sloping terrain	120-150	AGS
	49 Worthington	Tsina River	$61^{o}10'$	$145^{o}46'$	6	simple valley glacier	in lake	660	AGS
	50 Miles	Copper River	$60^{o}38'$	$144^{o}15'$	52	branched valley glacier	in lake, actively calving	30- 60	AGS
	51 Slide	Martin River	$60^{o}30.5'$	$144^{o}18.5'$	10	branched valley glacier	stagnant, heavy debris cover	205	NSF,UND
	52[1] Bering	Gulf of Alaska	$60^{o}15'$	$143^{o}30'$	200	branched valley glacier	active ice on flat terrain	10	USGS (T)
	53[1] Guyot	Icy Bay	$60^{o}10'$	$141^{o}40'$	26	branched valley glacier	in deep tidewater	0	USGS (T)
	54[1] Yahtse	Icy Bay	$60^{o}15'$	$141^{o}25'$	54	branched valley glacier	in deep tidewater	0	USGS (T)
	55[1] Tyndall	Icy Bay	$60^{o}10'$	$141^{o}10'$	29	branched valley glacier	in deep tidewater	0	USGS (T)
St. ELIAS MTS.	56 Geikie	Geikie Inlet[2]	$58^{o}35.5'$	$136^{o}36.5'$	10	branched valley glacier	active ice on gently sloping terrain	ca.30	AGS
	57 Hugh Miller	Hugh Miller Inlet[2]	$58^{o}43'$	$136^{o}40'$	11	simple valley glacier	active ice on gently sloping terrain, stagnant little debris cover	10- 30	AGS
	58 Reid	Reid Inlet[2]	$58^{o}48'$	$136^{o}48'$	21	outlet glacier	in deep tidewater, actively calving	0	AGS

1) Glaciers 52 to 55 are considered by some to be in the Chugach Mountains and by others to be in the St. Elias Mountains.

2) Nos. 56 to 82 inclusive are in the Glacier Bay National Monument.

Table 2.2. USA - Description of the observed glaciers.

Part 5 of 10

	No.	Name	Location	Lat.N	Long.W	Length km	Type of Glacier	Terminus: Type of Terminus	Terminus: Alt. m	Sponsoring Agency
ALASKA, St. ELIAS MOUNTAINS	59	Lamplugh	Johns Hopkins Inlet 2)	58°50'	136°54'	31	outlet, branched valley glacier	in shallow tidewater, negligible calving	0	AGS
	60	Unnamed	Johns Hopkins	58°52.5'	136°59.5'	2	cirque glacier	active ice on steep slope	370-400	AGS
	61	Kashoto	Inlet 2)	58°51.5'	137°01'	4.2	simple valley glacier	in shallow tidewater, negligible calving	0	AGS
	62	Hoonah	Johns Hopkins	58°50'	137°02.5'	11	simple valley glacier	in shallow tidewater, negligible calving	0- 30	AGS
	63	Gilman	Inlet 2)	58°49'	137°04'	13	branched valley glacier	in deep tidewater, actively calving	0	AGS
	64	Clark	Johns Hopkins	58°48'	137°07'	1.4	slope glacier	active ice on steep slope	670-700	AGS
	65	Johns Hopkins	Inlet 2)	58°48'	137°10'	26	branched valley glacier	in deep tidewater	0	AGS
	66	Tyeen	Johns Hopkins Inlet 2)	58°52'	137°09'	5.3 - 7.4	simple valley glacier when not surging	22.8.65 and 5.4.66 active ice on steep slope	700 (1961)	AGS
							branched valley glacier when surging	23.7.66 and 17.9.66 on tidal flat 14.7.67 and 17.7.68 on tidal flat, stagnant little debris cover	0	AGS
	67	Toyatte	Johns Hopkins Inlet 2)	58°54'	137°06'	5.4	simple valley glacier	active ice on steep slope	ca. 75 (1967)	
	68	Margerie	Tarr Inlet 2)	59°00'	136°10'	39	branched valley glacier	in deep tidewater	0	AGS
	69	Unnamed	Tarr Inlet 2)	59°02.5'	137°06.5'	3	niche glacier	active ice on steep slope	305-335	AGS
	70	Grand Pacific	Tarr Inlet 2)	59°10'	137°16'	46	branched valley glacier	in shallow tidewater	0	AGS
	71	Romer	Rendu Inlet 2)	58°59'	136°44'	3.5	cirque glacier	active ice on steep slope	240-270	AGS
	72	Rendu	Rendu Inlet 2)	59°04'	136°49'	ca. 17	branched valley glacier	stagnant, heavy debris cover	0- 30	AGS
	73	Unnamed	Rendu Inlet 2)	59°03'	136°44'	3.2	slope glacier	active ice on steep slope	460-490	AGS

2) Glaciers 56 to 82 inclusive are in the Glacier Bay National Monument

Table 2.2. USA - Description of the observed glaciers.

Part 6 of 10

	No.	Name	Location	Lat.N	Long.W	Length km	Type of Glacier	Terminus: Type of Terminus	Terminus: Alt. m	Sponsoring Agency
ALASKA, St. ELIAS MOUNTAINS	74	Carroll	Queen Inlet 2)	59°05'	136°40'	ca. 42	branched valley glacier	active ice on gently sloping terrain	0- 30	AGS
	75	Cushing	Queen/Wachusett Inlets 2)	59°04'	136°28'	19	branched valley glacier	active ice on gently sloping terrain	210-270	AGS
	76	Baldwin	Wachusett Inlet 2)	58°55.5'	136°23'	ca. 3.5	simple valley glacier	active ice on steep slope	215 (1968)	AGS
	77	Plateau		58°59'	136°24'	ca. 2	branched valley glacier	in deep tidewater	0	
	78	Burroughs	Wachusett Inlet 2)	58°59'	136°18'	ca. 6.5	simple valley glacier	stagnant little debris cover	50	IPS,NSF
	79	Muir	Muir Inlet 2)	59°06'	136°23'	25	branched valley glacier	in deep tidewater	0	AGS
	80	Riggs	Muir Inlet 2)	59°06'	136°12'	27	branched valley glacier	in deep tidewater	0	AGS
	81	McBride	Muir Inlet 2)	59°05	136°03'	26	branched valley glacier	in shallow tidewater	0	AGS
	82	Casement	Adams Inlet 2)	59°01'	135°56'	32	branched valley glacier	active ice on gently sloping terrain	0- 30	AGS
	83	Davidson	Chilkat Inlet	59°05'	135°30'	20	branched valley glacier	in lake	0- 30	AGS
	84	Garrison	Chilkat River	59°10'	135°40'	12	branched valley glacier	stagnant, heavy debris cover	220-250	AGS
	85	Bertha	Takhin River	59°12'	135°48'	12	branched valley glacier	active ice on gently sloping terrain	110-150	AGS
	86	Unnamed	Takhin River	59°12'	135°52'	7	branched valley glacier	active ice on gently sloping terrain	150-180	AGS
	87	Unnamed	Takhin River	59°13'	136°00'	11	branched valley glacier	active ice on steep slope	305-335	AGS
	88	Takhin	Takhin River	59°15'	136°10'	11	branched valley glacier	active ice on gently sloping terrain	245-275	AGS

2) Glaciers 56 to 82 inclusive are in the Glacier Bay National Monument

Table 2.2. USA - Description of the observed glaciers.

Part 7 of 10

	No.	Name	Locatior	Lat.N	Long.W	Length km	Type of Glacier	Terminus: Type of Terminus	Terminus: Alt. m	Sponsoring Agency
St. ELIAS MTS.	89	Le Blondeau	Tsirku Fiver	$59^{\circ}15.5'$	$136^{\circ}14'$	13	branched valley glacier	active ice on gently sloping terrain	245-275	AGS
	90	Unnamed	Tsirku River	$59^{\circ}16'$	$136^{\circ}21.5'$	3	simple valley glacier	active ice on steep slope	365-395	AGS
	91	Unnamed	Tsirku River	$59^{\circ}16'$	$136^{\circ}23'$	3.7	branched valley glacier	active ice on steep slope	365-395	AGS
	92	Unnamed	Tsirku River	$59^{\circ}16.5'$	$136^{\circ}25'$	5.8	slope glacier	active ice on gently sloping terrain	365-395	AGS
	93	Tsirku	Tsirku River	$59^{\circ}20'$	$136^{\circ}35'$	ca. 22	branched valley glacier	active ice on gently sloping terrain	430-460	AGS
ALASKA, COAST MOUNTAINS	94	Eagle	Favorite Channel	$58^{\circ}37'$	$134^{\circ}43'$	14	branched valley glacier	in lake, negligible calving, stagnant heavy debris cover	60 (1948)	AGS
	95	Herbert	Favorite Channel	$58^{\circ}33'$	$134^{\circ}37'$	19	branched valley glacier	active ice on steep slope	30 - 60 (1948)	AGS
	96	Mendenhall	Stephens Passage	$58^{\circ}29'$	$134^{\circ}32'$	22	branched valley glacier	active ice on steep slope, in lake	0- 30	AGS
	97	Lemon Creek	Gastineau Channel	$58^{\circ}24'$	$134^{\circ}22'$	7	simple valley glacier	active ice on steep slope	400 (1948)	AGS
	98	Norris	Taku Inlet	$58^{\circ}27'$	$134^{\circ}11'$	24	branched valley glacier	active ice on gently sloping terrain	0- 30	AGS
	99	Taku	Taku Inlet	$58^{\circ}33'$	$134^{\circ}08'$	48	branched valley glacier	active ice on gently sloping terrain	0- 30	AGS
	100	Hole in the wall	Taku River	$58^{\circ}28'$	$134^{\circ}03'$	3	explained under remarks	active ice on gently sloping terrain	0- 30	AGS
	101	West Twin	Taku River	$58^{\circ}35'$	$133^{\circ}58'$	8	explained under remarks	in lake, actively calving	9	AGS
	102	East Twin	Taku River	$58^{\circ}35'$	$133^{\circ}53'$	10	explained under remarks	in lake, negligible calving	9	AGS
	103	Wright	Taku River	$58^{\circ}28'$	$133^{\circ}30'$	33	branched valley glacier	in lake, negligible calving	15- 25	AGS

Table 2.2. USA - Description of the observed glaciers.

Part 8 of 10

	No.	Name	Location	Lat.N	Long.W	Length km	Type of Glacier	Terminus: Type of Terminus	Terminus: Alt. m	Sponsoring Agency
ALASKA, COAST MOUNTAINS	104	Unnamed	Speel River	$58^{o}20'$	$133^{o}38'$	18	branched valley glacier	active ice on gently sloping terrain	150-180	AGS
	105	Speel	Speel River	$58^{o}20'$	$133^{o}25'$	15	branched valley glacier	in lake	150-180	AGS
	106	Sawyer	Tracy Arm	$57^{o}57'$	$133^{o}04'$	35	branched valley glacier	in deep tidewater, actively calving	0	AGS
	107	South Sawyer	Tracy Arm	$57^{o}49'$	$133^{o}00'$	48	branched valley glacier	in deep tidewater, actively calving	0	AGS
	108	North Dawes	Endicott Arm	$57^{o}36'$	$132^{o}57'$	16	branched valley glacier	active ice on gently sloping terrain	180-240	AGS
	109	Dawes	Endicott Arm	$57^{o}28'$	$132^{o}43'$	41	branched valley glacier	in deep tidewater, actively calving	0	AGS
	110	Baird	Thomas Bay	$57^{o}09'$	$132^{o}40'$	48	branched valley glacier	active ice on gently sloping terrain	0- 30	AGS
	111	Patterson	Thomas Bay	$56^{o}58'$	$132^{o}36'$	23	branched valley glacier	active ice on gently sloping terrain	30- 60	AGS
	112	Unnamed	Muddy River	$56^{o}54.5'$	$132^{o}35'$	13	simple valley glacier	active ice on steep slope	150-180	AGS
	113	Le Conte	Frederick Sound	$56^{o}53'$	$132^{o}22'$	37	branched valley glacier	in deep tidewater, actively calving	0	AGS
	114	Blue	Olympic Mts.	$47^{o}49'$	$123^{o}41'$	4.2	branched valley glacier		1275	UW (AS)

Table 2.2. USA - Description of the observed glaciers.

Part 9 of 10

Region	No.	Name	Location	Lat.N	Long.W.	Length km	Type of Glacier	Terminus: Type of Terminus	Terminus: Alt. m	Sponsoring Agency
WASHINGTON, CASCADE MOUNTAINS	115	Coleman	North Cascade Range 1)	48°47.7'	121°51.4'	4.5	slope glacier	active ice on gently sloping terrain	1265	Field data etc
	116	Roosevelt	North Cascade 1)	48°47.9'	121°50.5'	3.6	slope glacier	active ice on steep slope	1450	UW (EE)
	117	Park	Range 1)	48°47'	121°46'	5	slope glacier	active ice on steep slope	1200	USGS (T)
	118	Boulder	North Cascade 1)	48°46'	121°47'	4	slope glacier	active ice on steep slope	1400	USGS (T)
	119	Deming	Range 1)	48°45'	121°51'	6	simple valley glacier	active ice on gently sloping terrain	1170	USGS (T)
	120	Boston	North Cascade 1)	48°30'	121°01'	5	slope glacier	in lake	1500	USGS (T)
	121	South Cascade	Range 1)	48°22'	121°03'	3.2	simple valley glacier	in lake, negligible calving	1610	USGS (T)
	122	North Guardian	North Cascade 2)	48°08'	121°05'	2	slope glacier	active ice on gently sloping terrain	1890	USGS (T)
	123	Dusty	Range 2)	48°08'	121°05'	3	slope glacier	active ice on steep slope	1860	USGS (T)
	124	Chocolate	North Cascade 2)	48°07'	121°05'	4	simple valley glacier	active ice on steep slope	1680	USGS (T)
	125	White Chuck	Range 2)	48°04'	121°07'	2.5	slope glacier	stagnant little debris cover	2010	USGS (T)
	126	Carbon	Middle Cascade Range 3)	46°56'	121°47'	9.7	simple valley glacier	active ice on gently sloping terrain	1130	USGS (T)
	127	Emmons	Middle Cascade 3)	46°52'	121°41'	6.9	simple valley glacier	active ice on gently sloping terrain	1690	USGS (T)
	128	Cowlitz	Range 3)	46°49'	121°42'	6.7	branched valley glacier	active ice on gently sloping terrain	1740	USGS (T)
	129	Nisqually	Middle Cascade 3)	46°48'	121°44'	6.5	branched valley glacier	active ice on gently sloping terrain	1410	USGS (T)
	130	Kautz	Range 3)	46°49'	121°47'	4.8	branched valley glacier	active ice on steep slope	1890	USGS (T)
	131	South Tahoma	Middle Cascade 3)	46°49'	121°49'	4.5	simple valley glacier	active ice on gently sloping terrain	1640	USGS (T)
	132	Eliot	Range 4)	45°23'	121°41'	3.6	simple valley glacier	stagnant, heavy debris cover	1920	USGS (T)

1) Mt. Baker ; 2) Glacier Peak ; 3) Mt. Rainier : 4) Mt. Hood

Table 2.2. USA - Description of the observed glaciers.

Part 10 of 10

	No.	Name	Location	Lat.N	Long.W	Length km	Type of Glacier	Terminus: Type of Terminus	Terminus: Alt. m	Sponsoring Agency
SIERRA NEVADA	133	Maclure	California; Yosemite National Park	37°45.1'	119°16.9'	0.4	cirque glacier	active ice on gently sloping terrain	3598 (1966)	USGS (S)
ROCKY MOUNTAINS	134	Grinnell	Montana: Glacier National Park	48°45'	113°43.6'	1.8	cirque glacier	in lake, actively calving	1945	USGS/NPS (1)
	135	Sperry		48°37.6'	113°45.1'	1.4	cirque glacier	active ice on gently sloping terrain	2250	
	136	Grasshopper	Beartooth Mts.	45°08'	109°53'	0.6	cirque glacier	in lake	3145	MSU
	137	Teton	Grand Teton National Park	43°44.5'	110°47.2'	0.9	simple valley glacier	stagnant heavy debris cover	3300	USGS (D)
ANTARCTICA	138	Hobbs	McMurdo Sound	77°54'	164°21'	ca. 18	simple valley glacier	active ice on gently sloping terrain	ca. 75	NSF (2)
	139	Taylor	McMurdo Sound	77°44'	162°09'	ca. 80	outlet glacier	in lake	ca.100-200	NSF (2)
Alaska	140	McCall	Brooks Range	69°18'	143°48'	7.5	branched valley glacier		1520	UC
Color.	141	Isabelle	Front Range	42°03'	105°39'		cirque glacier		3660	UC

1) Data at present in office of Arthur Johnson, University of North Dakota; eventually to be transferred to the U.S. Geological Survey, Helena, Montana.

2) Data in office of investigator at University of Wisconsin, Madison, Wisconsin.

Iceland

For the location of the glaciers see Index map of the observed glacier fronts in Iceland, in "Fluctuations of Glaciers", Vol. 1, Figure 4, or: Jökull, 17 ÁR, Reykjavik 1967, with the official IHD-designation.

Norway and Sweden

For the location of the glaciers see Inventory publications:

- Atlas over breer i Sør-Norge (Atlas of glaciers in South Norway) by G. Østrem og T. Ziegler. Norges Vassdrags- og Elektrisitetsvesen, Meddelelse Nr. 20 fra Hydrologisk avdeling, 1969 (see also annexed map of "Fluctuations of Glaciers", Vol. 2).
- Atlas over breer i Nord-Skandinavia (Glacier Atlas of Northern Scandinavia) by G. Østrem, N. Haakensen, O. Melander. Norges Vassdrags- og Elektrisitetsvesen og Stockholms Universitet, Meddelelse Nr. 22 fra Hydrologisk avdeling, Meddelande Nr. 46 från Naturgeografiska, 1973.

Austria

For the location of the glaciers see Index map of the observed glacier fronts in the Austrian Alps 1959/64, in "Fluctuations of Glaciers", Vol. 1, Figure 1.

In Austria aerial photographs of all glaciers were made in 1969 for mapping on a scale of 1 : 10 000. In addition, seismic depth measurements for the determination of the glacier volumes are being carried out in the preparation of a glacier inventory.

France

For the location of the glaciers see Index map of the observed glacier fronts in the French Alps 1959/64 in "Fluctuations of Glaciers", Vol. 1, Figure 2.

A contribution to a French glacier inventory has been published as "Fiches des glaciers français" (R. Vivian and others) in the periodical "Revue de Géographie Alpine", Grenoble. The "Fiches" for 29 glaciers were published in 1967 to 1970.

Germany

The Commission for Glaciology of the Bavarian Academy of Sciences, in co-operation with the Institute for Photogrammetry and Cartography of the Technical University Munich, is working on the "Nördlicher Schneeferner" in Germany as well as on various glaciers in Austria (Oetztal). Cf. Chapters 4, 5 and remark on the annexed maps.

Italy

Cf. reference (11): "Catasto dei ghiacciai italiani", Vol. I to IV, in "Fluctuations of Glaciers", Vol. 1, page 18.

The glaciers of the mountain group Ortler-Cevedale are described in the publication "I ghiacciai del gruppo Ortles-Cevedale" by Ardito Desio, XXIII + 876 pages of text + 207 pages of figures + 1 map, Torino, Palazzo Carignano, 1967.

Switzerland

Cf. Index map of the observed glacier fronts in the Swiss Alps 1959/64, in "Fluctuations of Glaciers", Vol. 1, Figure 3.

The mapping of the Swiss Alps on a scale of 1 : 25 000 will be completed in the near future. Independently of this work, all Swiss glaciers have been covered by aerial photography in 1973. Under the direction of Fritz Müller a complete inventory is being made and will be finished in a year or two.

USSR

For the report on the fluctuations of glaciers in the USSR 1959-1970 we are indebted to the Academy of Sciences of the USSR, Soviet Geophysical Committee. This report has been prepared by the working group for the study of the fluctuation of glaciers of the Section of Glaciology of the Soviet Geophysical Committee under the direction of K. G. Makarevich. It contains general information on the studied glaciers of the USSR (Table 2.3), data on the variations in the position of glacier fronts (Table 9.1.14), the mass balance values (Table 9.2.20), variations in the surface elevation and displacement of the glacier surface (Table 9.3.6) for glaciers of the Polar Ural, Caucasus, Pamir, Tien-Shan and Altai, hydrometeorological data (Table 9.5.2) and location maps. The location maps are given in Figures 2.3 to 2.9, followed by the references.

The Institute of Geography of the Academy of Sciences of the USSR began in 1965 to publish the series entitled "Catalogue of Glaciers in the USSR" based on a Russian manual: Rudovodstvo (1966): "Directions for Compiling a Glacier Catalogue of the USSR". The series, consisting of 100 volumes, is to be completed by 1975.

Table 2.3. USSR - General information on the studied glaciers.

Part 1 of 4

	Glacier	Geographical location	General description: Type of glacier	Area km2	Length km	Altitude Max. m	Altitude Min. m	Average height of firn line, accumulation area: Period of observation	Altitude m a.sl.	Accumulation area km2	Institution where data and photos are held
Polar Ural	MGU	M.Shchuch'ye river's valley	Cirque shaped valley	0.91	1.50	1150	693		900	0.61	Geographical Institute AN SSSR
	Southern Karskiy	M.Kara river's valley	Cirque	0.61	1.40	1000	662			0.33	
	Northern Karskiy	M.Kara river's valley	Cirque			725	550				
	Obrucheva	B.Khadat river's valley	Cirque	0.26	0.55	650	395		550	0.19	
	Chernova	B.Usa river's valley	Cirque	0.26	0.55	700	535		600	0.13	
	Bolshoy Usinskiy	B.Usa river's valley	Apron	0.63	0.45	1000	710		950	0.07	
	IGAN	B.Khadat river's valley	Cirque shaped valley	1.06	1.43	1180	827		980	0.25	
Caucasus	Mayli	Main mountain ridge of Caucasus	Valley glacier	7.0	6.2	4598	2360		3170	0.80	Hydrometeorological Service administration Gruz. SSR
	Devdoraki		Valley glacier	7.0	7.3	5034	2260		3260	0.86	
	Abano			2.0	4.1	5034	2950		3700	0.55	UGMS Gruzinskoy SSR
	Gergeti			8.3	8.5	5034	2870		3650	5.00	
	Mna			4.6	4.1	4598	2860		3480	0.87	
	Middle Suatisi			2.5	4.7	4760	2850		3520	0.76	
	Chalaat	Mestiachala river's valley	Valley glacier		7.0	4405	1900		2940	0.82	
	Koruldash	Koruldash river's valley	Valley glacier	5.5	4.5	4544	2220		3220	0.78	
	Kirtisho	Chashuri river's valley	Valley glacier	4.6	5.0	3837	2380		2950	0.78	
	Juzhnyi	Gora Addalashukhgel	Valley glacier	1.1	1.9	3850	2900		3360	0.46	UGMS Azerbaydzhanskoy SSR
	Lazg - Tsiti	Gora Lazg - Tsiti	Valley glacier	1.8	2.2	3860	3120		3300	0.91	UGMS Gruzinskoy SSR
	Dzhankuat	Basin of Terek river valley Adylsu	Valley glacier	2.8	3.0	4013	2700		3065		MGU, Geographical faculty

Table 2.3. USSR - General information on the studied glaciers.

Part 2 of 4

	Glacier	Geographical location	General description					Average height of firn line, accumulation area			Institution where data and photos are held
			Type of glacier	Area	Length	Altitude		Period of ob-servation	Alti-tude	Accumu-lation area	
						Max.	Min.				
				km^2	km	m	m		m a.sl.	km^2	
Pamiro Alai	Akbaytal	Mountain ridge Muzkol	Valley glacier	6.88	5.5	5790	4650				Hydrometeorological fund UGMS, Tadzh. SSR
	Malyi Oktyabr'skiy	Mountain ridge Zaalayskiy	Valley glacier	13.8	6.5	5450	4575				
	Bakchigir	Mountain ridge southern Alichurskiy	Valley glacier	11.6	6.0	5200	4480				
	Russian geographical society	Mountain ridge Darvazskiy	Valley glacier	129.2	20.1	6595	2600				
	Medvezhiy	Mountain ridge Darvazskiy	Valley glacier	29.5	13.0	5903	2900				
	Mushketova	Mountain ridge Petra Pervogo	Valley glacier	34.4	14.0	7105	2860				
	Mazarskiy	Mountain ridge Darvazskiy	Valley glacier	35.7	17.0	4500	3280				
	Zeravshanskiy	Between the mountain ridges Zeravshansky and Turke-stanskiy	Dentritic	121.2	27.6	4840	2775				
	Rama	Mountain ridge Turkestanskiy	Valley glacier	43.4	15.0	4800	2800				
	Dikhadant	Mountain ridge Zeravshansky	Cirque shaped valley	2.76	4.5	4600	3700				
	Tro	Mountain ridge Turkestanskiy	Valley glacier	4.13	6.5	4800	3100				
	Diakhandara	Northern slope of Gissar mountain ridges spur	Cirque shaped valley	1.44	2.3	4000	3600				
	Raygorodskogo	Basin of Sokh river, Turke-stanian mountain ridge	Valley glacier	6.0	6.4	4840	2750	1960-1970	3940	0.53	
	Kyrchin		Valley glacier	2.2	4.2	4800	3040	1964-1970	3830	0.48	
	Kokbeles		Valley glacier	3.2	5.6	5080	3120	1964-1968	4020	0.40	
	Tutek	Basin of Sokh river, mountain ridge Alay	Valley glacier	4.6	4.1	4600	3260	1961-1970	3840	0.50	
	Klyuyeva		Valley glacier	5.9		4400	2880	1960-1970	3540	0.50	
	Turamuz - 1		Valley glacier	1.4	2.8	4600	3000	1963-1970	3830	0.54	

Table 2.3. USSR - General information on the studied glaciers.

Part 3 of 4

	Glacier	Geographical location	General description: Type of glacier	Area km²	Length km	Altitude Max. m	Altitude Min. m	Average height of firn line accumulation area: Period of observation	Altitude m a.sl.	Accumulation area km²	Institution where data and photos are held
Pamiro Alai	Kyzylgorum	Basin of Sokh river, mountain ridge Alay	Valley glacier	2.1	3.7	4800	3200	1961-1970	4200	0.62	Sarnigmi Tashkent
	Bytyrbay	Basin of Sokh river Gissar Range	Cirque	1.6	2.4	3800	3280	1960-1970	3600	0.46	
	Severtsova	Basin of Kashdar'ya river	Apron	2.6	3.1	4000	3270	1960-1970	3500	0.62	
	Abramova	Kyzylsu river's basin mountain ridge Alay	Valley glacier	22.8	9.0	4960	3666	1968-1971	4150	16.40	
Tien - Shan	Turkpakbel-Nizhniy	Pskem river's basin western Tien-Shan	Cirque	0.6	1.4	3640	3200	1960-1970	3440	0.46	Sarnigmi Tashkent Astronomicheskaya 31
	Ayutor - 2		Cirque shaped valley	3.2	3.2	3810	3120	1960-1970	3520	0.56	
	Akbulakul'kun		Valley glacier	2.0	3.5	4140	3045	1962-1970	3720	0.58	
	Karabulak		Valley glacier	0.7	2.1	3990	3070	1960-1970	3420	0.34	
	Chotan - 2		Valley glacier	3.9	3.8	3910	3190	1960-1968	3610	0.54	
	Pakhtakor		Cirque shaped valley	3.5	3.7	4180	3455	1963-1968	3760	0.50	
	Tekeshsay - 1		Valley glacier	1.1	3.2	4020	3070	1962-1969	3530	0.48	
	Barkrak pravy		Cirque shaped valley	2.1	3.0	4130	3590	1962-1970	3870	0.42	
	Barkrak sredniy		Valley glacier	3.0	2.2	4180	3450	1968-1970	3720	0.46	
	Karabatkak	Chon-Kyzyl-Su Tereskey Ala-Too	Valley glacier	4.6				1956-1966	3745	3.10	Tyan'-Shan's physical-geographical station AN Kirg.SSR
	Tsentral'ny Tuyuksu	Ili river's basin Zailiyskiy Alatau	Valley glacier	3.2	3.5	4219	3400	1956-1970	3745	1.75	Geographical sector AN Kaz. SSR
	Igly Tuyuksu		Valley glacier	1.74	2.2	4216	3450	1957-1964	3682	1.26	
	Molodezhnyy		Apron	1.43	1.7	4147	3450	1957-1964	3661	0.77	
	Mametova		Hanging	0.35	1.4	4137	3610	1957-1964	3790	0.20	
	Konstitutsiya		Valley glacier	5.8	5.7	4520	3320	1961-1964	3970	3.70	

Table 2.3. USSR - General information on the studied glaciers.

Part 4 of 4

	Glacier	Geographical location	General description: Type of glacier	Area (km²)	Length (km)	Altitude Max. (m)	Altitude Min. (m)	Average height of firn line accumulation area: Period of observation	Altitude (m a.sl.)	Accumulation area (km²)	Institution where data and photos are held
Tien - Shan	Dmitriev	Ili river's basin Zailiyskiy Alatau	Short valley with large accumulation area	17.0	5.7	4520	3450	1961-1964	3940	8.40	Geographical sector AN Kaz. SSR
	Toguzak		↓	3.1	3.8	4510	3700	1961-1964	4030	1.20	
	Shokalskiy			10.8	4.7	4540	3370	1962-1964	3930	5.70	
	TEU - northern		Hanging valley	1.9	2.6	4260	3580	1962-1964	3840	1.20	
	TEU - southern		Hanging valley	1.8	2.5	4360	3560	1962-1964	3870	1.10	
	Talgar - southern		↓	1.8	3.0	5000	3320		3830	1.10	
	Talgar - northern			1.6	3.4	5020	3260		3790	1.00	
	Korzhenevsky		Valley glacier	37.5	11.5	5020	3300	1964-1970	3950- -4000	27.2	
	Bogatyr'		Short valley with large accumulation area	29.8	8.7	4580	3450	1965-1970	4020	20.1	
	Zhangyryk	Ili river's basin of the point of Zailiyskiy and Kungey Alatau	↓	17.3	8.0	4520	3510	1965-1970	3950	11.8	
	Zhangyryk-southern	Ili river's basin Kungey Alatau	Valley glacier	8.9	7.2	3640	3450	1965-1970	4000	5.6	
Altay	Malyi Berel	Buchtarma river's basin	Valley glacier	6.8	7.0	3830	2108	1962,1967, 1969-1970	2800	3.4	Geographical sector AN Kaz. SSR
	Bol'shoy Berel	Buchtarma river's basin	Valley glacier	9.5	8.7	4506	2050	1962,1967 1969-1970	3100	5.3	

Fig. 2.3.

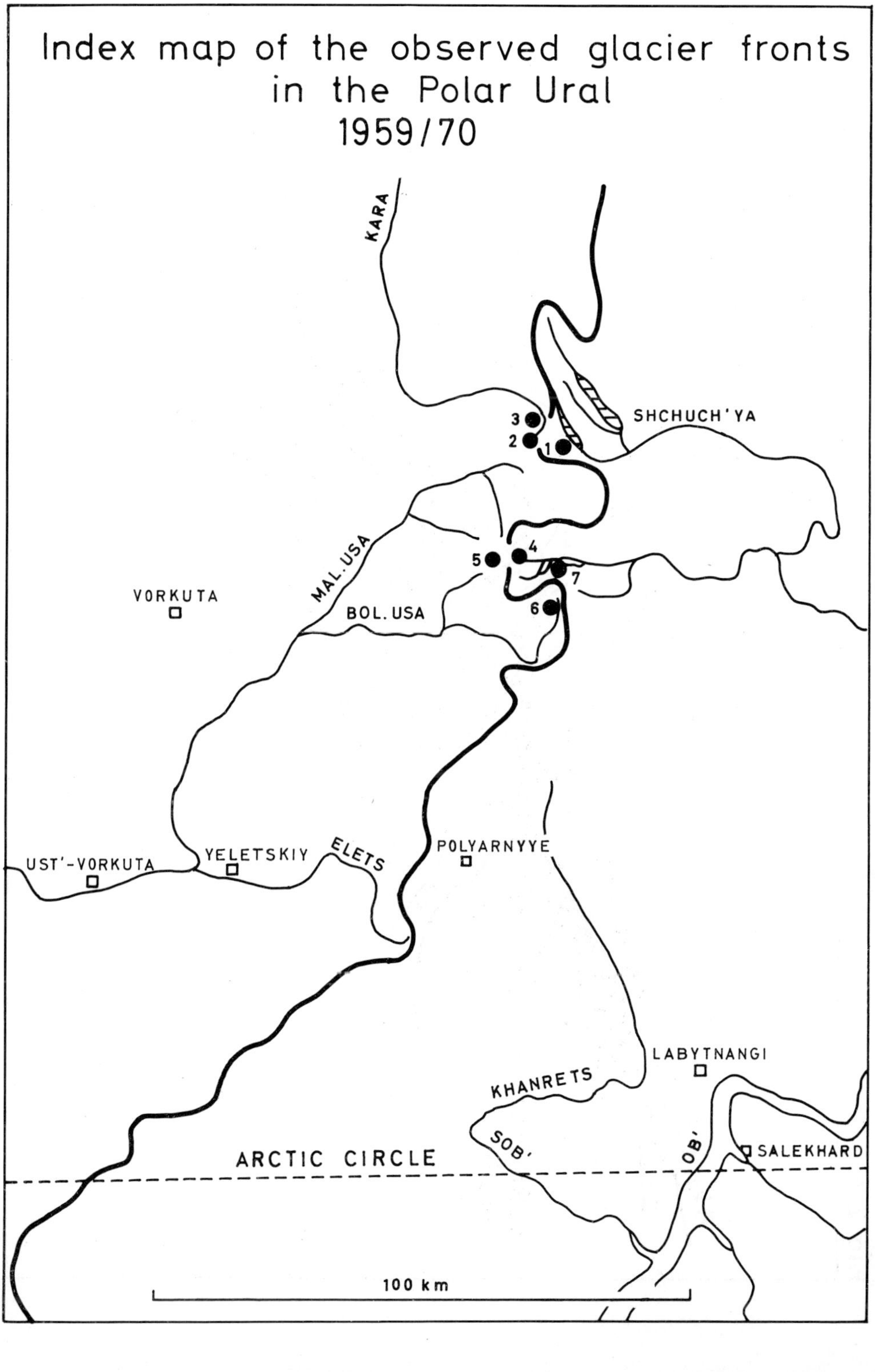
Index map of the observed glacier fronts in the Polar Ural 1959/70
KARA
SHCHUCH'YA
3
2
1
MAL. USA
4
5
7
6
VORKUTA
BOL. USA
UST'-VORKUTA
YELETSKIY
ELETS
POLYARNYYE
LABYTNANGI
KHANRETS
SOB'
OB'
SALEKHARD
ARCTIC CIRCLE
100 km

Fig. 2.4.

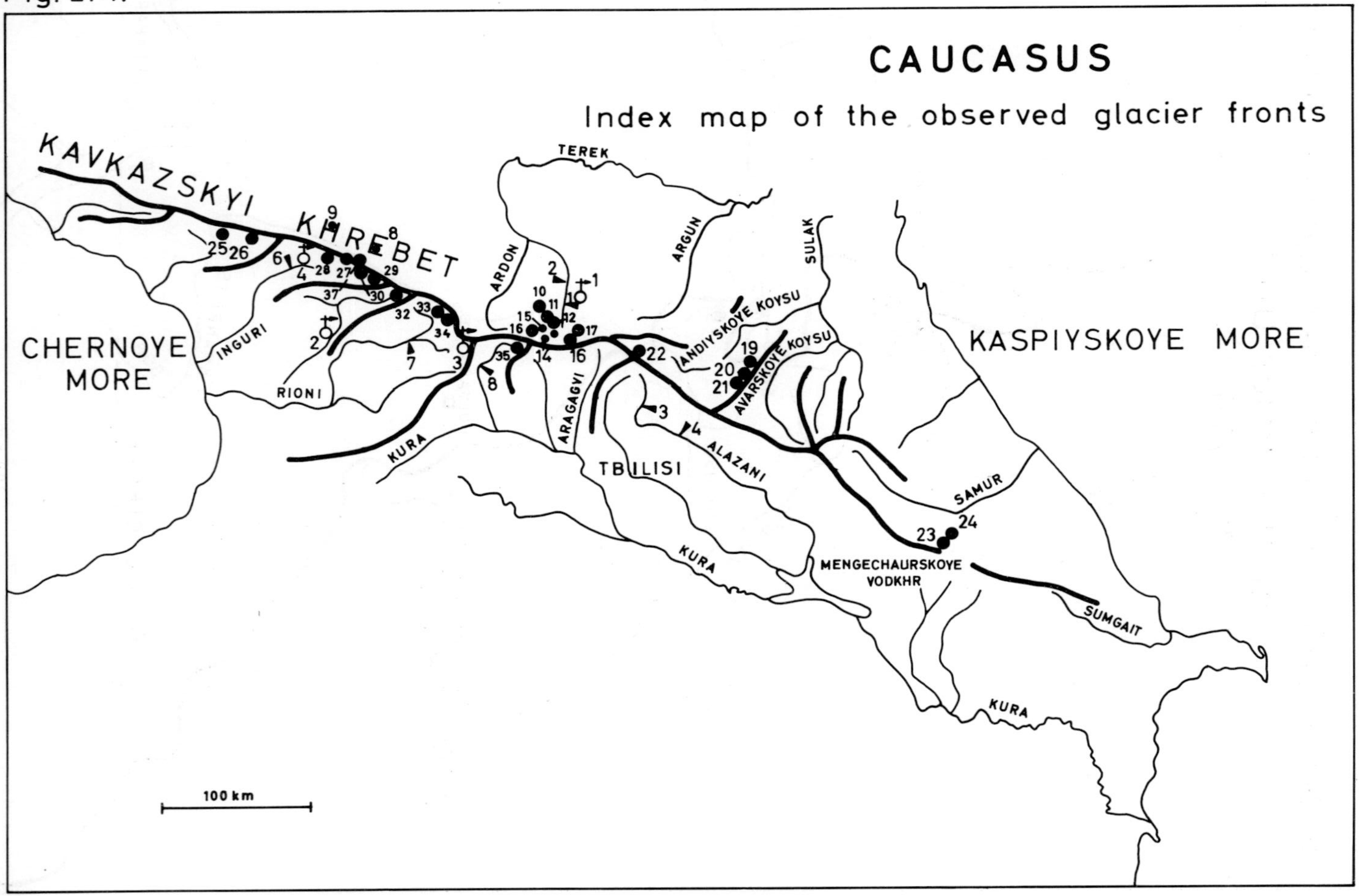
CAUCASUS
Index map of the observed glacier fronts
KAVKAZSKYI KHREBET
CHERNOYE MORE
KASPIYSKOYE MORE
TEREK
ARGUN
ARDON
SULAK
ANDIYSKOYE KOYSU
AVARSKOYE KOYSU
INGURI
RIONI
KURA
ARAGAGVI
TBILISI
ALAZANI
SAMUR
MENGECHAURSKOYE VODKHR
SUMGAIT
100 km

Fig. 2. 5.

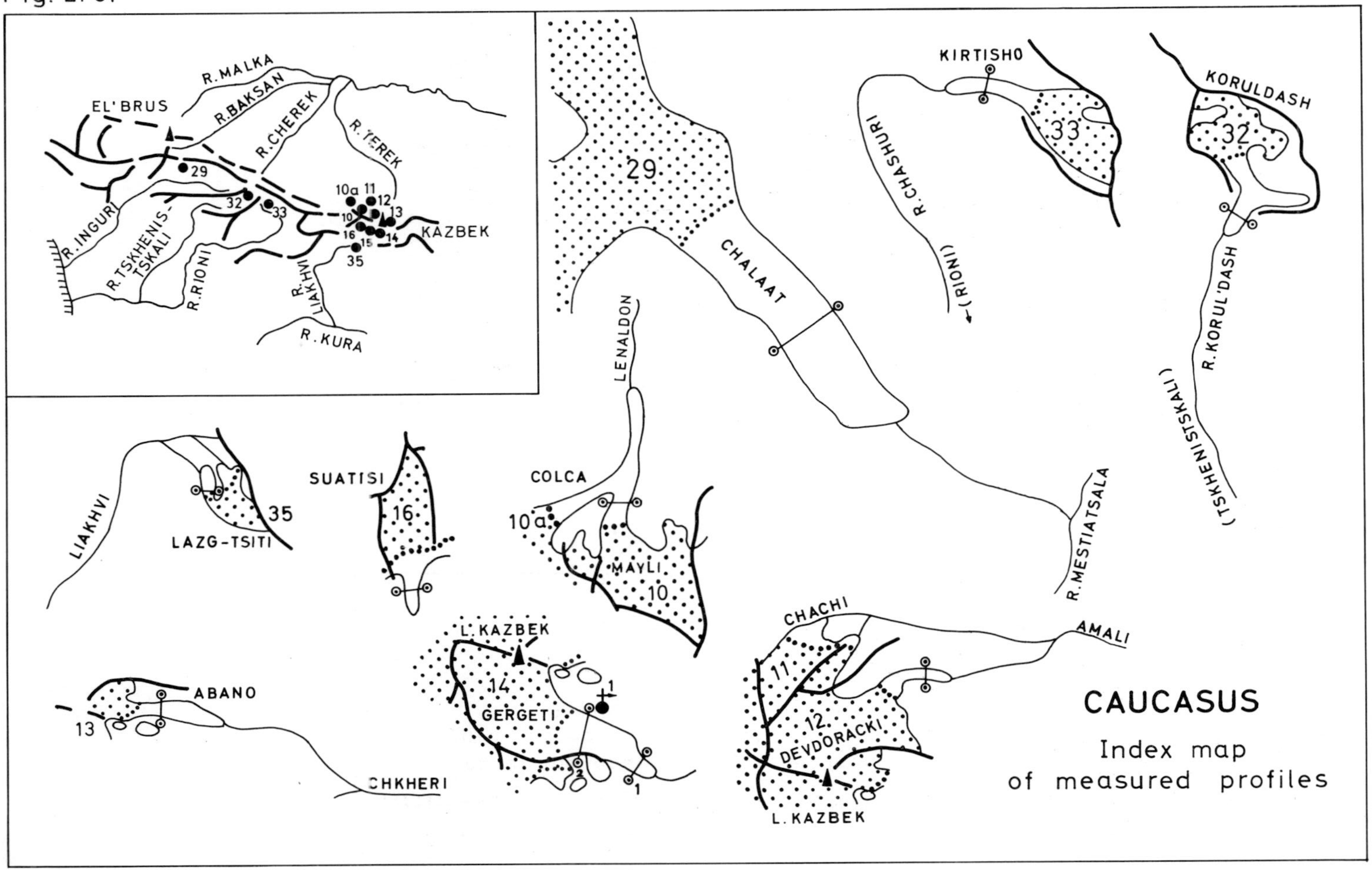
CAUCASUS
Index map
of measured profiles
EL'BRUS
R.MALKA
R.BAKSAN
R.CHEREK
R.TEREK
R.INGURI
R.TSKHENIS-TSKALI
R.RIONI
R.LIAKHVI
R.KURA
KAZBEK
29
32
33
10a
11
12
13
10
16
15
14
35
KIRTISHO
R.CHASHURI
(RIONI)
KORULDASH
R.KORUL'DASH
(TSKHENISTSKALI)
CHALAAT
LENALDON
R.MESTIATSALA
COLCA
10a
MAYLI
10
SUATISI
16
LIAKHVI
LAZG-TSITI
35
ABANO
13
CHKHERI
L.KAZBEK
14
GERGETI
1
2
CHACHI
AMALI
11
12
DEVDORACKI
L.KAZBEK

Fig. 2.6.

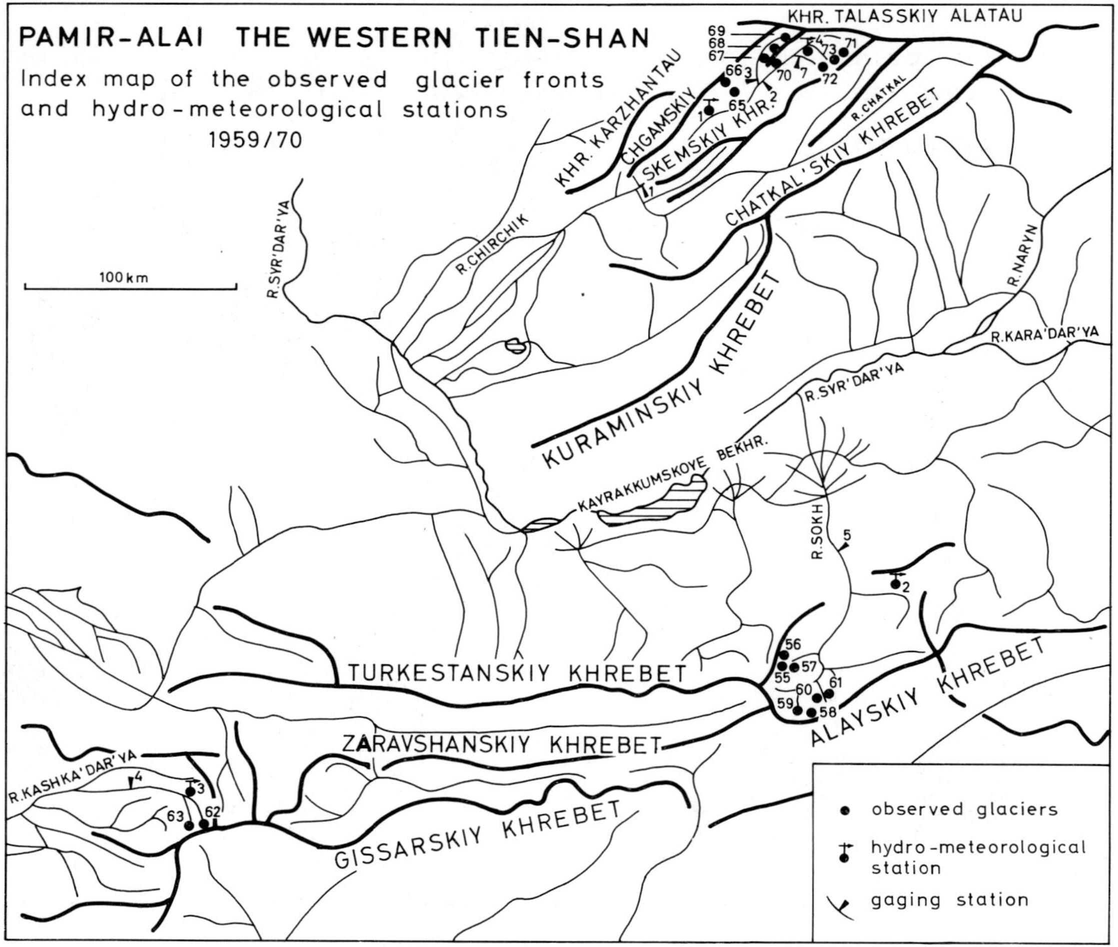
PAMIR-ALAI THE WESTERN TIEN-SHAN
Index map of the observed glacier fronts
and hydro-meteorological stations
1959/70
100 km
KHR. TALASSKIY ALATAU
KHR. KARZHANTAU
CHGAMSKIY
SKEMSKIY KHR.
CHATKAL'SKIY KHREBET
R. CHATKAL
R. CHIRCHIK
R. SYR'DAR'YA
R. NARYN
R. KARA'DAR'YA
KURAMINSKIY KHREBET
KAYRAKKUMSKOYE BEKHR.
R. SOKH
TURKESTANSKIY KHREBET
ALAYSKIY KHREBET
ZARAVSHANSKIY KHREBET
GISSARSKIY KHREBET
R. KASHKA'DAR'YA
observed glaciers
hydro-meteorological station
gaging station

Fig. 2.7.

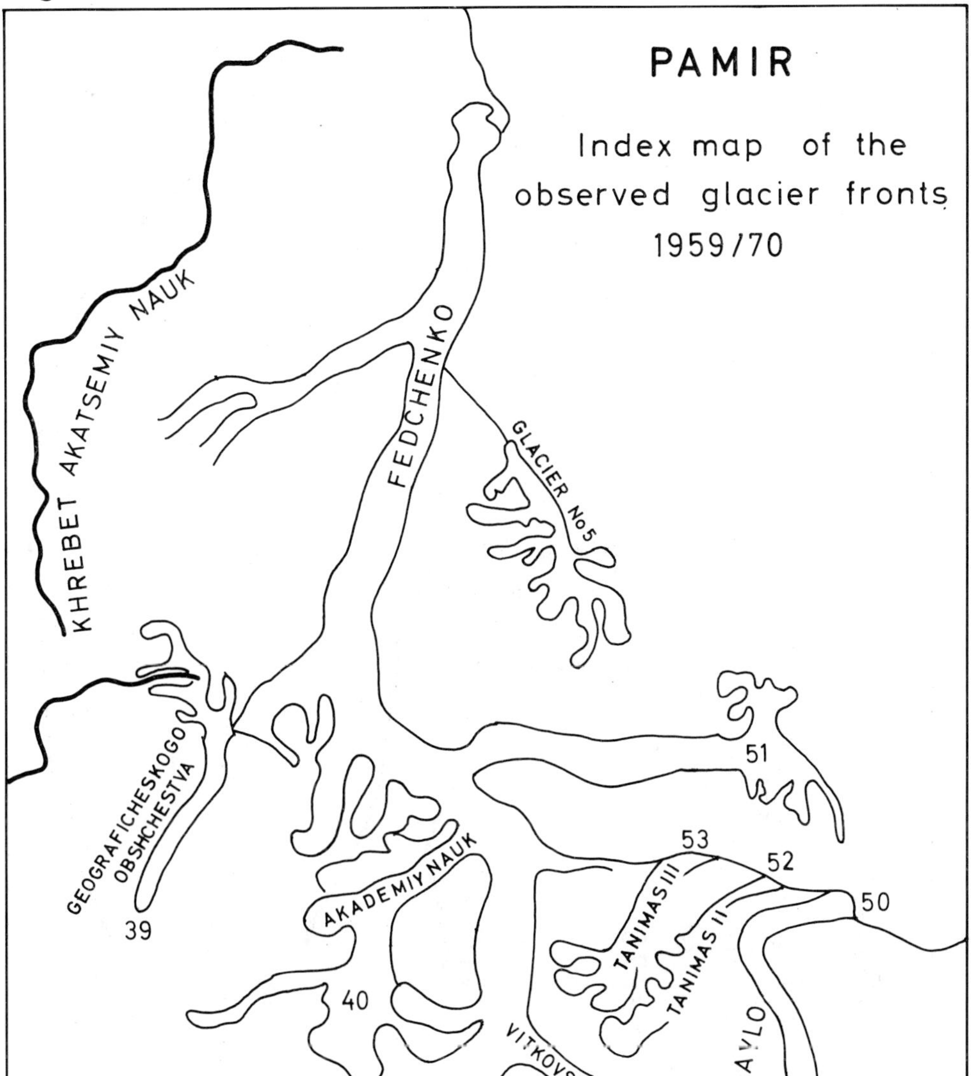
PAMIR
Index map of the observed glacier fronts 1959/70
KHREBET AKATSEMIY NAUK
FEDCHENKO
GLACIER No5
GEOGRAFICHESKOGO OBSHCHESTVA
39
51
53
52
50
AKADEMIY NAUK
TANIMAS III
TANIMAS II
40
VITKOVSKOGO
GRUMM GRZHIMAYLO
48
KHR. XAZ GULEMSKIY

Fig. 2.8.

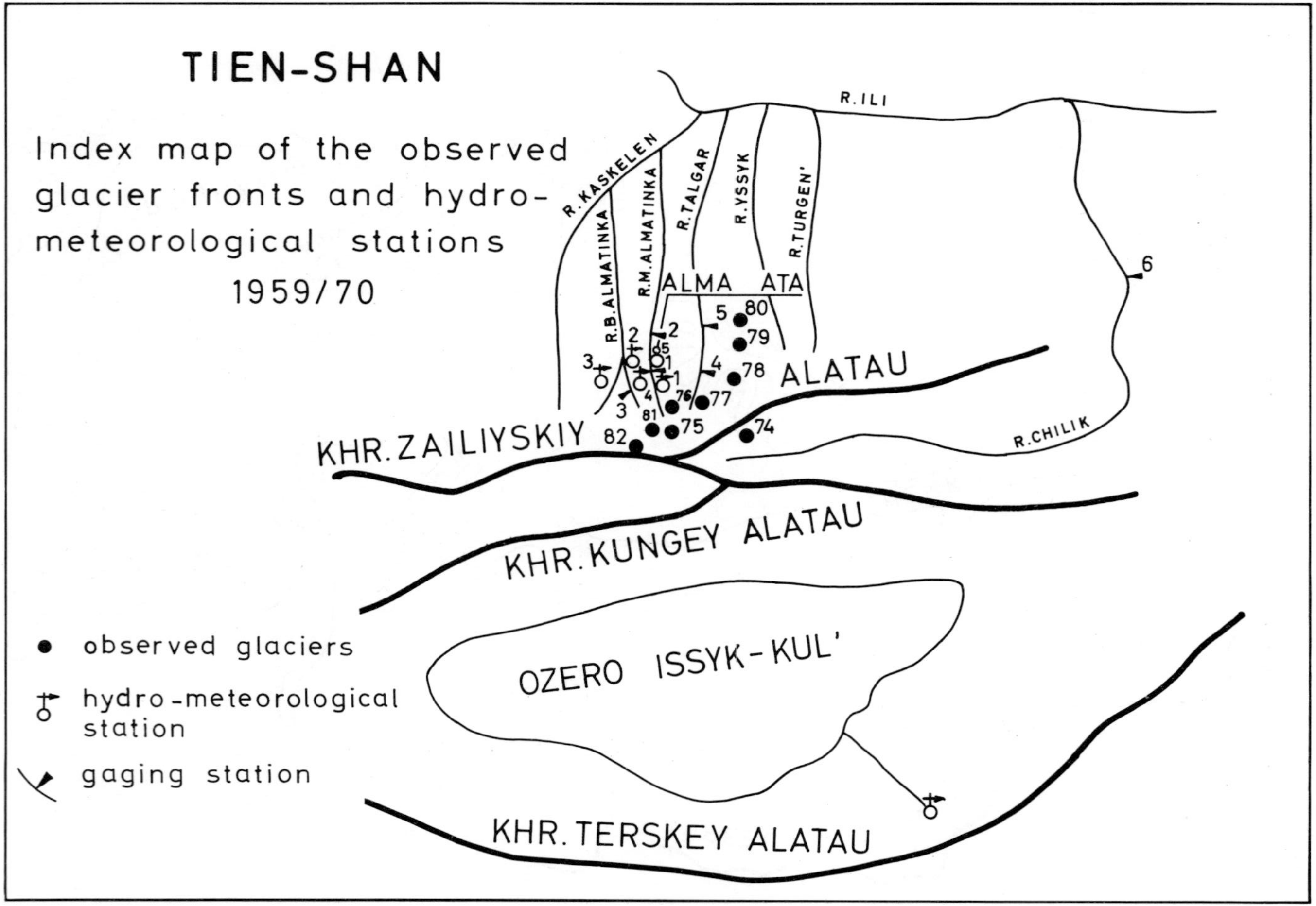
TIEN-SHAN
Index map of the observed glacier fronts and hydro-meteorological stations 1959/70
R. ILI
R. KASKELEN
R.B. ALMATINKA
R.M. ALMATINKA
R. TALGAR
R. YSSYK
R. TURGEN'
R. CHILIK
ALMA ATA
KHR. ZAILIYSKIY ALATAU
KHR. KUNGEY ALATAU
OZERO ISSYK-KUL'
KHR. TERSKEY ALATAU
observed glaciers
hydro-meteorological station
gaging station

Fig. 2.9. Part 1 of 3

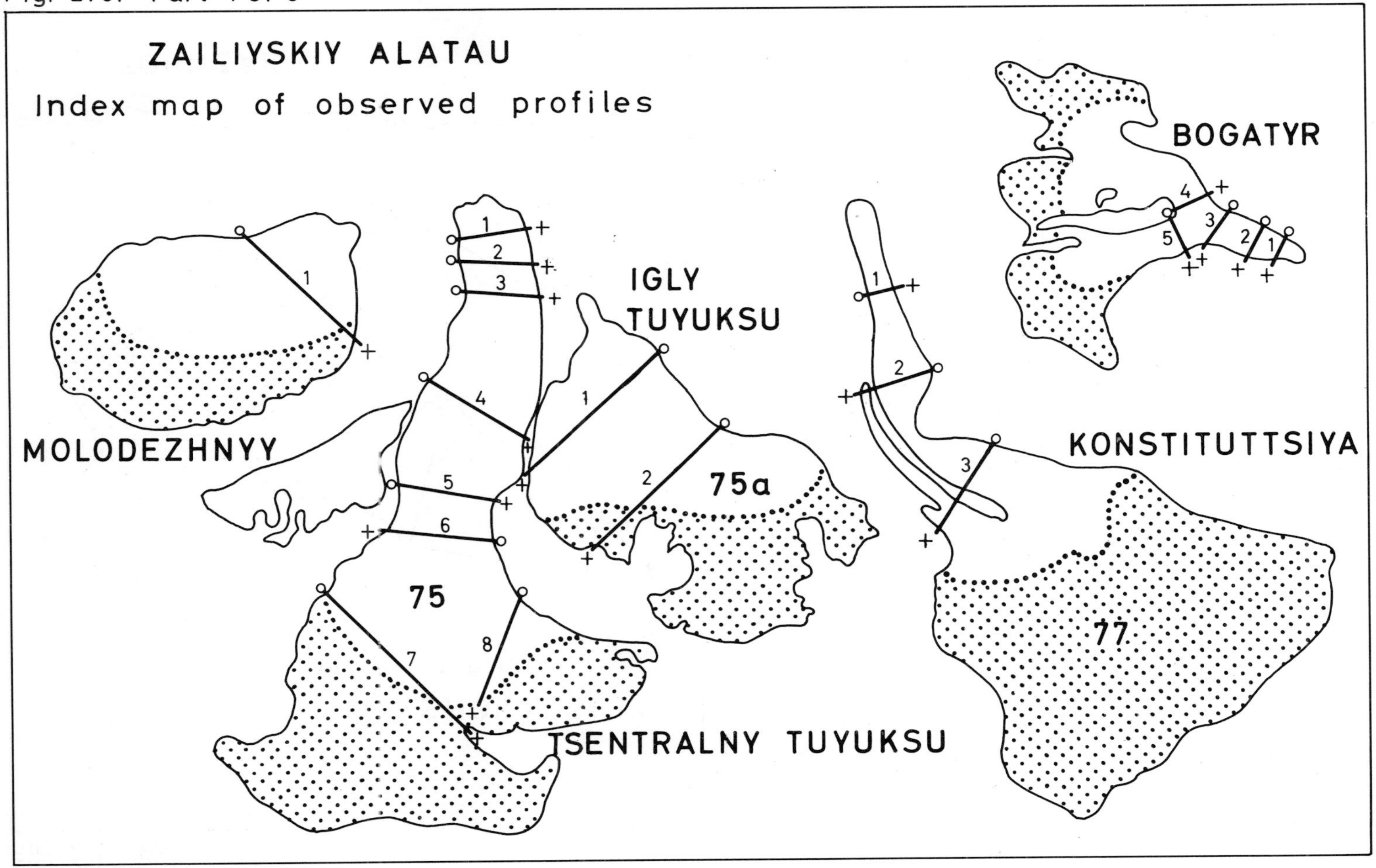

Fig. 2.9. Part 2 of 3

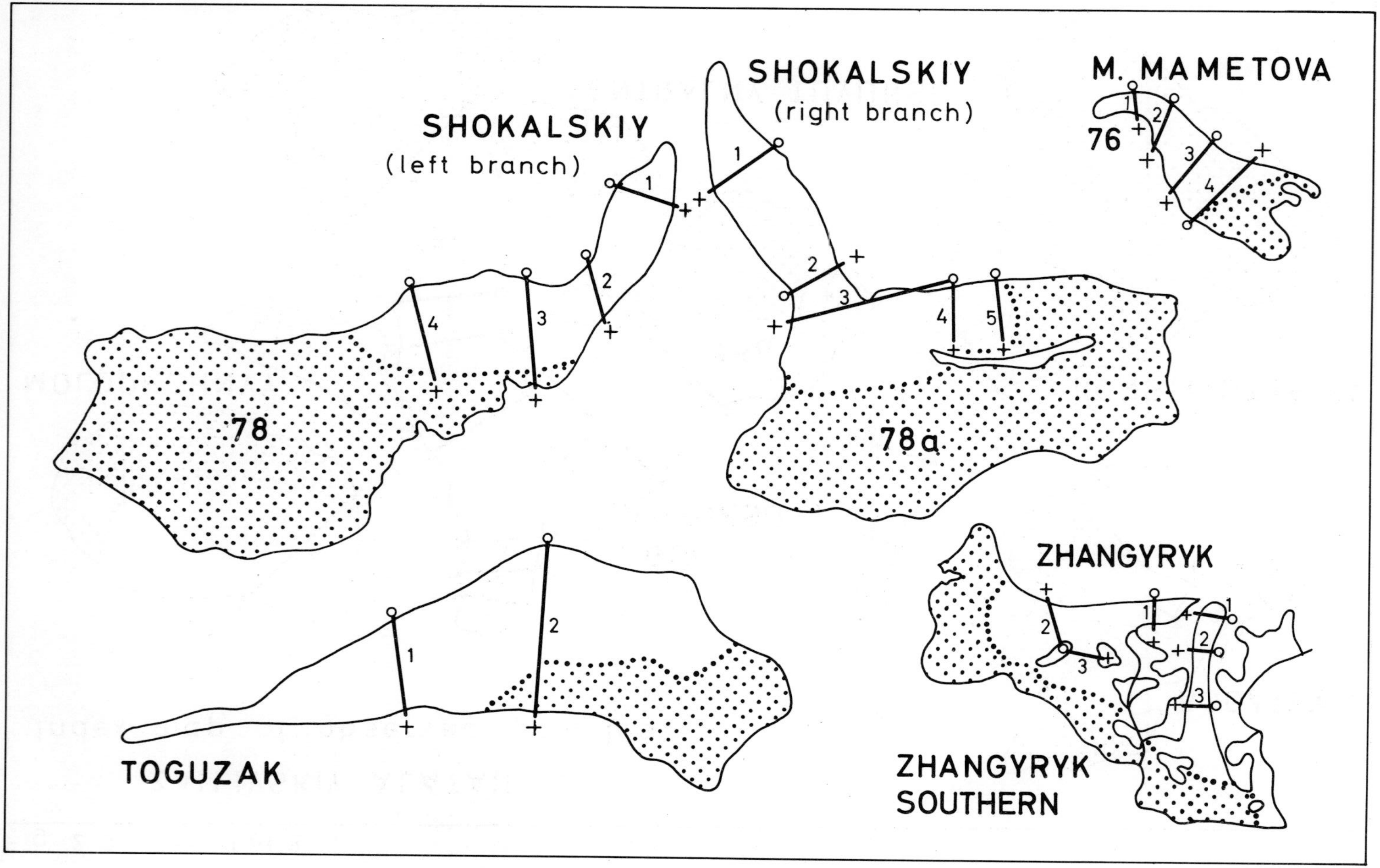

Fig. 2.9. Part 3 of 3

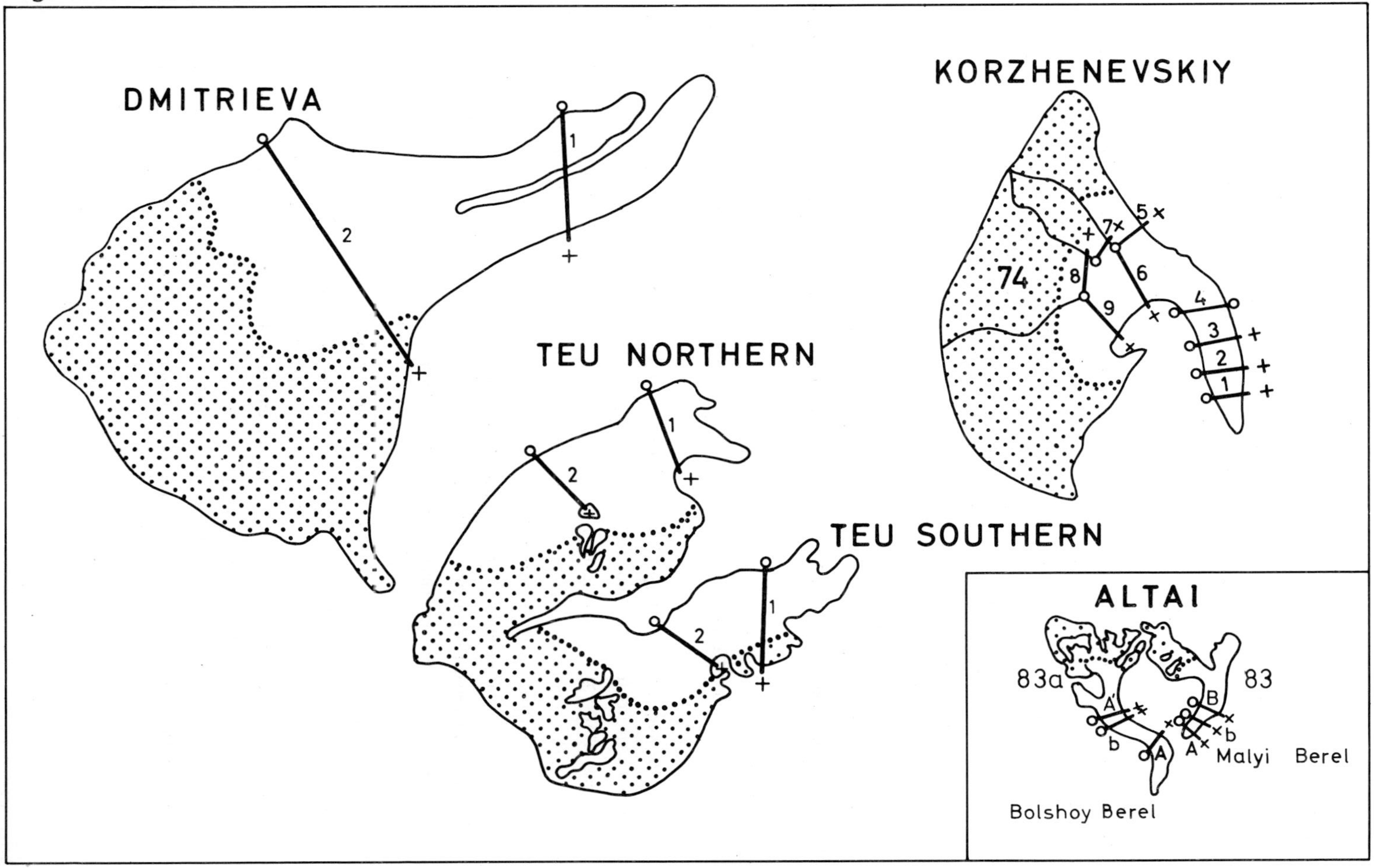

References.

Arctic Ural

Troitskiy L. S., Khodakov V. G. and others, Glaciation of Ural.
The results of research work according to program MGG, Glaciology 16, Moscow 1966.

Caucasus

Gigineyshvili V. M., Napetvaridze E. A., Papinashvili K. I., Atmospheric Processes as a Factor of the Fluctuation of Glaciers.
Bol'shoy Kavkaz. Trudy TbilNIGMI, 1961, issue 8.

Gudushauri Sh. L., The Changes in the Meteorological Elements and the Regime of Glaciers.
Gergeti Kazbekian Glaciation. Trudy TbilNIGMI, 1962, issue 9.

Panov V. D., The Glaciers of the Basin of River Terek.
Leningrad, 1971.

Tsomaya V. Sh., Kisin I. M., Retreat of Glaciers in the Caucasus.
Scientific Notes, Azerbaydzhanian State University, 1959, No. 6,

Tsomaya V. Sh., Kisin I. M., Recession of Glaciers in the Central and Eastern Caucasus in the last 90 - 100 Years.
Hydrology and Meteorology, 1959, No. 5.

Tsomaya V. Sh., The Change in the Speed of Recession of Caucasian Glaciers in Relation to Climatic Warming.
Meteorology and Hydrology, 1959, No. 10.

Tsomaya V. Sh., Glacier Movement in the Caucasus.
Trudy TbilNIGMI, 1960, issue 7.

Tsomaya V. Sh., Characteristic Traits of Retreat Regimen of Caucasian Glaciers.
Trudy TbilNIGMI, 1961, issue 9.

Tsomaya V. Sh., The Results of Levelling on Caucasian Glaciers.
Meteorology and Hydrology, 1969, No. 10.

Tsomaya V. Sh., The Present Growth of Kasbekian Glaciers.
Trudy Zak NIGMI, 1965, issue 19 (25).

Tsomaya V. Sh., The Materials of Glaciological Investigations.
Kasbek, Bagosskiy ridge, Bazar-Dyuzi, 1951-1963 g. g., I.

Tsomaya V. Sh., The Speed of Movement of Caucasian Glaciers and the Possibility of its Determination According to Preceding Meteorological Factors. (For example the Glacier Gergeti).
Trudy V. G. I., 1967, issue 12.

Tsomaya V. Sh., The Regime of Glaciers and Snowfields.
The supply of superficial coaters SSR, 1969, volume 9, issue I.

Tsomaya V. Sh. and Drobyshev O. A., The Results of Glaciological Observations on the Caucasian Glaciers 1963 - 1968.
Trudy ZakNIGMI, 1970, issue 45.

Tsomaya V. Sh. and Agayev M. M., The Glaciers on the Northern Slope of Bazar - Dyuzi Mountain.
Trudy ZakNIGMI, 1970, issue 45.

Tsomaya V. Sh. and Drobyshev O. A., Results of the Investigation into the Catastrophic Movements of the Kolka Glaciers.
Collection of works Tblissian GMO UGMS, Gruzian SSR, 1971, issue 6.

Pamir

Kreyter A. A. and Lesnik Yu. N., The Oscillations of Pamirian Glaciers.
Reports, Academy of Sciences Usbekian SSR, No. I, Tashkent, 1965.

Pamiro-Alay

Kanayev L. A., Podkopaeyva L. D., Shchetinnikov A. S. and Yazykov L. D., Some Results of the Observations of the State of Uzbekistan's Glaciers.
Trudy SANIGMI, 1967, issue 30 (45).

Suslov V. F. and Kidyayev A. M., The Glaciers of the Upper Part of River Koksu.
Trudy SANIGMI, 1969, issue 44 (59). The Glaciology of Central Asia. L. 1969.

Tien-Shan, Zailiyskiy Alatau

Makarevich K. G., The Dynamics of the Central Tuyuksu Glacier in the Past Decades.
Collection of Glaciological Investigations for the Period MGG, issue 2, Alma-Ata, 1962.

Makarevich K. G., Problems of the Trans Ili Alatau Left Talgar Glacier.
Collection "Glaciological Investigations", Vol. 4, Alma-Ata, 1969.

Makarevich K. G., The Mass Balance of Some Small Glaciers in Zailiyskiy Alatau.
Collection "Glaciological Investigations in Kazakhstan", Vol. 6, Alma-Ata, 1966.

Makarevich K. G. and others, The Glaciation of the Zailiyskiy Alatau.
Series "Results of the IGY Program", Moscow, 1969.

Makarevich K. G., Fedulov I. Ya. and Borovinskiy S. A., Electrical Sounding of the Korzhenevskiy Glacier, Its Runoff and Mass Budget.
Heat and Hydrological Regime of the Kazakhstan Glaciers, Alma-Ata, 1969, Vol. 8,

Makarevich K. G., Fedulov I. Ya., Initial Data on the Regime of the Korzhenevskiy Glaciers in Zailiyskiy Alatau.
Hydrological Regime of the Kazakhstan Glaciers, Alma-Ata, 1968.

Makarevich K. G. and Denisova T. Ya., Dependence of the Hydrological Regime of Glaciers on Climatic Conditions.
Regime of the Kazakhstan Glaciers, Alma-Ata, 1971.

Altai

Makarevich K. G., Regime and Mass Balance of the Berel Glaciers on Altay.
Heat and Hydrological Regime of the Kazakhstan Glaciers, Alma-Ata, 1969.

West Irian

Carstensz Glaciers

References:

Dozy J.J., 1938, Eine Gletscherwelt in Niederländisch-Neuguinea.
Zeitschrift für Gletscherkunde, Bd. XXVI, H. 1/2, 45-51

Champion C.R., Bennett, J.M. and Radok, U., Programs for Glaciology and Meteorology.
Carstensz Glaciers Expedition Progress Report, 1971.

Budd, W. F. and Allison, I., An empirical scheme for estimating the dynamics of unmeasured glaciers. - Presented at XV General Assembly IUGG, Moscow, July 1971, Symposium on Interdisciplinary Studies of Snow and Ice in Mountainous Areas.

Champion, C.R. and Radok, U., 1972. Preliminary Report of the Carstensz Glaciers Expedition 1971/1972. - Meteorology Department, University of Melbourne.

Hope, G.S. and Peterson, J.A. 1972, Late Pleistocene and Recent Fluctuations of the Carstensz Glaciers, West Irian. Mimeo - ANZAAS, Sydney, 1972, Section 21.

Peterson, J.A. and Hope, G.S. 1972, A lower limit and maximum age for the last major advance of the Carstensz Glaciers, West Irian. - Nature, London 240 : 36-37.

Galloway, R.W., Hope G.S., Loffler E. and Peterson J.A. (in press), Late Quaternary glaciation and periglacial phenomena in Australia and New Guinea.
Palaeoecology of Africa and Antarctica, Vol. 8, Ed. E.M. von Zinderen Bakker, Cape Town, Balkema.

Peterson, J.A. 1973, Carstensz Glaciers Expedition (CGE) 1973, Progress Report

New Zealand

References

Bell, J.M. 1910. A Geographical Report on the Franz Josef Glacier, with Topographical Maps and Data by R.P. Greville and Botanical Notes by L. Cockayne.
N.Z. Geological Survey Wellington, 14 pp.

Bowen, F.E. 1960, Advances and Retreats of the Franz Josef Glacier.
Journal of Glaciology, 3 (28): 742.

Cotton, C.A., 1958, Geomorphology.
7th ed. Whitecombe & Tombs, Christchurch, pp. 276-347.

Douglas, C.E. 1894, On the Westland Alps. - Append. J. House Repres. N.Z.C.-1:71-5.

Douglas, C.E. 1896, Report on Cook's River District. - Append. J. House Repres. N.Z.C.-1(8): 110-2

Douglas, C,E. and Harper A.P. 1895, The Westland Alps, New Zealand. - Geogr. J. 5(1):61-8.

Gunn, B.M. 1964, Flow Rates and Secondary Structures of Fox and Franz Josef Glaciers, New Zealand. - J. Glaciol. 5(38): 173-90.

Haast, J. 1879, Geology of the Provinces of Canterbury and Westland, New Zealand. Times Office, Christchurch, 486 pp.

Harper, A.P., 1893, Exploration and Character of the Principal New Zealand Glaciers. Geogr. J. 1 (1): 32-42.

Harper, A.P., 1894, The Franz Josef Glacier.- Append. J. House Repres. N.Z.C.-1(6): 75-9.

Harper, A.P., 1896, Pioneer Work in the Alps of New Zealand. - Unwin London, 336 pp.

Harper, A.P., 1926, Variations of the Franz Josef Glacier. - N.Z. alp. J. 3 (15):343-9.

Harper, A.P., 1934, Glacier Retreat. - N.Z. alp. J. 5 (21):322-6.

Harper, A.P., 1935, Glacier Retreat. - N.Z. alp. J. 6 (22) : 173.

Harrington, H.J. 1952, Glacier Wasting and Retreat in the Southern Alps of New Zealand. Journal of Glaciology, 2 (12) : 140-5.

von Kolb, A. 1958, Historic Variations in Glaciers in the Southern Hemisphere, Particularly in New Zealand. - Schlern-Schriften, Innsbruck 190 : 123-46.

Lawrence, D.B. and Lawrence, E.G., 1965, Glacier Studies in New Zealand. Mazama 47 (13) : 17-27.

McSaveney, K.J. and Gage, M. 1966, Ice Flow Measurements on Franz Josef Glacier, New Zealand in 1966. - N.Z. J. Geol. Geophys. 11 (3) : 564-92.

Mercer, J.H. 1962, Glacier Variations in New Zealand. - Glaciol. Notes 12 (IGY World Data Center A: Glaciology. American Geographical Society, New York) pp. 38-41.

Mercer, J.H. 1965, Glaciers in New Zealand. - American Geographical Society, New York.

Odell, N.E. 1955, Franz Josef Glacier, New Zealand, Speed of Flow. - J. Glaciol. 2 (18) : 598.

Odell, N.E. 1960, The Mountains and Glaciers of New Zealand. - J. Glaciol. 3 (28) : 739-41.

Sara, W.A., Soons, J. and Lawrence, D.B. 1966, The Glaciers. - Pp. 19-25 in "Handbook to the Westland National Park", Westland National Park Board, Westland.

Sara, W.A. 1968, Franz Josef and Fox Glaciers, 1951-1967. N.Z. J. Geol. Geophys. 11(3) :768-80

Sara, W.A. 1970, Glaciers of Westland National Park. A New Zealand Geological Survey Handbook, New Zealand Department of Scientific and Industrial Research, Information Series 75, 47 pp.

Speight, R. 1915, Recent Changes in the Position of the Terminal Face of the Franz Josef Glacier. Trans. N.Z. Inst. 47 : 353-4.

Speight, R. 1921, Recent Changes in the Terminal Face of the Franz Josef Glacier. Trans. N.Z. Inst. 53 : 53-7.

Speight, R. 1935, Notes on the Franz Josef Glacier, February, 1934.
Trans. R. Soc. N.Z. 64 (3) : 315-22.

Speight, R. 1940, Ice Wasting and Glacier Retreat in New Zealand. J. Geomorph. 3(2) : 131-43.

Speight, R. 1941, Note on the Franz Josef Glacier, December, 1940.
Trans. R. Soc. N.Z. 71 (2) : 128-33.

Suggate, R.P. 1950, Franz Josef and Other Glaciers of the Southern Alps, New Zealand.
J. Glaciol. 1 (8) : 422-9.

Suggate, R.P. 1952, Franz Josef Glacier, March, 1951. - N.Z. J. Sci. Tech. 33(4) : 299-304.

Wilson, J.M. 1966, Glacier Studies in New Zealand. - Cant. Mountaineer, N.Z. 35 (35) : 56-67.

Wilson, W. 1896, The Fox Glacier. - Append. J. House Repres. N.Z.C. - 1 (8) : 108-9.

Heard Island

References

Budd, G,M. 1964a. Heard Island expedition, 1963. - Polar Record, Vol. 12, No.77, p.193-195.

Budd, G.M. 1946b, The ANARE 1963 expedition to Heard Island.
ANARE Report No. 74, Department of Supply, Melbourne.

Polar Record, 1965, South Indian Ocean Expedition to Heard Island, 1964-65.
Polar Record, Vol. 12, No. 81, p. 744.

Budd, G.M. and Stephenson, P.J. 1970, Recent glacier retreat on Heard Island. In:
Gow, A.J. et al., eds. Proceedings of International Symposium on Antarctic Glaciological Exploration (ISAGE) Hanover, New Hampshire, USA, 3 -7 September 1968, Cambridge, p.449

Budd, G.M. 1970, Heard Island Reconnaissance, 1969. - Polar Record, Vol. 15 (96).

Deception Island

Reference

Orheim, O. 1972, A 200-year Record of Glacier Mass Balance at Deception Island, Southwest Atlantic Ocean, and its Bearing on Models of Global Climatic Change.
The Ohio State University, Institute of Polar Studies, Report No. 42.

3 VARIATIONS IN THE POSITIONS OF GLACIER FRONTS

The following information is given for each country listed:

Sources of data

References

Supplementary data and comments

Canada

Table 9.1.2

Source: Water Survey of Canada Division, Department of Energy, Mines and Resources, Glacier Survey Report 1970.

Data provided by Dr. O. H. Løken, Head of the Glaciological Subdivision, Inland Waters Branch Hydrologic Sciences Division, Ottawa, Ontario, Canada.

USA

Table 9.1.3

Sources: see Chapter 2: Report "Fluctuations of Glaciers, 1964-68", United States Report, by W.O. Field and Tables 2.1 and 2.2.

The investigator mentioned in Table 9.1.3 may be the field observer or the interpreter of the data from aerial photographs. When the responsibilities have been shared, the names of the participants are given. Space does not allow mention of the various other individuals and organizations who participated in the observational programs.

References for Table 9.1.3

Black, R. F. and C.J. Bowser (1968). "Salts and Associated Phenomena at the Termini of the Hobbs and Taylor Glaciers, Victoria Land, Antarctica", General Assembly of Berne, IUGG, Publ. n. 79, pp. 226-238.

Dodge, N.A. (1964). "Recent Measurements on the Eliot Glacier", Mazama, v. 46, p. 47

Field, W.O. (1969). "Current Observations on Three Surges in Glacier Bay, Alaska, 1965-1968", Can. Jour. Earth Sci., v. 6, n. 4, pp. 831-839.

Marangunic, C. (1968). "Effects of a Landslide on Sherman Glacier, Alaska". Ohio State Univ., Dept. Geol. (PhD. dissertation).

Reed, John C., Jr. (1964). "Recent Retreat of the Teton Glacier, Grand Teton National Park, Wyoming", U.S. Geol. Survey, Prof. Paper, 501-C, pp. C147-153.

Reed, John C., Jr. (1965). "Rate of Ice Movement and Estimated Ice Thickness in Part of the Teton Glacier, Grand Teton National Park, Wyoming", U.S. Geol. Survey, Prof. Paper, 525-B, pp. B137-141.

Reed, John C., Jr. (1967). "Observations on the Teton Glacier, Grand Teton National Park, Wyoming, 1965 and 1966", U.S. Geol. Survey, Prof. Paper, 575-C, pp. C154-159.

Reid, J.R. (1969). "Effects of a Debris Slide on "Sioux Glacier", South-Central Alaska", Jour. Glac., v. 8, n. 54, pp. 353-368.

Tuthill, S.J., W.O. Field, and L. Clayton (1968). "Postearthquake Studies at Sherman and Sheridan Glaciers", pp. 318-328 in: The Great Alaska Earthquake of 1964, v. 3, Hydrology, Nat'l. Acad. Sci., Washington.

U.S. Geological Survey. "Ice and Water Balance of Maclure Glacier, California, 1967 Balance Year", in Hydrological Atlas. (In preparation.)

Iceland

Table 9.1.4

Sources: Sigurjón Rist, Hydrological Survey, National Energy Authority, Reykjavík, Iceland: "Glacier variations", in Jökull, 17 Ár, 1967, 18 Ár, 1968, 20 Ár, 1970.

The glaciers in Table 9.1.4 are designated in two different ways. "No." is related to the index map, Figure 4 in "Fluctuations of Glaciers 1959-1965", Vol. 1, "IHD-No." is related to the above mentioned periodical "Jökull". The names of observers are mentioned in "Jökull".

Norway

Table 9.1.5

Sources: Olav Liestøl, in Årbok, volumes 1965, 1966, 1967, 1968, 1969 and 1970, Norsk Polar Institutt, Middelthunsgate 29, Oslo 3, Norway.

Sweden

Table 9.1.6

The data for 1963/64-1966/67 were taken from a personal communication from Dr. V. Schytt, Naturgeografiska Institutionen, Drottninggatan 120, Stockholm Va, Sweden.

Variations of the extent of Swedish glaciers
(Comments of V. Schytt for the years 1963/64 to 1966/67)

In the Kebnekaise massif:

a. Storglaciären. The front was surveyed on 27 Aug. 1965, 7 Aug. 1966 and 1 Sept. 1967. Five reference points were used, and both because of irregular rate of retreat and because of local snow patches along the front, it is difficult to express the annual retreat very accurately. Along the central parts of the front the retreat was:

1965 - 1966 - 10 m

1966 - 1967 - 5 m

whereas retreat as measured from the two outer reference points was approximately 50 % larger.

Since 1959 the average retreat has been 74 m (mean of 80 m, 60 m, 83 m and 74 m) and 249 m since 1945.

b. Isfallsglaciären. Observation dates: 28 Aug. 1965, 7 Aug. 1966 and 29 Aug. 1967.
The average retreat was 11 m in 1965-1966, 16 m in 1966-1967 and since 1959 the total retreat has been 90 m.

c. Rabots glaciär. Observation dates: 22 Aug. 1965 and 20 Aug. 1967. Average retreat 33 m over the period of 2 years.

Glaciers north of Kebnekaise:

a. Stuor-Räitaglaciären. The glacier front was not very well defined in 1967 (28 Aug.) - certain areas near fix point I-63 were covered by till and silt.
According to the observations the front had retreated 4 m at II-63 and advanced 5 m at I-63 since 23 Aug. 1965. Obviously the changes had been smaller than the natural spread in observations along this partly debris-covered ice front.

b. Unna-Räitaglaciären. This glacier was visited on 31 Aug. 1963, on 23 Aug. 1965 and on 28 Aug. 1967. In 1965 there was snow all along the ice edge and all measurements had to be made to the edge of the snow; the values are thus difficult to use. However, most of the 1963 and 1967 observations are made to the real glacier edge and they record a total retreat of 6 m over the 4-year period.

c. Mårmaglaciären. Four points were marked in 1963 and the snout was not visited until 1967. The front is surrounded by high ice-cored end moraines and it is quite probable that all the four reference points are in fact resting on ice - or on an ice-cored boulder deposit. The retreat of a glacier, the front of which is slowly melting down within its own "end moraine basin", is difficult to record by just measuring variations in its length. In this case a leveled longitudinal profile would serve the purpose much better.
The following values were obtained: Fix I-63, 7 m (7 m in 1963), Fix II-63, 20 m (9 m to icy, old firn in 1963), Fix III-63, 10-11 m (11 m in 1963) and Fix IV-63, 12 m (11 m in 1963). Apparently no frontal retreat can be recorded.

Glaciers south of Kebnekaise:

a. Ruopsokglaciären (67°21' N, 18°6' E). This glacier is the eastern-most large glacier in the

Äpar massif and is facing north. The southwest part of the front lies at 1170 m a.s.l. and the northeast part goes down to 1080 m. The glacier was visited on 31 July 1965 and on 12 July 1967. In 1965 the front was surveyed in relation to 9 fix points - five new and four older ones. Eight of these distances were re-measured in 1967 and the average retreat was 14 m. The variations were large - the actual observations were 20 m, 0 m, 10 m, 5 m, 23 m, 25 m, 23 m and 5 m at the eight points. The transient snow line on 12 July 1967 was at 1380 m.

b. Pårteglaciären (front at 67°10' N, 17°45' E). The largest glacier in Sweden. The front, facing east, goes down to 1075 m according to the new topogr. map. A survey of 4 new and 7 old fix points was made on 1 Aug. 1965 and again on 15 July 1967. There was much snow along the front on both these occasions and no meaningful value can be given for the retreat. It is, however, quite clear from the observations that retreat has taken place. The whole glacier was snow-covered on 15 July 1967.

c. Mikkajökeln (front at 67°24' N, 17°42' E). This glacier has been studied intensively by Axel Hamberg during the early part of the century and later by several scientists from the Department of Physical Geography at Uppsala University. It rests on the south and southwest slopes of Sarektjåkko, the next highest mountain massif in Sweden. The tongue ends at 950 m a.s.l.. The gradient over the first 100 m inside the front is 40 m/100 m and 27 m/100 m over the next 100 m.

When the glacier was visited on 4 Aug. 1965 no new reference points were established, only a few of the old ones re-painted. On 18 July 1967 the southeast part of the glacier front was still snow covered, but good observations were obtained from the rest of the ice edge. The average retreat along the exposed part of the front had been 25 m over the two years. The transient snow line of 18 July 1967 was at 1110 m.

d. Ruotesglaciären (front at 67°21' N, 17°30' E). This is the largest glacier in the Kuoper massif. It flows down to the north and its melt water, as well as the water from Mikkajökeln and Pårteglaciären, drains into Rapaälven. According to barometer readings made on 3 Aug. 1965 the front is at 935 m a.s.l. Six fix points were marked (I-65 to VI-65) on 3 August 1965 but the front was then snow-covered near four of them. On 20 July 1967 the snow conditions were rather similar. Fix points V-65 and VI-65 (the two lowest and most northerly) recorded a retreat of 2 m and 17 m respectively. The glacier front near the other four points was snow-covered, but the average distance from marker to snow edge had increased from 55 m to 91 m. This may, however, not indicate much retreat - the transient snow line was at 1070 m on 3 Aug. 1965 and at 1180 m on 20 July 1967.

e. Suotasglaciären (front at 67°27' N, 17°39' E). Lying on the north side of the Sarek massif, Suotasglaciären faces north and northeast. The elevation of the front is 1100 m a.s.l. When the glacier was visited on 30 July 1965 there was much snow left along the very front. 18 fix points were surveyed - most of them old ones. One distance was measured to glacier ice (EE 65), ten to the snow edge and seven to superimposed ice. On 21 July 1967 the whole glacier edge was hidden under a large snow bank. It was quite evident that retreat had taken place, but it is impossible to use the distances measured to the snow edge for computation of a quantitative measure of this retreat.

f. Vartasglaciären (front at 67°25' N, 17°42'E). Situated on the north side of the Sarek massif, three km SE of Suotasglaciären, but with its front approximately 160 m higher. The glacier's size variations have been studied by Uppsala geographers for several years. On 8 September 1964 and on 31 July 1965 some fix points were re-painted and a few new ones marked in order to get them identifiable on air photographs. At the time of the 1965 visit the whole ice front was snow-covered and since the 1967 observations (22 July) were also made to the snow edge little can be said about size variations. The glacier has not advanced, and any possible retreat has been small.

On 22 July 1967 the transient snow line was at 1500 m.

g. Hyllglaciären (front at 67°41'N, 17°29'E). This is a small glacier (about 2 km long) on the north slopes of the Akka. Its front is at approximately 1350 m a.s.l. Seven fix points (I-65 to VII-65) were marked and surveyed on 29 July 1965 and were visited again on 28 July 1967. On both occasions thick snow on and outside the toe of the glacier prevented a reliable survey of the ice front. We have, however, good air photographs from 1963 showing the exposed front and upon which the large boulders selected and marked as fix points in 1965 can easily be identified. The ice front of 1963 stood on an average 5 m outside the snow edge of 1967.

At the end of July 1965 only a few patches of the glacier were snow-free and in 1967 (28 July) no ice at all was exposed. The transient snow line was thus below 1350 m.

h. Salajekna (front at 67°08'N and 16°27' E). This is one of the largest glaciers in Sweden with an area of approximately 13.4 km^2 on the Swedish side. It is flowing south from a large accumulation area east and south of the high Sulitelma massif. Its front is about 2.5 km wide and reaches partly below 900 m elevation. Seven fix points were established along the eastern half of the front on 2 August 1965 (I-65 to VII-65) and were re-surveyed on 2-3 August 1967. The average retreat was 20 m for the two years.

The transient snow line was at 1110 m on 3 Aug. 1967 and at approximately the same elevation on 2 Aug. 1965.

The data for 1967/68 - 1970/71 were taken from a report by V. Schytt in "Ice", No. 38, 1st issue 1972, and from a personal communication from O. Melander.

Austria

Cf. Index map (Fig. 1) in "Fluctuations of Glaciers 1959-1965", Vol. 1, Tables 9.1.7 and 9.1.8.

Data provided by personal communications from Prof. Dr. H. Hoinkes and Dr. G. Patzelt, Institut für Meteorologie und Geophysik, Schöpfstrasse 41, A-6020 Innsbruck.

Reports on the variations in the positions of glacier fronts have also appeared in:

- Zeitschrift für Gletscherkunde und Glazialgeologie:

  Vol. V, No. 2, Innsbruck 1968: R. v. Klebelsberg: Observations for 1958/59-1962/63; H. Kinzl: 1963/64 + 1964/65, 1965/66.

  Vol. VI, Innsbruck 1970: H. Kinzl: Supplement 1962/63, 1966/67-1968/69.

  Vol. VII, Innsbruck 1971: H. Kinzl, 1969/70.

These publications give the names of the informants and additional particulars.

France

Cf. Index map (Fig. 2) in "Fluctuations of Glaciers 1959-1965", Vol. 1, Table 9.1.9.

Data provided by R. Vivian, Institut de Géographie Alpine, rue Maurice Gignoux, 38-Grenoble, France. Reference: R. Vivian, 1971: Les variations récentes des glaciers dans les Alpes françaises (1900-1970). Possibilité de prévision. Revue de Géographie Alpine, Vol. LIX, No. 2, Grenoble 1971.

The observations have been carried out by the following institutions (the list may conceivably be incomplete):

Institut de Géographie Alpine, Université de Grenoble
Laboratoire de Glaciologie, Université de Grenoble
Service des Eaux et Forêts
Société du Montenvers, Chamonix.

Prof. Dr. Paul Veyret, Director of the Institut de Géographie Alpine, further reports on the Glacier des Bossons, the Alpine glacier with the lowest terminus:

In 1952 (see the Aiguille du Midi sheet of the 1 : 10 000 map of Mont Blanc) the Glacier des Bossons stopped at an altitude of 1340 m; in 1972 it reached down to a datum of 1180. The difference of 160 m in altitude, measured on the map, corresponds to a distance of 600 m. The most marked advance took place in 1956-1958, when the front reached the bottom of the alluvial basin. Since then, the advance of the glacier has been slower, perhaps for general reasons but also no doubt because the flattening of the slope slows its movement, which also expresses itself in the almost complete absence of crevasses in the lower section and the formation of prows of ice that are often of a considerable height. After a standstill or even a slight retreat in 1960-1964, the advance was resumed from 1965 to 1970 over a distance of some 100 m. It has been continuously observed since 1968, and this has revealed the peculiar behaviour of the front. The glacier is of the slope type, of moderate thickness, and does not flow in a straight line but describes a curve from its left towards its right bank. The curve causes the front to abut against the bed (a bank of scree with large boulders) of the massive old right-hand moraine, which hinders its progress. The rhythm at the present time is as follows: in the cold season (October-April) the front advances (about 30 m in 1969-70); then it remains stationary in the warm season in spite of intense melting, becoming thinner but not withdrawing if account is taken of the blocks of ice concealed in the moraine; in spring (May-June) the left-hand flank pushes forwards towards the alluvial plain and towards the old moraine on the left-hand bank; in the right-hand bank local thrusts occur at various levels and various times and erode the rocky slope (through the intermedium of the many large blocks they carry with them) or else deposit lateral moraines on it.

In 1971 and 1972 the advance of the front was 10 m per year.

Italy

Tables 9.1.10 and 9.1.11

The results shown in Table 9.1.10 are due to Prof. Manfredo Vanni, Secretary of the Comitato Glaciologico Italiano in Turin up to 1971 and Honorary Secretary since 1972.

The data for Table 9.1.11 were taken from the Bollettino del Comitato Glaciologico Italiano, No. 16, Parte seconda, Turin 1970. In No. 20 (1972) of this yearbook Prof. Corrado Lesca gives revised figures for 1964/65 t0 1971/72 which do not correspond to the figures given for the years 1964/65-1967/68. A note by C. Lesca in "Ice", No. 41, 1973, page 8, states: "The reports given for the past years are partially wrong, and therefore not comparable."

Switzerland

Tables 9.1.12 and 9.1.13

Cf. index map (Fig. 3) in "Fluctuations of Glaciers 1959-1965", Vol. 1.

The data are taken from the reports of the Commission des Glaciers of the Société Helvétique des Sciences Naturelles, which have been published up to and including 1969/70.

References:

"Les variations des glaciers suisses, rapports annuels de la Commission des Glaciers de la Société helvétique des Sciences naturelles":

86e rapport 1964/65

87e rapport 1965/66

88e rapport 1966/67

89e rapport 1967/68

90e rapport 1968/69

91e rapport 1969/70

USSR

Table 9.1.14

See remarks in Chapter 2.

The measurements used in Table 9.1.14 are taken from the following authors:

Region	Number of Glaciers	Table 9.1.14 Part	Number of Figure	Author
Polar Ural	1 - 7	1	2.3	D. G. Tsvetkov, V. I. Mikhalev, Geographical Institute, Academy of Sciences, SSR
Caucasus	8	1	2.4	V. D. Panov, Northern-Caucasian Administration of Hydrometeorological Service, SSR
	9	1	2.4	N. G. Golubev, V. S. Freydlin, Moscow State University
	10 - 30 32 - 33 35	1 and 2	2.4 and 2.5	V. Sh. Tsomaya, O. A. Drobyshev, Zak. NIGMI and Administration of Hydrometeorological Service Gruzin, SSR
	31 and 34	2	2.4 and 2.5	R. G. Shengeliya, Geographical Institute, Academy of Sciences Gruzin, SSR
Pamir Alai	36 - 47	2 and 3		A. G. Sannikov, L. N. Sokolov, Administration of Hydrometeorological Service Tadzhikian, SSR
	48 - 54	3	2.7	A. A. Kreyter, Iu. N. Lesnik, Institute of Geology and Geophysics, Academy of Sciences, Uzbekian, SSR
	55 - 64	3 and 4	2.6	A. S. Shchetinnikov, A. T. Moshkin, B. V. Kislov, Sarnigmi
Tien-Shan	65 - 73	4 and 5	2.6	A. S. Shchetinnikov, A. T. Moshkin, Sarnigmi
	74 - 82	5	2.8	K. G. Makarevich, I. Ia. Fedulov, Geographical Sector of Academy of Sciences, Kazakhstan, SSR
Altai	83	5	2.9 Part 3	K. G. Makarevich, N. V. Ierasov, N. Ospanov, G. A. Tokmagambetov

West Irian

Table 9.1.15

We owe the results to personal communications from Dr. Uwe Radok, Meteorology Department, University of Melbourne, Parkville, 3052 Victoria, Australia. See also references, Chapter 2.

New Zealand

All information on the glaciers of New Zealand was contributed by Dr. P.W. Anderton, Ministry of Works, Glaciology Section, P.O. Box 1479, Christchurch, New Zealand.

The references are given in detail in Chapter 2.

Table 9.1.16
Franz Joseph Glacier: 1951-1967
References: Sara, W.A. 1968
Sara, W.A. 1970

Figure 9.1.1
Franz Joseph Glacier: 1894-1971
References: Sara, W.A. 1970 and personal communication of Dr. P.W. Anderton, 31.1.72

Figure 9.1.2
Fox Glacier: 1894-1971
References: Sara, W.A. 1970 and personal communication of Dr. P.W. Anderton, 31.1.72

In a letter dated 31.1.1972 Dr. P.W. Anderton supplied the following supplementary information:

Recent Fluctuations of Glaciers in New Zealand:

Westland National Park

The Fox and Franz Josef Glaciers are the largest glaciers on the western side of the Southern Alps. Both glaciers appear to respond to mass balance variations with time lags of only a few years. Fluctuations of the termini of the two glaciers have been recorded by the Geological Survey since 1951. Observations to 1967 are summarised in the enclosed publications by W.A. Sara (1. Franz Josef and Fox Glaciers 1951-1967 and 2. Glaciers of Westland National Park).

Franz Josef Glacier

In June 1965 the glacier commenced a major advance which continued until October 1967. This advance was followed by retreat of the terminus until September 1971. However, the glacier is currently showing signs of another advance, which probably represents a response to heavy snowfalls in late 1967 and 1968. The terminal positions of the glacier since 1967 are sketched on the enclosed map (copy of figure 2, publication 1).

Fox Glacier

A slow advance of the glacier occurred from 1965 until late 1968, after which the terminus retreated steadily. By May 1970 the terminus had returned almost to its position of June 1965. However, the glacier has since re-advanced to a position ahead of the limit of its previous advance. As in the case of the Franz Josef Glacier, this re-advance is probably a response to the heavy snowfalls of late 1967 and 1968. The terminal positions of the Fox Glacier since 1967 are sketched on the accompanying map (copy of figure 24, publication 2).

Mt Cook National Park

The glaciers on the eastern side of the Southern Alps do not respond rapidly to mass balance variations. Termini of these glaciers have shown little or no fluctuation in recent years, and are characterised by inactive ice with a thick cover of moraine.

Photographs of the Tasman Glacier have been taken annually since 1955 by the Geological Survey, primarily for study of erosion processes and rates. Accumulation and ablation measurements have been carried out on the Tasman Glacier between 1958 and 1965 by the Geological Survey and since 1965 by the Water and Soil Division, Ministry of Works.

No regular observations have been made on other major glaciers in the park, but vertical aerial photographs, taken by the Lands and Survey Department in 1965 and by the Ministry of Works in 1971, have been used to assess major changes in the intervening period. Maps of the area have not yet been published but copies of the provisional photogrammetric plots (Mt Cook and Godley sheets) are enclosed, showing terminal positions of the major glaciers.

Tasman Glacier

The terminal face is well defined and has shown no significant retreat since 1965. The lower 10 km of the glacier is covered by moraine, which is approximately 1 m thick at the terminus, and consequently inhibits ablation. Downwasting of the glacier tongue is occurring, and the aerial photographs show an increase in the extent of subsidence, and up-glacier migration of the limit of white ice.

Hooker Glacier

The terminus is narrow and well defined and has shown no significant retreat since 1965. Moraine covers the lower 4.5 km of the glacier tongue. Aerial photographs indicate an increase in the extent of subsidence near the terminus, and an up-glacier migration of the limit of white ice.

Mueller Glacier

As shown on the enclosed map, the Mueller Glacier is no longer a unified glacier system. The lower Mueller tongue is now entirely covered by moraine, and fed primarily by the Tuckett Glacier. The terminus is wide and fairly well defined, and has shown no significant retreat since 1965, but an increase in extent of sudsidence is evident. Ice from the upper Mueller Glacier now terminates in a narrow moraine-covered tongue against active ice of the Tuckett Glacier.

Murchison Glacier

The lower 7.5 km of the glacier tongue is covered by moraine. The terminus of the glacier is not well defined, and an active alluvial fan is extending across the terminal area. On the enclosed map, the approximate boundary of the partly buried, partly eroded terminal ice is shown by a dashed line. The boundary of the main ice mass in 1965 is shown by a solid line. Some minor retreat of this boundary had occured by 1971 but was obscured by general subsidence and down - wasting.

Godley Glacier

The Godley Glacier is no longer confluent with the Grey and Maud Glaciers. Retreat of the terminus from the former medial moraine was evident between 1965 and 1971 and is shown on the enclosed map. In 1965 white ice was present almost to the terminus, but the boundary has since retreated up-glacier.

The combined Grey and Maud Glaciers terminate in a proglacial lake. Significant retreat of the terminus has occurred since 1965, and is shown on the enclosed map.

Heard Island

Figs. 9.1.3 and 9.1.4

The particulars on Heard Island were contained in a personal communication dated 21.12. 1972 from Dr. Grahame M. Budd, Department of Tropical Health and Public Medicine, University of Sydney, New South Wales, Australia.

The references are given in Chapter 2.

In "Budd, G, M., 1970" the terminus changes are summarized as follows:
Glaciology: Major re-advance was obvious in the glaciers of Corithian Bay which, after the striking retreat observed in 1963 (Budd and Stephenson, 1970), had now regained their 1954 status. The sea cliffs of the eastern part of Baudissin Glacier were 20 - 40 m in height although in 1963 they had been only 20 - 25 m; and the adjacent Little Challenger Glacier, which in 1963 had had a sloping terminus behind a broad beach, had spread laterally and had again developed sea cliffs 20 m high.

In other areas, especially Jacka Glacier of the Laurens Peninsula and Brown Glacier of the north-east coast, further retreat from the positions of 1963 and 1965 was apparent.

G.M. Budd gives the following commentary on Figs. 9.1.3 and 9.1.4:

Abbotsmith Glacier

This is one of the two major ice streams (the other being the Gotley) on the western side of Heard Island. Henderson Bluff, the promotory at the southern end of its sea cliffs, was photographed from the sea in 1948 but has been visited only twice. From my notes and sketches of 23rd December , 1954, and measurements and photographs made on 7th February, 1971, it is clear that the ice edge above Henderson Bluff has retreated about 800 metres since 1954. The distance was paced out in 1971 by Ian Holmes, Iain Dillon and myself, and the results ranged from 878 to 900 paces. My impression was that the glacier was then re-advancing, but as this has not been confirmed its present trend must be regarded as uncertain. However, as we made careful measurements to a marked reference block and took many photographs the question could readily be clarified on the next visit.

Winston Glacier

As Fig. 4 and Plate 2 of the ISAGE paper show, since 1954 the front of this glacier has been divided into two parts, a southwestern (SW) and a northeastern (NE), by a central rock promontory. The SW part has continued to terminate in ice cliffs which calve into deep water in Winston Lagoon. However, the NE part has retreated inland and now terminates above high bluffs, except for a short ice-tongue which in 1963 and 1965 ended about 50 metres from the water, on the northern side of the central rock promontory.

In my re-examination of all available photographs of this glacier, I have been able to fix successive positions of the ice front in relation to recognizable features of the adjacent rock. The results, plotted on the enclosed copy of Fig. 4 of the ISAGE paper, show that from 1963 to 1965 the SW stream not only increased in thickness, as previously reported, but also advanced more than 200 metres, although no change was apparent in the NE part of the glacier. From 1965 to 1971, as a comparison between the enclosed panoramic photograph (Photo 1) and Plate 2 of the ISAGE paper will show, the SW stream advanced about the same distance again. The NE part also advanced, as shown by (a) the ice above the bluffs, which in 1963 and 1965 had appeared stagnant but in 1971 was thicker, had developed sharply defined and overhanging ice cliffs, and showed considerable avalanche activity; and (b) the central ice tongue, which had thickened, spread laterally to cover some of the adjacent bluff, and advanced into the lagoon on a broad front.

In Photo 1 dirt on the ice unfortunately tends to hide the left and right thirds of the ice tongue, although the original colour transparency is clearer; and the advance of the SW stream is also less obvious in the photos than it might have been, owing to the fact that the camera view - point in 1965 was some distance to the right (further from the sea) of that in 1963, and in 1971 was a little further to the right again. This explains how the re-advance in 1965 happened to be overlooked until a close examination was made of the adjacent rock landmarks. Perhaps I should mention that the reason for this variation in the camera positions is that the observations each year were made under adverse conditions, by travelling parties who could not afford to spend much time either building large cairns or looking for those made on previous visits. For example, the 1971 photographs were taken during a long forced march to get help for Ian Holmes after he had broken his leg. It is ironic that one of the main objectives that year, which was the first in which we have had sufficient time for such work, had been to establish large paint-marked cairns at all photo sites, in order to put future observations on a better basis. I hope it will prove possible to do this on some future visit.

Stephenson

Further retreat of this glacier is suggested by the finding, on 13th February, 1971, that on the SE margin of the glacier a lake several hundred metres wide had formed since 1965, at the point where a large meltwater stream flows through the Dovers moraine to the northern beach of Spit Bay. No evidence was found of a dam in the stream.

Brown

The ISAGE paper (Plates 4 and 5) shows the retreat of this glacier between 1947 and 1963. Continuing retreat and decrease in thickness have been apparent in 1965, 1969 and 1971. Photo 2, which was taken in 1969 from a lower viewpoint on Round Hill than was the 1963 photo, shows that in those 6 years the middle of the ice front retreated by about 300 metres. The ice front retreated another 50 metres between 1969 and 1971.

Little Challenger and Baudissin

In the Postscript to the ISAGE paper, and in a paper entitled "Heard Island Reconnaissance, 1969" (Polar Record 15 (96), 335-336, September 1970) I commented that "major re-advance was obvious" in these glaciers by 1969. Photos 3 and 4, taken in 1963 and 1971 from virtually the same viewpoint, provide some of the evidence for this conclusion, by showing that Little Challenger, which in 1963 had sloped down to the sand behind a broad beach, had re-advanced into the sea and developed ice cliffs conservatively estimated to be more than 20 metres high. It had also spread laterally onto the bluff separating it from the Baudissin, and the appearance of both glaciers was generally similar to that of 1954. No clear changes were apparent from 1969 to 1971.

Schmidt

This glacier is really a small westerly tongue of the Baudissin. Although the position of the terminus cannot be seen in any of the photographs, Photos 5 to 12 show the following changes in the ice just above the terminus:

1954 - 1963 Narrowing and apparent decrease in thickness.

1963 - 1969 No clear change, other than slight wasting of southern side.

1969 - 1971 Marked decay of whole southern half of glacier tongue. This is particularly obvious in Photos 13 and 14, which were taken in 1963 and 1971 from virtually identical viewpoints (with 55 mm and 50 mm lenses respectively).

Vahsel

Photos 15 to 18, taken in 1954, 1963, 1969 and 1971 from similar viewpoints (immediately south of the "Erratic Boulder" at Erratic Point), appear to show the following changes:

1954 - 1963 Marked recession, as shown by lower and round-edged ice cliffs, retreat at north end, and development of extensive moraines at the southern end of the sea cliffs, near Cape Gazert.

1963 - 1969 Further recession, as shown by still lower, and more rounded, sea cliffs.

1969 - 1971 Marked re-advance, as shown by development of high and sharp-edged sea cliffs, which extend closer to Cape Gazert and to Erratic Point. The general appearance approaches that of 1954.

Photos 19 to 22, taken from near the north end of the beach at Southwest Bay in 1954, 1969 and 1971, also show these fluctuations. Cape Gazert on the right, and the large and angular Erratic Boulder (which had slipped downhill by 1971) on the left, can be seen at each end of the ice front. A possible decrease in thickness of the body of the glacier, some hundreds of metres behind the ice front, is suggested by changes in the appearance of the rock outcrops on the far side of the Vahsel, which can be seen above the Erratic Boulder in Photos 19 (although snow-covered), 20, and 21 - but not in Photo 22, which was taken from a viewpoint somewhat nearer the glacier. However, variations in camera position, height, and focal length make it difficult to confirm this impression.

Jacka

Continuing recession from its 1963 status was apparent in 1969 and 1971.

4 MASS BALANCE STUDY RESULTS

The following information is given for each country listed:

Sources of data

References

Supplementary data and comments

Canada

All data were supplied by Dr. O. Løken, Head of the Glaciological Subdivision, Hydrologic Sciences Division, Inland Waters Branch, Ottawa, Ontario, Canada.

References to Table 9.2.2:

Koerner, R.M.: The Mass Balance of the Devon Island Ice Cap, Northwest Territories, Canada, 1961-66. Journal of Glaciology, Vol. 9, No. 37, 1970, and personal communication, 1971.

Hattersley-Smith, G. and Serson, H.: Mass Balance of the Ward Ice Rise and Ice Shelf: A 10 Year Record. Journal of Glaciology, Vol.9, No. 56, 1970.

Reference to Table 9.2.3:

Glacier Mass Balance Data 1965-1971, Department of the Environment, Inland Waters Branch, Ottawa, Canada, Technical Bulletin No. 56, 1972.

USA

All data were transmitted by Dr. William O. Field.

Cf. remark concerning the investigators mentioned in the tables in Chapter 3.

References to Tables 9.2.4 and 9.2.5:

Alford, D. and R. Clark (1968). "The 1967 Mass Balance of Grasshopper Glacier, Montana", Northwest Science, v. 42, n. 3, pp. 115 - 122.

Bull, C. and Marangunic, C. (1967). "The Earthquake-Induced Slide on Sherman Glacier, South-Central Alaska, and its Glaciological Effects", pp. 395 - 408, in: Physics of Snow and Ice, H. Oura, ed. Int'l. Conf. Low Temp. Sci., Sapporo, Japan, v. 1.

Handewith, H. (1959). "Recent Glacier Variations on Mt. Hood", Mazama, v. 41, pp. 23 - 28.

LaChapelle, E. (1965). "The Mass Budget of Blue Glacier, Washington", Jour. Glac., v. 5, n. 41, pp. 609 - 623.

Lloyd, D. (1970). "The Isabelle Glacier, Front Range, Colorado, during the 1968-69 Budget Year", M.A. Thesis, Univ. Colorado Dept. of Geography.

Matthes, F. and Phillips, K.N. (1943). "Surface Ablation and Movement of Ice on Eliot Glacier", Mazama, p. 17.

Meier, M.F. and Tanborn, W.V. 1965). "Net Budget and Flow of South Cascade Glacier Washington", Jour. Glac., v. 5, n. 41, pp. 547 - 566.

Meier, M. F. (1966). "Some Glaciological Interpretations of Remapping Programs on South Cascade, Nisqually, and Klawatti Glaciers, Washington", Can. Jour. Earth Sci., v.3, n.6, pp. 811 - 818.

Meier, M. F. (1968). "Calculations of Slip of Nisqually Glacier on its Bed: No Simple Relation of Sliding Velocity to Shear Stress", General Assembly of Berne, IUGG, Publ. n. 79, pp. 49 - 57.

Reger, R. D. (1968). "Recent History of Gulkana and College Glaciers, Central Alaska Range, Alaska", Jour. Geol., v. 76, n. 1, pp. 2 - 16.

Veatch, F. M. (1970). "Analysis of a 24-Year Photographic Record of Nisqually Glacier, Mount Rainier National Park, Washington", U.S. Geol. Survey, Prof. Paper 631, 52 pp.

Wendler, G. (in press). "Mass Balance Studies on the McCall Glacier", General Assembly of Moscow, IAHS-IUGG.

Norway

All data were supplied by Dr. Gunnar Østrem, Glaciology Section, Vassdrags- og Elektrisitetsvesen, Hydrologisk Avdeling, P.O. Box 5091, Majorstua, Oslo 3, Norway.

References to Tables 9.2.6 and 9.2.7:

Glazio-Hydrologiske Undersøkelser: Norge Vassdragsdirektoratet, Hydrologisk avdeling, for the year 1965: by Pytte and Liestøl, Oslo 1966

1966: by Pytte, Oslo 1967

1967: by Østrem and Pytte, Oslo 1968

1968: by Pytte with others, Oslo 1969

1969: by G. Østrem, T. Ziegler, S.R. Ekman, Oslo 1970

1970: by T. Ziegler with others, Oslo 1972

Sweden

References to Tables 9.2.8 and 9.2.9

All data were supplied by Dr. V. Schytt, Department of Physical Geography, University of Stockholm.

The following extracts are from the report of V. Schytt entitled "Notes on glaciological activities in Kebnekaise, Sweden, during 1966 and 1967":

Mass balance studies on Storglaciären (Table 9.2.8)

1965 - 1966 winter balance

The inventory of the winter balance was started in April 1966. W. Karlén first measured the snow depth along evenly spaced profiles across the glacier. The end points of these profiles have been marked in such a way as to permit the measurements to be repeated in the same place year after year - with 100 m between each reading and 100 m between the profiles. Based on these

323 snow depth observations and on the density distribution in 9 pits the winter balance was calculated for 17-18 April. Additional surveys of 64 accumulation stakes were made on 28 April (with 2 pits), on 6 May (3 pits) and on 27 May, and the winter balance was found to reach its maximum on 27 May and then it stayed unchanged for 10 days. The stake readings of 27 May werde used for a correction of the value obtained on 17-18 April.

The total winter balance, as calculated for 27 May, amounted to $3.73 \cdot 10^6$ m^3 of water ($1.21 \cdot 10^6$ m^3/km^2 or 121 g/cm^2) or 7 % less than the average, $4.03 \cdot 10^6$ m^3, for the 20-year period 1945-1965.

1966 summer balance

As stated above the accumulation season lasted until 27 May, but after that a cold period delayed the beginning of the ablation season until 7 June. Melting was then studied continuously during the summer by means of 57 ablation stakes and the last melting was recorded on 7 September. The total length of the ablation season was thus 93 days; the number of positive degree-days at the Tarfala station (1130 m a.s.l.) was 545.5 and there were only two days during the ablation season, both in September, which had slightly negative mean temperatures. The accumulation which took place during the ablation period was quite negligible.

The summer balance agrees very nearly with the total ablation. The winter balance also agrees this year with the total accumulation. The total net loss during the ablation season was $5.36 \cdot 10^6$ m^3 of water ($1.74 \cdot 10^6$ m^3/km^2 or 174 g/cm^2). The 1945-1965 average was $5.65 \cdot 10^6$ m^3. The summer balance gradient was 50 cm/100 m.

1965 - 1966 net balance

The net balance of the year 1965 - 1966 increased from a minimum of - 275 cm in the interval 1130 - 1140 m a.s.l. to 0 cm at 1500 m, the equilibrium line of 1966, and to + 165 cm at 1680 - 1700 m. The net balance gradient was approximately 90 cm/100 m.

The glacier surface below 1200 m is of such small extent that the large figures for summer balance as well as net balance do not influence the total result very much. Between 1340 and 1400 m, however, there are large areas between the contours and in this quite narrow interval we record 31 per cent of the total summer balance. The same part of the glacier is of equal importance to the total net balance. Out of the net loss below the equilibrium line, a total of $2.54 \cdot 10^6$ m^3, 42 per cent was due to the negative balance between 1340 and 1400 m.

The final result of the year 1965 - 1966 was a net balance of $- 1.63 \cdot 10^6$ m^3 of water ($- 0.53 \cdot 10^6$ m^3/km^2 or 53 g/cm^2) - very close to the 1945 - 1965 average of $- 1.62 \cdot 10^6$ m^3.

The equilibrium line was situated at 1500 m a.s.l.

1966 - 1967 winter balance

The map of the winter balance is based on 312 observations of snow depth. The same rectangular network with 100 m squares, which was used in 1966, could be used again, since the transverse profiles are now well marked along the sides of the glacier. The observations were made during 19-22 May and this coincided with the end of the accumulation season. During the days following 22 May very small changes took place on the glacier; some surface melting occurred on the lower parts of the glacier, but the water refroze either in the snow or onto the ice surface. Runoff measurements in the main stream draining Storglaciären on 24, 25, 26 and 27 May gave 0.3 l/sec., 0.5 l/sec., 0.5 l/sec. and 0.9 l/sec., respectively. On 28 May the whole runoff channel was filled with water, dammed up by the snow, and no measurement could be made - the runoff was then estimated at 50-100 l/sec. At the outlet of the lake below Isfallsglaciären, Storglaciären's nearest neighbour, the runoff was less than 0.1 l/sec. through 31 May, it was 1.3 l/sec on 1 June and more than 20 l/sec on the 2nd. Since the runoff during these first days of the melting season comes mainly from snow-covered areas outside the glacier, it is reasonable to assume that the runoff from the lower parts of the glacier started on 1 June. There was thus a 10-day period separating the winter season from the summer.

The winter balance of 1966 - 1967 was more influenced by wind conditions and topography than it was during the previous year. The great amount of snow deposited on the snout - below 1300 m - is very striking. Above 1300 m the irregular winter balance curve has an average gradient of 45 cm/100 m, not very different from the 50 cm/100 m valid for the 1965 - 1966 curve.

When it comes to the areal winter balance, 1966 and 1967 show more similar conditions. Because of the small area below 1300 m great snow depths on this part of the glacier add little to the total winter balance, which per 22 May amounted to $4.19 \cdot 10^6$ m^3 of water ($1.36 \cdot 10^6$ m^3/km^2 or 136 g/cm^2).

1967 summer balance

The summer melt season started on 1 June and lasted until 21 September, 113 days. The number of positive degree-days recorded at the station (1130 m a.s.l.) was 577. Very little snow fell during the summer and there was not much difference between summer balance and summer ablation.

The total summer balance was $4.89 \cdot 10^6$ m^3 of water ($1.59 \cdot 10^6$ m^3/km^2 or 159 g/cm^2); the summer balance gradient was 45 cm/100 m.

1966 - 1967 net balance

The net balance curve shows very similar variations to the winter balance curve and its gradient (above 1300 m) is naturally the sum of the gradients of the other two curves, i.e. in this particular case 90 cm/100 m.

The area distribution is the dominant factor in the areal mass balance. Out of the total summer balance 29 per cent was recorded between 1340 and 1400 m (31 per cent in 1966), and 46 per cent (42 in 1966) of the total net loss below the equilibrium line came from the same height interval with 26 per cent of the glacier area. The areal net balance of the year 1966 - 1967 was

Temperature data (°C) from the Tarfala Station (1130 m a.s.l.), 1966

Month	Mean monthly temp.	Highest daily mean (date)	Lowest daily mean (date)	Maximum (date)	Minimum (date)
January	- 13.7	+ 0.8 (7)	- 21.9 (3)	+ 2.8 (7)	- 24.7 (29)
February	- 18.7	- 12.1 (19)	- 30.9 (2)	- 10.0 (19)	- 33.0 (1, 2)
March	- 13.2	- 4.0 (20)	- 20.0 (23)	- 1.0 (20)	- 24.0 (14)
April	- 10.1	+ 1.4 (29)	- 18.6 (11)	+ 2.7 (29)	- 23.0 (12)
May	- 3.1	+ 4.2 (19)	- 11.7 (10)	+ 7.8 (19)	- 17.5 (10)
June	+ 5.6	+ 14.9 (20)	- 2.5 (5)	+ 19.5 (20)	- 5.0 (5)
July	+ 7.1	+ 12.5 (26)	+ 4.0 (30)	+ 16.0 (26)	+ 1.2 (1)
August	+ 4.5	+ 10.2 (4)	+ 0.6 (31)	+ 11.8 (4)	- 1.2 (24)
September	- 1.8	+ 6.0 (6)	- 8.8 (27)	+ 8.0 (6)	- 11.3 (27)
October	- 7.0	- 3.1 (13)	- 10.5 (29)	- 2.0 (13)	- 13.0 (30)
November	- 6.8	- 0.8 (19)	- 12.4 (1)	+ 1.4 (19)	- 14.5 (27)
December 1)	- 13.1	- 1.4 (3, 7)	- 32.8 (16)	-	-
Year	- 5.86	+ 14.9 (20/6)	- 32.8 (16/12)	+ 19.5 (20/6)	

1) Because of instrument failure no temperature recordings were obtained at the station during December. Temperatures for this month have been computed from recordings made at Nikkaluokta, 20 km ESE of Tarfala.

Temperature data (°C) from the Tarfala Station (1130 m a.s.l.), 1967

Month	Mean monthly temp.	Highest daily mean (date)	Lowest daily mean (date)	Maximum	Minimum
January	- 13.4	- 1.2 (16)	- 23.4 (29)	± 0.0 (15, 16)	- 26.2 (29)
February	- 9.5	- 2.5 (11)	- 18.0 (1)	- 0.3 (11)	- 20.8 (3)
March	- 7.3	- 1.2 (5)	- 13.5 (23)	+ 2.0 (8)	- 17.3 (15)
April	- 5.1	+ 0.8 (27)	- 11.1 (21)	+ 3.5 (27)	- 16.5 (21)
May	- 0.5	+ 5.5 (29, 30)	- 6.6 (1)	+ 7.0 (29,30,31)	- 9.5 (1)
June	+ 2.5	+ 8.1 (2)	- 2.8 (9)	+ 10.5 (20)	- 5.2 (8)
July	+ 5.9	+ 11.1 (19)	+ 1.4 (1)	+ 15.7 (19)	- 0.5 (1)
August	+ 6.8	+ 10.9 (4)	+ 3.7 (20)	+ 13.3 (3)	+ 0.7 (20)
September	+ 3.7	+ 7.1 (13)	- 3.1 (27)	+ 11.1 (17)	- 4.1 (26)
October	- 4.7	+ 1.2 (3, 5)	- 12.0 (21)	+ 3.2 (3)	- 15.1 (21)
November	- 4.7	+ 0.7 (20)	- 11.4 (22)	+ 2.8 (19)	- 16.7 (22)
December	- 14.1	- 6.3 (1)	- 21.1 (23)	- 1.3 (14)	- 24.0 (23)
Year	- 3.4	+ 11.1 (19/7)	- 23.4 (29/1)	+ 15.7 (19/7)	- 26.2 (29/1)

$-0.69 \cdot 10^6$ m^3 ($-0.23 \cdot 10^6$ m^3/km^2 or 23 g/cm^2). With the positive results of 1961/62, 1963/64 and 1964/65 the net balance for the years 1961/67 is still slightly positive, $+ 0.9 \cdot 10^6$ m^3 (3g/cm^2) for the whole period.

The equilibrium line of 1966 - 1967 was at 1500 m a.s.l.

Spot observations of mass balance 1965-67 on 6 glaciers south of Kebnekaise (Table 9.2.9.)

Between 29 July and 5 August 1965 a group from the Department of Physical Geography, University of Stockholm, visited the eight glaciers south of Kebnekaise which are listed in the previous section. The group was supported by two Army helicopters and one day was devoted to each glacier. Two members selected suitable fix points and made various general observations, one student painted markers, two surveyors fixed the position of all fix points and measured longitudinal ice surface profiles on Mikkajökeln, Ruotesglaciären and Hyllglaciären, and four students, finally, drilled down two ablation stakes on each 100 m contour in the ablation areas (occasionally other spacing had to be used).

Most stakes were drilled to a depth of between 5 and 6 m; the drilling was done with a hand-operated drill developed for the Tarfala investigations in 1947 and most 5-m holes were finished in less than 25 min.

When O. Melander with two assistants visited these glaciers in 1967 they also tried to remeasure the stakes. This visit had to be made in July (Salajekna only on 2-3 Aug.) and there was much snow left, but 1-3 stakes were found on six of the eight glaciers. Since the observations were made at approximately the same time in 1965 and 1967 we get approximate values of the net balance for a period of two years. The results of this study are reported in Table 9.2.9, which also includes the net balance at various elevations on Storglaciären as recorded per 1 Aug. 1965 to 23 July 1967, i.e. from the middle of the first study period to the middle of the next.

The values reported in Table 9.2.9 are of course too few to permit any safe conclusions about variations in net balance gradient, but they stress the great differences between glaciers lying on the eastern and northern, more continental sides of the Kebnekaise and Sarek massifs and those lying in Sulitelma and south of Sarek. It is quite natural that the net balance curves for the maritime glaciers are situated lower in the diagram than those for the more continental ones - the curves intercept the ordinate at the equilibrium line, and the equilibrium line as well as the glaciation limit rises with the degree of continentality. There exists a definite relationship between summer balance gradient and the degree of continentality (Schytt, 1967, p. 331), i.e. a maritime glacier has a large gradient and a continental glacier a small gradient. The few Swedish data now available on variations of net balance gradient suggest that even this gradient is larger in maritime areas and smaller where the climate is drier.

Acknowledgements: Several students have helped to collect the data presented above. Mr. Wibjörn Karlén has been responsible for the field work connected with the mass balance studies and has also compiled the accumulation and ablation data. Mr. Olle Melander made the 1967 survey of the glaciers south of Kebnekaise and Mr. Åke Fleetwood the observations at three glaciers further north. Mrs. E. Kuldver drew the maps and Mrs. B. Hansson the diagrams. Financial support was received from the Swedish Natural Science Research Council and from the city of Kiruna. The Army Helicopter School helped with transportation.

References:

Schytt, V., 1966. Notes on glaciological activities in Kebnekaise, Sweden - 1965, Geogr. Ann. 48A (1), p. 43-50.

Schytt, V., 1967. A study of "ablation gradient". Geogr. Ann. 49A (2-4), p. 327-332.

Austria

Tables 9.2.10 and 9.2.11

The data were supplied by the following correspondents:

For Hintereisferner and Kesselwandferner:

Prof. Dr. H. Hoinkes, Institut für Meteorologie und Geophysik, Schöpfstrasse 41, A-6020 Innsbruck.

For Vernagtferner and Langtalerferner:

O. Reinwarth, Kommission für Glaziologie der Bayerischen Akademie der Wissenschaften, Marstallplatz 8, D-8 München 22.

For Stubacher Sonnblickkees:

Dr. H. Slupetzky, Geographisches Institut der Universität, Akademiestrasse 20, A-5020 Salzburg

Abbreviations:

S_c	accumulation area	B_a	net ablation
S_a	ablation area	b_a	specific net ablation
S	total area of glacier	B	mass balance
B_c	net accumulation	b	specific mass balance
b_c	specific net accumulation	E	equilibrium line

The values for Stubacher Sonnblickkees in Table 9.2.10 for 1963/64 and 1964/65 differ from those given in "Fluctuations of Glaciers 1959-1965", Vol. 1, Table 10, having been recalculated from improved data.

The following note, which is important for the understanding of the values for the Langtalerferner in Tables 9.2.10 and 9.2.11, is translated from the report of the Kommission für Glaziologie in the "Jahrbuch der Bayerischen Akademie der Wissenschaften 1970":

Special difficulties were encountered in determining the glacier area of the Langtalerferner. The previous uncertainties in fixing the orographically right-hand edge of the glacier in particular were eliminated in autumn 1969 by a completed photogrammetric survey. The result of planimetering was a figure of 3.049 km^2, a glacier area 5 % larger than the figure evaluated in 1966 and used so far. It is not possible, however, to apply the improved area figure to earlier mass balance analyses, since the corrected value was subject to changes in time which are not sufficiently well known. In order to be able to compare the mass balance values for 1968/69 for the Langtalerferner exactly with those of earlier years, the mass balance of the difference in area, which belongs mainly to the ablation region, was evaluated separately and the mass balance thus calculated for the glacier area used hitherto. The following figures were then obtained:

Langtalerferner - Mass Balance 1968/69

	Survey 1969	Survey 1966	
Accumulation Area	1.354	1.348	km^2
Net Accumulation	+ 0.353	+ 0.352	10^6 m^3
Ablation Area	1.695	1.555	km^2
Net Ablation	- 1.908	- 1.761	10^6 m^3
Total Area	3.049	2.903	km^2
Net Balance	- 1.555	- 1.409	10^6 m^3
Specific Net Balance	- 51.0	- 48.5	cm

France

The data on the Glacier de Sarennes (Table 9.2.12) were supplied by L. de Crécy, Ingénieur en Chef of the C.E.R.A.F.E.R., Division Nivologie, B.P. 114, 38 St-Martin-d'Héres, France.

L. de Crécy's commentary on the measurements on the Glacier de Sarennes is here given in excerpt form:

Extract from a note presented on 5 July 1972 at the 1972 meeting on glaciology and hydro - electric documentation of the Section de Glaciologie of the Société Hydrotechnique de France (SG/SHF).

The first measurements of the terminal tongue date back to 1891, the year in which Prince Roland Bonaparte stated, according to the guide Rodaron, that the glacier was retreating slightly. But the first complete study of the glacier, accompanied by a remarkable contour map, was the work of Flusin, Jacob, Offner and Lafay in "Etudes Glaciologiques", Vol. 1, in 1905-1906. Some measurements of the tongue carried out by the administration of the department Eaux et Forêts in 1927- 1932 - 1935 showed only insignificant changes in the positinn of the tongue.

The fact is that in reality the Sarennes Glacier is of the type of the cirque glaciers. Since the nineteen-forties it no longer crosses the two barriers that enclose its valley. There is accordingly no further observable movement of its tongue. It has, moreover, a moderate gradient only and shows no crevasses that might indicate internal movement (the only irregularities are "glacier mills" belonging to an intraglacial karstic network and a very inconspicuous crevasse of circular shape which sometimes appears at the centre of the glacier and suggests the initial melting of the keystone of an arch above a hypothetical sub-glacial lake). There thus appears to be no appreciable movement of the ice, but only level variations due solely to the cycle of ablation and accumulation.

The work of Ahlmann and of the school of Scandinavian glaciologists had established, by the thirties of this century, the measurement of ablation and accumulation, of mass balance and régime, by means of coring and stakes over the whole surface of a glacier. Their methods are universally adopted for determining one of the essential factors of the fluctuation of glaciers, that of the exchanges between the surface and the atmosphere (precipitation and melting). The peculiarity of the Sarennes Glacier is that this factor is practically the only one involved. For this reason Cherrey had the happy idea, in 1948, of using it for the application of the Scandinavian methods, so as to be able to read the indications of the vast natural level gauge constituted by the cirque of Sarennes.

The Ahlmann method, applied to the Sarennes Glacier in an extremely simplified form to permit easy and absolutely continuous use by non-specialized personnel, is based on five measuring points on a long axis of the glacier at altitudes of approximately 2870, 2920, 2970, 3000 and 3050 m, the figures for each point being weighted by a coefficient proportional to the glaciated surface area in the altitude sector represented by each point (these coefficients are 0.5, 1, 2.5, 2.5 and 3.5 respectively). These measurements are made at intervals of about three weeks from 1 June till 30 September. They consist in measuring the water equivalent of the sonw cover by taking cores and weighing them, then measuring the lowering of the surface by ablation, as the season advances, by means of stakes driven into the ice. It is then possible to calculate the annual accumulation (stock of snow found on 1 June) and the ablation (sum of the amounts of snow and ice melted between 1 June and 1 October). Thence an annual balance and a régime are obtained for the glacier. The special behaviour of the Sarennes Glacier enables these balances to be checked very easily. since they express themselves solely as a thinning or thickening on the spot, without any lateral movement. Topographical surveys with level measurements carried out at intervals from 1948 to 1959 showed good agreement with the balances and justified the use of this method.

The Sarennes Glacier registers only climatic variations and registers them with sensitivity, since it extends in practice only between 2850 and 3100 m. In fact, a variation of the altitude of the firn limit of only 250 m may either uncover the glacier completely, thus favouring very rapid ablation, or may leave it entirely covered with snow and protected against melting by the high albedo of the snow. This explains the importance of the yearly variations of the balances and of their factors (standard deviation for accumulation 0.45 m on a mean of 1.50 m of water, for ablation 0.75 on a mean of 2.33, for the balance 0.86 on a mean of 0.6, for the régime 0.92 on a mean of 3.65). It would have been tempting to bring these values measured on the glaciers into relation with meteorological figures proper. Unfortunately the nearest long series of measurements available were those for the climate of Grenoble, a region rather distant from Sarennes.

It was on the basis of the climatic factors of the Grenoble area, however, that Garavel ("Comportement glaciaire et fluctuations climatiques", Revue Forestière No. 1, 1955) and de Crecy ("Le glacier de Sarennes et le climat Grenoblois", Annale de l'Ecole Nationale des Eaux et Forêts, Vol. XX, No. 3, 1963) tried to find some correlation with the fluctuations of the glacier. These endeavours met with mixed success, but they permitted a reconstruction of the old balances for the Sarennes Glacier from 1906 to 1948 which is in good agreement with the thinning of the glacier in the same period as revealed by the contour map by Flusin, Jakob, Offner and Lafay dating from 1906.

The Sarennes balance figures, which have now been recorded for 23 years by the same method and are therefore strictly comparable, are the subject of an annual report of the SG/SHF. That of 1969 for the season 1967-1968 included a chart of the variations of the balances for 1948 to 1968 and showed that the years 1964-1968 had been characterized by slightly positive balances and weak régimes. Since then, in 1968 and 1969, and more particularly in 1970 and 1971, there has been a return to negative balances (-0.36; -0.41; -1.10) and to régimes of normal amplitude (3.5 to 4 m of water).

The balance measurements of Sarennes are very simple, and their only merit is that they have been unchanged for 23 years. They are pratically the only ones in France permitting comparison with the balances measured in the other countries of the Alpine arc. They may have extensions into other fields and in particular they may be compared with the growth of trees in the nearby forest regions: since the glacier régime is higher when there is a large acculumation of snow and therefore large water reserves, and when there is more ablation and therefore summer heat, one may assume that high régimes correspond to better growing conditions for the forests. It would be worth while to make a comparison between the régimes of the glaciers and the width of the age-rings of the forest trees.

But a new development now begins to threaten the Sarennes balance measurements: the progressive advance of skiers is changing the situation at the two upper measuring points on the glacier. It would be desirable to find a location that is not disturbed in this way and where the indications of point balances will be the same.

References

Cf. "Fluctuations of Glaciers, 1959-1965", Vol. 1, page 38, ref. 24 and 25, and mimeographed communications to the SG/SHF:

Communications by M. de Crécy at the meeting of

23/24 February 1967
7 / 8 March 1968
4 / 5 March 1969
17/18 March 1970
4 / 5 March 1971.

Germany

All data shown in Tables 9.2.13 and 9.2.14 were supplied by O. Reinwarth, Glaciology Commission of the Bayerische Akademie der Wissenschaften (Bavarian Academy of Sciences), Marstallplatz 8, D-8 Munich 22.

Italy

All data shown in Table 9.2.15 were supplied by Dr. Giorgio Zanon, Geographic Institute, University of Padua.

Further information will be found in:

Bollettino del Comitato Glaciologico Italiano No. 15, Parte Prima, Turin 1969 (Ghiacciaio Marmolada 1964/65 and 1965/66), No. 18, Turin 1970 (Ghiacciaio del Caresèr 1966/67 and 1967/68).

Switzerland

Tables 9.2.16 to 9.2.19, Figures 9.2.1 to 9.2.3.

References:

"Les variations des glaciers suisses", annual reports of the Glacier Commission of the Société Helvétique des Sciences Naturelles (Swiss Academy of Natural Sciences),

86th Report 1964/65	89th Report 1967/68
87th Report 1965/66	90th Report 1968/69
88th Report 1966/67	91st Report 1969/70.

These reports, which are also published in German under the title "Die Gletscher der Schweizer Alpen", can be obtained from: Sekretariat der Gletscherkommission der SNG, c/o Versuchsanstalt für Wasserbau, Hydrologie und Glaziologie an der ETH-Z, Voltastrasse 24, CH-8044 Zurich.

USSR

The results given in Table 9.2.20 were contributed by the following authors and organizations:

Region	Name of Glacier	Table 9.2.20 Part	Authors
Polar Ural	Obrucheva Igan	1 1	A. Gus'kov, Geographical Institute, Academy of Sciences, SSR
Caucasus	Dzhankuat	1	G.N. Golubev, V.S. Freydlin, Moscow State University
	Bezingi Kelbashi Zeiskiy	2 2 2	V.D. Panov, Northern Caucasian Administration of Hydrometeorological Service, SSR

Region	Name of Glacier	Table 9.2.20 Part	Authors
Tien-Shan	Karabatkak Tsentralny Tuyuksu Igly Tuyuksu	2 1 and 3 3	A. Dikikh, Tien-Shan Physical-Geographical Station, Academy of Sciences, Kirgizian, SSR
	Molodezhnyy Shokalskiy Mametova Teu-Northern Teu-Southern Korzhenevsky	3 3 3 3 3 3	K. G. Makarevich Geographical Sector, Academy of Sciences, Kazakhstan, SSR
Altai	Bolshoy Berel Malyi Berel	3 3	K. G. Makarevich, Geographical Sector, Academy of Sciences, SSR

New Zealand

The information in Table 9.2.21 and Figure 9.2.4 was supplied by Dr. Peter W. Anderton, Glaciology Section, Ministry of Works, P.O. Box 1479, Christchurch, New Zealand.

Dr. Anderton comments as follows on Table 9.2.21:

Tasman Glacier: Snow surveys.

Since 1965 the Ministry of Works has carried out specific balance measurements at six locations on the upper glacier (Fig. 9.2.4). The surveys have encountered difficulties due to the excessive snowfalls, the high rates of ablation, the high initial density and rapid densification of the snowpack, and the lack of easily identifiable annual layering in the snowpack. Consequently, the records are incomplete.

Data for the glacier budget years 1966-67 to 1970-71 are enclosed. Winter balance was low during 1966-67, high during 1967-68 and 1968-69, and moderate during 1969-70 and 1970-71. The high total accumulation during 1967-68 was largely the result of an unusually heavy freak snowfall in November 1967.

Summer balance appears to have been fairly constant from 1966 to 1969, so that net balance variations were controlled by winter balance. However, the summer balance during 1969-70 was very high and remained high during 1970-71. No satisfactory estimate of mass balance of the glacier can be made, but the estimated altitudes of the equilibrium line can be used as an index of mass balance variations.

Year	Altitude of equilibrium line (m)
1966-67	1970
1967-68	1630
1968-69	1690
1969-70	2200
1970-71	1930

Personal communication, P. W. Anderton, January 1972.

Deception Island

The information for Table 9.2.22 and Figure 9.2.5 was supplied by Dr. O. Orheim, Institute of Polar Studies, Ohio State University, 125 South Oval Drive, Columbus, OH 43210, USA.

Cf. also Reference in Chapter 2.

5. CHANGES IN THICKNESS, AREA AND VOLUME

The following information is given under each of the countries listed:

Sources of data

References

Supplementary data and comments.

USA

All data for Tables 9.3.1 were supplied by Dr. William O. Field.

Cf. the note in Chapter 3 concerning the Investigator mentioned in the tables.

References

Alden W.C. 1914. Glaciers of Glacier National Park, Dept. of Interior, Washington.

Bengtson, K.B. 1956. Activity of the Coleman Glacier, Mt. Baker, Washington, USA, 1949-1955. Jour. Glac., V. 2, No. 20, pp. 708-713.

Bull, C. and Marangunic, C. 1967. The Earthquake - Induced Slide on Sherman Glacier, South-Central Alaska, and its Glaciological Effects, pp. 395-408 in: Physics of Snow and Ice, H. Oura, ed. Int. 1, Conf. Low Temp. Sci., Sapporo, Japan Vol. 1.

Dyson, J.L. 1948. Shrinkage of Sperry and Grinnell Glaciers, Glacier National Park, Montana. Geog. Rev., Vol. 38, pp. 95-103.

Meier. M.F. 1968 Calculations of Slip of Nisqually Glacier on its Bed; No Simple Relation of Sliding Velocity to Shear Stress.- General Assembly of Bern, IUGG, Publ. n.79, pp. 49-57.

Meier M.F. and Tangborn, W.V. 1965. Net budget and flow of South Cascade Glacier, Washington. Jour. Glac., Vol. 5, No. 41, pp. 547-566

Reid, J.R. 1969. Effects of a debris slide on "Sioux Glacier", south-central Alaska. Jour. Glac., Vol. 8, No. 54, pp. 353-368.

Sperry, A.I. 1938. Avalanche, Christopher Publishing House, Boston.

Austria

Table 9.3.2 . The data were communicated by the following correspondents:

For Vernagtferner and Guslarferner: O. Reinwarth, Glaciology Commission of the Bayerische Akademie der Wissenschaften, Marstallplatz 8, D-8 Munich 22.

For Hintereisferner: Prof. Dr. H. Hoinkes, Institut für Meteorologie und Geophysik, Schöpf-strasse 41, A-6020 Innsbruck.

For Vernagtferner and Guslarferner cf. also the chapter "Remarks on the Annexed Maps" and the annexed maps of the Vernagtferner.

France

Table 9.3.3 and Figure 9.3.1

The data were supplied by the Electricité de France, Equipement Hydraulique Alpes Nord, Section Topographie, Albertville, France.

Germany

Table 9.3.4

The data were supplied by o. Reinwarth, Munich.

Switzerland

Table 9.3.5

References

" Les variations des glaciers suisses en 1969-1970", 91st report of the Glacier Commission of the SHSN, Berne, 1972.

Cf. also the annexed map of the Mattmark Glaciers, scale 1 : 10 000, surveys 1956 and 1967.

USSR

Table 9.3.6

The figures given in Table 9.3.6 were contributed by the following authors and organizations:

Region	Name of Glacier	Table 9.3.6 Part	Number of Figure	Authors
Caucasus	Mayli Gergeti Suatisi Koruldash Kirtisho Abano Chalaat Lazg-Tsiti	1 2 and 3 4 4 6 6 5 2	2.5 2.5 2.5 2.5 2.5 2.5 2.5 2.5	V. Sh. Tsomaya, O.A. Drobyshev, Zak. NIGMI and administration of hydrometeorological Service Gruzin, SSR
Tien-Shan Zailiyskiy Alatau	Tsentralny Tuyuksu Igly Tuyuksu Manshuk Mametova Konstitutsiya Dimitrieva Toguzak Shokalskiy Korzhenevskiy	7 - 14 15-16 17-19 21-23 24-25 26-27 28-36 39-43	2.9 P.1 2.9 P.1 2.9 P.2 2.9 P.1 2.9 P.3 2.9 P.2 2.9 P.2 2.9 P.3	K.G. Makarevich Geographical Section, Academy of Sciences, Kazakhstan, SSR
Tien-Shan Zailiyskiy Alatau	Bogatyr	44-47	2.9 P.1	K.G. Makarevich I. Ia. Fedulov, Geographical Section, Academy of Sciences Kazakhstan, SSR
Zailiyskiy and Kuney Alatau	Zhangyryk	48-49	2.9 P.2	
Kungey Alatau	Southern Zhangyryk	50-51	2.9 P.2	
Altai	Bolshoy Berel	52	2.9 P.3	
	Malyi Berel	53-54	2.9 P.3	K.G. Makarevich, G.A. Tokmagambetov, N.V. Ierasov, N. Ospanov
Tien-Shan Zailiyskiy Alatau	Teu-Southern Molodezhny	37-38 20	2.9 P.3 2.9 P.1	K.G. Makarevich

6 VARIOUS AND COMBINED GLACIOLOGICAL OBSERVATIONS

The following information is given for each country listed:
Sources of data
References
Supplementary data and comments.

France

Table 9.4.1 and Figure 9.4.1

The Glacier of Saint Sorlin, d'Arves, is entered in "Fluctuations of Glaciers 1959-65", Vol.1 as No. 15 in the index map, Figure 2. The glacier is 3.5 km^2 in size and has been thoroughly investigated since 1957 by the Institute of Glaciology of the C.N.R.S. in Grenoble (Director: Prof. Dr. L. Lliboutry). The data used in Table 9.4.1 and Figure 9.4.1 were contributed in part in a personal communication by C. Carle and in part were taken from the following dissertation: Garcin, Jean-Paul, 1971: Etude de la zone d'ablation de Glacier de Saint Sorlin, Bilan annuel et débit solide (Thèse de 3e cycle), Publication No. 143 of the Glaciology Laboratory of the C.N.R.S., Grenoble.

Switzerland

Table 9.4.2 and Figure 9.4.2

The measurements were carried out by A. Flotron, Dipl. Ing., who has developed a new automatic method for the purpose. Reference: "Les variations des glaciers suisses en 1969-1970", 91st report.

7 HYDROMETEOROLOGICAL DATA

The present volume, "Fluctuations of Glaciers 1965-1970", contains only such data as are not given in meteorological yearbooks or are difficult of access for language reasons.

USA

Cf. report by W.O. Field in Chapter 2 and Table 2.1.

Particulars of Investigators, References and Sponsoring Agencies are given in the headings of Tables 9.5.1.

Sweden

Temperature figures for Tarfala are given in Chapter 4, in the report by V. Schytt.

USSR

Table 9.5.2

The figures contained in Tables 9.3.6 were supplied by the following authors and organizations:

Region	Table 9.5.2 Part	Authors
Polar Ural	1 - 3	A. Gus'kov, Geographical Institute, Academy of Sciences
Caucasus	4 - 16	V. Sh. Tsomaya, O.A. Drobyshev, Zak. NIGMI and Administration of Hydrometeorological Service, Gruzin
		V.S. Freydlin, Moscow State University
Pamir Alai	17 - 28	A.S. Sachetinnikov, A.T. Mashkin, Sarnigmi
Tien-Shan Terskey Alatau	29 - 31	A. Dikikh, Tien-Shan Physical-Geographical Station, A.N. Kirgizian SSR
Tien-Shan Zailiyskiy Alatau	32 - 43	R.G. Golovkova, Geographical Sector, Academy of Sciences, Kazakhstan SSR
Altay	44 - 46	N.V. Ierasov, R.R. Bekten'yarov, Geographical Sector, Academy od Sciences, Kazakhstan, SSR

8 FINAL COMMENTS AND REMARKS

8.1. Objectives and limits of the periodical publication of the PSFG

In "Fluctuations of Glaciers 1959-1965" the following requirements were laid down for the periodical publications of the PSFG:

The data should be so reproduced that :

they afford a general view of development;

they encourage more extensive measurements;

they invite further processing of the results;

they can serve as a basis for such investigations;

they facilitate consultation of the sources.

Analysis of the measurements is beyond the scope of the present publication, which is primarily conceived as a collection of data and source indications qualified to facilitate further investigation for those interested. The available funds make it necessary to accept the restriction to data concerning glacier variations, to the exclusion of almost all climatic data, which can be found in meteorological yearbooks.

8.2 Summary of the measurements for the period 1965-1970

Taken as a whole, the changes in the positions of glacier fronts for the period 1965-1970 present a very heterogeneous picture. In a few mountain ranges, as for instance in the Cascade Mountains in the Rockies or in the Pamir Alai, most glaciers advanced. In the Tien-Shan the numbers of advancing and retreating glaciers are roughly equal. The glaciers of Alaska and the Coast Mountains (USA) differ considerably from one area to another, the general trend being if anything towards retreat, In Iceland, the Alps, the Caucasus, New Zealand and Heard Island the receding tendency is growing less pronounced, while in Scandinavia it is as strong as ever. The glaciers of West Irian are still shrinking. The distribution of retreat and advance over the earth, as mirrored in the observations collected here, reveals no systematic differences on a global scale.

The mean annual net balances in 1965/70 were mostly negative for the observed glaciers in Canada, USA and Spitsbergen. In Scandinavia the unfavourable growth year 1965/66 was followed by two years with mass balances that were mostly positive, while in the Alps four consecutive years, from 1964/65 to 1967/68, brought mass growth that was in some cases considerable. The mass balances for 1968/69 and 1969/70 were again almost wholly negative both in Scandinavia and in the Alps. In the Polar Ural only 1966/67 and 1967/68 of the five years under review favoured growth, while the corresponding years for the Caucasus and the Tien-Shan were 1966/67 and 1969/70. In the other years positive and negative mass balances are approximately equally distributed.

On Deception Island the mass balance of Glacier G 1 was roughly zero in 1968/69, and distinctly negative in 1969/70.

The changes in the glacier termini and the mass balances for the period 1965/70 seem to imply that the beginning of a general advance is not to be expected in the next few years. It is more likely that the increased frequency of advances observed in some areas is only of episodic significance and that the general receding trend will continue, possibly in a weaker form. This assessment of the situation should be qualified by the reminder that information on the polar regions is still very scanty and that some important areas, such as the Karakorum, Himalayas and Andes, are missing from our collected data.

8.3 Notes on the observation network and measuring programme

The scientific object of glacier observations is the study of the mechanism of glacier variations and their relation to climatic variations. Among the numerous parameters determining the reaction of a glacier to climatic changes there are some that are time-dependent and some that are not. Among those not dependent on time is the morphology of the glacier bed, which decides what possibilities of flow are open to the ice stream. The time-dependent parameters include the climatic factors, the mass balances and the changes taking place in the shape of the glacier. Long and if possible unbroken series of measurements are required for all time-dependent factors to permit the investigation of the underlying laws. The termini of glaciers are in many cases measured only every few years. It would be desirable, for the later evaluation of the observations, if the position of the terminus were known every year for as many glaciers as possible. Very often a commentary would also be very valuable for a better understanding of the measurement figures. Unfortunately an indication of the degree of accuracy of the measurements is likewise still missing in the majority of reports.

Regular annual observations must be aimed at in the various observation networks. This will of course hardly be possible in areas that are rarely visited and are difficult of access. From these areas even less complete observations are nevertheless of great value.

8.4 Relationship of the PSFG to the snow and ice projects of the IHD

On pages 43 to 49 of "Fluctuations of Glaciers, 1959-1965", Vol. 1, are given Resolutions I-13, I-14 and I-15 of the first session of the Co-ordinating Council of the IHD relating to the "World Inventory of Perennial and Annual Ice and Snow Masses", the "Measurement of Glacier Variations on a World-Wide Basis" and the "Combined Water, Ice and Heat Balance Measurements at Selected Representative Glacier Basins".

On the basis of these resolutions, which are still in force, a number of guides have been prepared by the ICSI and have been published jointly by the Unesco and IAHS as "Technical Papers in Hydrology". They are:

No. 1: Perennial Ice and Snow Masses. A Guide for Compilation and Assemblage of Data for a World Inventory, 1970.

No. 3: Variations of Existing Glaciers. A Guide to International Practices for their Measurement, 1969.

No. 5: Combined Heat, Ice and Water Balances at Selected Glacier Basins. Part I: A Guide for Compilation and Assemblage of Data for Glacier Mass Balance Measurements, 1970 Part II: Specifications, Standards and Data Exchange, 1973.

Close contact with the two working groups attached to the International Commission on Snow and Ice (ICSI), viz. that studying the combined balances and that engaged on the inventory of perennial ice and snow masses, is essential for the future activities of the PSFG. On the basis of Resolution VIII-3 of the eighth session of the Co-ordinating Council of the IHD of May 1973, the ICSI is planning to set up a temporary Technical Secretariat to handle the world inventory.

The three World Data Centres for Glaciology are also of great importance to the PSFG. Their addresses are:

WDC-A Glaciology: US Geological Survey,
1305 Tacoma Avenue South,
Tacoma, Washington 98402, USA

WDC-B Glaciology: Molodezhnaya 3,
Moscow B-926, USSR

WDC-C Glaciology: c/o Scott Polar Research Institute
Cambridge CB2 1ER, England

T A B L E S

Table 9.1.1. Variations in the positions of glacier fronts - Compilation.

n number of observed glaciers, number of glaciers : in advance (+), stationary (st) or in retreat (-)

	1964/65				1965/66				1966/67				1967/68				1968/69				1969/70			
	n	+	st.	-	n	+	st.	-	n	+	st.	-	n	+	st.	-	n	+	st.	-	n	+	st.	-
Iceland	32	12	4	16	36	11	5	20	37	9	-	28	34	8	-	26	31	5	-	26	33	11	-	22
Norway	14	1	3	10	10	3	-	7	13	5	-	8	11	4	-	7	10	2	-	8	10	2	1	7
Sweden	2	-	-	2	2	-	-	2	2	-	-	2	11	-	-	11	5	-	-	5	3	-	-	3
Austria	85	16	31	38	87	22	25	40	90	19	23	48	88	13	23	52	95	13	17	65	92	19	4	69
France	18	4	-	14	18	3	-	15	18	3	-	15	18	2	3	13	18	2	1	15	18	2	2	14
Italy *)	86	12	10	64	106	25	12	69	101	16	14	71	88	15	28	45	70	19	12	39	111	16	26	69
Switzerland	90	22	10	58	90	34	3	53	100	24	3	73	98	35	6	57	102	29	4	69	98	31	6	61
USA Cascade Mts.	10	8	1	1	12	8	-	4	16	11	3	2	11	6	4	1								
USSR Caucasus	5	2	-	3	8	1	-	7	10	5	-	5	11	4	-	7	11	4	-	7	9	5	-	4
USSR Pamir Alai	11	7	1	3	11	4	2	5	13	5	1	7	11	4	1	6	10	7	1	2	10	6	1	33
USSR Tien-Shan	8	3	-	5	10	4	-	6	12	6	1	5	11	4	1	6	11	6	-	5	8	3	1	4

*) From C. Lesca, La campagna ghiaciologica 1972. Bolletino del Comitato Glaciologico Italiano, No. 20, 1973, p. 103.
These values are controlled and partly do not correspond with the data presented in table 9. 1. 11.

Table 9.1.2. Canada - Variations in the positions of glacier fronts

Athabasca Glacier, Alta 52° 12' N, 117° 15' W

Saskatchewan Glacier Alta 52° 12' N, 117° 08' W

Period	Mean rate of terminus change in metres / year	
	Athabaska Glacier	Saskatchewan Glacier
1945 - 1946	- 23.8	- 46.6
1946 - 1947	- 26.6	- 76.7
1947 - 1948	- 24.7	- 44.5
1948 - 1949	- 27.2	- 48.1
1949 - 1950	- 33.2	- 48.7
1950 - 1952	- 23.8	- 38.4
1952 - 1954	- 19.2	- 45.7
1954 - 1956	- 19.8	- 36.0
1956 - 1958	- 36.0	- 39.6
1958 - 1960	- 36.9	- 33.2
1960 - 1962	- 0.6	- 40.5
1962 - 1964	- 17.1	- 26.8
1964 - 1966	- 2.1	- 13.4
1966 - 1968	- 8.5	- 20.4
1968 - 1970	- 11.9	- 7.6

\- retreat

Table 9.1.3. USA - Variations in the positions of glacier fronts

Part 1 of 19 Changes of terminus length 1964 - 1968

( )* estimated value

a.

	No.	Name	Dates of Observation from	to	Change of Terminus Length m	error m	m/yr	how determined
ALASKA, KENAI MOUNTAINS	2	Portage	-8.65	-8.66	S: -23 N: - 6		- 23 - 6	single linear measurement
			-8.66	-8.67	S: -17 N: - 5		- 17 - 5	
			-8.67	-8.68	S: -15 N: - 5		- 15 - 5	
			-8.65	-8.68	S: -55 N: -16		- 18 - 5	
	3	Spencer	7.9.64	11.8.66		5 - 20	-(10-50)*	interpretation of photos taken from the same point
	4	Bartlett	6.9.64	13.8.66			-(10-50)*	interpretation of aerial photos
	5	Trail	11.7.57	12.8.66	- 410	5-20	- 45	ground survey
	7	Taylor	12.9.64	25.8.66	- 80	2-5	- 40	ground survey
	8	Tebenkof	28.8.61 14.9.64	14.9.64 26.8.66	- 80	5-20	- 27 -(10-50)*	ground survey interpretation of aerial photos
	9	Lawrence	14.9.64	26.8.66	(0)*	2-5		interpretation of photos taken from the same point
	10	Marquette	14.9.64	26.8.66	(0)*	2-5		interpretation of photos taken from the same point
	11	Beloit	14.9.64	26.8.66		2-5	+(0-10)*	interpretation of photos taken from the same point

Table 9.1.3. cont.

Part 1 of 19 Changes of terminus area 1964 - 1968
for references and abbreviations see chapter 3

b.

Change of Terminus Aerea		Previous Trends	Remarks	Investigators (References)
m^2	error (m^2)			
$- 135 \cdot 10^5$	$10^3 - 10^6$	Recession of lake terminus in deep water has been going on since 1914; average rate 1950-1964:-50m/yr	S : south side of terminus N : north side of terminus	USFS
		Steady recession over the last few decades; average rates : 1931-1950: -18m/yr 1950-1964: -11m/yr	Station photos: AGS 1964,1966.	M.T.Millett and W.O.Field
		Steady recession since advance in 1916, average rate ca. -9m/yr. 1931-1957.	Photos: AGS 1964, 1966; aerial photos AGS 1966.	M.T.Millett and W.O.Field
$- 431 \cdot 10^3$	$10^2 - 10^4$	Steady recession; average rate 1925-1957: -22m/yr	Surveys:AGS 1957, 1966. Station photos: AGS 1961, 1966	M.T.Millett and W.O.Field
$- 38 \cdot 10^3$	$10^2 - 10^4$	Recession since early 1900's average rate: 1950-1964: -23m/yr	Average of five points along terminus. Surveys: AGS 1964, 1966	W.O.Field
		Recession; average rates: 1910-1961: -18m/yr 1961-1964: -27m/yr	Surveys:AGS 1961, 1964	W.O.Field
		Relatively stable since early 1900's; slight recession 1957-1964	Station photos: AGS 1964, 1966	W.O.Field
		Relatively stable; Recession of 200 - 250m since 1910	Station photos: AGS 1964, 1966	W.O.Field
		Relatively little change since 1909, but slight recession in recent years.	Station photos: AGS 1964, 1966	W.O.Field

Table 9.1.3. USA - Variations in the positions of glacier fronts.

Part 2 of 19 Changes of terminus length 1964 - 1968
( )* estimated value

a.

	No. Name	Dates of Observation		Change of Terminus Length			
		from	to	m	error m	m/yr	how determined
ALASKA, KENAI MOUNTAINS	12 Blackstone	14.9.64	26.8.66		2-5	+(0-10)*	interpretation of photos taken from the same point
ALASKA, KENAI MOUNTAINS	13 Unnamed	16.8.57	23.8.66			-(0-10)*	interpretation of photos taken from the same point
ALASKA, KENAI MOUNTAINS	14 Tiger	16.8.57	23.8.66	(0)*			interpretation of aerial photos
ALASKA, KENAI MOUNTAINS	15 Tigertail	16.8.57	23.8.66			-(0-10)*	interpretation of aerial photos
ALASKA, KENAI MOUNTAINS	16 Chenega	25.8.64	23.8.66	(0)*	5-20		interpretation of photos taken from the same point
ALASKA, KENAI MOUNTAINS	17 Nellie Juan	12.9.64	25.8.66	-45	5-20	-22	ground survey
ALASKA, KENAI MOUNTAINS	18 Falling	31.8.61	25.8.66	-(60)*	5-20	-(12)*	interpretation of photos taken from the same point
ALASKA, KENAI MOUNTAINS	19 Langdon	29.8.57	13.8.66			-(10-50)*	interpretation of aerial photos
CHUGACH MTS.	20 Toboggan	22.9.64	29.8.66			-(10-50)*	interpretation of photos taken from the same point
CHUGACH MTS.	21 Dirty	22.9.64	29.8.66			-(10-50)*	interpretation of photos taken from the same point

Table 9.1.3. USA - Variations in the positions of glacier fronts.

Part 2 of 19 Changes of terminus area 1964 - 1968

for refernces and abbreviations see chapter 3

b.

Change of Terminus Aerea		Previous Trends	Remarks	Investigators (References)
m^2	error (m^2)			
		Relatively little change since 1909, but slight recession in recent years	Station photos: AGS 1964,1966	W.O.Field
		Slow recession since 1935	Photos : AGS 1957,1966	W.O.Field
		Relative stability 1909-1957,followed by recession of 100-200m 1957-1964	Station photos: AGS 1957,1966. Aerial photos: USGS 1964,1965, 1966; AINA 1964	W.O.Field
		Minor oscillations, but no appreciable change since 1909. Slight recession in last two decades	Station photos: AGS 1957,1966. Aerial photos: USGS 1964,1965	W.O.Field
		No appreciable change since 1908	Station photos: AGS 1957,1966. Aerial photos: USGS 1964,1965, 1966; AINA 1964	W.O.Field
$20 \cdot 10^3$	$10^2 - 10^4$	Recession from maximum in 1880's totalled 1750m in 1964. Rate 1935-1964: -40m/yr	Average of five points along terminus. Surveys: AGS 1964, 1966	W.O.Field
		Outermost point of terminus has receded less than 200m since 1908	Measurement at lowest tongue which may just reach high tide. Station photos: AGS 1961, 1966	W.O.Field
		Recession 1950-1966 averaged about 15m/yr	Photos: AGS 1957 Aerial photos: AGS 1966	W.O.Field
		Recession 1935-1961 at average rate of 18m/yr seems to have continued to 1966	Station photos: AGS 1961,1964, 1966	W.O.Field
		Recession 1935-1961 at average rate of 17m/yr apparently continuing to 1966	Station photos: AGS 1961,1964 Aerial photos: USGS 1965	W.O.Field

Table 9.1.3. USA - Variations in the positions of glacier fronts.

Part 3 of 19 Changes of terminus length 1964 - 1968

( ) * estimated value

a.

	No. Name	Dates of Observation from	to	Change of Terminus Length m	error m	m/yr	how determined
ALASKA, CHUGACH MOUNTAINS	22 Harriman	16.9.64	27.8.66	-26	2-5	-13	single linear measurement
		24.8.65	24.8.68		2-5	-(0-10)*	interpretation of aerial photos
	23 Roaring	15.9.64	27.8.66	(0)*	5-20		interpretation of photos taken from the same point
	24 Cataract	21.9.64	29.8.66	(0)*	5-20		interpretation of photos taken from the same point.
		3.9.66	24.8.68	(0)*	5-20		interpretation of aerial photos
	25 Surprise	21.9.64	29.8.66	(0)*	5-20		interpretation of photos taken from the same point
		29.8.66	24.8.68	-(200)*	20-50	-(100)*	interpretation of aerial photos
	26 Baker	20.9.64	27.8.66	(0)*			interpretation of photos taken from the same point
	27 Serpentine	22.9.64	29.8.66	-(100)*	5-20	-(50)*	interpretation of photos taken from the same point
		24.8.65	24.8.68			-(0-10)*	interpretation of aerial photos

Table 9.1.3. USA - Variations in the positions of glacier fronts.

Part 3 of 19 Changes of terminus area 1964 - 1968

for references and abbreviations see chapter 3

b.

Change of Terminus Aerea		Previous Trends	Remarks	Investigators (References)
m^2	error (m^2)			
		Terminus advanced from 1900 to 1964 at following average rates: 1910-1935: 23m/yr; 1935-1961: 12m/yr; 1961-1964: 10m/yr. Glacier seems to have reached a maximum in mid-1960's.	Measurement on land at south margin. Measurements and station photos: AGS 1964,1966. Aerial photos: USGS 1965,1966,1968	W.O.Field
		Sligth recession in recent decades.	Station photos: AGS 1964,1966.	W.O.Field
		Recession from tidewater since 1935. Has been fairly stable in last few years.	Station photos: AGS 1964,1966. Aerial photos: USGS 1964,1966, 1968	W.O.Field
		Terminus has been relatively unchanged since 1909 except for minor fluctiations.Large slides occurred on upper glacier in early 1960's and during 1964 earthquake.	Station photos: AGS 1964, 1966. Aerial photos: USGS 1964,1965, 1966,1968.	W.O.Field
		Substantial recession in recent decades up to 1961-1964 period	Station photos: AGS 1964,1966	W.O.Field
		Slight recession since 1910, marked by readvance in late 1940's and subsequent recession of 100-200m up to 1964.	Station photos: AGS 1964,1966. Aerial photos: 1965,1968.	W.O.Field

Table 9.1.3. USA - Variations in the positions of glacier fronts

Part 4 of 19 Changes of terminus length 1964 - 1968

( )* estimated value

a.

	No. Name	Dates of Observation		Change of Terminus Length			
		from	to	m	error m	m/yr	how determined
ALASKA, CHUGACH MOUNTAINS	28 Cascade	15.9.64	27.8.66		5-20	+(0-10)*	interpretation of photos taken from the same point
		27.8.66	24.8.68		5-20	+(0-10)*	interpretation of aerial photos
	29 Barry	15.9.64	27.8.66		5-20	+(10-50)*	interpretation of photos taken from the same point
		27.8.66	24.8.68		5-20	+(10-50)*	interpretation of aerial photos
	30 Coxe	15.9.64	27.8.66		2-5	-(0-10)*	interpretation of photos taken from the same point
		27.8.66	24.8.68		2-5	-(0-10)*	interpretation of aerial photos
	31 Wellesley	19.9.64	24.8.65	(0)*	5-20		interpretation of photos taken from the same point
		1.9.66	24.8.68			-(0-10)*	interpretation of aerial photos
	32 Vassar	19.9.64	24.8.65	(0)*			interpretation of aerial photos
		1.9.66	24.8.68	(0)*			
	33 Bryn Mawr	19.9.64	1.9.66	+(30)*	5-20		interpretation of photos taken from the same point
		3.9.66	24.8.68			+(0-10)*	interpretation of aerial photos

Table 9.1.3. USA - Variations in the positions of glacier fronts.

Part 4 of 19 Changes of terminus area 1964 - 1968

for references and abbreviations see chapter 3

b.

Change of Terminus Aerea		Previous Trends	Remarks	Investigators (References)
m^2	error (m^2)			
		Terminus has not changed apprecia-bly since 1914; current advance is where terminus coelesces with Barry glacier.Such an oscillation has occurred previously	Coelesces at ter-minus with Barry Glacier. Station photos: AGS 1964,1966. Aerial photos: USGS 1964,1965, 1966,1968;USC&GS 1964	W.O.Field
		Terminus has not changed appreciably since 1914.Current advance is similar to previous ones and may be only a minor oscillation	Coelesces at ter-minus with Cascade Glacier. Station photos: AGS 1964,1966. Aerial photos: USGS 1964,1965, 1966,1968; USC & GS 1964	W.O.Field
		Terminus has re-mained relatively stable since 1914.	Station photos: AGS 1964,1966. Aerial photos: USGS 1964,1965, 1966,1968; USC & GS 1964	W.O.Field
		Slight net reces-sion in last few decades but little change 1957-1964	Station photos: AGS 1964,1966. Aerial photos: USC & GS 1964; USGS 1964,1965, 1966,1968	W.O.Field
		Little change in this terminus du-ring last few deca-des. Moraine-cove-red area appears to be shrinking slowly	On site and station photos: AGS 1964, 1966. Aerial photos: USGS 1965,1968; AGS 1966	W.O.Field
		Oscillating glacier Last maximum ca. 1950 and latest minimum ca.1961	Survey:AGS 1964. Station photos: AGS 1964,1966. Aerial photos: USC & GS 1964; USGS 1965,1966, 1968	W.O.Field

Table 9.1.3. USA - Variations in the positions of glacier fronts

Part 5 of 19 Changes of terminus length 1964 - 1968

( )* estimated value

a.

	No. Name	Dates of Ob- servations		Change of Terminus Length			
		from	to	m	error m	m/yr	how determined
ALASKA, CHUGACH MOUNTAINS	34 Smith	17.9.64	1.9.66	(0)*	5-20		interpretation of photos taken from the same point
	35 Harvard	17.9.64	1.9.66	+(100)*	5-20	+(50)*	interpretation of photos taken from the same point
		3.9.66	24.8.68		20-50	+(10-50)*	interpretation of aerial photos
	36 Yale	19.9.64	31.8.66	- 100	5-20	- 50	ground survey
		3.9.66	24.8.68	-(500)*	>50	-(250)*	interpretation of aerial photos
	37 Meares	23.9.64	2.9.66	+ 50	5-20	+ 25	ground survey
		24.8.65	24.8.68			+(0-10)*	interpretation of aerial photos
	38 Brilliant	23.9.64	2.9.66			- ?	interpretation of aerial photos

Table 9.1.3. USA - Variations in the positions of glacier fronts.

Part 5 of 19 Changes of terminus area 1964 - 1968

for references and abbreviations see chapter 3

b.

Change of Terminus Aerea		Previous Trends	Remarks	Investigators (References)
m^2	error (m^2)			
		Terminus has not changed appreciably since 1930's	Station photos: AGS 1964,1966	W.O.Field
		Steady advance totalling 1300m since 1900. Average rate 1957-1964: 25m/yr	Observations at both margins of tidal ice front. Station photos: AGS 1964,1966; Aerial photos: USC & GS 1964; USGS 1964,1965, 1966, 1968	W.O.Field
$-75 \cdot 10^3$ ($-350 \cdot 10^3$)*	10^2-10^4 10^4-10^6	Recession since maximum in 1935. Greatest recession on south side (point of measurement) averaged 80m/yr.1950-1964	Measurement at southern part of tidal ice front. Surveys:AGS 1964, 1966; Aerial photos: USC & GS 1964; USGS 1964,1965, 1966,1968	W.O.Field
$15 \cdot 10^3$	10^2-10^4	Glacier has been advancing with some interruptions since 1910. Rate at point of measurement on north side 1910-1964: 11m/yr. There has been little change in the middle and south side since 1957.	Measurement at north margin of ice front. No appreciable change elsewhere. Surveys:AGS 1964, 1966. Aerial photos: USC & GS 1964; USGS 1964,1965, 1966,1968	W.O.Field
		Considerable shrinkage of lower end of glacier since 1931	Photos: AGS 1964, 1966	W.O.Field

Table 9.1.3. USA - Variations in the positions of glacier fronts

Part 6 of 19 Changes of terminus length 1964 - 1968

( )* estimated value

a.

	No. Name	Dates of Observation from	to	Change of Terminus Length m	error m	m/yr	how determined
ALASKA, CHUGACH MOUNTAINS	39 Columbia	26.9.64	4.9.66	a) (0)*	2-5		interpretation of photos taken from the same point
				b) -13.5	2-5	- 7	ground survey/ single linear measurement
				c) -14	2-5	- 7	single linear measurement
				d) -72	2-5	-36	single linear measurement
		4.9.66	. .68	(0)*			interpretation of aerial photos
	40 Shoup	24.8.64	24.8.68			-(10-50)*	interpretation of aerial photos
	41 Valdez	24.8.64	24.8.68			-(10-50)*	interpretation of photos taken from the same point/ interpretation of aerial photos
	42 Sheridan	24.8.64	12.7.68	a) -(30)*	5-20		ground survey/ interpretation of photos taken from the same point
				b) -(75)*	5-20		

Table 9.1.3. USA - Variations in the positions of glacier fronts.

Part 6 of 19 Changes of terminus area 1964 - 1968

for references and abbreviations see chapter 3

b.

Change of Terminus Aerea		Previous Trends	Remarks	Investigators (References)
m^2	error (m^2)			
		Tidal part of terminus fluctuates seasonally but has not changed position appreciably since 1900. Measurements where terminus rests on land are better indicators of the glaciers' fluctuations. Total recession along the terminus since a maximum about 1920 varies from 375 to 900m. Minor advances were recorded in 1935, 1949-1950 and 1957-1961	Measurements 1964-1966 at four points: a) west margin on land; b) west side of Heather Island near middle of terminus; c) east side of Heather Island; d) eastern land terminus Surveys: AGS 1964, 1966. Aerial photos: USC & GS 1964; USGS 1964, 1965,1966,1968; AGS 1968	W.O.Field
		Terminal part of glacier has been shrinking for decades with little change in the position of the terminus until 1957. Since then recession of several hundred meters has occurred in one area.	Photos: AGS 1964; Aerial photos: AGS 1964; USGS 1964,1968	W.O.Field
		Recession 1909-1961 at average rate of 12m/yr appears to be continuing	Station photos: AGS 1964, 1966. Aerial photos:USC & GS 1964; USGS 1964, 1968	W.O.Field
		Gradual recession for last few decades. Rate which from 1950 to 1965 was 7 m/yr on land and 10m/yr in lakes seems to be continuing	a) Average of change at seven points on land,1950-1965 b) Average of change at seven points in lakes, 1950-1965	W.O.Field (Ref.:Tuthill, at.al.,1968)

Table 9.1.3. USA - Variations in the positions of glacier fronts.

Part 7 of 19 Changes of terminus length 1964 - 1968

( )* estimated value

a.

	No. Name	Dates of Observation from	to	Change of Terminus Length m	error m	m/yr	how determined
ALASKA, CHUGACH MOUNTAINS	43 Sherman	. .64 .7.65 .7.66 - .7.67	.7.65 .7.66 .7.67 .8.68	- 25 -5 to -10 0 to + 20 0 to + 20	? < 2 < 2 < 2	- 25 - 7 + 10 + 10	ground survey
	44 Saddlebag	25.8.65	10.7.68		20-50	-(10-50)*	interpretation of aerial photos
	45 Childs	24.8.64	10.7.68	+(100)*	5-20	+ 25	ground survey/ single linear measurement/ interpretation of photos taken from the same point
	46 Allen	27.8.63	8.9.66			+ ?	interpretation of aerial photos
	47 Schwan	25.8.64	8.9.66			-(0-10)?*	interpretation of aerial photos
	48 Woodworth	25.8.64	8.9.66		5 -20	-(0-10)*	interpretation of aerial photos

Table 9.1.3. USA - Variations in the positions of glacier fronts.

Part 7 of 19 Changes of terminus area 1964 - 1968

for references and abbreviations see chapter 3

b.

Change of Terminus Aerea		Previous Trends	Remarks	Investigators (References)
m^2	error (m^2)			
		Recession 1890's to 1941 at an average rate of 11-15m/yr and from 1941 to 1964 at 28m/yr	North side of terminus has advanced at rate of 20 m/yr from 1965 to 1969. South side has remained stationery. Advance is caused by landslide debris cover emplaced by the earthquake of 1964. Mass balance remains slightly negative	M.McSavency C.Bull C.Marangunic (Ref.: Marangunic,1968)
		Recession 1950-1965 at average rate of 65m/yr. Little change 1965-1968	On site photos: AGS 1968; Aerial photos: USGS 1965,1966; Glacier affected by extensive debris slides during 1964 earthquake	W.O.Field
		Oscillating glacier; surge 1909-1912; recession of 400m to 1959, then advance of ca.75m to 1964	Station photos: AGS 1961, 1968;W.M.Laird (NDGS) 1964; S.J.Tuthill (MC) 1965; Aerial photos:USGS 1964,1966,1968; AGS 1964,1966	W.O.Field
		Recession of 1-2km since small advance in 1912. Recent recession up to 1963	Aerial photos: AGS 1964,1966; USGS 1966. Large debris slides occurred on upper glacier in 1964 earthquake	W.O.Field
		Slow recession for several decades	Aerial photos:USGS 1964; AGS 1966, Large debris slide fell on upper glacier in 1964 earthquake	W.O.Field
		Slow recession of active terminus during last few decades. Total recession since 1898 about 1.3 km	Aerial photos:USGS 1964; AGS 1966	W.O.Field

Table 9.1.3. USA - Variations in the positions of glacier fronts.

Part 8 of 19 Changes of terminus length 1964 - 1968

( )* estimated value

a.

	No. Name	Dates of Observation from to	Change of Terminus Length m	error m	m/yr	how determined
ALASKA, CHUGACH MOUNTAINS	49 Worthington	28.9.64 17.8.66 17.8.66 24.8.68	- 35	2-5	-17 -(10-50)*	ground survey interpretation of aerial photos
	50 Miles	16.4.64 24.8.68			- ?	interpretation of aerial photos
	51 Slide	15.6.65 25.6.68		2-5	+ 45	ground survey
	52 Bering	27.8.63 2.9.66 2.9.66 8.9.67 8.9.67 24.8.68	+110 - 20 - 20	5-20 5-20 5-20	+ 40 - 20 - 20	interpretation of aerial photos
	53 Guyot	23.8.64 23.8.65 23.8.65 17.9.66 17.9.66 24.8.68	-150 -200 -300	> 50 > 50 > 50	-150 -200 -150	interpretation of aerial photos

Table 9.1.3. USA - Variations in the positions of glacier fronts.

Part 8 of 19 Changes of terminus area 1964 - 1968

for references and abbreviations see chapter 3

b.

Change of Terminus Aerea		Previous Trends	Remarks	Investigators (References)
m^2	error (m^2)			
		Recession since 1930's. Average rate 1957-1964: 18m/yr	Surveys: AGS 1964 1966; Aerial photos USC & GS 1964, USGS 1964,1968	W.O.Field
		Recession since the 19th century marked by occasional advances. Total recession 1910-1957 ca.3 km	Aerial photos: AGS 1964,1966; AINA 1964; USGS 1964, 1966,1968	W.O.Field
		Evidence of some recession from the 1930's to 1964.	The lower third of the glacier was covered by a debris slide during the 1964 earthquake. Active push moraine advancing over vegetation; margin very steep, ca. 38°. Activity noted in aerea ca. 500m above terminus; calculations based on transverse profile ca. 700m from terminus	J.R.Reid (Ref.: Reid 1969)
$+3.2 \cdot 10^6$ $-0.5 \cdot 10^6$ $-0.5 \cdot 10^6$	10^2-10^4 10^2-10^4 10^2-10^4		Major surge advanced along 30km long southeast terminus between 1957 and 1960 with an advance up to 5 km. Glacier was retreating from this advance between 1960 and 1964. A new smaller surge occurred in late 1965 and early 1966 moving glacier terminus slightly in advance of the 1960 surge. Since 1966 terminus has been retreating.	A.Post
		Continuous retreat since about 1910. Formerly joined to Yahtse Glacier		A.Post

Table 9.1.3. USA - Variations in the positions of glacier fronts.

Part 9 of 19 Changes of terminus length 1964 - 1968

( )* estimated value

a.

Region	No. Name	Dates of Observation from	to	Change of Terminus Length m	error m	m/yr	how determined
CHUGACH MTS.	54 Yahtse	23.8.64 23.8.65 17.9.66	23.8.65 17.9.66 24.8.68	 -150	> 50 > 50 > 50	-(0-10)* -(0-10)* - 80	interpretation of aerial photos
CHUGACH MTS.	55 Tyndall	23.8.64 23.8.65 17.9.66	23.8.65 17.9.66 24.8.68	(Ø)* -1800 -2000	> 50 > 50 > 50	 -1800 -1000	aerial photogrammetry interpretation of aerial photos
ALASKA, ST. ELIAS MOUNTAINS	56 Geikie	23.8.64	23.7.68	- 275	5-20	- 70	ground survey
ALASKA, ST. ELIAS MOUNTAINS	57 Hugh Miller	22.8.64	22.7.68	- 280	5-20	- 70	ground survey
ALASKA, ST. ELIAS MOUNTAINS	58 Reid	22.8.64 19.9.66	19.9.66 18.7.68	(0)* (0)*	5-20 5-20		interpretation of photos taken from the same point
ALASKA, ST. ELIAS MOUNTAINS	59 Lamplugh	21.8.64 19.9.66	19.9.66 16.7.68	ca.+60 ca.+60	5-20 5-20	ca.+30 ca.+30	single linear measurement/ interpretation of photos taken from the same point
ALASKA, ST. ELIAS MOUNTAINS	60 Unnamed	20.8.64	17.7.68	(0)*			interpretation of aerial photos/ interpretation of photos taken from the same point
ALASKA, ST. ELIAS MOUNTAINS	61 Kashoto	20.8.64	17.8.68	(0)*			interpretation of photos taken from the same point/ interpretation of aerial photos
ALASKA, ST. ELIAS MOUNTAINS	62 Hoonah	20.8.64 17.9.66 12.7.67	17.9.66 12.7.67 17.7.68		5-20 5-20 5-20	+(0-10)* -(0-10)* -(0-10)*	interpretation of photos taken from the same point

Table 9.1.3. USA - Variations in the positions of glacier fronts.

Part 9 of 19 Changes of terminus area 1964 - 1968

for references and abbreviations see chapter 3

b.

Change of Terminus Aerea		Previous Trends	Remarks	Investigators (References)
m^2	error (m^2)			
		Continuous retreat since about 1910		A. Post
		Continuous retreat since about 1910	Changes 1964-1965 along terminus cliff nearly balance.	A. Post
$-92 \cdot 10^3$	10^2-10^4	Steady recession since minor advance in early 1920's; average rate ca. 100m/yr	Surveys: AGS 1964, 1968	W.O.Field
$-160 \cdot 10^3$	10^2-10^4	Steady recession since the 1890's. Average rate 1941-1964, ca. 95m/yr	Surveys: AGS 1964 1966,1968	W.O.Field
		Recession of 5750m 1919-1941 and about 250m 1941-1964.	Station photos:AGS 1964,1966,1968	W.O.Field
		Steady advance since minimum in 1941. Average rate 1941-1964 20m/yr. No evidence of positive regime; advance more likely due to reduced calving because of shoaling.	Measurement on land at east margin; little change along rest of terminus	W.O.Field
		Little change during last two decades	Photos: AGS 1964, 1966; Station photos: AGS 1967, 1968	W.O.Field
		Minor fluctuations over the past few decades but no appreciable net change.	Photos: AGS 1964, 1966; Station photos: AGS 1967, 1968	W.O.Field
		Except for minor fluctuations this terminus has remained in essentially the same position since the 1920's.	Station photos: AGS 1964,1966,1967,1968	W.O.Field

Table 9.1.3. USA - Variations in the positions of glacier fronts.

Part 10 of 19 Changes of terminus length 1964 - 1968

( )* estimated value

a.

	No. Name	Dates of Observation from	to	Change of Terminus Length m	error m	m/yr	how determined
ALASKA, ST. ELIAS MOUNTAINS	63 Gilman	20.8.64	17.7.68		20-50	+(0-10)*	ground survey/ interpretation of photos taken from the same point
	64 Clark	20.8.64	17.7.68	(0)*	5 -20		interpretation of photos taken from the same point
	65 Johns Hopkins	20.8.64	17.9.66	+ 140	5 -20	+ 70	ground survey
		17.9.66	14.7.67	+ 60	5 -20	+ 30	
		14.7.67	17.7.68		20-50	+(10-50)*	interpretation of photos taken from the same point
	66 Tyeen	29.8.64	22.8.65	+(1000)*	>50	+(1000)*	interpretation of aerial photos
		22.8.65	5.4.66	+(800)*	>50	+(1275)*	
		5.4.66	23.7.66	+(350)*	>50	+(1200)*	
		23.7.66	17.9.66	-(0)*	5 -20		
		17.9.66	14.7.67	-(25)*	2 - 5	-(25)*	interpretation of photos taken from the same point
		14.7.67	17.7.68	-(25)*	2 - 5	-(25)*	
	67 Toyatte	20.8.64	17.9.66	+(100)*	20-50	+(50)*	interpretation of photos taken from the same point
		17.9.66	14.7.67	+(50)*	20-50	+(50)*	
		14.7.67	17.7.68	-(25)*	5 -20	-(25)*	
	68 Margerie	19.8.64	18.7.68	(0)*			interpretation of photos taken from the same point
	69 Unnamed	19.8.64	18.7.68	(0)*			interpretation of photos taken from the same point

Table 9.1.3. USA - Variations in the positions of glacier fronts.

Part 10 of 19 Changes of terminus area 1964 - 1968

for references and abbreviations see chapter 3

b.

Change of Terminus Aerea		Previous Trends	Remarks	Investigators (References)
m^2	error (m^2)			
		Little net change for several decades. Slight advance may be due to reduced calving because of shoaling	Station photos: AGS 1964, 1966, 1967, 1968	W.O.Field
		No appreciable change since the 1930's	Station photos: AGS 1964, 1966, 1967, 1968	W.O.Field
$+155 \cdot 10^3$ $+57 \cdot 10^3$	10^2-10^4 10^2-10^4	Steady advance since 1929 at average rate of 40m/yr	Advance measured at outermost point of terminus. Surveys: AGS 1964, 1966, 1967; Station photos: AGS 1967, 1968	W.O.Field
$+500 \cdot 10^3$ $+300 \cdot 10^3$ $+250 \cdot 10^3$	10^2-10^4 10^2-10^4 10^2-10^4	Surge similar to that of 1964-1966 occurred in 1946-1948.	Surge occurred between 1964 and 1966, following maximum in 1966, terminus became stagnant in 1967. Surveys: AGS 1966, 1967. Station photos: AGS 1967, 1968; Aerial photos: USGS 1964, 1965, 1966, NPS 1965, 1966	W.O.Field (Ref.: Field, 1969 pp.833-834)
		An oscillating glacier. Former maximum occurred in 1935; 1946-1948 and 1958	An oscillating glacier; Photos: AGS 1964, 1966; Station photos: 1967, 1968	W.O.Field
		Terminus oscillates but has remained in essentially the same position since 1912	Minor oscillations. Station photos: AGS 1964, 1966, 1967, 1968	W.O.Field
		Little change for several decades	Station photos: AGS 1964, 1966, 1967, 1968	W.O.Field

Table 9.1.3. USA - Variations in the positions of glacier fronts.

Part 11 of 19 Changes of terminus length 1964 - 1968

( )* estimated value

a.

	No. Name	Dates of Observation from	to	Change of Terminus Length m	error m	m/yr	how determined
ALASKA, ST. ELIAS MOUNTAINS	70 Grand Pacific	19.8.64	18.9.66	W + 50	2-5	+ 25	interpretation of photos taken from the same point
				E +110		+ 55	
		18.9.66	18.7.68	W ca +15	5-20	+ 7	
				E ca +35	5-20	+ 17	
	71 Romer	18.9.64	15.7.68	(0)*	5-20		interpretation of photos taken from the same point
	72 Rendu	18.8.64	22.8.65	0	2-5		interpretation of photos taken from the same point/ interpretation of aerial photos
		22.8.65	16.9.66	+ 450	5-20	+ 450	ground survey
		16.9.66	11.7.67	+ 20	2- 5	+ 20	single linear measurement
		11.7.67	15.7.68	0	2- 5		interpretation of photos taken from the same point
	73 Unnamed	16.9.66	15.7.68	(0)*			interpretation of photos taken from the same point
	74 Carroll	21.8.64	22.8.65			-(10-50)*	interpretation of aerial photos
		22.8.65	16.9.66			+(>50)*	
		16.9.66	10.7.67			+(>50)*	
		10.7.67	19.7.68			+(10-50)*	

Table 9.1.3. USA - Variations in the positions of glacier fronts.

Part 11 of 19 Changes of terminus area 1964 - 1968

for references and abbreviations see chapter 3

b.

Change of Terminus Aerea		Previous Trends	Remarks	Investigators (References)
m^2	error (m^2)			
$+144 \cdot 10^3$	10^2-10^4	Surges 1912-1913 and 1931. Minimum S in mid 1920's and mid 1930's. Advance since late 1930's at average rate of 48 m/yr 1948 - 1964	Measurements on margins of tidal front:E = east W = west Surveys: AGS 1964, 1966; Station photos: AGS 1966, 1967, 1968	W.O.Field
		Relatively stable since a surge between 1907 and 1911. Slight advance 1961 - 1964	Station photos: AGS 1961, 1966,1967, 1968	W.O.Field
0 $+612 \cdot 10^3$ ca. 140 0	 10^2-10^4 10 -100	Surge about 1908 and other minor advances since then. Terminus in 1966 - 1968 was in advance of its position in 1892. From 1958 to 1964 terminus became progressively more inactive,but remained in same position	Terminus was virtually stagnant in 1964 and 1965. Surge reached terminus 1966 and ended by 1967. Terminus began to stagnate in 1967. Surveys: AGS 1958, 1966, 1967; Station photos:AGS 1967, 1968	W.O.Field (Ref.: Field,1969)
		Except for a surge between 1911 and 1916 this terminus has not changed appreciably since 1892	Station photos:AGS 1961,1966,1968	W.O.Field
		Surge in 1919 and minor advance in 1943 followed by recession and thinning until 1964 Lower margin of active ice receded up to 1000 m 1929-1958	Precise measurements difficult because terminus of active ice was poorly defined until 1968.Advance of 1965 to 1967 reactivated the terminal stagnant area and then thrust it forward from 1967 to 1968. Station photos:AGS 1964,1966,1967,1968 Aerial photos: USGS 1964,1966,1967; NPS 1966	W.O.Field (Ref.: Field, 1969 pp. 838)

Table 9.1.3. USA - Variations in the positions of glacier fronts.

Part 12 of 19 Changes of terminus length 1964 - 1968

( )* estimated value

a.

	No. Name	Dates of Observation		Change of Terminus Length			
		from	to	m	error m	m/yr	how determined
ALASKA, ST. ELIAS MOUNTAINS	75 Cushing	30.8.64	21.7.68		>50	-(10-50)*	interpretation of aerial photos
	76 Baldwin	14.8.64	21.7.68	- 510	5-20	- 125	ground survey
	77 Plateau	14.8.64 17.7.67	14.7.67 21.7.68	- 865 - 350	5-20 5-20	- 290 - 350	ground survey
	78 Burroughs	30.8.64	27.8.69	a) -158 b) -164 c) -235 d) -243	5-20 5-20 5-20 5-20	- 39,6 - 41,0 - 58.8 - 60,8	aerial photogrammetry
	79 Muir	16.8.64 20.7.67	20.7.67 15.7.68	- 2050 - 600	20-50 20-50	- 680 - 600	ground survey
	80 Riggs	15.8.64	15.7.68	- 175	5-20	- 45	ground survey
	81 Mc Bride	22.8.64	20.7.68			-(10-50)*	interpretation of photos taken from the same point

Table 9.1.3. USA - Variations in the positions of glacier fronts.

Part 12 of 19 Changes of terminus area 1964 - 1968

for references and abbreviations see chapter 3

b.

Change of Terminus Aerea		Previous Trends	Remarks	Investigators (References)
m^2	error (m^2)			
		Steady recession in recent decades. Average rate 1948 to 1966 was about 50 m/yr	Aerial photos: USGS 1964,1966,1967 Station photos: AGS 1968	W.O.Field
		A tributary of Plateau Glacier until 1956-1957. Recession since 1957 at rate of about 80 m/yr	Surveys: AGS 1964, 1967,1968	W.O.Field
$-1410 \cdot 10^3$ $-619 \cdot 10^3$	10^2-10^4 10^2-10^4	Steady recession since 1890's. Average rate of tidal front 1948-1964 was -500m/yr	This is a remnant glacier:average of recession at three points along terminus. Surveys: AGS 1964,1967,1968	W.O.Field
		This has been a remnant glacier since the 1890's	This is a stagnant remnant glacier. Measurements at four points: (a) and (b) land slope toward ice on a nunatak (c) land-	D.M.Mickelson R.P.Goldthwaite -slope nearly horizontal.(d) landslope away from ice
$-875 \cdot 10^3$	10^2-10^4	Recession since at least early 19th century. Total 1892 -1966 32.25 km average rate 1892-1948 -436 m/yr and from 1948-1966 -430 m/yr	Point of measurement for 1964 to 1967 is medial moraine near middle of terminus. Recession of 1967 to 1968 is measured in middle of terminus. Surveys: AGS 1964, 1967,1968	W.O.Field
$-209 \cdot 10^3$	10^2-10^4	Tributary of Muir Glacier until 1961. Advanced slightly 1961 to 1964	Recession figure is average of five points along terminus. Surveys: AGS 1964,1966,1967,1968	W.O.Field
		Tributary of Muir Glacier until 1945, Terminus formerly in deep water is now on tidal flat. Slight advance around 1960 probably due to reduced calving	Station photos:AGS 1964,1966,1967,1968	W.O.Field

Table 9.1.3. USA - Variations in the positions of glacier fronts.

Part 13 of 19 Changes of terminus length 1964 - 1968

( )* estimated value

a.

	No. Name	Dates of Observation		Change of Terminus Length			
		from	to	m	error m	m/yr	how determined
ALASKA, ST. ELIAS MOUNTAINS	82 Casement	17.8.64	22.8.65			-(10-50)*	interpretation of aerial photos
	83 Davidson	23.8.64	4.8.67			-(> 50)*	interpretation of aerial photos
	84 Garrison	11.8.58	3.8.67	0*			interpretation of aerial photos
	85 Bertha	11.8.58	3.8.67			-(0-10)*	interpretation of aerial photos
	86 Unnamed	11.8.58	3.8.67	0*			interpretation of aerial photos
	87 Unnamed	11.8.58	3.8.67			-(0-10)*	interpretation of aerial photos
	88 Takhin	11.8.58	3.8.67			-(0-10)*	interpretation of aerial photos
	89 Le Blondeau	11.8.58	3.8.67			-(0-10)*	interpretation of aerial photos
	90 Unnamed	11.8.58	3.8.67			+(10-50)*	interpretation of aerial photos
	91 Unnamed	11.8.58	3.8.67	0*			interpretation of aerial photos

Table 9.1.3. USA - Variations in the positions of glacier fronts.

Part 13 of 19 Changes of terminus area 1964 - 1968

for references and abbreviations see chapter 3

b.

Change of Terminus Aerea		Previous Trends	Remarks	Investigators (References)
m^2	error (m^2)			
		Tributary of Muir Glacier until 1911 Recession 1911 - 1962 about 4000 m	Photos: AGS 1964; Aerial photos:USGS 1964,1965,1968	W.O.Field
		Recession and shrinkage of terminus since the 19th century	Aerial photos:USGS 1964,1966;AGS 1967	W.O.Field
		Terminus is covered with moraine and has receded only slightly in last several decades. However, lower end of glacier has thinned appreciably	Aerial photos:AGS 1958,1967	W.O.Field
		Maximum in 1880's Since then lower end of glacier has shrunk but there has been relatively little recession of the terminus	Aerial photos:AGS 1958,1967	W.O.Field
		Recession of an estimated 300 m from 1894-1948 and little change 1948-1958	Aerial photos:AGS 1958,1967	W.O.Field
		Considerable recession 1948-1958	Aerial photos:AGS 1958,1967	W.O.Field
		Only very slight recession since 1910	Aerial photos:AGS 1958,1967	W.O.Field
		Recession of about 200 m 1910 to 1948 and about 50 m 1948 to 1958	Aerial photos:AGS 1958,1967	W.O.Field
		No appreciable change in recent decades	Aerial photos:AGS 1958,1967	W.O.Field
		No appreciable change in recent decades	Aerial photos:AGS 1958,1967	W.O.Field

Table 9.1.3. USA - Variations in the positions of glacier fronts.

Part 14 of 19 Changes of terminus length 1964 - 1968

( )* estimated value

a.

	No. Name	Dates of Observation		Change of Terminus Length			
		from	to	m	error m	m/yr	how determined
ST.ELIAS MTS.	92 Unnamed	30.8.64	3.8.67	0*			interpretation of aerial photos
ST.ELIAS MTS.	93 Tsirku	30.8.64	3.8.67	0*			interpretation of aerial photos
ALASKA, COAST MOUNTAINS	94 Eagle	23.8.64	29.8.67		5 -20	-(10-50)*	interpretation of aerial photos
ALASKA, COAST MOUNTAINS	95 Herbert	23.8.64	29.8.67		5 -20	-(0 -10)*	interpretation of aerial photos
ALASKA, COAST MOUNTAINS	96 Mendenhall	23.8.64	31.7.68		2 - 5	-(0 -10)*	interpretation of aerial photos/ interpretation of photos taken from the same point
ALASKA, COAST MOUNTAINS	97 Lemon Creek	30.8.64	4.8.67		5 -20	-(0 -10)*	interpretation of aerial photos
ALASKA, COAST MOUNTAINS	98 Norris	23.8.64	30.7.68		5 -20	-(10-50)*	interpretation of aerial photos
ALASKA, COAST MOUNTAINS	99 Taku	20.7.65	29.7.68	a) +75 b) +70 c) 0*	5 -20 5 -20	+25 +23	ground survey/ interpretation of photos taken from the same point interpretation of photos taken from the same point

Table 9.1.3. USA - Variations in the positions of glacier fronts.

Part 14 of 19 Changes of terminus area 1964 - 1968

for references and abbreviations see chapter 3

b.

Change of Terminus Aerea		Previous Trends	Remarks	Investigators (References)
m^2	error (m^2)			
		Little change since before 1948	Aerial photos:USGS 1964; AGS 1967	W.O.Field
		Slow advance 1910-1958 at average rate of 5 - 10 m/yr	Aerial photos: USGS 1964; AGS 1967	W.O.Field
		Terminus appears to be receding at rate of about 35-40 m/yr which is similar to the rate during the 1948-1967 period	Station photos: AGS 1964;Aerial photos USGS 1964,1965,1967	W.O.Field M.T.Millett
		Terminus has receded to ledge and current rate of recession appears less than the rate of 25 - 30 m/yr from 1948 to 1965	Station photos: AGS 1964; Aerial photos USGS 1964,1965,1967	W.O.Field M.T. Millett
		Current rate of recession slower than average of 30 - 35 m/yr from 1945 - 1964	Surveys: AGS 1964; Station photos: AGS 1964,1965,1966,1967 1968	W.O.Field
		Current rate of recession about 10 m/yr which is less than the rate of 30 - 35 m/yr from 1929 - 1958	Aerial photos:USGS 1964,1965,1966; AGS 1967	W.O.Field
		Slow recession since maximum about 1915. Rate of recession has averaged about -10m/yr for last four decades	Aerial photos:USGS 1964,1965,1967; AGS 1968	W.O.Field
		Almost continuous advance since early 1900's. Average rate 1948 - 1961 was +70 - 75 m/yr	a) South side of terminus b) Near middle of terminus c) North side of terminus Surveys: AGS 1961, 1965,1968; Station photos: AGS 1961, 1965, 1967, 1968	W.O.Field

Table 9.1.3. USA - Variations in the positions of glacier fronts.

Part 15 of 19 Changes of terminus length 1964 - 1968

( )* estimated value

a.

	No. Name	Dates of Observation		Change of Terminus Length			
		from	to	m	error m	m/yr	how determined
ALASKA, COAST MOUNTAINS	100 Hole in the wall	23.8.64	30.7.68		5 -20	+(10-50)*	interpretation of photos taken from the same point/ interpretation of aerial photos
	101 West Twin	23.8.64	29.7.68		5 -20	-(0 -10)*	interpretation of aerial photos/ interpretation of photos taken from the same point
	102 East Twin	23.8.64	29.7.68	0*	5 -20		interpretation of photos taken from the same point/ interpretation of aerial photos
	103 Wright	23.8.64	30.7.68	- 400	20-50	- 100	interpretation of aerial photos/ ground survey
	104 Unnamed	23.8.64	25.7.68		20-50	-(10-50)*	interpretation of aerial photos
	105 Speel	26.8.60	25.7.68			-(> 50)*	interpretation of aerial photos
	106 Sawyer	23.8.64	18.9.66	0*	20-50		interpretation of aerial photos
		18.9.66	25.7.68	- 25	5 -20	- 12	ground survey
	107 South Sawyer	23.8.64	25.7.68	0*	20-50		interpretation of aerial photos/ ground survey

Table 9.1.3. USA - Variations in the positions of glacier fronts.

Part 15 of 19 Changes of terminus area 1964 - 1968

for references and abbreviations see chapter 3

b.

Change of Terminus Aerea		Previous Trends	Remarks	Investigators (References)
m^2	error (m^2)			
		Reflects advance of Taku Glacier . Steady advance since early 1900's. Current rate less than the ca.+40m/yr 1948-1964	Distributary tongue of Taku Glacier. Station photos: AGS 1965, 1967, 1968	W.O.Field
		Continued slow re-cession of 15 - 20 m/yr since 1948	Distributary tongue of Taku Glacier. Station photos: AGS 1965, 1967, 1968	W.O.Field
		Present ralative stability was pre-ceded by slow re-cession of about 20 m/yr 1948-1965	Distributary tongue of Taku Glacier. Station photos: AGS 1965, 1967, 1968	W.O.Field
$-173 \cdot 10^3$	10^2-10^4	This period marks final breakup of semi-stagnant ter-minus in lake ba-sin. Recession at average rate of 130m/yr 1948-1967	Survey: AGS 1967; Aerial photos: USGS 1964,1965,1966; AGS 1968	W.O.Field
		Recession for se-veral decades; average rate 1948-1965 ca. -40 m/yr	Aerial photos: USGS 1964,1965; AGS 1968	W.O.Field
		Recession for se-veral decades;	Aerial photos: UW (AS) 1960; AGS 1968	W.O.Field
		An oscillating glacier. Terminus has fluctuated within limits of 700 m from 1948-1968. In 1967 it was only 800 m back of its 1889 position	1966-1968 recession is estimated avera-ge for whole termi-nus. Surveys: AGS 1958, 1967; Aerial photos: USGS 1964, 1965,1966; AGS 1968	W.O.Field
		Recession inter-rupted by still-stands since 1889 at average rate of 47 m/yr. Current stillstand follows recession at avera-ge rate of 185m/yr from 1958-1964	Surveys: AGS 1958, 1967; Aerial photos: USGS 1964,1965,1966, 1967; AGS 1968	W.O.Field

Table 9.1.3. USA - Variations in the positions of glacier fronts.

Part 16 of 19 Changes of terminus length 1964 - 1968

( )* estimated value

a.

	No. Name	Dates of Observation		Change of Terminus Length			
		from	to	m	error m	m/yr	how determined
ALASKA, COAST MOUNTAINS	108 North Dawes	23.8.64	25.7.68	-(200)*	20-50	-(50)*	interpretation of aerial photos
	109 Dawes	23.8.64	25.7.68	0*	5 -20		interpretation of aerial photos/ ground survey
	110 Baird	23.8.64	20.8.68		5 -20	+(0-10)*	interpretation of aerial photos
	111 Patterson	23.8.64	20.8.68	-(200)*	20-50	-(50)*	interpretation of aerial photos
	112 Unnamed	8.8.48	25.7.68	-(200)*	20-50	-(10)*	interpretation of aerial photos
	113 Le Conte	23.8.64	20.8.68	0*	20-50		interpretation of aerial photos

Table 9.1.3. USA - Variations in the positions of glacier fronts.

Part 16 of 19 Changes of terminus area 1964 - 1968

for abbreviations and references see chapter 3

b.

Change of Terminus Aerea		Previous Trends	Remarks	Investigators (References)
m^2	error (m^2)			
		Steady recession at average rate of 45-55 m/yr since a small advance about 1920.	Aerial photos: USGS 1964,1966,1967; AGS 1968	W.O.Field
		Recession interrupted by still-stand since 1889. Average rate of recession -95 m/yr. The present still-stand follows period of rapid recession which averaged about 180m/yr from 1958 to 1963	Surveys: AGS 1958,1967; aerial photos: USGS 1964,1965,1967; AGS 1968	W.O.Field
		Advance 1887-1935 at average rate of 12 m/yr and since then at 0-5 m/yr. Terminus was advancing slowly when visited in July 1967. The glacier is now at greatest length since 16th century but since 1930 marginal zones above the terminus indicate a shrinkage in volume	On site observations: AGS 1967; Aerial photos: USGS 1964,1965,1966, 1967; AGS 1968	W.O.Field
		Steady recession since maximum in 1890's. Total recession to 1964 ca. 1.65 km	Aerial photos: USGS 1964,1965,1966, 1967; AGS 1968	W.O.Field
		Slow recession for several decades	Aerial photos: USN 1948; AGS 1968	W.O.Field
		Recession interrupted by still-stands since 1890's. Average rate of recession 1893-1964 ca. - 50 m/yr	Aerial photos: USGS 1964,1965,1967; AGS 1968	W.O.Field

Table 9.1.3. USA - Variations in the positions of glacier fronts.

Part 17 of 19 Changes of terminus length 1964 - 1968

( )* estimated value

a.

	No. Name	Dates of Observation		Change of Terminus Length			
		from	to	m	error m	m/yr	how determined
WASHINGTON, CASCADE MOUNTAINS	115 Coleman	19.9.64	19.9.65	+ 35	<2	+ 35	terrestrial photogrammetry
		19.9.65	.9.66	+ 2	2 - 5	+ 2	interpretation of photos taken from the same point
		.9.66	19.9.67	+ 17	2 - 5	+ 17	terrestrial photogrammetry/ interpretation of photos taken from the same point
		19.9.67	.9.68	+ 1	2 - 5	+ 1	aerial photogrammetry
	116 Roosevelt	26.9.64	12.9.65	+ 5	<2	+ 5	terrestrial photogrammetry
		12.9.65	.9.66	+ 3	2 - 5	+ 3	aerial photogrammetry
		.9.66	20.9.67	+ 6	2 - 5	+ 6	terrestrial photogrammetry/ interpretation of photos taken from the same point
		20.9.67	.9.68	+ 6	2 - 5	+ 6	terrestrial photogrammetry
	117 Park	9.9.64	23.9.65	+ 15	2 - 5	+ 15	interpretation of aerial photos
		23.9.65	22.9.66	+ 40	2 - 5	+ 40	
		22.9.66	21.9.67	+ 40	5 -20	+ 40	
		21.9.67	22.8.69	+ 40	5 -20	+ 20	
	118 Boulder	9.9.64	23.9.65	+ 60	2 - 5	+ 60	interpretation of aerial photos
		23.9.65	22.9.66	+ 40	2 - 5	+ 40	
		22.9.66	21.9.67	+ 60	2 - 5	+ 60	
		21.9.67	22.8.68	+ 81	5 - 20	+ 40	
	119 Deming	7.9.62	23.9.65	+ 137	20-50	+ 46	interpretation of aerial photos
		23.9.65	22.9.66	+ 6	2 - 5	+ 6	
		22.9.66	21.9.67	+ 15	2 - 5	+ 15	
	120 Boston	3.9.63	23.9.65	+ 15	2 - 5	+ 8	interpretation of aerial photos
		23.9.65	21.9.67	+ 15	2 - 5	+ 8	
	121 South Cascade	.10.64	.9.65	- 12	2 - 5	- 12	ground survey
		.9.65	.9.66	- 16	2 - 5	- 16	
		.9.66	.9.67	- 17	2 - 5	- 17	
		.9.67	.7.68	- 12	2 - 5	- 12	
	122 North Guardian	23.9.65	22.9.66	+ 6	2 - 5	+ 6	interpretation of aerial photos
		27.9.66	21.9.67	0	2 - 5	0	
		21.9.67	27.9.68	0	2 - 5	0	

Table 9.1.3. USA - Variations in the positions of glacier fronts.

Part 17 of 19 Changes of terminus area 1964 - 1968

for references and abbreviations see chapter 3

b.

Change of Terminus Aerea		Previous Trends	Remarks	Investigators (References)
m^2	error (m^2)			
			Terminus reached glacier creek in 1965 advancing perpendicular to stream valley	A.E.Harrison
			Reached top of waterfall in 1967	A.E.Harrison
		Marked retreat till about 1953 then strong advance		A.Post
		Marked retreat till about 1953 then strong advance		A.Post
		Retreat 1940-1956, advance 450 m 1959-1965		
		Retreat before 1945 advance 400 m 1947-1954		A.Post
3.02 - 4.06 - 5.23 - 3.08	10^2 10^4 10^2-10^4 10^2-10^4 10^2-10^4	Rapid retreat by calving 372 m 1952-1963, slower retreat since due to land-based terminus	Values from combination of photogrammetry (aerial), photo interpretation and ground surveying	W.V.Tangborn R.A.Krimmel
		Retreat 900 m 1900-1939, retreat 300 m 1939-1947, advance 150 m 1949-1955	Thickening	A.Post

Table 9.1.3. USA - Variations in the positions of glacier fronts.

Part 18 of 19 Changes of terminus length 1964 - 1968

( )* estimated value

a.

	No. Name	Dates of Observation from	to	Change of Terminus Length m	error m	m/yr	how determined
WASHINGTON, CASCADE MOUNTAINS	123 Dusty	27. 9.64	23. 9.65	0	2 - 5	0	interpretation of aerial photos
		23. 9.65	27. 9.66	- 6	2 - 5	- 6	
		22. 9.66	21. 9.67	0	2 - 5	0	
		21. 9.67	27. 9.68	0	2 - 5	0	
	124 Chocolate	27. 9.64	23. 9.65	+ 25	2 - 5	+ 24	interpretation of aerial photos
		23. 9.65	21. 9.67	0	2 - 5	0	
		21. 9.67	27. 9.68	0	2 - 5	0	
	125 White Chuck	16.10.63	23. 9.65	- 50	5 -20	- 25	interpretation of aerial photos
		23. 9.65	22. 9.66	- 25	5 -20	- 25	
		22. 9.66	22. 9.67	- 25	5 -20	- 25	
	126 Carbon	25. 9.64	2. 9.65	+ 17	<2	+ 17	ground survey
		2. 9.65	2. 9.66	+ 20	<2	+ 20	
		2. 9.66	27. 9.67	+ 16	<2	+ 16	
	127 Emmons	20. 8.64	22. 9.66	+ 70	5 -20	+ 35	interpretation of aerial photos
		22. 9.66	18. 9.67	+ 9	2 - 5	+ 9	
	128 Cowlitz	20. 8.64	22. 9.66		5 -20	+(10-50)*	interpretation of aerial photos
		22. 9.66	18. 9.67	+ 15	2 - 5	+ 15	
	129 Nisqually	31. 8.64	25. 8.65	+ 52	2 - 5	+ 52	ground survey
		25. 8.65	29. 8.66	+ 44	2 - 5	+ 44	
		29. 8.66	31. 8.67	+ 25	2 - 5	+ 25	
		31. 8.67	24. 9.68	+ 8	2 - 5	+ 8	
	130 Kautz	20. 8.64	22. 9.66		5 -20	+(10-50)*	interpretation of aerial photos
		22. 9.66	18. 9.67	+ 15	5 -20	+ 15	
	131 South Tahoma	20. 8.64	22. 9.66	+ 140	20-50	+ 70	interpretation of aerial photos
		22. 9.66	18. 9.67	+ 100	20-50	+ 100	
		18. 9.67	15. 7.69	+ 170	20-50	+ 85	

Table 9.1.3. USA - Variations in the positions of glacier fronts.

Part 18 of 19 Changes of terminus area 1964 - 1968

for references and abbreviations see chapter 3

b.

Change of Terminus Aerea		Previous Trends	Remarks	Investigators (References)
m^2	error (m^2)			
		Retreat 1100m 1900-1939, retreat 300m 1939-1947, advance 150m 1949-1955		A. Post
		Retreat 900m 1900-1939, retreat 1200m 1939-1947, advance 370m 1949-1956, retreat 30m 1956-1959, no change 1959-1964	Greatly increased crevassing thickening. Extensive crevassing, smooth in lower glacier	A. Post
		Retreat 90m 1949-1950, retreat 240m 1956-1960		A. Post
		Retreat 276m 1932-1958, advance 16m 1959-1961		D.Richardson
		Retreat 705m 1930-1952, advance 200m 1953-1957, advance continued slowly 1958-1963		M.F. Meier
		Retreat until about 1960		M.F. Meier
		Sudden recession of more than 1000m during outburst flood 1938, slow recession 1938-1959, advance 1959-1964		M.F. Meier
		Sudden recession of more than 1000 m during outburst flood 1938 slow recession 1938-1959, advance 1959-1964		M.F. Meier
		Recession until 1962, advance since	Outburst flood 31.8.67. Terminus stagnant 15.7.69	M.F. Meier

Table 9.1.3. USA - Variations in the positions of glacier fronts.
Part 19 of 19 Changes of terminus length 1964 - 1968
( )* estimated value

a.

Region	No. Name	Dates of Observation from	to	Change of Terminus Length m	error m	m/yr	how determined
WASHINGTON, CASCADE MOUNTAINS	132 Eliot	4.9.63	5.9.69	+ 105		+ 105	interpretation of photos taken from the same point
		5.9.64	1.9.65	+ 8		+ 8	ground survey
		1.9.65	22.9.66	- 16		- 16	interpretation of photos taken from the same point
		22.9.66	19.9.67	0		0	
		19.9.67	28.9.68	0*		0	
SIERRA NEVADA	133 Maclure	18.10.66	15.10.67		2 - 5	+(0-10)*	interpretation of aerial photos
		15.10.67	8.10.68		2 - 5	-(0-10)*	
ROCKY MOUNTAINS	134 Grinnell	28. 8.64	27. 8.68	- 61	2 - 5	- 15.2	ground survey
	135 Sperry	16. 9.61	23. 8.69		5 -20	-(0-10)*	ground survey
	136 Grasshopper	. 6.67	. 9.69	0*			terrestrial photogrammetry
	137 Teton	. 8.65	. 7.66	+ 10	2 - 5		ground survey
ANTARCTICA	138 Hobbs	1965	1966		< 2	-(0-10)*	interpretation of aerial photos
	139 Taylor	1965	1966		<2	+(0-10)*	interpretation of aerial photos

Table 9.1.3. USA - Variations in the positions of glacier fronts.

Part 19 of 19 Changes of terminus area 1964 - 1968

for abbreviations and references see chapter 3

b.

Change of Terminus Aerea		Previous Trends	Remarks	Investigators (References)
m^2	error (m^2)			
			The kinematic wave responsible for the terminus change in 1963-1964 was also detected by a cross-sectional profile previously established 145 m up-glacier from the September 1964 terminus position. The kinematic wave was first detected in 1958 by these profile measurements. Profiles for most years are available since 1940.	N.A. Dodge (Ref.: Dodge, 1964)
				D.R. Scully (Ref.: Maclure,Gl.)
$10.5 \cdot 10^3$	10 - 100		Terminal recession primarily due to ice calving into lake	A. Johnson (Ref.: Annual progress reports in open file of the U.S. Geological Survey)
			No measurements in 1964 or 1968. Terminal recession very irregular. Estimated overall annual average 3 to 5 m	A. Johnson (Ref.: s. 134)
				D.L. Alford
negligible	negligible			J.C. Reed jr. (Ref.: Reed 1964, 1965,1967)
			Inspection of moraine VS ice face	R.F. Black (Ref.: Black and Bowser,1968)
			Inspection of moraine VS ice face	R.F. Black (Ref.: Black and Browser,1968)

Table 9.1.4. Iceland - Variations in the positions of glacier fronts.

Part 1 of 2 - retreat + advance n: not measured na: not available

Main glacier	No	IHD No	Name	1964 -1965	1965 -1966	1966 -1967	1967 -1968	1968 -1969	1969 -1970	1964	1965	1966	1967	1968	1969	1970
			Observed glacier snout	Variations in meters in the years						Day and month of the surveys in the years						
II Snaefellsjökull	2	1	Jökulhals	?	- 20	+ 49	- 13	←	+ 29			20.09.	25.09.	23.09.	n	18.09.
	1		Hyrningsjökull	- 8	- 14	+ 3	+ 1	- 10	-			20.09.	25.09.	23.09.	15.10.	18.09.
I Drangajökull	1	2	Kaldalonsjökull	- 46	- 12	←	- 66	←	- 12			27.11.	-	22.10.	n	23.09.
	3		Reykjarfjardarj.		←	- 23	- 13	- 44	na			na	29.09.	20.10.	17.10.	na
	2		Leirufjardarj. 63/64 ←	←	- 156	n	n	- 24	n			25.09.	8.09.	7.09.	13.09.	27.08.
X Nordurlandsjöklar	1	3	Gljufurarjökull	- 12	- 4	na	←	- 30	+ 5						25.08.	31.08.
	3	4	Baegisarjökull			1)	- 5	na	na							
VII Langjökull	2	5	Hagafellsjökull W	←	- 175	←	←	←	- 292	.08.		13.08.	-	-	-	10.10.
	3	6	Hagafellsjökull E 63/64 ←	- 228	- 50	n	←	- 185	- 77			27.08.	29.10.	-	2.08.	30.08.
		7	Jökulkrokur, Pjofad.		←	- 54	na	na	- 69		1.08.	-	26.08.			29.08.
IX Kerlingarfjöll	1	8	Lodmundarjökull	- 3	na	na	na	na	na							
VI Hofsjökull	4	9	Lambahraunsjökull		na	na	na	na	na			na	na			
	2	10	Nauthagajökull	- 1	- 22	- 5	←	- 13	- 11			8.09.	8.09.		20.09.	3.10.
	3	11	Mulajökull W	} - 33	+ 71	- 13	←	- 85	n			8.09.	8.09.		20.09.	3.10.
	3	11	Mulajökull S			- 78	←	- 26	+ 4			8.09.	8.09.		20.09.	3.10.
III Eyjafjallajökull	1	12	Gigjökull	+ 12	- 24	- 12	- 25	n	na			20.08.	28.10.	29.09.	7.12	
IV Myrdalsjökull	1	13	Solheimajökull W	+ 50	± 0	- 11	- 22	- 42	+ 26			13.11.	29.10.	21.11.	27.09.	31.10.
	2	13	Solheimajökull E	+ 22	+ 3	+ 4	- 8	- 9	+ 8			13.11.	29.10.	21.11.	27.09.	31.10.
	3	13	Solheimaj. Jökulhöfud	+ 18	± 0	+ 1	- 1	- 4	+ 5			13.11.	29.10.	21.11.	27.09.	31.10.
		13	Myrdalsj. E (Merkig.)			- 9	←	←	- 116			30.08.	2.08.			19.07.
V Vatnajökull	22	14	Tungnaarj. (Jökulh.)	- 50	- 200	- 77	- 190	- 55	- 192			8.10.	17.09.	14.09.	14.09.	20.09.
		15	Sidujökull E (M-175)	- 25	- 77	na	na	na	na		.09.	-	17.09.			
	1	16	Skeidararjökull W	+ 30	+ 37	- 9	- 14	- 77	- 10			7.10.	.09.	na	3.11.	29.11.
	2	17	Skeidararj. E_1 (saeluhus)	}		- 5	- 11	+ 13	+ 39			23.10.	26.10.	2.11.	9.10.	2.10.
	2	17	Skeidararj. E_2	+ 51	- 6	- 19	- 12	- 11	+ 4			23.10.	26.10.	2.11.	9.10.	2.10.
	2	17	Skeidararj. E_3 (farvegur)	}		- 7	- 3	- 3	- 4			23.10.	26.10.	2.11.	9.10.	2.10.
	3	18	Morsarjökull	- 18	- 24	- 7	- 8	- 12	- 11			16.10.	18.10.	29.10.	25.10.	18.11.
	4	19	Skaftafellsjökull	- 12	- 18	- 30	+ 11	- 5	- 55			30.11.	30.11.	30.10.	25.11.	15.11.
	5	20	Svinafellsj. N-side	- 5	- 1	- 6	- 5	n	- 5			30.11.	30.11.	3.11.	25.11.	15.11.
	6	20	Svinafellsj. S-side	- 12	- 2	- 18	- 14	- 3	- 7			30.11.	30.11.	3.11.	25.11.	15.11.
	7	21	Virkisjökull	- 4	+ 18	+ 5	- 15	- 5	- 10			30.11.	30.11.	3.11.	25.11.	20.11.
	10	21	Falljökull	- 5	+ 12	+ 3	- 10	- 3	- 7			30.11.	30.11.	3.11.	25.11.	20.11.
	8	22	Kviarjökull	- 7	- 7	- 46	+ 28	n	- 5			6.10.	16.10.	30.10.	26.10.	10.10.
	9	23	Hrutarjökull	0	- 27	- 3	n	n	n			7.10.	16.10.	22.10.	10.10.	15.10.
	10	24	Fjallsj. (Gamla-Sel)	}		- 24	- 9	- 14	- 5			7.10.	30.10.	23.10.	10.10.	15.10.
	10	24	Fjallsj. Fitjar	+ 11	- 2	- 24	n	+ 12	n			7.10.	30.10.	23.10.	10.10.	15.10.
	10	24	Fjallsj. Breidamerkurfj.	}		- 5	- 3	+ 4	+ 4			7.10.	30.10.	23.10.	24.10.	20.10.

Table 9.1.4. Iceland - Variations in the positions of glacier fronts.

Part 2 of 2 - retreat + advance n: not measured na: not available

Main glacier	Observed glacier snout No	IHD No	N a m e	Variation in meters in the years 1964 -1965	1965 -1966	1966 -1967	1967 -1968	1968 -1969	1969 -1970	Day and month of the surveys in the years 1964	1965	1966	1967	1968	1969	1970
V Vatnajökull	11	25	Breidamj. W (Brmfj.)	- 51	- 54	- 29	- 20	- 34	- 30			15.10.	21.10.	3.10.	24.10.	20.10.
	11	25	Breidamj. W (Brm.skali)			- 57	- 45	- 41	- 30			15.10.	18.10.	24.10.	20.10.	28.10.
	11	25	Breidamj. W (Nygr.bakki)			- 75	- 70	- 132	- 85			24.10.	18.10.	24.10.	20.10.	28.10.
	11	25	Breidamj. W (Nygraedur)			- 150	na	na	na			24.10.	17.10.			
	12	26	Breidamj. E (Jökulsa/St.)	- 35	- 48	- 45	- 20	+ 5	+ 80			na	6.11.	12.11.	29.10.	20.10.
	12	26	Breidamj. E (Fellsfjall)			- 80	- 17	+ 23	n			na	6.11.	12.11.	29.10.	20.10.
	14	27	Brokarjökull	± 0	± 0	n	+ 36	- 50	+ 31			2.10.	30.10.	1.11.	24.10.	9.10.
	15	27	Birnudalsjökull	± 0	± 0	n	n	n	n			30.09.	30.10.	5.10.	24.10.	9.10.
	16	27	Eyvindartungnakollur	± 0	± 0	n	n	n	n			14.10.	1.10.	2.10.	25.09.	9.10.
	17	28	Skalafellsj. W (Skalf.)	- 35	- 25	+ 10	+ 6	n	- 5			16.10.	2.10.	31.10.	25.09.	27.10.
	17	28	Skalafellsj. E (Hafrafell)			n	n	n	n			16.10.	16.10.	31.10.	26.09.	8.10.
	18	29	Heinabergsj. (Hafraf.)	- 15	- 40	n	- 28	- 10	n			1.10.	16.10.	20.10.	26.09.	8.10.
	18	29	Heinabergsj. (Geitakinn)			n	- 50	- 50	n			1.10.	16.10.	20.10.	26.09.	8.10.
	19	30	Flaaj. (W Holmsa) J. 152	- 28	- 6	+ 9	+ 8	- 10	n			4.10.	16.10.	20.10.	26.09.	8.10.
	19	30	Flaaj. (E Holmsa) J. 150			- 1	n	- 1	- 6			4.10.	16.10.	20.10.	26.09.	8.10.
	19	30	Flaaj. (E Holmsa) J. 148			- 5	+ 40	- 11	- 19			4.10.	16.10.	20.10.	26.09.	8.10.
	19	30	Flaaj. (E Holmsa) J. 146			+ 30	+ 9	- 1	- 3			4.10.	16.10.	31.10.	26.09.	8.10.
	20	31	Hoffellsjökull W	←	- 85	na	←	- 150	- 18			15.12.	.12.		20.03.	1.04.
	21	32	Hoffellsjökull E	←	+ 20	na	na	n	+ 5			15.12.	.12.		20.03.	1.04.
		33	Bruarjökull	1)		na	na	na	na			na	na			

1) special mapping

Table 9.1.5. Norway - Variations in the positions of glacier fronts.

Glacier	Variations in metres in the years					
	1964/65	1965/66	1966/67	1967/68	1968/69	1969/70
Jotunheimen						
Storbreen	- 8	- 6	- 5	- 8	- 14	- 18
Styggedalsbreen	- 3	- 9	- 3	- 1	- 16	- 9
Jostedalsbreen						
Briksdalsbreen	- 2	+ 3	+ 15	+ 17	+ 8	- 19
Lodalsbreen	- 120	- 145	- 93	- 131	- 127	- 116
Fåbergstølbreen	- 103	- 138	- 101	- 137	- 135	- 121
Austerdalsbreen	- 8	- 31	- 1	- 22	- 35	- 13
Stegholtbreen	- 67	- 70	- 69	- 110	- 99	
Nigardsbreen	- 54		- 20			
Abrekkebreen	+ 7					
Tunsbergdalsbreen	- 28					
Bersetbreen			+ 10	+ 12		
Store Supphellebre			+ 1			
Folgefonni						
Buarbreen						0
Bandhusbreen	0	+ 5	- 1	- 1	- 3	
Møre						
Trollkyrkjebreen	0	- 15	+ 5	+ 5	- 15	- 8
Finnebreen	0					
Veslebreen	- 1					
Svartisen						
Engabreen		+ 5	+ 10	+ 10	+ 8	+ 34
Østerdalsisen						- 11

Table 9.1.6. Sweden - Variations in the positions of glacier fronts.

\- retreat

Glacier	Position N	Position E	Variations in m in the years 1963–1964	1964–1965	1965–1966	1966–1967	1967–1968	1968–1969	1969–1970	1970–1971	Day and month of surveys in the years 1963	1964	1965	1966	1967	1968	1969	1970	1971
Salajekna	67°08'	16°27'			←	- 20	-11	←	-34	- 5			2.8.	-	2-3.8.	28.8.	-	21.8.	30.7.
Hyllglaciären	67°41'	17°21'				snow	-10*	←	-10*	- 2			29.7.	-	28.7.	22.8.	-	31.7.	8.8.
Ruotesglaciären	67°26'	17°30'			←	- 10	-13.5	←	-42	-10			3.8.	-	20.7.	4.8.	-	6.8.	10.8.
Suotasglaciären	67°27'	17°39'				snow	-15	←	-53	- 5			30.7.	-	21.7.	6.8.	-	3.8.	3.8.
Vartasglaciären	67°21'	17°41'				snow	- 1	←	-26	- 1		8.9.	31.7.	-	22.7.	6.8.	-	4.8.	3.8.
Mikkaglaciären	67°24'	17°42'			←	- 25	-12	←	-39	-12			4.8.	-	18.7.	30.7.	-	8.8.	12.8.
Pårteglaciären	67°10'	17°43'				snow	←	←	-65	- 7			1.5.	-	15.7.	-	-	17.8.	15.8.
Ruopsokglaciären	67°21'	17°59'			←	- 14	-10	←	-50	- 1			31.7.	-	12.7.	21.7.	-	11.8.	30.7.
Rjukojietna	68°05'	18°05'	←	←	←	←	-64	-12	←	-30*	31.8.					13.8.	8.8.	-	23.7.
Kårsajökeln	68°22'	18°21'							←	-50*							29.7.	-	31.7.
Stuor-Räitaglaciären	67°58'	18°23'	←	-15	←	0	-20	←	-10*	-20	31.8.	-	23.8.	-	28.8.	10.8.	-	3.9.	22.7.
Västra Påssus	68°04'	18°24'						-17	←	-23						12.8.	7.8.	-	24.7.
Östra Påssus	68°04'	18°25'						-14	←	-14						12.8.	7.8.	-	24.7.
Rabots glaciär	67°55'	18°27'	←	-10	←	- 33	-13	-17	-10*	-15	31.8.	-	22.8.	-	20.8.	9.8.	11.8.	26.8.	11.8.
Unna-Räitaglaciären	67°59'	18°27'	←	←	←	- 6	- 5*	←	←	- 8*	31.8.	-	23.8.	-	28.8.	11.8.	-	-	22.7.
Kuototjåkkaglaciären	68°09'	18°34'							←	-31						20.8.	-	-	27.7.
Isfallsglaciären	67°55'	18°35'	0	- 5	-11	- 16	←	- 7	-34	-15		18.8.	28.8.	7.8.	29.8.	-	?	17.9.	2.9.
Storglaciären	67°54'	18°36'	0	- 5	-10	- 5	- 5	- 9*	-13	- 6	24.8.	30.8.	27.8.	7.8.	1.9.	22.8.	?	17.9.	2.9.

* Figures less reliable because of partly snow covered fronts or obtained from aerial photographs

Table 9.1.7. Austria - Variations in the positions of glacier fronts.

Part 1 of 2 + advance n not measured sn under snow
- retreat x not measurable () interpolated

Mountain group	Glacier No	Glacier Name	Variations in metres in the years 1964 -1965	1965 -1966	1966 -1967	1967 -1968	Day and month of the surveys in the years 1965	1966	1967	1968
I Silvretta-gruppe	1	Litzner NE	- 0.1	+ 0.2	+ 1.7	0.0	8.9.	13.9	8.9.	18.9.
	2	Klostertaler	0.0	+ 7.0	+ 4.0	0.0	9.9.	12.9.	8.9.	18.9.
	3	Ochsental	- 6.8	-16.8	-29.0	-16.0	9.9.	11.9.	7.9.	17.9.
	4	Fermunt	0.0	- 0.8	- 5.3	- 6.5	9.9.	11.9.	7.9.	17.9.
	5	Bieltal	sn	sn	sn	sn	9.9.	11.9.	7.9.	17.9.
	6	Jamtaler	- 1.0	+ 1.1	- 6.6	- 9.1	8.9.	11.9.	7.9.	17.9.
	7	Larain	sn	+ 2.0	0.0	- 6.5	8.9.	10.9.	7.9.	17.9.
II Oetztaler Alpen	1	Weissee	-30.3	- 4.8	-17.3	- 8.8	24.8.	12.8.	24.8.	21.8.
	2	Gepatsch	-47.0	-10.7	-15.0	-42.0	24.8.	13.8.	25.8.	21.8.
	3	Hinterer Oelgruben	←	←	-48.0	0.0	25.8.	12.8.	24.8.	22.8.
	4	Sexegerten	-24.5	- 8.0	-22.0	-17.7	29.8.	11.8.	24.8.	22.8.
	5	Taschach	-19.0	- 4.0	- 6.0	- 5.0	28.8	11.8	23.8.	22.8.
	6	Mittelberg	0.0	0.0	-24.0	- 5.0	27.8.	11.8.	23.8.	23.8.
	7	Karles	0.0	0.0	- 4.5	- 9.0	26.8.	10.8.	22.8.	23.8.
	8	Hochjoch	-25.0	- x	-22.5	-15.0	19.9	8.8.	1.10.	30.9.
	9	Hintereis	-10.0	-18.0	-30.0	-13.0	23.8.	7.8.	6.9.	4.9.
	10	Guslar	0.0	- 7.0	-24.0	-15.0	26.8	9.8.	6.9.	10.10.
	11	Vernagt	-30.0	-20.0	-45.0	-25.0	25.8.	9.8.	6.9.	10.10.
	12	Rettenbach	- 3.0	0.0	- 7.0	0.0	26.8.	10.8.	22.8.	23.8.
	13	Pitztaler Jöchl	sn	sn	sn	sn	26.8.	10.8.	22.8.	23.8.
	14	Mitterkar	←	←	←	+37.0	-	-	-	24.9.
	15	Rofenkar	←	←	←	+31.6				
	16	Taufkar	sn	sn	sn	sn	-	-	-	-
	17	Niederjoch	-11.5	- 7.5	- 9.6.	-12.0	17.10.	10.-13.9.	18.10.	22.9.
	18	Marzell	- 1.3	- 6.9.	←	-13.4	17.10.	10.-13.9.	-	22.9.
	19	Schalf	n	-19.1		-26.3	17.10.	10.-13.9.	-	22.9.
	20	Diem	←	-28.4	←	-16.4	-	10.-13.9.	-	23.9.
	21	Spiegel	←	-11.4	←	-18.0	-	10.-13.9.	-	15.9.
	22	Gurgler 62/63←	←	-67.2	+ 6.5	n	5.10.	10.-13.9.	17.10.	20.9.
	23	Langtaler	-12.5	-12.4	-11.9	-13.6	5.10.	10.-13.9.	17.10.	20.9.
	24	Rotmoos	- 6.7	-11.8	- 8.5	-23.3	5.10.	10.-13.9.	16.10.	21.9.
	25	Gaisberg	- 9.7	-18.8	-27.3	-21.4	5.10.	10.-13.9.	16.10.	19.9.
	26	Kesselwand	n	0.0	+ 6.0	+ 5.0	-	15.10.	30.9.	9.10.
III Stubaier Alpen	1	Sulztaler	←	-25.5	-29.5	-15.0	3.8.	30.8.	24.8.	21.8.
	2	Bockkogel	n	n	n	n	3.8.	3.8.	24.8.	21.8.
	3	Schwarzenberg	←	+11.0	-26.5	- 7.0	3.8.	30.8.	24.8.	21.8.
	4	Bachfallen	+ 0.5	+ 1.5	+ 1.5	- 8.0	4.8.	30.8.	26.8.	22.8.
	5	Längentaler	sn	sn	sn	sn	4.8	28.8.	26.8.	22.8.
	6	Lisenser	- 4.5	sn	- 2.3	+ 0.5	5.8.	25.9.	27.8.	23.8.
	7	Alpeiner	-15.0	- 3.0	-25.0	- 3.0	5.8.	28.8.	28.8.	25.8.
	8	Berglas	- 1.5	+ 9.0	+ 7.0	+ 1.0	5.8	28.8	28.8.	24.8.
	9	Hochmoos	sn	sn	sn	sn	5.8.	27.8.	25.8.	26.8.
	10	Daunkogel	+ 2.0	- 1.0	- 1.6	- 3.5	4.8.	26.8.	24.8.	26.8.
	11	Schaufel	sn	sn	sn	sn	4.8.	27.8.	24.8.	27.8.
	12	Fernau	←	+ 2,0	+ x	+ 0.7	4.8.	27.8.	24.8.	27.8.
	13	Grawawand	sn	sn	sn	sn	5.8.	27.8.	25.8.	26.8.
	14	Sulzenau	+ 5.0	0.0	- 5.0	+ x	4.8.	25.8.	23.8.	27.8.
	15	Grünau	←	+ 7.0	+ 3.0	0.0	3.8.	24.8.	23.8.	27.8.
	16	Grübl W	sn	sn	sn	sn	3.8.	23.8.	22.8.	28.8.
	17	Grübl E	sn	sn	sn	sn	3.8.	23.8.	22.8.	28.8.
	18	Simming	sn	- 5.5	- 3.0	- 3.5	2.8.	23.8.	21.8.	28.8.

Table 9.1.7. Austria - Variations in the positions of glacier fronts.

Part 2 of 2

+ advance	n not measured	sn under snow	
- retreat	x not measurable	() interpolated	

Mountain group		Glacier No	Name	Variations in metres in the years 1964 -1965	1965 -1966	1966 -1967	1967 -1968	Day and month of the surveys in the years 1965	1966	1967	1968
IV	Zillertaler Alpen	1	Waxegg	+14.5	+ 9.0	+20.3	+10.3	4.- 5.9.	10.-11.9.	21.-24.9.	19.-20.8.
		2	Horn	-10.0	-39.5	+ 1.0	-22.5	4.- 5.9.	10.-11.9.	21.-24.9.	19.-20.8.
		3	Schwarzenstein	-30.0	-436.0	-35.0	-35.0	4.- 5.9.	10.-11.9.	21.-24.9.	19.-20.8.
V	Venediger-gruppe	1	Krimmler	+10.0	- 2.5	- 2.0	- 5.0	7.-11.9.	13.-21.9.	15.-24.9.	8.-26.9.
		2	Obersulzbach	-16.0	-24.5	- 6.5	-33.0	7.-11.9.	13.-21.9.	15.-24.9.	8.-26.9.
		3	Untersulzbach	- 6.0	- 9.1	- 2.5	-16.8	7.-11.9.	13.-21.9.	15.-24.9.	8.-26.9.
		4	Habach	n	n	+ x	n	7.-11.9.	13.-21.9.	15.-24.9.	8.-26.9.
		5	Viltragen	- 6.0	- 2.3	+ 0.7	-10.7	24.-28.8	13.-21.9	15.-24.9.	8.-26.9.
		6	Schlaten	+ 1.0	-10.2	- 1.0	- 1.4	24.-28.8.	13.-21.9.	15.-24.9.	8.-26.9.
		7	Frosnitz	+ 5.5	- 3.5	+ 1.1	- 2.3	24.-28.8	13.-21.9.	15.-24.9.	8.-26.9.
		8	Zetalunitz	- 1.3	-13.3	- 3.0	-12.3	24.-28.8.	13.-21.9.	15.-24.9.	8.-26.9.
		9	Dorfer	+16.0	- 5.5	- 0.8	-13.2	24.-28.8	13.-21.9.	15.-24.9.	8.-26.9.
		10	Maurer	sn	-13.0	sn	sn	24.-28.8.	13.-21.9.	15.-24.9.	8.-26.9.
		11	Simony	- 8.7	- 6.0	- 1.0	- 7.8	13.9.	13.-21.9.	15.-24.9.	8.-26.9.
		12	Umbal	- 8.0	- 7.5	- 1.5	-13.8	14.9.	8.9.	15.-24.9.	8.-26.9.
VI	Granatspitz-gruppe	1	Sonnblick	+ 0.6	+ 1.0	+ 1.4	+ 0.3	25.9.	13.9.	14.10.	11.9.
		2	Kalser Tauern	n	n	n	n	-	-	-	-
		3	Prägrat	n	n	n	n	-	-	-	-
VII	Grossglockner-gruppe	1	Karlinger	n	+ x	0.0	+17.0	-	8.9.	25.9.	11.9.
		2	Bärenkopf	n	n	n	n	-	-	3.9.	8.9.
		3	Klockerin	sn	sn	+ 4.1	- 7.7	-	7.9.	23.9.	17.9.
		4	Pasterzen	- 5.5	- 1.2	- 8.5	-16.5	25.-31.8.	26.-30.8.	27.-31.8.	19.-22.8.
		5	Wasserfall	←	+ 6.0	- 1.6	+ 0.6	-	3.9	2.9	11.9.
		6	Freiwand	←	- 0.9	+15.8.	+ 2.6	-	26.-30.8.	27.-31.8.	19.-22.8.
		7	Pfandlscharten	n	n	n	n	-	26.-30.8.	27.-31.8.	-
		8	Eiser	←	+ 9.8	sn	sn	-	7.10.	24.9.	18.9.
		9	Grieskogel	←	+ 4.6	+ 4.0	+ 3.0	-	7.10.	24.9.	18.9.
		10	Schwarzköpfl	←	+ 4.0	- 1.4	- 5.2	-	8.9.	25.9.	17.9.
		11	Schmiedinger	63/64 ←	←	- 1.3	+ x	-	16.-19.9.	27.9.	20.9.
		12	Oedenwinkel	- 5.9	- 7.2	-11.0	-27.4	18.9.	15.9.	26.9.	10.9.
		13	Unteres Riffl	- 2.6	- 1.5	- 2.6	- 3.7	18.9.	14.9.	15.9.	9.9.
		14	Maurer	(+ 2.5)	(+ 4.2)	(+ 5.9)	(+ 1.4)	-	-	-	-
		15	Wurfer	n	n	n	n	-	-	-	--
		16	Schwarzkarl	n	n	n	n	-	-	-	-
		17	Kl. Eiser	(+ 3.4)	(+ 5.0)	(+ 7.0)	(+ 1.2)	-	-	-	-
		18	Riffl	n	n	n	n	-	-	-	-
		19	Vd. Kasten	n	n	n	n	-	-	-	-
VIII	Goldberg-gruppe	1	Kl. Fleiss	←	+ 4.2	- 8.3	- 0.5	-	8.9.	11.9.	6.9.
		2	Gr. Goldberg	←	+ 6.1	- 6.3	- 4.1	-	8.9.	9.9.	2.9.
		3	Kl. Sonnblick	sn	sn	sn	sn	-	8.9.	9.9.	2.9.
		4	Wurten	←	- 3.0	- 0.4	- 4.1	-	10.9.	6.9.	1.9.
IX	Hochkönig	1	Uebergossene Alm	- 1.8	- 0.3	- 0.9	- 1.2	17.8.	8.-10.9.	23.-24.9.	15.-16.9.
X	Ankogel	1	Grosselend	←	+ 6.8	- 2.4	+ 0.5	5.9.	10.9.	3.9.	3.9.
		2	Kleinelend	←	+ 3.6	- 1.0	+ 2.4	-	4.9.	2.10.	4.9.
		3	Kälberspitz	←	+ 7.0	-14.6	- 7.3	6.9.	5.9.	1.10.	6.9.
		4	Tripp - W	sn	sn	sn	n	4.9.	5.9.	3.9.	-
		5	Hochalm	←	←	- 8.5	+ 0.2	4.9.	2.9.	30.8.	30.8.
		6	Winkel	←	+ 3.5	- 1.1	+ 1.0	4.9.	9.9.	10.9.	7.9.
XI	Dachstein	1	Gr. Gosau	←	- 4.4	+ 3.2	- 4.7	13.9.	8.9.	26.8.	30.-31.8.
		2	Hallstätter	- 4.6	- 9.2	- 2.2	- 9.1	17.9.	27.9.	27.-28.8.	30.8.-2.9.
		3	Schladminger	n	n	n	0.0	-	-	-	2.9.

Table 9.1.8. Austria - Variations in the positions of glacier fronts

Number and percentage of glaciers in advance or in retreat 1889/90 - 1968/69

year	number of observed glaciers	number of glaciers in retreat	number of glaciers stationary	number of glaciers in advance	percentage of observed glaciers in retreat	percentage of observed glaciers stationary	percentage of observed glaciers in advance
1889/90	16	12	0	4	75	0	25
1890/91	26	15	0	11	58	0	42
1891/92	39	27	1	11	69	3	28
1892/93	38	28	2	8	74	5	21
1893/94	30	21	1	8	70	3	27
1894/95	49	26	4	19	53	8	39
1895/96	35	22	4	9	63	11	26
1896/97	51	28	3	20	55	6	39
1897/98	56	30	7	19	54	12	34
1898/99	45	29	1	15	64	2	34
1899/00	30	15	1	14	50	3	47
1900/01	56	41	2	13	73	4	23
1901/02	49	37	4	8	76	8	16
1092/03	18	11	0	7	61	0	39
1903/04	45	36	3	6	80	7	13
1904/05	59	45	5	9	76	9	15
1905/06	37	30	1	6	81	3	16
1906/07	26	25	0	1	96	0	4
1907/08	25	23	0	2	92	0	8
1908/09	37	33	3	1	89	8	3
1909/10	35	23	5	7	66	14	20
1910/11	34	31	3	0	91	9	0
1911/12	41	33	4	4	80	10	10
1912/13	38	26	2	10	68	5	27
1913/14	20	11	3	6	55	15	30
1914/15	40	21	7	12	53	17	30
1915/16	27	13	3	11	48	11	41
1916/17	33	7	4	22	21	12	67
1917/18	32	8	3	21	25	9	66
1918/19	28	5	2	21	18	7	75
1919/20	33	8	1	24	24	3	73
1920/21	27	13	3	11	48	11	41
1921/22	30	18	0	12	60	0	40
1922/23	32	20	4	8	63	12	25
1923/24	55	42	5	8	76	9	15
1924/25	50	34	4	12	68	8	24
1925/26	57	30	7	20	53	12	35
1926/27	56	37	6	13	66	11	23
1927/28	56	50	2	4	89	4	7
1928/29	56	54	1	1	96	2	2

year	number of observed glaciers	number of glaciers in retreat	number of glaciers stationary	number of glaciers in advance	percentage of observed glaciers in retreat	percentage of observed glaciers stationary	percentage of observed glaciers in advance
1929/30	61	58	2	1	95	3	2
1930/31	62	60	1	1	97	2	1
1931/32	58	51	6	1	88	10	2
1932/33	67	64	3	0	96	4	0
1933/34	63	61	1	1	97	2	1
1934/35	66	56	6	4	85	9	6
1935/36	69	64	4	1	93	6	1
1936/37	66	65	0	1	98	0	2
1937/38	62	59	1	2	95	2	3
1938/39	52	50	1	1	96	2	2
1939/40	43	42	0	1	98	0	2
1940/41	58	56	1	1	96	2	2
1941/42	55	52	2	1	95	3	2
1942/43	43	43	0	0	100	0	0
1943/44	55	53	0	2	96	0	4
1944/45	41	39	2	0	95	5	0
1945/46	71	69	2	0	97	3	0
1946/47	53	52	0	1	98	0	2
1947/48	71	61	5	5	86	7	7
1948/49	51	48	1	2	94	2	4
1949/50	76	75	1	0	99	1	0
1950/51	58	52	3	3	90	5	5
1951/52	80	80	0	0	100	0	0
1952/53	63	58	2	3	92	3	5
1953/54	83	77	5	1	93	6	1
1954/55	56	44	7	5	79	12	9
1955/56	79	67	8	4	85	10	5
1956/57	62	52	8	2	84	13	3
1957/58	86	81	5	0	94	6	0
1958/59	61	53	4	4	87	6	7
1959/60	83	77	5	1	93	6	1
1960/61	71	54	13	4	76	18	6
1961/62	82	74	7	1	90	9	1
1962/63	92	90	2	0	98	2	0
1963/64	90	87	2	1	97	2	1
1964/65	85	38	31	16	45	36	19
1965/66	87	40	25	22	46	29	25
1966/67	90	48	23	19	53	26	21
1967/68	88	52	23	13	59	26	15
1968/69	95	65	17	13	68	18	14

Table 9.1.9. France - Variations in the positions of glacier fronts.

Drainage basin	Glacier No.[1	Glacier Name	Variations in meters in the years 1964-1965	1965-1966	1966-1967	1967-1968	1968-1969	1969-1970
Arve	1	Tour	+ 10	←	- 11	←	- 8	← 0
	2	Argentière	- 7	- 5	- 5	0	0	- 3
	3	Bois Mer de Glace	- 8	- 10	- 10	- 15	- 90[2	- 5
	4	Bossons	←	+ 40	+ 10	←	+ 15	+ 20
	5	Taconnaz	←	←	+ 50	←	+ 10	+ 5
	6	Bionnassay	←	←	- 20	←	- 10	- 5
	7	Tré la Tête	←	- 15	- 3	← 0	- 2	- 5
Isère (en amont de l'Arc)	8	Sources de l'Isère	←	←	- 35	- 15	←	- 10
	9	Gébroulaz	←	- 20	- 25	←	←	st?
Arc	10	Sources de l'Arc	- 10	←	- 25	- 10	←	- 20
	11	Mulinet	←	←	0	- 5	←	- 10
	12	Grand Méan	+ 8	←	+ 10	0	- 3	- 5
	13	Evettes	←	←	- 50	- 15	- 10	- 5
	14	Arnès	←	←	- 20	- 5	←	- 10
	15	St. Sorlin d'Arves	- 5	←	- 25	←	- 20	- 10
Vénéon	25	Pilatte	- 20	- 5	←	- 20	- 10	- 5
	26	Chardon	- 5	←	- 25	←	- 30	- 10
Romanche	29	Sarennes						
Torrent del'Ailefroide (Bassin de la Durance)	31	Blanc de Pelvoux-1963	- 45	- 10	←	- 40	- 10	- 5
	32	Noir du Pelvoux	- 2	- 5	←	- 5	- 3	0

(1 cf. index map Fig. 2 Fluctuations of Glaciers 1959-1965.

(2 artificial retreat

Table 9.1.10. Italy - Variations in the positions of glacier fronts.

Part 1 of 6 + advance, - retreat, st stationary, n not measured, x value unknown, ? not measurable, sn under snow

Drainage basin		Glacier		Variations in metres				Day and month of the surveys					Altitude of the snout	
				in the years				in the years						
I. Order	II. Order	No	Name	1964 -1965	1965 -1966	1966 -1967	1967 -1968	1964	1965	1966	1967	1968	metres above sea level	year
Po	Stura di Demonte	1	Clapier	- 1.0	- 2.0	- 3.0	- 0.9	10.08.	15.09.	10.08.	9.08.	10.08.		
		2	Peirabroc	- 0.8	- 1.5	- 2.0	- 1.0	10.08.	15.09.	11.08.	10.08.	10.08.		
		3	Maledia	- 2.0	- 2.0	- 4.0	- 1.3	10.08.	15.09.	11.08.	10.08.	11.08.		
		4	Muraion	- 1.0	- 1.0	- 2.0	- 0.5	10.08.	15.09.	10.08.	10.08.	11.08.		
		5	Ciafraion	- 1.0	- 1.0	- 2.0	- 0.5	10.08.	15.09.	10.08.	10.08.	12.08.		
		6	Gelas	- 0.7	- 1.0	- 2.0	- 0.5	10.08.	15.09.	11.08.	12.08.	12.08.		
	Alto Po	21	Viso NE											
	Dora Riparia	25	Vallonetto											
		26	Galambra	- 4.0	sn	n	n	30.08.	06.08.					
		27	Fourneaux		sn				06.08.					
		29	Agnello	- 8.0	sn	n	n	30.08.	.08.					
		30	Muttet		sn				06.08.					
		32	Bard	n	sn	n	n			.08.				
		34	Lamet	sn	sn	sn	n	20.08.	14.09.	17.09.	5.09.		2960	1960
	Stura di Lanzo	40	Bessanese	- 4.0	st	- 2.0	- 3.0	27.08.	4.08.	18.08.	30.08.	28.08.	2580	1960
		43	Ciamarella	- 4.0	st	- 1.0	st	28.08.	5.08.	19.08.	31.08.	29.08.	3180	1968
	Orco	57	Nel centrale	- x	- x	- x	n		26.09.	4.09.	17.09.			
		59	Carro											
		61	Capra	- 2.0	- 1.0	- x	- 0.8	11.10.	12.09.	25.09.		22.09.		
		64	Basei	sn	- 15.0	n	st		12.09.	10.09.		6.10.		
		69	Breuil	- 10.0	sn	n	sn	10.08.	25.08.		10.09.	22.08.	2975	1962
		72	Noaschetta	sn	n	st	sn	10.08.	22.08.		10.09.			
	Dora Baltea	110	Money					13.09.						
		111	Grand Croux	- 12.0	sn	n	n	19.09.						
		112	Tribolazione	- x	sn	n	n	13.09.						
		121	Trajo	- x	n	- x	n	27.09.	3.10.		31.09.			
		129	Lavacciau					8.09.						
		130	Gran Paradiso	- 2.2	st	- 10.3	n	6.09.	15.09.	27.08.	27.08.			
		131	Moncorvé	- 43.0	- 6.6	- 22.4	n	6.09.	10.09.	27.08.	27.08.			
		132	Monciair	- 18.1	- 2.9	- 6.2	n	6.09.	10.09.	27.08.	27.08.			
		134	Grand Etrét											
		144	Lavassey	sn	- 16.0	- 5.6	- 18.0	6.09.	3.09.	31.08.	16.08.	5.09.		
		145	Fond orientale	sn	- x	- 4.0	n	6.09.	3.09.	31.08.				
		147	Soches Tsanteleina	sn	- 9.1	- 4.7	- 10.5	7.09.	4.09.	1.09.	16.08.	4.09.		
		148	Goletta	sn	sn	n	n							
		155	Torrent	sn	- 0.8	+ 6.0	- 8.0	5.09.		29.08.	15.09.	3.09.		

Table 9.1.10. Italy - Variations in the positions of glacier fronts.

Part 2 of 6

Drainage basin		Glacier		Variations in metres				Day and month of the surveys					Altitude of the snout	
				in the years				in the years						
I. Order	II. Order	No	Name	1964 -1965	1965 -1966	1966 -1967	1967 -1968	1964	1965	1966	1967	1968	metres above sea level	year
Po	Dora Baltea	168	Gliaretta Vaudet											
		177	Ormelune											
		180	Morion or. le	sn	sn	- 12.0	- 11.1	7.09.		.09.	5.09.	15.09.		
		181	Chateau Blanc	sn	sn	- 22.4	sn	10.09.		.09.	8.09.	15.09.		
		189	Rutor	st	+ 4.0	n	n	25.09.						
		202	Breuil merid. le	n	n	n	sn							
		204	Chavannes	- 5.0	sn	- x	sn	14.09.	5.08.	10.09.	11.09.	15.09.		
		208	Estellette	- 5.0	- x	- 3.0	st	5.09.	4.09.	1.09.	10.09.	4.08.		
		209	Lex Blanche	- 2.0	- x	- 1.0	st	20.09.	5.09.	1.08.	10.09.	4.08.		
		213	Miage	- 4.0	- x	- 3.0	+ 1.0	20.09.	6.09.	1.08.	10.09.	6.08.		
		216	Brouillard	- 1.5	- x	n	n	20.09.	6.09.	1.08.				
		218	Freynay	- 1.5	n	n	n	20.09.	6.09.					
		219	Brenva	?	n	n	n	20.09.						
		221	Toula	+ 10.0	+ 17.0	+ 12.0	+ 5.0	18.08.	.08.	20.08.	.08.	15.09.	2637	1968
		222	Mon Frety	+ 2.0	+ x	n	n							
		223	Col de Gigante	+ x	+ x	n	n							
		224	Rochefort	st	+ x	n	n	.08.	25.08.					
		225	Planpincieux	st	+ x	n	n	.08.	25.08.				2350	1968
		226	Grandes Jorasses	+ 2.0	+ 3.0	n	n	25.08.	.08.	.08.			2550	1968
		229	Frébouzie	st	+ x	n	n	.08.	.08.					
		234	Triolet	st	+ x	n	n	.08.	.08.					
		235	Pré de Bar	+ 5.0	+ 6.0	+ 20.0	+ 10.0	12.08.	.08.	.08.	.08.	.08.	2040	1968
		244	Mont Gelé	+ 2.0	+ 5.0	+ 1.0	- 4.0		19.09.	18.09.	17.09.	6.10.		
		257	Col Collon	n	- 52.0	st	- 6.0			4.09.	20.09.	6.10.	2800	1968
		258	Mont Braulé	n	n						20.09.		3138	1967
		259	Tza de Tzan	- 20.0	- 13.0	- 15.0	- 17.0	27.10.	20.08.	21.08.	4.10.	6.09.	2620	1968
		260	Grandes Murailles	- 18.0	-130.0	- 15.0	+ 6.5	27.10.	20.08.	21.08.	4.10.	6.09.	2465	1968
		267	Chavacour	- 30.0	- x	n	n							
		272	Roisetta	- 3.0	- x	+ 5.0	+ 0.6	13.08.	31.08.	11.09.	20.08.	8.09.	2900	1968
		281	Mon Tabel	- 5.0	- 6.0	- x	- 12.0	29.08.	24.08.	29.08.	23.09.	4.09.		
		282	Chérillon	sn	+ 4.0	- 4.5	- 4.5	28.08.	25.08.	28.08.	23.08.	5.09.	2900	1959
		283	Leone	sn	sn	sn	st	.08.	25.08.	.08.	23.08.	21.08.		
		284	Tyndall	st	st	st	st	24.08.	24.08.	.08.	15.08.	21.08.		
		285	Cervino	st	st	st	st	24.08.	24.08.	.08.	15.08.	21.08.		
		286	Forca	- 10.6	- 1.5	+ 11.0	- 10.0	.08.	25.08.	.08.	14.08.	21.08.		
		289	Valtournanche	- 7.5	st	- 20.0	- 10.0	17.08.	.08.	.08.	.08.	.08.	2805	1967

Table 9.1.10. Italy - Variations in the positions of glacier fronts.

Part 3 of 6

Drainage basin		Glacier		Variations in metres				Day and month of the surveys					Altitude of the snout	
I. Order	II. Order	No	Name	in the years 1964 -1965	1965 -1966	1966 -1967	1967 -1968	in the years 1964	1965	1966	1967	1968	metres above sea level	year
Po	Dora Baltea	290	Gran Sometta	sn	+ x	+ x	+ x	17.08.	.08.	.08.	.08.	16.08.	3180	1968
		297	Verra (grande di)	sn	- 5.0	- 2.0	- 4.0	30.08.	11.09.	27.08.	26.08.	7.09.		
		298	Verra (piccolo di)	sn	+ 6.0	st	st	28.08.	11.09.	27.08.	26.08.	7.09.	2790	1968
		299	Castore	sn	st	st	- 2.0	28.08.	11.09.	27.08.	26.08.	2.09.	2870	1968
		304	Lys	- 30.0	- 7.0	- 8.3	- 22.0	.10.	3.11.	.08.	.10.	15.10.	2355	1959
		306	Jndren	- x	sn	sn	sn	.10.	15.09.	.08.	15.11.	.11.	3050	1965
		308	Netschio	- 6.2	sn	sn	- 2.7		1.08.	.08.	15.11.	18.10.		
	Sesia	314	Sesia	- x	- x	?	- x		17.08.	.08.	15.11.	21.10.	2600	1905
	Ticino	324	Nordend	- 10.0	st	+ 2.0	+ 3.0	28.08.	25.					
		325	Belvedère	- 18.0	- 10.0	- 5.0	st	27.08.	25.09.	3.09.	15.11.	13.09.		
		337	Leone	sn	- x	- x	- x	25.09.				.10.		
		338	Aurona	sn	sn	- 32.0	-274.0	25.09.				.10.		
		339	Rebbio	- x	- x	st	st	25.09.				.10.		
		340	Taramona	- x	st	st	st					.10.		
		341	Mottiscia	- x	- x	st	st					.10.		
		342	Boccareccio	st	- x	n	st		.10.			.10.		
		354	Gemelli di Ban	sn	- x	n	st	.10.	.10.	.10.	.10.	.10.		
		355	Costone	sn	sn	sn	sn	.10.	.10.	.10.	.10.	.10.		
		356	Osand mer. le.	- 50.0	- 40.0	- 25.0	n	.10.	28.10.	.10.	.10.	.10.		
		357	Osand sett. le	?	- x	?	?	.10.	28.10.	.10.	.10.	.10.		
		360	Blindenhorn	sn	sn	sn	sn	.10.	28.10.	.10.	.10.	.10.		
		361	Camosci	- x	?	n	n		28.10.					
		363	Basodino occid. le	st	?	n	st	25.08.	.10.					
	Adda	365	Pizzo Ferré	- 8.0	- 6.0	- 10.0	n	25.08.	7.09.	5.09.			2490	1967
		367	Val Loga	sn	sn	sn	n		2.09.	1.09.	3.09.		2830	1962
		368	Passo Zoccone	sn	sn	n	n							
		371	Suretta	- 3.0	sn	sn	n		27.08.	1.09.	4.09.		2725	1962
		372	Orsareigls	sn	sn	sn	n		27.08.	1.09.	4.09.		2700	1964
		388	Cengalo		- ?	- 8.5							2559	1968
		408	Predarossa	sn	+ 5.0	st	st	15.09.	9.09.	10.09.	10.09.	1.09.		
		409	Corna Rossa	st	+ 5.0	st	st	15.09.	9.09.	10.09.		1.09.	2997	1968
		416	Ventina	- 90.0	- 81.0	n	n			10.09.				
		432	Scerscen inf. re.											
		435	Caspoggio											
		439	Fellaria occid. le	- 10.0	- 12.0	- 8.0	n			10.09.	17.09.			
		440	Fellaria orientale											

Table 9.1.10. Italy - Variations in the positions of glacier fronts.

Part 4 of 6

Drainage basin: I. Order	Drainage basin: II. Order	Glacier: No	Glacier: Name	Variations in metres in the years: 1964 -1965	1965 -1966	1966 -1967	1967 -1968	Day and month of the surveys in the years: 1964	1965	1966	1967	1968	Altitude of the snout: metres above sea level	year
Po	Adda	466	Rinalpi											
		467	Val Lia	n	n	n	- 21.0					20.08.	2500	1968
		468	Cardonnè or. le	- 3.0	- 15.0	- 2.0	- 3.0				.09.	.09.	2320	1959
		469	Cardonnè occid. le	- 1.5	- 1.0	- 1.5	- 1.0	26.08.	.08.	.08.	.09.	.09.		
		471	Verva maggiore di	+ 1.0	st	st	st	14.09.	.08.	.08.		.09.	2630	1959
		473	Dosdè orient. le	- 8.0	- 46.0	- 12.0	- 10.0	29.08.	.08.	.08.		.10.	2530	1968
		475	Dosdè occid. le											
		476	Val Viola or. le											
		477	Val Viola occid. le											
		483	Vitelli	- 0.7	- 14.0	- 4.0	st	12.09.	21.09.	8.10.	23.09.	29.09.	2518	1968
		488	Campo	- 5.0	+ 12.0	st	st				.09.			
		503	Cedéch	← 1951	- 12.5	- 8.5	+ 29.4			25.09.	12.10.	5.10.	2687	1968
		506	Rosole	←	- 15.0	- 9.0	+ 30.1			25.09.	12.10.	5.10.	2735	1968
		507	Forno	←	- 30.5	- 65.0	?	29.08.			10.10.	5.10.	2430	1968
		511	Treséro	- 6.7	- 4.5	- 2.5	- 3.8	10.09.	24.09.	9.10.	24.09.	28.09.		
		512	Dosegu	+ 40.0	+ 20.0	+ 14.0	+ 10.0	11.09.	23.09.	9.10.	25.09.	27.09.		
		516	Sforzellina	- 1.5	sn	+ 1.5	+ 6.5	12.09.	23.09.	8.10.	25.09.	27.09.	2780	1964
		517	Lago Bianco	- 19.0	+ 47.5	+ 5.5	- 2.0	13.09.	22.09.	8.10.	23.09.	26.09.	2855	1965
		518	Gavia	sn	sn	sn	?		22.09.	8.10.	23.09.	27.09.	3000	1966
		549	Porola	sn	?	+ 52.0	- 5.0	9.09.		10.08.	19.09.	18.09.	2260	1968
		550	Scais	sn	- 4.0	+ 10.0	- 20.0	9.09.		10.08.	19.09.	18.09.	2350	1968
		566	Gleno	sn	sn	n	n			23.08.				
		567	Trobio	sn	+ 17.0	st	n	13.09.		23.08.	10.09.		2620	1966
	Oglio	577	Pisgana occid. le	- 7.0	- 19.0	- 13.0	- 13.0			8.09.	8.10.	6.09.		
		581	Venerocolo	- 21.0	+ 7.0	+ 1.0	st			8.09.	8.10.	6.09.		
		583	Avio	n	+ 23.0	- 2.0	sn			8.09.	8.10.	6.09.		
		604	Salarno	← 1958	- 15.0	- 15.0	st			8.09.	8.10.	6.09.		
	Minico (Sarca)	632	Carè Alto	st	?	sn	sn	18.08.	8.08.	2.09.	27.08.	11.08.		
		633	Niscli	- 4.7	?	- 4.0	st	19.08.	9.08.	2.09.	27.08.	10.08.		
		634	Lares	- 2.6	- 7.0	- 5.5	- 2.0	19.08.	9.08.	2.09.	28.09.	10.08.		
		637	Lobbia	+ 3.0	+ 3.0	+ 1.0	st	14.09.	27.08.	14.09.	17.09.	28.08.		
		639	Mandrone	- 5.1	- 4.0	- 6.0	- 0.2				31.08.	27.08.		
		640	Nardis	- 8.7	- 8.0	- 18.0	- 12.0				24.08.			
		644	Amola	- 16.7	+ 1.9	- 2.4	- 5.3	24.08.	4.09.	23.08.	23.08.	13.09.		
		646	Cornisello	- 23.0	- x	- x	- 22.0				23.08.	13.09.		
		649	Vallesinella	sn	?	- 8.0	st	16.08.	16.09.	10.08.	23.08.	13.09.	2390	1968

Table 9.1.10. Italy - Variations in the positions of glacier fronts.

Part 5 of 6

Drainage basin I. Order	Drainage basin II. Order	Glacier No	Glacier Name	Variations in metres in the years 1964 -1965	1965 -1966	1966 -1967	1967 -1968	Day and month of the surveys in the years 1964	1965	1966	1967	1968	Altitude of the snout metres above sea level	year
Po	Minico (Sarca)	650	Tuckett	sn	?	?	+ x	18.08.	16.09.	10.08.	23.08.	18.08.	2360	1968
		652	Brentei	sn	sn	- x	+ 2.0	20.08.	19.09.	12.08.	25.08.	20.08.	2550	1968
		653	Sfulmini	sn	- 2.0	- 3.0	st	20.08.	19.09.	12.08.	25.08.	20.08.	2610	1968
		655	Crozzon	sn	sn	- x	sn	22.08.	21.09.	15.08.	27.08.	22.08.	2290	1968
		657	Lagol	sn	- 6.0	- x	st	24.08.	24.09.	17.08.	29.08.	24.08.	2550	1968
		658	Prà Fiori	sn	- 2.0	- x	st	24.08.	24.09.	17.08.	29.08.	24.08.	2582	1968
		659	Dodici Apostoli	sn	?	?	- 4.0	25.08.	24.09.	17.08.	29.08.	25.08.	2585	1968
		678	Presanella	st	- 0.3	- 7.2	st	13.08.	15.08.	11.08.	10.09.	8.09.		
Adige	Noce	699	Mare	- 8.0	- 11.0	n	n	28.08.						
		700	Marmotte	sn	sn	n	n							
		701	Caresèr	+ 13.0	st	n	n							
		702	Cavaion	sn	sn	n	n							
		706	Saént di Mezzo											
		710	Sternai											
	Adige Alto	729	Ultima	←	- 4.5	- 5.5	n			27.09.	10.10.		2760	1966
		730	Alta											
		731	Forcola	←	- 22.5	- 17.0	n		27.09.	27.09.	12.10.		2620	1966
		732	Cevedale	←	- 11.0	+ 3.0	n			27.09.	10.10.		2630	1966
		733	Lunga	←	- 14.0	- 14.0	n			27.09.	10.10.		2630	1966
		762	Solda	←	- 11.0	- 7.5	- 3.5						2260	1968
		777	Vallelunga	- 5.0	- 7.0	←	- 10.0	8.09.	11.09.		29.08.			
		778	Barbadorso di dentro	←	—	—	+ 44.0	8.09.				28.09.		
		779	Barbadorso di fuori	sn	←	—	+ 41.0	29.09.	11.09.	11.09.		29.08.		
		780	Fontana	sn	- 4.5	←	- 1.5	28.09.	11.09.	11.09.		29.08.		
		794	Saldura merid.	sn	+ 10.0	n	n	13.09.	12.09.	25.09.				
		795	Ramudla	sn	+ 3.0	n	n	13.09.	12.09.	25.09.			2760	1966
		813	Giogo alto	- 2.5	- 4.5	←	- 12.0	16.09.	10.09.	10.09.		11.09.	2751	1968
	Isarco	937	Cristallo	- 0.4	sn	n	sn							
Piave		963	Cresta Bianca	sn	sn	- 1.5	- 30.0				17.09.	.09.		
		973	Sorapis or. le	sn	sn	- 3.0	- 50.0				17.09.	.09.		
		974	Sorapis centrale	sn	sn	- 1.3	- 20.0				17.09.	.09.		
		975	Sorapis occ. le	- 0.7	sn	- 1.0	n				17.09.	.09.		
Tagliamento		979	Montasio minore	sn	sn	sn	sn	11.09.	8.09.	3.09.	9.08.	15.08.		
		980	Montasio or. le	sn	sn	sn	sn	11.09.	8.09.	3.09.	9.08.	15.08.		
		981	Montasio occ. le	sn	sn	sn	sn	11.09.	8.09.	3.09.	9.08.	15.08.		
		983	Ursic	sn	sn	sn	sn	10.09.	7.09.	2.09.	8.08.	14.08.		

Table 9.1.10. Italy - Variations in the positions of glacier fronts.

Part 6 of 6

Drainage basin I. Order	Drainage basin II. Order	Glacier No	Glacier Name	Variations in metres 1964 -1965	1965 -1966	1966 -1967	1967 -1968	Day and month of the surveys 1964	1965	1966	1967	1968	Altitude of the snout metres above sea level	year
Tagliamento	Isarco	984	Canin or. le	sn	sn	sn	sn	10.09.	7.09.	2.09.	8.08.	14.08.		
		985	Canin occ. le	sn	sn	sn	sn	10.09.	7.09.	2.09.	8.08.	14.08.		
Danubio	Inn	996	Val Nera occ. le	- 2.0	- 1.0	- 1.0	- 1.0	10.09.	7.09.	12.09.		12.09.	2700	1968
		997	Campo	- 1.0	+ 6.0	st	st	11.09.	14.09.	7.09.	6.09.	11.09.		
Vomano		1006	Calderone	st	←	- 3.0	n	27.09.			3.09.		2676	1967

Table 9.1.11. Italy - Variations in the positions of glacier fronts.

Percentage of glaciers in advance or in retreat 1924/25 - 1967/68

Year	Number of observed glaciers	Percentage of observed glaciers in advance %	stationary %	in retreat %
1924/25	74	3	3	94
1925/26	84	24	11	65
1926/27	108	11	8	81
1927/28	126	0	6	94
1928/29	147	0	4	96
1929/30	150	15	14	71
1930/31	179	4	7	89
1931/32	209	9	22	69
1932/33	258	2	5	93
1933/34	252	4	5	91
1934/35	228	5	3	92
1935/36	133	22	14	64
1936/37	149	7	10	83
1937/38	186	7	4	89
1938/39	112	21	7	72
1939/40	57	22	11	67
1940/41	110	20	13	67
1941/42	94	7	4	89
1942/43	25	14	11	75
1943/44	25	12	10	78
1944/45	25	13	9	78
1945/46	63	6	10	84
1946/47	71	4	4	92
1947/48	59	6	14	80
1948/49	96	2	3	95
1949/50	95	1	4	95
1950/51	105	10	4	86
1951/52	100	2	11	87
1952/53	102	12	9	79
1953/54	102	9	9	82
1954/55	123	4.1	15.4	80.5
1955/56	119	8.4	17.7	73.9
1956/57	118	8.5	13.5	78
1957/58	125	4	9	87
1958/59	112	4.5	15.5	80
1959/60	101	6.9	42.6	50.5
1960/61	107	10.4	30.0	59.0
1961/62	119	4.1	27.9	68.0
1962/63	115	9.6	45.2	45.2
1963/64	119	8	13	79
1964/65	142	7	49	44
1965/66	141	15	36	49
1966/67	91	11	35	54
1967/68	113	15	42	43

Table 9.1.12. Switzerland - Variations in the positions of glacier fronts.

Part 1 of 8 + advance, - retreat, st stationary, n not measured, x value unknown

Drainage basin		Glacier		Variations in metres				Day and month of the surveys					Altitude of the snout		see remark
				in the years				in the years					metres above sea level	year	
I. Order	II. Order	No	Name	1964 -1965	1965 -1966	1966 -1967	1967 -1968	1964	1965	1966	1967	1968			No
Rhone	Rhone	1	Rhone	- 11.5	+ 10.7	- 5.1	- 11.1	3.10.	19.09.	19.09.	19.09.	15.09.	2125	1965	
		2	Mutt	←	+ 11.7	- 4.8	- 4.3	20.09.	19.09.	18.09.	23.08.	16.09.	2626	1968	
		3	Gries (Aegina)	- 11.5	-159.0	- 36.5	- 33.7	16.10.	9.10.	28.06.	13.10.	12.10.	2364	1968	3
		4	Fiescher	- 23.1	- 14.8	- 27.5	- 10.0	11.09.	20.10.	9.09.	14.09.	14.09.	1636	1968	4
		5	Grosser Aletsch	- 62.7	- 74.0	- 36.2	- 7.9	12.09.	5.10.	10.09.	12.10.	15.09.	1503	1968	
		6	Ober Aletsch	- 8.8	- 6.1	- 19.3	- 7.5	5.10.	5.10.	14.09.	2.10.	3.10.	2129	1968	
		7	Kaltwasser	- 5.9	←	- 36.8	st	17.10.	?	n	19.10.	18.10.	2650	1968	
		8	Tälliboden	←	+ 5.2	- 6.2	- 2.6	24.09.	18.09.	26.09.	27.09.	30.09.	2629	1968	
		9	Ofental	←	- 12.9	- 6.2	- 3.5	28.09.	20.09.	30.09.	20.09.	30.09.	2628	1968	
		10	Schwarzberg	- 3.8	- 3.6	- 3.0	- 5.9	2.10.	16.09.	27.09.	28.09.	1.10.	2659	1968	
		11	Allalin	-219.5	+ 7.7	+ 10.9	+ 93.1	4.10.	4.10.	18.09.	26.09.	13.09.	2416	1968	11
		12	Kessjen	sn/st	+ 10.3	- 3.8	st	22.09.	12.09.	20.09.	19.09.	25.09.	2848	1968	
		13	Fee (Nord)	+ 8.8	←	+ 12.9	+ 7.4	24.08.	25.10.	12.09.	16.10.	14.10.	2038	1968	13
		14	Gorner	- 22.8	←	- 54.1	- 40.0	28.08	27.10.	n	11.10.	11.10.	2056	1968	
		15	Z'Mutt	- 23.8	- 12.9	- 16.8	- 12.6	27.08.	26.10.	23.08.	22.08.	26.08.	2232	1968	
		16	Findelen	- 21.4	- 27.2	- 10.9	- 11.9	2.10.	25.09.	5.10.	18.10.	19.09.	2482	1961	
		17	Ried	- 22.7	- 10.8	- 20.7	- 9.5	6.10.	11.10.	29.09.	1.10.	2.10.	2045	1968	
		18	Lang	1962 ←	←	←	- 81.3	n	n	n	n	24.10.	2010	1961	18
		19	Turtmann - West	- 6.0	- 6.3	- 9.2	- 10.4	2.10.	12.10.	21.10.	7.10.	14.09.	2262	1968	
		20	Turtmann - Ost (Brunegg)	+ 20.7	- 28.6	- 45.4	- 4.1	2.10.	12.10.	21.10.	7.10.	14.09.	2456	1968	
		21	Bella Tola	- 1.4	+ 18.5	+ 21.4	+ 20.8	1.10.	16.10.	29.09.	30.10.	18.09.	2763	1967	21
		22	Zinal	- 18.9	st	- 1.6	- 16.0	19.09.	19.10.	14.10.	22.09.	14.09.	1995	1968	
		23	Moming	- 24.3	- 25.7	-167.4	- 3.3	19.09.	19.10.	14.10.	29.09.	14.09.	2327	1968	
		24	Moiry	- 37.6	- 21.9	- 10.8	- 9.8	24.09.	29.10.	8.10.	17.10.	11.10.	2438	1967	
		25	Ferpècle	1963 ←	- 23.2	- 4.5	- 2.7	n	n	5.10.	12.10.	26.10.	1990	1957	25
		26	Mont Miné	1963 ←	st	st	- 10.0	n	n	5.10.	12.10.	26.10.	1965	1955	26
		27	Bas d'Arolla	- 29.2	- 8.0	- 2.4	- 2.4	26.09.	23.10.	22.10.	13.10.	10.10.	2130	1959	
		28	Tsidjiore Nuove	←	←	- 13.6	- 5.8	26.09.	n	n	13.10.	10.10.	2251	1959	
		29	Cheillon	- 2.0	- 3.9	- 6.8	- 2.9	26.09.	4.10.	29.09.	28.09.	4.10.	2620	1964	
		30	L'En Darrey	←	←	- 9.6	- 1.9	25.09.	5.10.	28.09.	27.09.	5.10.	2445	1961	
		31	Grand Désert	←	←	- 8.8	n	7.10.	29.09.	23.09.	2.10.	n	2800	1964	
		32	Mont Fort	1963 ← - 11.0	←	- 30.0	n	n	25.09.	?	29.08.	n	2740	1967	32
		33	Tsanfleuron	- 3.0	+ 10.0	- 20.9	- 3.0	31.10.	4.10.	8.10.	28.09.	19.10.	2420	1961	
		34	Otemma	- 11.5	- 13.0	- 9.5	- 1.0	26.09.	25.09.	1.10.	30.09.	28.09.	2408	1963	34
		35	Mont Durand	- 7.0	- 2.4	st	- 18.0	27.09.	25.09.	2.10.	1.10.	28.09.	2265	1960	
		36	Breney	- 3.0	- 10.0	- 13.5	- 9.5	26.09.	25.09.	1.10.	30.09.	28.09.	2570	1965	

Table 9.1.12. Switzerland - Variations in the positions of glacier fronts.

Part 2 of 8

Drainage basin		Glacier		Variations in metres				Day and month of the surveys					Altitude of the snout		see remark
I. Order	II. Order	No	Name	in the years 1964 -1965	1965 -1966	1966 -1967	1967 -1968	in the years 1964	1965	1966	1967	1968	metres above sea level	year	No
Rhone	Rhone	37	Giétro	←	+ 11.7	+ 1.2	+ 7.5	25.08.	5.10.	27.09.	22.08.	13.09.	2500	1968	37
		38	Corbassière	- 27.0	+ 1.0	- 13.0	- 17.0	25.09.	4.10.	4.10.	23.10.	19.09.	2188	1968	
		39	Valsorey	- 5.0	- 4.6	- 0.7	+ 1.0	23.10.	13.10.	7.10.	9.10.	15.10.	2395	1968	
		40	Tseudet	st	- 3.2	+ 2.5	- 13.5	23.10.	13.10.	7.10.	9.10.	15.10.	2422	1968	
		41	Boveyre	+ 9.0	- 3.2	+ 6.0	+ 2.0	24.10.	14.10.	7.10.	9.10.	16.10.	2606	1968	
		42	Saleina	st	←	+ 10.5	+ 5.0	24.10.	14.10.	n	10.10.	17.10.	1738	1968	
		43	Trient	+ 8.1	+ 23.7	+ 2.7	+ 23.8	4.10.	20.09.	2.10.	22.10.	21.10.	1775	1968	43
		44	Paneyrosse	n	n	n	n	n	n	n	n	n	2370	1958	44
		45	Grand Plan Névé	1963 ←	←	- 1.4	+ 2.0	n	7.10.	12.10.	5.10.	3.10.	2345	1963	45
		46	Martinets	n	?	?	n	n	6.10.	11.10.	1.10.	n	2095	1961	
		47	Scex Rouge	+ 15.0	st	- 2.5	+ 3.0	.10.	15.10.	.09.	27.08.	7.10.	2660	1968	
		48	Prapio	+ 24.0	- 39.0	- 6.0	+ x	18.09.	15.10.	25.10.	26.10.	1.11.	2370	1968	48
Rhein	Aare	49	Pierredar	+ 15.0	+ 8.0	- 6.0	+ x	17.09.	16.10.	25.10.	26.10.	1.11.	2400	1968	49
		50	Oberaar	- 29.5	- 5.8	- 23.8	- 18.4	15.07.	22.07.	14.07.	21.07.	30.07.	2296	1968	
		51	Unteraar	- 18.6	- 17.0	- 21.2	- 14.6	29.07.	5.08.	27.07.	24.07.	1.08.	1908	1968	
		52	Gauli	- 4.0	- 8.0	- 7.0	- 11.0	25.09.	14.10.	24.09.	12.10.	24.09.	2220	1958	
		53	Stein	- 6.0	- 9.0	- 1.0	- 0.4	23.09.	13.09.	23.09.	6.10.	28.09.	1935	1968	
		54	Steinlimmi	- 1.0	- 0.4	- 1.4	st	23.09.	13.09.	23.09.	6.10.	28.09.	2092	1968	
		55	Trift	1963 ← - 18.0	- 5.0	- 9.0	n	n	21.09.	6.10.	7.10.	n	1650	?	55
		56	Rosenlaui	1963 ← + 18.0	+ 24.6	+ 12.0	+ x	n	16.09.	13.09.	1.10.	27.09.	1900	?	56
		57	Oberer Grindelwald	+ 10.0	+ 15.0	+ 30.0	+ 65.0	4.10.	13.10.	6.10.	20.10.	7.10.			57
		58	Unterer Grindelwald	- x	- x	- x	- x	4.10.	31.10.	12.08.	22.10.	18.10.	1220	1961	58
		59	Eiger	- 10.0	- 11.9	+ 42.7	+ 10.8	1.10	26.10.	20.10.	26.09.	8.09.	2050	1968	59
		60	Tschingel	←	- 6.8	←	- 10.8	25.09.	n	25.09.	n	20.09.	2150	1968	60
		61	Gamchi	- 4.3	+ 8.4	+ 2.8	+ 5.1	24.09.	9.10.	13.09.	30.09.	28.09.	1990	1968	
		62	Schwarz	0.0	- 0.4	- 0.7	- 2.0	7.10.	4.10.	24.09.	30.09.	20.09.	2220	1968	
		63	Lämmern	- 3.5	- 0.9	- 26.0	- 8.2	7.10.	2.10.	9.10.	6.10.	21.09.	2490	1968	
	Reuss	64	Blümlisalp	n	n	n	n	n	n	n	n	27.09.	2200	1957	64
		65	Rätzli	1963 ← - 46.0	- 6.0	- 3.0	- 4.9	n	6.10.	24.09.	19.09.	4.10.	2310	1968	65
		66	Tiefen	- 19.5	+ 3.0	- 16.5	- 5.5	13.09.	28.10.	2.09.	24.09.	13.10.	2490	1968	
		67	St. Anna	+ 4.0	+ 15.0	- 7.5	- 1.0	2.10.	23.09.	2.09.	7.10.	25.10.	2592	1968	
		68	Chelen	- 31.0	+ 21.0	- 35.7	+ 0.7	12.09.	15.09.	23.09.	18.09.	25.10.	2134	1968	
		69	Rotfirn	- 9.5	+ 4.0	+ 13.5	+ 10.5	13.09.	15.09.	23.09.	18.09.	25.10.	2031	1968	
		70	Damma	1963 ← - 10.3	- 0.5	- 22.0	- 1.7	13.09.	21.09.	1.09.	17.09.	27.09.	2044	1968	
		71	Wallenbur	- 1.2	- 5.0	- 2.0	- 6.5	n	24.09.	15.09.	4.10.	20.09.	2240	1968	71
		72	Brunni	- 1.7	- 2.0	- 1.7	+ 4.0	16.09.	24.09.	27.09.	11.10.	22.10.	2305	1968	

Table 9.1.12. Switzerland - Variations in the positions of glacier fronts.

Part 3 of 8

Drainage basin: I. Order	Drainage basin: II. Order	Glacier: No	Glacier: Name	Variations in metres in the years 1964-1965	1965-1966	1966-1967	1967-1968	Day and month of the surveys in the years 1964	1965	1966	1967	1968	Altitude of the snout: metres above sea level	year	see remark No
Rhein	Reuss	73	Hüfi	- 6.0	- 8.0	+ 4.5	- 24.0	25.09.	22.09.	4.10.	14.10.	22.10.	1740	1968	
		74	Griess (b. Unterschächen)	- 1.0	- 0.5	- 1.0	+ 1.0	24.08.	27.08.	3.10.	21.09.	11.09.	2217	1968	
		75	Firnälpli - E	+ 48.5	←	+ 2.6	+ 3.0	7.10.	11.10.	n	15.10.	28.09.	2151	1956	
		76	Griess (Griessengl.) 1962 ←	+ 78.0	+ 38.6	st	+ 44.3	n	14.10.	23.09.	.10.	21.10.	2500	1968	76
	Limmat	77	Biferten	+ 3.4	- 1.2	- 3.3	+ 3.7	2.10.	22.09.	21.09.	6.09.	14.10.	1945	1967	
		78	Limmern	- 1.1	- 1.7	- 1.8	+ 0.6	16.09.	20.09.	17.09.	15.09.	11.09.	2245	1968	78
		79	Sulz	+ 1.0	+ 4.7	+ 3.1	+ 11.9	30.09.	14.10.	22.09.	10.10.	22.10.	1785	1968	79
		80	Glärnisch	+ 18.1	- 6.0	- 5.1	- 2.2	6.10.	7.10.	5.10.	27.09.	12.10.	2298	1968	
		81	Pizol	←	+161.0	- 26.2	- 1.2	16.09.	6.10.	5.10.	24.10.	11.10.	2500	1968	
	Rhein	82	Lavaz	+ 11.0	+ 2.0	- 21.3	0.0	1.10.	9.10.	24.09.	11.10.	5.10.	2280	1968	
		83	Punteglias	+ 2.5	+ 7.1	- 3.8	- 3.6	25.09.	5.10.	30.08.	13.09.	14.09.	2330	1968	
		84	Lenta	+ 0.7	- x	- 24.5	- 48.4	5.10.	27.10.	26.09.	28.09.	9.10.	2275	1968	
		85	Vorab	←	←	+ 25.8	st	4.09.	n	27.09.	11.10.	21.10.	2535	1968	
		86	Paradies 1963 ←	- 48.0	- 25.0	- 2.0	+ 11.5	n	7.10.	25.10.	10.10.	21.10.	2365	1968	86
		87	Suretta 1963 ←	- 8.6	+ 44.0	+ 10.0	+ 58.0	n	4.10.	8.10.	15.10.	22.10.	2178	1968	87
		88	Porchabella	- 31.0	- 8.5	- 9.0	- 13.0	15.09.	23.10.	23.09.	23.09.	5.10.	2577	1968	
		89	Verstankla	- 2.0	- 4.0	- 1.0	+ 2.0	1.09.	5.10.	6.10.	28.09.	22.10.	2360	1968	
		90	Silvretta	+ x	+ 6.3	- 9.4	- 0.4	18.09.	17.09.	23.09.	30.09.	27.09.	2426	1968	90
Inn		91	Sardona	+ 10.5	+ 18.2	+ 4.2	+ 1.5	17.09.	7.10.	6.10.	11.10.	30.09.	2500	1968	
		92	Rosegg 1963 ←	←	- 50.8	- 27.8	- 50.0	27.08.	16.10.	15.10.	12.10.	9.10.	2170	1968	92
		93	Tschierva	- 40.9	- 12.6	- 10.0	+ 23.7	27.08.	16.10.	15.10.	12.10.	9.10.	2160	1968	
		94	Morteratsch	- 37.6	- 40.8	- 31.4	- 5.1	16.06.	14.10.	10.10.	9.10.	8.10.	2000	1968	94
		95	Calderas	- 2.4	- 8.5	- 3.8	- 4.0	27.08.	17.10.	9.10.	16.10.	19.10.	2675	1968	
		96	Tiatscha 1963 ←	-120.0	- 5.0	- 1.0	0.0	n	23.09.	22.09.	24.09.	6.10.	2615	1968	96
		97	Sesvenna	←	- 19.5	+ 20.5	+ 0.4	23.09.	n	21.09.	28.09.	25.10.	2735	1968	
		98	Lischana	←	+ 55.3	- 7.0	+ 8.5	15.09.	n	8.09.	3.10.	23.10.	2800	1968	
Adda		99	Cambrena	- 0.6	+ 3.0	+ 3.8	+ 5.3	27.08.	19.09.	21.09.	18.10.	6.10.	2491	1964	
		100	Palü	- 16.2	←	- 15.0	n	27.08.	22.10.	n	20.10.	n	2360	1967	
		101	Paradisino	- 5.6	+ 6.5	- 4.0	- 2.2	27.08.	26.09.	2.10.	21.10.	27.08.	2805	1968	
		102	Forno 1963 ←	- 40.9	- 27.7	- 20.9	- 29.1	n	23.10.	11.10.	19.10.	18.10.	2200	1968	102
Tessin		103	Bresciana	- 14.5	- 14.8	+ 5.5	+ 1.2	15.09.	21.09.	29.09.	12.10.	17.10.	2570	1967	
		104	Basodino	←	- 3.3	+ 16.0	+ 5.5	24.09.	31.10.	16.09.	11.10.	13.10.	2590	1964	
		105	Rossboden	- 5.6	- 3.3	- 6.6	- 1.0	30.09.	15.10.	8.10.	6.10.	10.10.	1928	1962	

Table 9.1.12. Switzerland - Variations in the positions of glacier fronts,

Part 4 of 8 Remarks

3 Gries (Aegina)

The glacier was first impounded in summer 1966, when the new Gries Storage Lake operated by Kraftwerke Aegina AG was created. The front is about 400 metres across, the terrain below it has a gradient of about 8 % (1 in 12). Calving of the glacier on 16th August and 19th September 1966 and additional melting due to the lake waters were the cause of the extremely marked retreat (159.0 m) of the glacier tongue, which prior to the filling of the lake had run out flat but on 28th June 1967 ended in a steep ice wall averaging 20 metres height and reaching 30 metres at its highest point. (The date of measurement 28.6. in Table 9.1.12, Column 1966, actually belongs to 1967). The maximum level of the lake is to be 2386.5 m above sea level. On 12th October 1968 the lowest point of the glacier was at 2364 m, while the foot of the vertical ice wall, with an average height of 26 m, lay at a mean altitude of about 2370 m. The Gries Glacier is the subject of special investigations (mass balance, changes of shape, movement, melting of the tongue in contact with the waters of the lake).

4 Fiescher

In its lowest part the tongue of the glacier, which flows southward through a deep gorge, is bounded orographically on the right by a ridge which at about 1800 m above sea level projects only a few metres above the surface of the ice. Beyond this ridge lies a small lateral valley running parallel to the main valley and collecting the waters of the Galtjinen and the Seebach in the socalled Glingelwasser. The latter joins the Weisswasser coming from the Fiescher Glacier at an altitude of about 1240 m. Up to about twenty years ago a lateral excrescence of the glacier covered the ridge just mentioned at 1800 m above sea level. The changes in the position of this lobe, which flowed down into the valley of the Glingelwasser, were recorded up to the autumn of 1959; since that time measurements have been made on the main tongue in the gorge.

11 Allalin

The tongue was measured on 4th October 1964 by theodolite, then in the following years by aerial photogrammetry. The figures given below are the horizontal components of the positional changes and displacements. As a result of the ice-fall of 30th August 1965, 17.15 hours, the tongue of the glacier, measured over a strip 280 metres in width, was on the average 219.05 metres shorter on 4th October 1965 than it had been on 4th October 1964.

After the ice fall (measurement of 4th September 1965) the tongue was even 70 metres shorter than this. Between 30th August and 4th October 1965 the bottom of the tongue slid forward by this mean distance, the front of the recess produced by the fracture moving about 3 to 4 metres per day while the foot of the south pillar, which had remained standing, changed its position only very little. The whole volume of ice which came down when the collapse occured was between $1.2 \cdot 10^6$ and $1.6 \cdot 10^6$ m^3. Of this, about one third came to rest on the ledges between the end of the glacier and the foot of the incline, while two thirds reached the valley bed and there caused the catastrophe in which 88 lives were lost. The normal surface velocities of the tongue are a few decimetres per day. The bottom of the tongue began to slide for a few weeks in 1966, 1967 and 1968 as it had in the autumn of 1965, the displacement amounting to a few metres per day, but without any major falls of ice occuring. The many small ice avalanches that came down in the period of slip never exceeded a few thousand cubic metres and came to a stop on the rock ledges below the glacier snout. The substantial advance of 93.1 m recorded in the year 1967/68 was prepared by the formation of these ice deposits below the glacier. Although the slip that took place in the autumn of 1968 began only after 13th September, the advancing front had at this time already reached the ice deposits on the lower ledge. These were consequently regarded as forming part of the glacier in the measurement of 13th September 1968. The next ledge valleywards is the big rock terrace upon which the glacier tongue ended before the great ice-fall of 30th August 1965. On this terrace there lay, on 13th September 1968, a number of large ice pyramids resulting from avalanches, which were separated from the glacier mountainwards by steeper rocks.

Table 9.1.12. Switzerland - Variations in the positions of glacier fronts.

Part 5 of 8 Remarks

13 Fee (Nord)
The glacier is advancing over the debris of the mountain slide which covered the lowest part of the tongue in summer 1954.

18 Lang
Retreat from 24.9.1962 till 24.10.1968 = 81.3 metres. This retreat presumably continued in 1967/68.

21 Bella Tola
The advance of this gorge-type glacier is mainly due to the fact that in the last four years the winter snow in front of the tongue has persisted through the summer, forming firn and thus becoming part of the glacier.

25 Ferpècle
Retreat from 25.10.63 till 5.10.66 = 23.2 metres.

26 Mont Miné
In the period 25.10.63 to 5.10.66 the glacier snout was stationary.

32 Mont Fort
Change between 11.10.63 and 25.9.65 = 11.0 metres.

34 Otemma
The neighbouring Epicoun Glacier has been advancing at least since 1963/64. Cf. the illustrations in the 87th Report of the Swiss Glacier Commission for 1965/66.

37 Giétro
Determination of the changes in length by planimetering aerial photographs to a scale of 1 : 2000, width of front = 260 metres. Tape measurements from measuring marks to the edge of the glacier on 24.9.64, 5.10.65 and 10.11.66 gave the following results: 1964/65 = + 3.0 m, 1965/66 = + 4.0 m. The Giétro Glacier is being supervised. Since 1968 the velocity of the bottom part of the tongue is continuously registered during the summer.

43 Trient
The glacier is in the process of an advance originating in the feed area and acting through the mechanism of glacier movement.

44 Paneyrosse
Has not been measured since autumn 1958.

45 Grand Plan Névé
The retreat of 1.4 m holds good for the interval from 6.10.1963 till 5.10.1967.

48 Prapio
The glacier ends in a small, narrow, rocky valley running westwards and opening at the top of a scree slope at 2340 metres above sea level. Comparison of aerial photographs shows that the tongue in the thalweg was about 4 metres shorter on 22nd October 1968 than on 18th August 1959. This figure and the annual observations from 1959/60 till 1966/67 would indicate an "advance" of 12 metres for 1967/68. About 330 m upwards from the tip, teh glacier passes a bottle-neck. It is so thin and narrow at this point that the bottom part of it is not fed by glacier movement but only from the side by avalanches from the southern

Table 9.1.12. Switzerland - Variations in the positions of glacier fronts.

Part 6 of 8 Remarks

slope of the Sex Rouge. These avalanches decide the position of the front. The narrow tongue is almost interrupted at several points. There are probably at most only small remnants of ice from the glacier proper under the snow from the avalanches. The bottle-neck mentioned above can be made out on an aerial photograph dating from as early as 6th September 1954.

49 Pierredar

The glacier is exposed to the north and extends in a coherent mass orographically on the right only down to the ledge at about 2410 metres altitude on which the measuring base is located. In its central and left-hand thirds the glacier is interrupted by a step in the rock. Above this the ice breaks away from a front about 30 metres high, the upper edge of which is at about 2590 metres in the west, while it drops in the east to about 2520 metres. The falling ice masses form a regenerated glacier below the rock step, the lower margin of this part of the glacier lying on the ledge where the measuring base is located and merging to the east into that of the coherent section of the glacier. In 1893 the glacier was only about 10 to 30 metres longer than it is today. At that time it reached the ledge where the measuring base is situated in a coherent mass covering its full width, and ended on the outer brink of the ledge in an ice wall. On 6th September 1954 the glacier tapered out on the ledge. The rock face which today interrupts the glacier above the ledge was already visible through two small windows in the central and western sections. In its present state the glacier threatens tho alp of Creux de Champs, at an altitude of about 1400 metres, with a potential mass of at most 10^6 m^3 of ice. For this reason it has been watched more closely since 1968.

55 Trift

The retreat of 18.0 metres holds good for the interval from 28.9.63 till 21.9.65.

56 Rosenlaui

The advance of 18.0 metres holds good for the interval from 11.10.63 till 16.9.65.

57 Oberer Grindelwald

This valley glacier, which is about 5 kilometres long measured from the Lauteraarsattel (3144 m above sea level), flows northwest. Direct access to the main valley is blocked by a rock barrier, which is circumvented by the glacier in a gorge first running southwest, then turning north. The measurements apply to the advance of the tongue in this gorge. Since measuring year 1966/67 a lateral excrescence is advancing over the rock barrier just mentioned.

58 Unterer Grindelwald

In the vicinity of the end of the tongue, a narrow gorge over 100 metres deep is cut in the bed of the U-shaped valley. At the top it is only about 40 metres wide and is bridged by the front of the glacier in an impressive ice vault. The front was last measured at the bottom of the gorge on 4.10.1964. As a result of the constant retreat of the glacier, its end was no longer accessible and no longer visible in the gorge on 31.10.1965. From this date onwards the observations have therefore been restricted to the end of the tongue visible above the gorge. The positional change of this vault and its edge zones has been followed since 1965 in the photographic records of Hans Boss (31.10.65, 22.10.67, 6.6., 18.6. and 18.10.68) and of the Swiss Federal Survey Administration (12.8. and 15.9.66). In the measuring year 1965/66 the retreat on the left bank, viz. 17 metres, was somewhat less than on the right bank. In 1966/67 the tongue was protected from melting factors on the left bank by ice deposited by the Schlosslaui avalanche, which came down beyond the end of the glacier. On the right the glacier became somewhat shorter. The mean retreat over the full width of the front in 1966/67 was probably less than 5 metres. In the year 1967/68 now reported on, the glacier freed a further 5 - 10 metres of its bed.

Table 9.1.12. Switzerland - Variations in the positions of glacier fronts.

Part 7 of 8 Remarks

59 Eiger

The series of observations, which had been made regularly from 1893 onwards, was interrupted from 1938 till 1962. In the summer of 1968 the marked growth of heaps of avalnache debris in the Trümmeltal indicated that the glaciers of Kühlauenen and Giessen are increasing in size and thickness.

60 Tschingel

In the course of the last 27 years the Tschingel Glacier has become about 500 metres shorter and continues to retreat. The following observations were made on the neighbouring glaciers on 6th October 1968: The Rottal Cabin is no longer visible from Hotel Tschingelhorn (1685 metres above sea level) because the Rottal Glacier has grown thicker in the last few years. The cabin was previously visible from this point, and as the glacier shrank it could even be seen for a time from the stable situated about 40 metres lower in altitude. The Schmadri Glacier is also advancing; this is confirmed by the fact that the edge of the ice is bordered by a freshly created moraine. The main tongue of the Breithorn Glacier shows no sign of an advance at its snout. On its northwest edge below a point 2285 metres above sea level, however, and along the stream of ice coming from the Schmadrijoch (to the south of the Schmadri Cabin) freshly raised moraines have been observed. The western arm, to the east of the chancel, is clearly advancing in the direction of the Tschingel Glacier.

64 Blümlisalp

Between 1893 and 1951 the end of the glacier withdrew about 280 metres, between 1951 and 1968 another 300 - 400 metres. According to the observations of 26th and 27th September 1968, it seems that the glacier will soon be reaching the end of a long shrinkage phase.

65 Rätzli

The retreat of 46.0 metres holds good for the interval 16.9.63 till 6.10.65.

71 Wallenbur

The retreat of - 1.2 metres holds good for the interval 25.9.63 - 24.9.65.

76 Griess (Griessen)

The advance of 78.0 metres holds good for the three-year period from 26.10.62 till 14.10.65.

78 Limmern

The series of measurements begins in 1964. The end of the tongue is about 240 metres across, lies on rock with a very gentle gradient and can be clearly distinguished. Cross-sectional records were begun as early as 1946, mass balance measurements on stakes in autumn 1959. The glacier is of importance for the Linth-Limmern Power Station.

79 Sulz

The glacier is only about 0.3 km^2 in size and extends over an altitude interval of 2100 to 1800 m above sea level. It is exposed to the north and is fed almost exclusively by avalanches which, however, hardly interfere with measurements at the end of the glacier. J. Becker, formerly Chief Forester of the Canton of Glarus, carried out his 42nd annual measurement on 22.10.68 and as usual prepared an informative and well documented report.

86 Paradies

The value of - 48.0 metres holds good for the interval 23.10.63 - 7.10.65. A report on measurements carried out on the Paradies Glacier over a period of 100 years ("100 Jahre Gletschermessung Paradies") is being compiled by F. Juvalta.

Table 9.1.12. Switzerland - Variations in the positions of glacier fronts.

Part 8 of 8 Remarks

87 Suretta

The value of - 8.6 metres holds good for the interval 20.10.63 - 4.10.65.

90 Silvretta

Mass balance measurements have been made since autumn 1960, velocity measruements on stakes since 1961.

92 Roseg

The value of - 50.8 metres holds good for the interval 12.11.63 - 15.10.66.

94 Morteratsch

The measurement. for 1964 was carried out on 16.6.1965

96 Tiatscha

The value of - 120.0 metres holds good for the interval 19.10.63 - 23.9.65.

102 Forno

The value of - 40.9 metres holds good for the interval 14.11.63 - 23.10.65.

Table 9.1.13. Switzerland - Variations in the positions of glacier fronts.

Number and percentage of glaciers in advance or in retreat 1890/91 - 1969/70

year	number of observed glaciers	number of glaciers			percentage of observed glaciers			year	number of observed glaciers	number of glaciers			percentage of observed glaciers		
		in retreat	stationary	in advance	in retreat	stationary	in advance			in retreat	stationary	in advance	in retreat	stationary	in advance
1890/91	29	16	3	10	55	10	35	1930/31	92	60	12	20	66	12	22
1891/92	28	18	2	8	64	7	29	1931/32	108	87	7	14	80	7	13
1892/93	30	21	0	9	70	0	30	1932/33	98	79	4	15	81	4	15
1893/94	44	32	3	9	73	7	20	1933/34	94	84	3	7	89	3	8
1894/95	52	40	4	8	77	8	15	1934/35	83	70	3	10	84	4	12
1895/96	38	30	4	4	78	11	11	1935/36	69	58	5	6	84	7	9
1896/97	58	43	5	10	74	9	17	1936/37	71	54	8	9	76	12	12
1897/98	61	48	4	9	79	6	15	1937/38	81	71	7	3	88	8	4
1898/99	62	56	0	6	90	0	10	1938/39	57	46	3	8	81	5	14
1899/00	68	58	4	6	85	6	9	1939/40	60	47	4	9	78	7	15
1900/01	61	52	6	3	85	10	5	1940/41	72	52	3	17	72	4	24
1901/02	59	38	9	12	65	15	20	1941/42	65	61	0	4	94	0	6
1902/03	51	37	5	9	73	10	17	1942/43	71	63	3	5	89	4	7
1903/04	57	53	2	2	92	4	4	1943/44	55	48	3	4	87	6	7
1904/05	47	42	2	3	89	5	6	1944/45	63	56	3	4	89	5	6
1905/06	55	47	1	7	85	2	13	1945/46	76	61	5	10	80	7	13
1906/07	59	48	1	10	81	2	17	1946/47	84	82	2	0	98	2	0
1907/08	57	42	2	13	74	3	23	1947/48	78	60	8	10	77	10	13
1908/09	56	46	1	9	82	2	16	1948/49	81	75	2	4	93	2	5
1909/10	51	36	0	15	71	0	29	1949/50	80	79	0	1	99	0	1
1910/11	60	56	0	4	93	0	7	1950/51	80	63	7	10	79	9	12
1911/12	50	29	2	19	58	4	38	1951/52	49	46	2	1	94	4	2
1912/13	52	32	5	15	61	10	29	1952/53	76	70	4	2	92	5	3
1913/14	47	26	4	17	55	9	36	1953/54	73	65	3	5	89	4	7
1914/15	38	21	4	13	56	10	34	1954/55	76	50	7	19	66	9	25
1915/16	56	16	4	36	29	7	64	1955/56	70	56	3	10	80	4	14
1916/17	70	31	4	35	44	6	50	1956/57	88	75	3	10	85	3	12
1917/18	58	23	8	27	40	14	46	1957/58	89	83	0	6	93	0	7
1918/19	82	22	3	57	27	4	69	1958/59	76	69	0	7	91	0	9
1919/20	101	33	6	62	33	6	61	1959/60	63	56	0	7	89	0	11
1920/21	132	80	8	44	61	6	33	1960/61	85	72	3	10	85	3	12
1921/22	82	43	12	29	51	14	35	1961/62	92	76	3	13	83	3	14
1922/23	118	66	23	29	56	19	25	1962/63	93	77	6	10	83	6	11
1923/24	100	66	11	23	66	11	23	1963/64	84	78	1	5	93	1	6
1924/25	98	69	11	18	70	11	19	1964/65	90	58	10	22	65	11	24
1925/26	95	37	8	50	40	8	52	1965/66	90	53	3	34	59	3	38
1926/27	96	68	6	22	71	7	22	1966/67	100	73	3	24	73	3	24
1927/28	103	83	6	14	81	5	14	1967/68	98	57	6	35	58	6	36
1928/29	102	79	12	11	77	12	11	1968/69	103	69	6	28	67	6	27
1929/30	96	77	11	8	80	12	8	1969/70	98	61	6	31	62	6	32

Table 9.1.14. USSR - Variations in the positions of glacier fronts.

Part 1 of 5 - retreat, + advance, n not measured, ( ) changes in area of the frontal and lateral parts of the front

High region	River system	Valley	No	Glacier	1959 1960	1960 1961	1961 1962	1962 1963	1963 1964	1964 1965	1965 1966	1966 1967	1967 1968	1968 1969	1969 1970
					Variations m (m2) Year										
Polar Ural															
	Ob'	M.Shchuch'ye	1	MGU 53/54 ←									-575		
	Kara	M.Kara	2	Karskiy Southern 53/54 ←									0		
			3	Karskiy Northern 53/54 ←									- 80		
	Ob'	B. Khadat	4	Obrucheva 53/54 ←	- 85	←		- 20	←		- 15	←		- 7	
	Pechora	B.Usa	5	Chernova 53/54 ←									- 80		
			6	Bol'shoy Usinakiy 53/54 ←									0		
	Ob'	B.Khadat	7	IGAN Northern 58/59 ←									0		
Caucasus															
Nothern slope of main ridge	r.Terek	Dykhsu	8	Dykhsu 53/54 ←		+ 134	←				+ 62				
		Adylsu	9	Dzhankuat								- 13	- 6	- 17	
		Genal'don	10	Mayli		- 1	+ 3	+ 9	+ 1	+ 6	- 15	+ 15	+ 3	- 8	
		Amali	11	Chachi						←			+ 44		
		Amali	12	Devdoraki		- 25	- 7	+ 12	- 6	+ 20	- 3	+ 3	- 14	- 2	+ 4
		Blota	13	Abano	+ 8	- 14	- 1	+ 5	+ 9	- 8	+ 9	+ 1	+ 2	+ 1	+ 1
		Chkheri	14	Gergeti	- 18	←	- 34	- 5	←	+ 10	- 6	+ 2	+ 6	+ 7	+ 4
		Mnaisidon	15	Mna					←		+ 32	←	+ 8		
		Suatisi don	16	Middle Suatisi							- 18	- 4	+ 3	+ 4	+ 5
		Dzhuta	17	Chaukhi						←			-140		
		Kistinka	18	Kibisha						←			- 57		
	Sulak	Pirikitel'skaya Alazani	19	Tebulo						←			- 32		

Table 9.1.14. USSR - Variations in the positions of glacier fronts.

Part 2 of 5 - retreat, + advance, n not measured, ( ) changes in area of the frontal and lateral parts of the front

High region	River system	Valley	No	Glacier	1959 1960	1960 1961	1961 1962	1962 1963	1963 1964	1964 1965	1965 1966	1966 1967	1967 1968	1968 1969	1969 1970
					Variations m (m2) Year										
					Caucasus										
Nothern slope of main ridge	Sulak	Kila	20	Juzhnyi	- 2	- 2	- 4	- 1	- 8	- 9	- 42				
		Kila	21	Jugo-Vostochnyi	- 5	- 1	- 12	- 1	- 3	- 2	- 19				
		Kila	22	Belengi	←								+ 19		
	Samur	Sel'di	23	Tikhitsar	←								- 82		
		Sel'di	24	Murkar	+690					←			- 87		
	Kadari	Klych	25	Klych						←			- 10		
		Chkhalta	26	Kvish						←			-118		
	Inguri	Dolra	27	Ushba						←			- 72		
		Tviber	28	Tsanner						←			-104		
		Mestia-chala	29	Chalaat	- 21	- 15	- 17				- 8	- 7	- 10	- 5	- 6
		Mestiachala	30	Lekzyr						←			-168		
		Adish-chala	31	Adishi			- 3	- 2	- 4						
	Tskhenistskali	Korul'dash	32	Korul'dash								- 12	- 7	+ 1	- 1
	Rioni	Chashuri	33	Kirtisho								- 13	- 7	- 10	- 5
		Chanchakhi	34	Tbilisi									- 2	- 1	- 1
	Bol. Liakhvi	Kalasamidon	35	Lazg - Tsiti							- 4	+ 1	- 3	- 10	+ 1
					Pamir - Alai										
Khrebet Muzkol	Lake Karakul' (nothern)	Akbaytal	36	Akbaytal		←	- 27	+ 31	- 12	+ 8	- 24	- 61			
Khrebet Zaalayskiy	Lake Karakul'	Karadzhilgasay	37	Malyi Oktyabr'skiy					+ 35	+ 17	- 10	- 29			
Khrebet Yuzhno Alichurskiy	Bartang	Bakchigir	38	Bakchigir		←		+ 15	←		- 17				

Table 9.1.14. USSR - Variations in the positions of glacier fronts.

Part 3 of 5 - retreat, + advance, n not measured, ( ) changes in area of the frontal and lateral parts of the front

High region	River system	Valley	No	Glacier	Variations m (m²) Year 1959 1960	1960 1961	1961 1962	1962 1963	1963 1964	1964 1965	1965 1966	1966 1967	1967 1968	1968 1969	1969 1970
				Pamir - Alai											
Khrebet Darvazskiy	Vanch	Vanch	39	Russian geographical society				+ 45		- 3	- 14	- 12	+ 38	- 3	
	Vanch	Khyrsdara	40	Medvezhiy				+1400	←	+ 75	-105	-153	-233	- 66	- 65
Khrebet Petra Pervogo	Muksu	Karasel	41	Mushketova				+ 12	- 1	+ 12	+ 6	+ 8	+ 18	+ 13	+ 15
Khrebet Darvazskiy	Obikhingou	Obimazar	42	Mazarskiy 58/59	←	←	- 62	←	+127	- 38	+ 9	-178	←	←	+ 11
Between Khrebet Zeravshansky and Turkestanskiy	Zarafshan	Zarafshan	43	Zeravshanskiy 57/59	←	←	←	- 18	←	←	-384	+ 45	- 12	←	-102
Khrebet Turkestanskiy	Zarafshan	Rama	44	Rama					- 65	+ 14	0	+ 35	- 54		- 14
Khrebet Zaravshanskiy	Zarafshan	Dikhadant	45	Dikhadant	←	←	+ 9	←	+ 17	←	←	←	+ 88		
Khrebet Turkestanskiy	Zarafshan	Tro	46	Tro	0	0	0	0	0	0	0	0	0	0	0
Khrebet Gissarskiy	Karatag	Diakhandara	47	Diakhandara						- 46	+ 41	←	- 37		
Pamir	Amu-Dar'ya	Sel'dara	48	Fedchenko	- 38	- 45	- 34	0							
			49	Ulugbeka		+ 72	+141								
		Bol'shoi' Tanymas	50	Grumm-Grzhimaylo	- 41	- 41	+ 42	+103							
			51	Northern Tanymas	+290	+ 10	+ 87	+ 61							
			52	II. Tanymas	- 16	- 17	- 4	- 5							
			53	III. Tanymas	+ 38	+ 7	- 6	+ 1							
			54	V. Tanymas		- 2	- 3	0							
Turkestanskiy Khrebet	Sokh	Kalaymakhmund	55	Raygorodskiy		+ 34 (+ 4980)	- 6 (+2860)	- 5 (-2080)	+ 21 (+4430)	+ 17 (+ 4380)	+ 2 (- 2730)	+ 4 (+ 390)	- 3 (- 700)	+ 17 (+ 4610)	+ 11 (+ 4270)

Table 9.1.14. USSR - Variations in the positions of glacier fronts.

Part 4 of 5 - retreat, + advance, n not measured, ( ) changes in area of the frontal and lateral parts of the front

High region	River system	Valley	No	Glacier	1959 1960	1960 1961	1961 1962	1962 1963	1963 1964	1964 1965	1965 1966	1966 1967	1967 1968	1968 1969	1969 1970
									Variations m (m2) Year						
									P a m i r - A l a i						
			56	Kyrchin						+ 13 (+ 2420)	- 6 (- 570)	- 5 (- 760)	- 23 (- 7480)	+ 33 (+10630)	- 6 (- 1790)
			57	Kokbeles						+ 26 (+ 6260)	←	- 45 (-10030)	+ 72 (+16830)		
Alayskiy Khrebet	Sokh	Tuteksu	58	Tutek			←	- 51 (-4600)	←	←	←	←	- 19 (- 3840)	+ 5 (+ 1000)	+ 4 (+ 1640)
			59	Klyuyeva		- 12 (- 2510)	- 6 (-2530)	0 (- 300)	+ 1 (+ 20)	←	←	←	+ 41 (+ 2080)	+ 16 (+ 3250)	+ 1 (+ 40)
		Iashil'kul'	60	Turamuz - 1					←	←	←	←	+ 19 (+ 7770)	+ 17 (+ 6260)	+ 5 (+ 3600)
			61	Kyzylgorum		←	←	- 3 (- 270)	←	←	←	←	+ 69 (+ 9410)	+ 13 (+ 1380)	+ 1 (+ 220)
Gissarskiy Khrebet	Kashka-dar'ya	Aksu	62	Severtsova		- 10 (+ 1880)	- 31 (-4000)	- 17 (-4620)	←	←	+ 85 (+15440)	+ 2 (+ 220)	- 10 (- 1500)	←	+ 16 (+ 2830)
		Batyrtay	63	Batyrbay		- 6 (- 1900)	+ 9 (+ 590)	- 33 (-3980)	←	←	+ 8 (+ 1380)	- 40 (- 5380)	+ 18 (+ 1030)	←	+ 1 (+ 140)
Alayskiy Khrebet	Kyzylsu	Koksu	64	Abramova 54/59	←	←	←	←	←	←	←	-500			
									T i e n - S h a n						
Ugamskiy Khrebet	Pskem	Anaul'gen	65	Turkpakbel- -Nizhniy		+ 32 (+ 6740)	←	- 6 (-1720)	←	- 11 (- 1860)	+ 27 (+ 2290)	+ 3 (+ 520)	+ 1 (+ 30)	←	+ 4 (+ 280)
		Ayutormaydan-tal'skiy	66	Ayutor - 2		- 4 (- 1200)	- 8 (-3650)	+ 1 (+ 620)	+ 2 (+1180)	- 4 (- 650)	- 1 (- 580)	+ 4 (+ 280)	- 12 (- 4400)	+ 23 (+10350)	+ 4 (+ 1250)
Maydantal'skiy Khrebet	Pskem	Akbulakul'kun	67	Akbuul'kun				- 6 (-1180)	+ 26 (+4200)	+ 38 (+ 7530)	+ 19 (+ 4650)	+ 22 (+ 2480)	+ 5 (+ 1030)	+ 1 (+ 120)	- 7 (- 500)
		Karabulak	68	Karabulak		←	←	←	←	←	←	←	+ 50 (+10500)	- 40 (- 9030)	- 44 (-41820)
		Chotan	69	Chotan - 2		←	←	+ 44 (+21650)	←	←	- 2 (- 320)	←	- 12 (- 710)		

Table 9.1.14. USSR - Variations in the positions of glacier fronts.

Part 5 of 5 - retreat, + advance, n not measured, ( ) changes in area of the frontal and lateral parts of the front

High region	River system	Valley	No	Glacier	1959 1960	1960 1961	1961 1962	1962 1963	1963 1964	1964 1965	1965 1966	1966 1967	1967 1968	1968 1969	1969 1970
					Variations m (m2) Year										
Tien - Shan															
		Tekeshsay	70	Tekeshsay - 1				+ 8 (+ 1720)	←		+ 27 (+ 5310)	- 18 (+ 7230)	+ 33 (+ 8130)	- 14 (+ 3260)	+ 14 (+ 2710)
Pskemskiy ridge	Pskem	Tokmaksaldy	71	Pakhtakor					←				- 62 (- 7550)		
		Barkrak	72	Barkrak sredniy										←	- 30 (- 1106)
		Barkrak	73	Barkrak pravy				- 20 (+ 3720)	- 5 (-1520)	+ 3 (+ 450)	- 22 (- 5500)	+ 2 (+ 100)	- 16 (- 2050)	- 1 (- 120)	+ 2 (+ 420)
Zailiyskiy Alatau	Ili	Chilik	74	Korzhenevskiy						- 22 (- 7406)	- 30 (- 9828)	- 25 (- 9378)	- 53 (-26234)	- 21 (-11500)	- 15 (- 9226)
		Malaya Almatinka	75	Tsentralny Tuyuksu	- 20 (n)	←	- 37 (-5290)	+ 12	←	- 65 (-17449)	- 21 (- 8236)	- 14 (- 5822)	- 18 (- 5218)	- 24 (- 7356)	- 16 (- 4663)
			76	Manshuk Mametova 58/59	←	(- 5005)	- 96 (-1800)	←		(0)		- 17 (- 336)	- 21 (- 1030)	+ 19 (+ 614)	0 (n)
		Levyy Talgar	77	Konstitutsiya			- 12 (n)	n	0 (0)	- 6 (- 1365)	- 5 (- 2313)	← ((0)	+ 5 (+ 1579)	+ 1 (+ 1579)	n
		Sredniy Talgar	78	Shokalskiy			- 8 (-1531)	+ 2 (n)	- 4 (n)	- 2 (- 980)	- 12 (- 2622)	- 10 (- 3158)	- 2 (- 756)	n	n
			79	Southern Talgar			+ 5 (n)	0 (n)	+ 6 (n)	+ 7 (+ 1193)	+ 2 (+ 309)	+ 2 (+ 299)	0 (+ 961)	+ 6 (+ 1176)	n
			80	Northern Talgar				- 8 (n)	- 4 (n)	- 4 (- 730)	+ 5 (+ 690)	+ 3 (+ 445)	+ 17 (+ 5142)	+ 11 (+ 2291)	n
		Bol'shaya Almati Almatinka	81	Glavnyy		- 8 (n)	←							- 15 (- 6291)	n
			82	Chernyy		- 3 (n)	←							- 8 (- 7678)	n
Altai															
Altai	Irtysh	Berel	83	Malyy Berel					←			- 26	←	- 24	- 10

Table 9.1.15. West Irian - Variations in the positions of glacier fronts.

Carstensz glaciers 1936, 1942, 1962 and 1972

4° 10' S, 137° 10' E Green.

Glacier	Variations in metres in the years				
	1936 -1942	1942 -1962	1962 -1972	1936 -1972	1972 -1973
Meren	- 325	- 665	- 440	-1430	
Carstensz	- 300	- 250	- 120	- 670	- 9

1936 : Colijn expedition

1942 : U.S. Army aerial photographs

1962 : Harrer expedition

1972 : Carstensz glaciers expedition

1973 : Carstensz glaciers expedition

Altitude interval of the Carstensz glaciers :

ca. 4 800 m a.sl. to 4 300 m a.sl.

Table 9.1.16. New Zealand - Variations in the positions of glacier fronts.

Franz Joseph Glacier 1950 - 1967

Record of retreat

Date of Survey or Observations	Distance from Peg 3 on Harper Rock		Retreat		Rate of Retreat Since Previous Survey		Height of Terminal Face		Width of Terminal Face		Altitude of Terminal Face at River Level	
	ft	m	ft	m	ft/day	m/day	ft	m	ft	m	ft	m
25.08.51	2.500	765							900	275	742	226
04.07.56	3.500	1.065	1.000	300			40-50	12-15				
08.05.58	4.500	1.370	1.000	300					700	210		
29.05.59	5.000	1.500	500	150	1.3	0.39			400	120	760	232
26.05.60	5.450	1.660	450	140	1.2	0.38			800	240	768	234
08.06.61	5.750	1.750	300	90	0.8	0.24			600	180	830	253
01.05.62	6.750	2.060	1.000	300	3.1	0.90			500	150		
30.04.63	7.050	2.150	300	90	0.8	0.24	90	27	500	150		
18.11.64	7.300	2.225	250	75	0.4	0.13						
05.05.65	7.500	2.300	200	60	1.2	0.35	50	15	250	75	900	274

Record of advance

Date of Survey	Distance from Peg 3		Advance		Rate of Advance Since Previous Survey		Height of Terminal Face		Width of Terminal Face	
	ft	m	ft	m	ft/day	m/day	ft	m	ft	m
05.05.65	7.500	2.300					50	15	250	75
18.10.65	7.130	2.175	370	115	2.5	0.75	300	90	600	180
07.12.65	6.850	2.090	280	85	5.5	1.6	300	90	600	180
08.02.66	6.650	2.030	200	60	3.0	0.9	300	90	600	180
04.04.66	6.590	2.010	60	18	1.0	0.3	100	30	700	210
08.06.66	6.490	1.980	100	30	1.5	0.45	100	30	900	275
09.08.66	6.410	1.950	80	24	1.3	0.4	250	76	1.000	300
10.10.66	6.340	1.930	70	27	1.2	0.37	150	45	1.000	300
06.02.67	6.270	1.910	70*	21	0.6	0.2	200	60	1.000	300
04.04.67	6.250	1.905	20	6	0.4	0.1	150	45	1.000	300
04.10.67	6.230	1.900	20	6	0.1	0.02	150	45	1.000	300

* This is the average advance over the whole face. On the western side the advance was 120 ft (36 m).

Fig. 9. 1. 1.

New Zealand - Franz Joseph Glacier

Variations in the positions of the glacier front

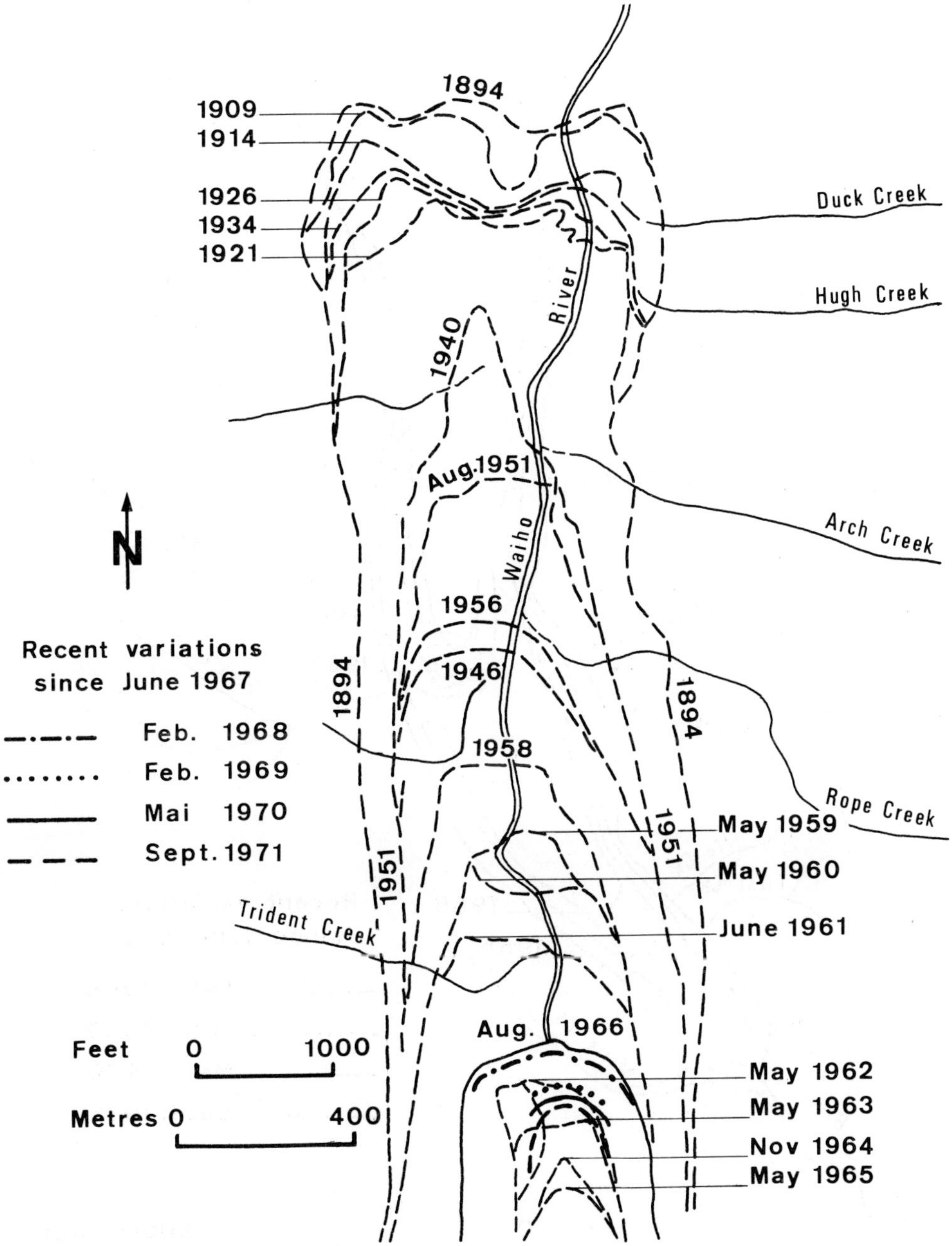

Fig. 9. 1. 2.
New Zealand - Fox Glacier
Variations in the positions of the glacier front

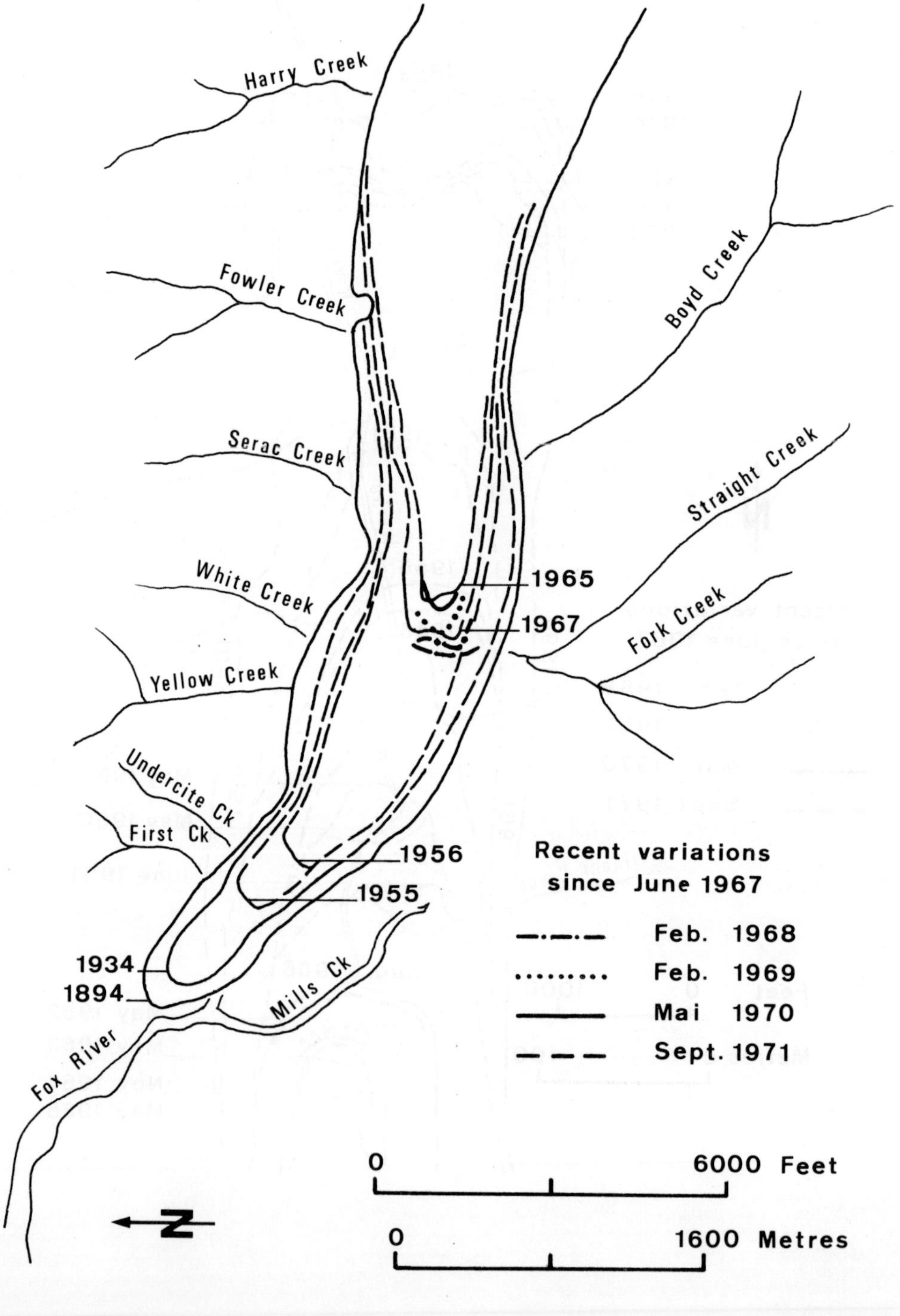

Fig. 9.1.3.

Topographical features of
HEARD ISLAND
(from Budd and Stephenson, 1970)

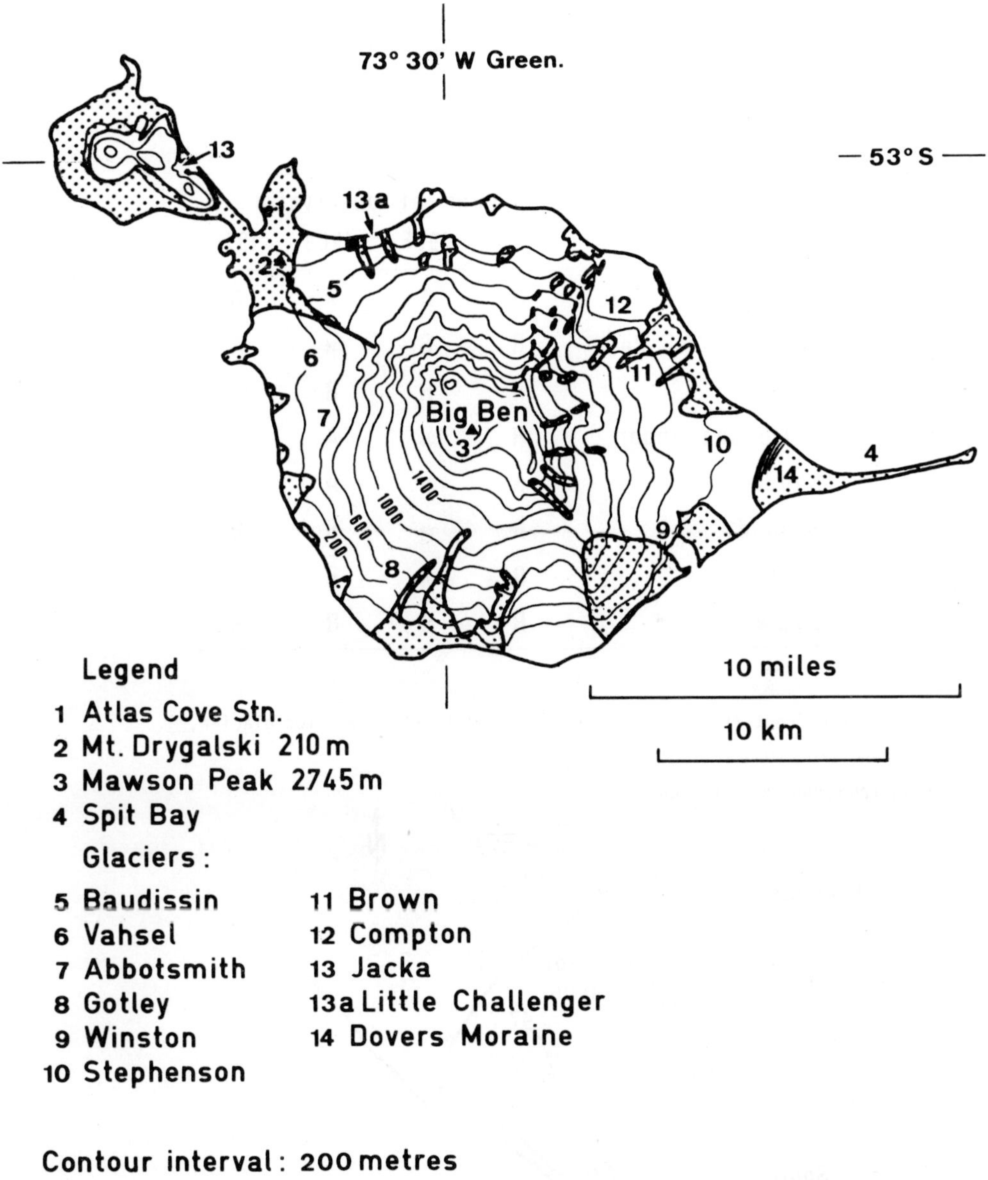

Fig. 9. 1. 4. and Table 9.1.17.
Heard Island - Winston Glacier
Variations in the positions of the glacier front

Glacier fluctuations at Heard Island from 1954 to 1971. Changes shown are those occurring since the previous year of observation, beginning with 1954, and are denoted by A (advance), R (recession) and S (stationary). Glaciers not seen in a particular year are indicated by asterisks.
(from Budd and Stephenson, in preparation)

Nr.	Glacier	year of observation 1963	1965	1969	1971
9 sw	Winston SW Part	R	A	*	A
9 ne	Winston NE Part	R	S	*	A
11	Brown	R	R	R	R
13 a	Little Challenger	R	*	A	S
5	Baudissin	R	*	A	S
	Schmidt	R	*	S	R
6	Vahsel	R	*	R	A
13	Jacka	R	*	R	R

Position of the front (shown by broken or dotted line) of Winston glacier, Heard Island from 1947 ("47") to 1971 ("71").
(Redrawn from Budd and Stephenson, 1970).

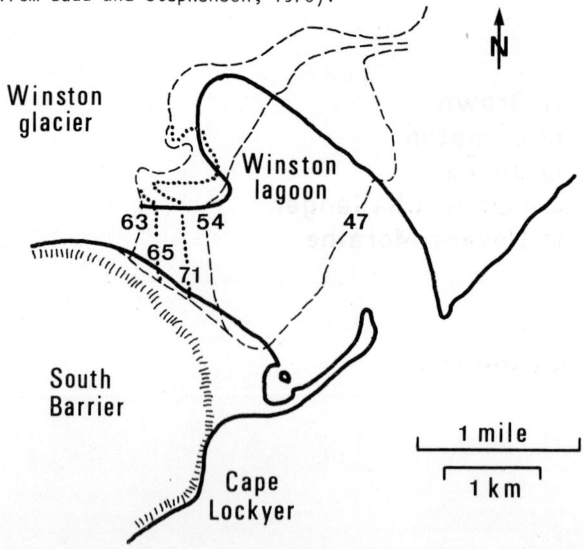

Table 9.2.1. Mass balance study results - Compilation

Part 1 of 3 $\bar{b}$: mean net mass balance in cm of water S: glacier area in km^2
E: altitude of equilibrium line, m a.sl.

G l a c i e r	Country	1965 / 66			1966 / 67			1967 / 68			1968 / 69			1969 / 70		
		$\bar{b}$	S	E	$\bar{b}$	S	E	$\bar{b}$	S	E	$\bar{b}$	S	E	$\bar{b}$	S	E
Devon Ice Cap	Canada	- 13.5												+ 3.9		900
Ward Hunt Ice Shelf	Canada	- 13.7			- 9.1			- 0.7								
Sentinel	Canada	+ 15	1.985	1820	- 18	1.985	1875	+ 38	1.985	1720	+ 12	1.985	1800	-130	1.927	2080
Place	Canada	+ 11	3.96	2020	-121	3.967	2370	- 13	3.957	2030	- 21	3.957	2090	-151	3.957	2380
Woolsey	Canada	- 21	3.865	2200	+ 12	3.865	2270	+ 17	3.865	2200	- 81	3.865	2380	-190	3.869	2600
Peyto	Canada	- 11	13.836	2610	- 80	13.936	2720	- 20	13.836	2620	- 65	13.836	2700	-140	13.836	2780
Per Ardua	Canada							- 32	4.682	1350						
Ram River	Canada	- 6	1.850	2760	-115	1.850	3000	+ 29	1.850	2760	- 29	1.850	2760	-124	1.850	<3000
Decade	Canada	- 71	8.654	*)	- 2	8.654	930	+ 24	8.654	770	- 75	8.654	*)	+ 8	8.654	865
Gulkana	USA	0	23.7	1770	- 50	23.7	1800	- 80	23.7	1850	-120	23.7	2000	+ 30	23.7	1700
Wolverine	USA	- 26	18.0	1250	-200	18.0	1460	-150	18.0	1350	-100	18.0	1250	+100	18.0	1000
Sherman	USA	+ 20	54.1	560	- 80	54.1	600	0	54.1	575		54.1			54.1	
Blue	USA	+ 38	4.2	1655	+ 59	4.2	1675	+ 27	4.2	1850		4.2			4.2	
South Cascade	USA	-130	2.8	2140	- 60	2.8	1870	0	2.8	2080	- 70	2.8	1910	-130	2.8	2050
Eliot	USA	- 30	1.8	2380	-160	1.8	2670	-260	1.8	2710	+ 70	1.8	2320	- 30	1.8	2380
Maclure	USA				+121	0.17	3640	- 76	0.17	3740	+ 70	0.17	3670		0.17	
Grasshopper	USA				+ 81	0.41	3180	+ 53	0.41	3160						
McCall	USA										- 28	6.2	2210			
Isabelle	USA										+120	0.12	3700			
Alfotbreen	Norway	-161	4.76		+128	4.76	950	+ 95	4.76	1075	-217	4.82	1550*)	-123	4.82	
Vesledalsbreen	Norway				+ 35	4.22	1400	+ 64	4.22	1320	-218	4.22	1850*)	-114		
Tunsbergdalsbreen	Norway	-109	43.77		+179	43.77	1160	+ 4	43.77	1270	-169	50.11	1700	- 84	50.11	
Nigardsbreen	Norway	- 92	40.86		+216	40.86	1310	+ 22	47.03	1550	-131	47.03	1850	- 56	47.03	
Folgefonni	Norway	-140			+136	19.74	1350	+ 64	19.52	1365				- 62		
Hardangerjøkulen	Norway	- 64	17.62		+119	17.62	1540	+ 53	17.55	1600	-190	17.53	1950*)	- 60	17.44	
Omnsbreen	Norway							- 18	1.52	1530	-259	1.50	?			

*) Equilibrium line above the highest point of the glacier

Table 9.2.1. Mass balance study results - Compilation

Part 2 of 3 $\bar{b}$: mean net mass balance in cm of water gl;S: glacier area in km^2
E: altitude of equilibrium line in m a.sl.

Glacier	Country	1965 / 66 $\bar{b}$	S	E	1966 / 67 $\bar{b}$	S	E	1967 / 68 $\bar{b}$	S	E	1968 / 69 $\bar{b}$	S	E	1969 / 70 $\bar{b}$	S	E
Storbreen	Norway	- 61			+ 72	5.45	1570	+ 5	5.45	1700	-142	5.36	2020*)	- 72		
Hellstugubreen		- 67	3.38		+ 55	3.38	1800	- 11	3.33	1875	-128	3.33	2130	-101	3.33	
Vestre Memurubre								+ 24	9.06	1820	-106	9.05	2170	- 79	9.00	
Austre Memurubre								+ 1	8.86	1960	-146	8.86	2130	- 90	8.71	
Grasubreen		- 29	2.39		+ 71	2.39		- 8	2.53	2140	-137	2.53	2350*)	- 66	2.53	
Store Supphellebre					+122	11.99	1190									
Blaisen		-127	2.18		- 97	2.18	1175	+ 26	2.18	1010						
Storsteinsfjellbreen		- 83	6.12		- 40	6.12	1450	+ 45	6.12	1275						
Cainhavarre		- 95	0.68		- 16	0.68	1450	+ 26	0.68	1290						
Spitzbergen																
Midre Lovenbreen								- 3	6.03	295	- 84	6.03	650*)	- 53		
Austre Brøggerbreen					+ 65		400	- 10	6.08	295	- 93	6.08	650*)	- 54		
Størglaciären	Sweden	- 53	3.07(1	1500	- 23	3.07(1	1500	- 11	3.07(1	1480	-103	3.07(1	1570	-153	3.07(1	1610
Hintereisferner	Austria	+ 34.5	9.05	2850	+ 20	9.03	2920	+ 34.0	9.03	2850	- 43.1	9.01	2960	- 55.2	9.01	3030
Kesselwandferner		+ 59	4.06	3040	+ 30	3.94	3070	+ 45	3.94	3060	- 15	3.94	3090	0	3.94	3080
Vernagtferner		+ 93.3	9.553	2935	+ 8.3	9.522	3015	+ 30.1	9.522	2995						
Langtalerferner		+ 47.8	2.902	2815	- 6.3	2.901	2820	+ 19.0	2.900	2795	- 51.5	3.050	2917	- 70.6	3.049	2975
Stubacher Sonnblickkees		+ 73.6	1.772		+ 16.0	1.772		+ 23.6	1.772		- 24.7	1.772		+ 14.4	1.772	
Glacier de Sarennes	France	+ 42	0.78(2		- 41	0.78(2		+ 34	0.78(2		- 36	0.78(2		- 41	0.78(2	
Nördl. Schneeferner	Germany	+ 99.7	0.325	<2560	+ 57.6	0.325	<2560	+ 62.3	0.335	<2560						
Marmolada	Italy	+ 36	1.542													
Ghiacciaio del Caresèr					- 39	4.7175		+ 26	4.7040		~ 0	4.7235		- 63	4.6820	

(1 Value for the year 1964/65; (2 Value for the year 1958/59; *) Equilibrium line above the highest point of the glacier;

Table 9.2.1. Mass balance study results - Compilation

Part 3 of 3 $\bar{b}$: mean net mass balance in cm of water gl S: glacier area in km^2
E: altitude of equilibrium line in m a.sl.

G l a c i e r	Country	1965 / 66 $\bar{b}$	S	E	1966 / 67 $\bar{b}$	S	E	1967 / 68 $\bar{b}$	S	E	1968 / 69 $\bar{b}$	S	E	1969 / 70 $\bar{b}$	S	E
Gries	Switzerland	- 27.9	6,69	2780	+ 26.0	6.54	2800	+ 33.2	6.38	2710	+ 26.9	6.38	2740	- 51.9	6.38	3040
Aletsch		+ 59.1	124,60		+ 30.7	124.15		+ 70.5	124.00		+ 32.1	123.71		- 12.9	123.26	
Limmern		+ 50.5	3.29	2420	- 36.3	3.29	2860	+ 39.7	3.29	2530	- 14.0	3.29	2740	- 15.8	3.29	2820
Silvretta		+109.6	3,33	2510	+ 26.2	3.33	2715	+ 45.6	3.33	2645	- 29.4	3.33	2800	+ 7.2	3.33	2730
Tsentralny Tuyuksu	USSR							- 78	3.15							
Obrucheva		+ 28	0,30		+ 20	0.30		+ 18	0.30		- 55	0.30		+ 34	0.30	
Igan		- 20	0,81		+ 31	0.81		+249	0.81		- 58	0.81		+ 1	0.81	
Dzhankuat								0		3090	- 10.5		3280	+ 5.2		3050
Bezingi		+ 0,4			+ 2,9			- 9.1			- 5.3			+ 1.3		
Kelbashi		- 0,6			+ 9,4			+ 0.7			+ 2.9			+ 7.4		
Zeiskiy		+ 0.4			+ 14,1			- 2.0			- 5.0			+ 3.6		
Karabatkak		- 70	4.58		+ 5	4.58		-279	4.58		+ 99	4.58		+120	4.58	
Glacier Gl	Deception Island										- 1	0.418		- 30	0.418	

Table 9.2.2. Canada - Mass balance study results.

Devon Ice Cap 75° 15' N, 82° W

Year	Mean net mass balance (kg $m^{-2}a^{-1}$) North-west sector	South-west sector	North-east sector	South-east sector	Equilibrium line North-west sector (m a.s.l.)
1960 - 61	-197	-	-	-	1323
1961 - 62	-359	-383	-	-370	1510
1962 - 63	+ 44	+ 76	-16	- 3	860
1963 - 64	+125	+195	+86	+201	610
1964 - 65	+ 64		0	-238	700
1965 - 66	-135				
1966 - 67	-103	-	-	-	-
1967 - 68		-	-	-	-
1968 - 69		-	-	-	-
1969 - 70	+ 38.8	-	-	-	900

Ward Hunt Ice Shelf [1] 83° 07' N 74° 10' W

Year	Winter balance in metres W.eq.	Winter balance date of Survey	Net balance metres W. eq.
1958 - 59	0.173	Late June 59	-0.110
1959 - 60	0.255	Late June 60	-0.068
1960 - 61	-	- -	-0.576
1961 - 62	-	- -	
1962 - 63	0.251	20/6/63	+0.117
1963 - 64	0.171	25/6/64	+0.104
1964 - 65	0.177	12/6/65	+0.202
1965 - 66	-	-	-0.137
1966 - 67	0.152	17/6/67	-0.091
1967 - 68	0.174	5/5/68	-0.007
1968 - 69	0.146	15/5/69	-

1) Measurements taken from stake network on ice rise north of Ward Hunt Island. For details see reference chapter 4.

Table 9.2.3. **Canada - Mass balance study results.**

Part 1 of 8 Mass balance versus altitude [1]

Sentinel Glacier, British Columbia 49° 54' N, 122° 59' W

Year	Altitude interval	Area	Winter balance	Summer balance	Net balance	Balance period Start	Winter to	Summer to	Equilibrium line
	m a.sl.	km²	m	m	m				m a.sl.
1965-1966	> 2100	0.023	3.40	2.25	+ 1.15	1.10.65.	16.06.66.	6.09.66.	1820
	2100 - 2000	0.346	3.62	2.32	+ 1.30				
	2000 - 1900	0.490	3.68	2.53	+ 1.15				
	1900 - 1800	0.515	3.32	2.98	+ 0.34				
	1800 - 1700	0.300	3.01	3.66	- 0.65				
	1700 - 1600	0.219	2.45	4.86	- 2.41				
	< 1600	0.092	2.44	4.64	- 2.20				
	Total	1.985	3.28	3.13	+ 0.15				
1966-1967	> 2100	0.023	4.48	2.26	+ 2.22	6.09.66.	6.06.67.	8.10.67.	1875
	2100 - 2000	0.346	4.40	3.00	+ 1.40				
	2000 - 1900	0.490	4.54	3.94	+ 0.60				
	1900 - 1800	0.515	4.42	4.65	- 0.23				
	1800 - 1700	0.300	3.80	5.14	- 1.34				
	1700 - 1600	0.219	3.53	5.50	- 1.97				
	< 1600	0.092	3.21	6.00	- 2.79				
	Total	1.985	4.21	4.39	- 0.18				
1967-1968	> 2100	0.023	3.55	2.26	+ 1.29	8.10.67.	23.05.68.	1.10.68.	1720
	2100 - 2000	0.346	3.54	2.89	+ 0.65				
	2000 - 1900	0.490	3.43	2.71	+ 0.72				
	1900 - 1800	0.515	3.71	2.77	+ 0.94				
	1800 - 1700	0.300	3.47	3.01	+ 0.46				
	1700 - 1600	0.219	2.97	4.12	- 1.15				
	< 1600	0.092	2.14	4.66	- 2.52				
	Total	1.985	3.42	3.04	+ 0.38				
1968-1969	> 2100	0.023	4.22	2.74	+ 1.48	1.10.68.	30.05.69.	6.10.69.	1800
	2100 - 2000	0.346	3.70	2.84	+ 0.86				
	2000 - 1900	0.490	3.85	3.08	+ 0.77				
	1900 - 1800	0.515	3.74	3.04	+ 0.70				
	1800 - 1700	0.300	3.24	3.96	- 0.72				
	1700 - 1600	0.219	2.80	4.44	- 1.64				
	< 1600	0.092	2.23	5.06	- 2.83				
	Total	1.985	3.52	3.40	+ 0.12				
1969-1970	> 2100	0.023	3.62	3.26	0.00	6.10.69.	18.05.70.	30.09.70.	2080
	2100 - 2000	0.346	2.94	3.23	- 0.29				
	2000 - 1900	0.490	2.78	3.10	- 0.32				
	1900 - 1800	0.515	2.56	4.02	- 1.46				
	1800 - 1700	0.298	2.47	4.81	- 2.34				
	1700 - 1600	0.186	2.14	5.14	- 3.00				
	< 1600	0.069	1.89	5.42	- 3.53				
	Total	1.927	2.62	3.92	- 1.30				

[1] mass balance in water equivalent

Table 9.2.3. Canada - Mass balance study results.

Part 2 of 8 Mass balance versus altitude

Place Glacier, British Columbia 50° 26' N, 122°36' W

Year	Altitude interval m a.sl.	Area km²	Winter balance m	Summer balance m	Net balance m	Balance period Start	Winter to	Summer to	Equilibrium line m a.sl.
	> 2500	0.23	2.17	1.38	+ 0.79				
	2500 - 2400	0.29	1.60	1.42	+ 0.18				
	2400 - 2300	0.26	1.93	1.71	+ 0.22				
1964-1965	2300 - 2200	0.18	2.17	2.06	+ 0.11	.09.64.	18.04.65.	10.09.65.	2200
	2200 - 2100	0.79	1.69	1.86	- 0.17				
	2100 - 2000	1.24	1.45	2.13	- 0.68				
	2000 - 1900	0.75	1.32	3.08	- 1.76				
	< 1900	0.24	1.32	3.84	- 2.52				
	Total	3.98	1.58	2.23	- 0.65				
	> 2500	0.05	2.50	0.74	+ 1.76				
	2500 - 2400	0.20	2.50	1.07	+ 1.43				
	2400 - 2300	0.47	2.31	1.37	+ 0.94				
1965-1966	2300 - 2200	0.29	2.02	1.87	+ 0.15	10.09.65.	15.06.66.	3.09.66.	2020
	2200 - 2100	0.64	2.19	1.91	+ 0.28				
	2100 - 2000	1.34	2.12	2.03	+ 0.09				
	2000 - 1900	0.76	1.59	2.04	- 0.45				
	< 1900	0.21	1.25	2.98	- 1.73				
	Total	3.96	2.02	1.91	+ 0.11				
	> 2500	0.040	3.25	1.75	+ 1.50				
	2500 - 2400	0.200	2.83	1.75	+ 1.08				
	2400 - 2300	0.467	1.93	2.21	- 0.28				
1966-1967	2300 - 2200	0.296	1.76	2.40	- 0.64	3.09.66.	31.05.67.	16.10.67.	2370
	2200 - 2100	0.637	2.16	3.30	- 1.14				
	2100 - 2000	1.345	2.30	3.61	- 1.31				
	2000 - 1900	0.762	1.78	3.92	- 2.14				
	< 1900	0.210	1.75	4.76	- 3.00				
	Total	3.967	2.10	3.31	- 1.21				
	> 2500	0.040	3.25	1.25	+ 2.00				
	2500 - 2400	0.200	3.15	1.25	+ 1.90				
	2400 - 2300	0.467	2.88	1.80	+ 1.08				
1967-1968	2300 - 2200	0.296	2.88	1.82	+ 1.06	16.10.67.	28.05.68.	1.10.68.	2030
	2200 - 2100	0.637	2.43	2.20	+ 0.23				
	2100 - 2000	1.345	2.29	2.36	- 0.07				
	2000 - 1900	0.762	1.73	3.43	- 1.70				
	< 1900	0.210	1.25	3.88	- 2.63				
	Total	3.957	2.32	2.45	- 0.13				
	> 2500	0.040	3.25	1.25	+ 2.00				
	2500 - 2400	0.200	3.25	1.25	+ 2.00				
	2400 - 2300	0.467	2.64	1.75	+ 0.89				
1968-1969	2300 - 2200	0.296	2.38	1.85	+ 0.53	1.10.68.	8.06.69.	1.10.69.	2090
	2200 - 2100	0.637	2.32	1.98	+ 0.34				
	2100 - 2000	1.345	1.89	2.16	- 9.27				
	2000 - 1900	0.762	1.41	3.07	- 1.66				
	< 1900	0.210	1.36	3.72	- 2.36				
	Total	3.957	2.05	2.26	- 0.21				
	> 2500	0.040	2.25	1.25	+ 1.00				
	2500 - 2400	0.200	2.25	1.25	+ 1.00				
	2400 - 2300	0.467	1.73	1.96	- 0.23				
1969-1970	2300 - 2200	0.296	1.54	2.11	- 0.57	1.10.69.	23.05.70.	19.09.70.	2380
	2200 - 2100	0.637	1.59	2.72	- 1.13				
	2100 - 2000	1.345	1.36	3.15	- 1.79				
	2000 - 1900	0.762	1.08	3.76	- 2.68				
	< 1900	0.210	1.02	4.68	- 3.66				
	Total	3.957	1.43	2.94	- 1.51				

Table 9.2.3. Canada - Mass balance study results.

Part 3 of 8 Mass balance versus altitude

Woolsey Glacier, British Columbia 51° 07' N, 118° 12' W

Year	Altitude interval m a.sl.	Area km²	Winter balance m	Summer balance m	Net balance m	Balance period Start	Winter to	Summer to	Equilibrium line m a.sl.
1965-1966	> 2600	0.019	3.10	1.26	+ 1.84	.09.65.	14.06.66.	.09.66.	2200
	2600 - 2500	0.208	3.24	1.25	+ 1.99				
	2500 - 2400	0.555	3.25	1.75	+ 1.50				
	2400 - 2300	0.676	2.95	2.20	+ 0.75				
	2300 - 2200	0.630	2.72	2.60	+ 0.12				
	2200 - 2100	0.606	2.40	3.15	- 0.75				
	2100 - 2000	0.671	2.09	3.70	- 1.61				
	< 2000	0.500	1.97	3.95	- 1.98				
	Total	3.865	2.56	2.77	- 0.21				
1966-1967	> 2600	0.019	3.74	1.74	+ 2.00	.09.66	21.06.67.	3.09.67.	2270
	2600 - 2500	0.208	3.76	2.00	+ 1.76				
	2500 - 2400	0.555	3.96	2.47	+ 1.49				
	2400 - 2300	0.676	3.83	2.27	+ 1.56				
	2300 - 2200	0.630	3.29	3.00	+ 0.29				
	2200 - 2100	0.606	2.97	3.27	- 0.30				
	2100 - 2000	0.671	2.54	3.99	- 1.45				
	< 2000	0.500	2.25	3.77	- 1.55				
	Total	3.865	3.13	3.01	+ 0.12				
1967-1968	> 2600	0.019	3.78	1.74	+ 2.05	3.09.67.	25.05.68.	4.10.68.	2200
	2600 - 2500	0.208	3.47	1.75	+ 1.72				
	2500 - 2400	0.555	2.47	1.75	+ 1.72				
	2400 - 2300	0.676	3.23	1.75	+ 1.48				
	2300 - 2200	0.630	2.88	3.10	- 0.22				
	2200 - 2100	0.606	2.61	2.75	- 0.14				
	2100 - 2000	0.671	2.37	3.30	- 0.93				
	< 2000	0.500	2.12	3.84	- 1.72				
	Total	3.865	2.84	2.67	+ 0.17				
1968-1969	> 2600	0.019	2.60	1.90	+ 0.70	4.10.68.	16.05.69.	4.10.69.	2380
	2600 - 2500	0.208	2.60	2.00	+ 0.60				
	2500 - 2400	0.555	2.60	2.20	+ 0.40				
	2400 - 2300	0.676	2.70	2.80	- 0.10				
	2300 - 2200	0.630	2.50	3.17	- 0.67				
	2200 - 2100	0.606	2.25	3.46	- 1.21				
	2100 - 2000	0.671	1.80	3.53	- 1.73				
	< 2000	0.500	1.74	3.94	- 2.20				
	Total 3.865	3.865	2.30	3.10	- 0.81				
1969-1970	> 2600	0.019	2.50	3.10	- 0.60	4.10.69.	11.05.70.	13.09.70.	2600
	2600 - 2500	0.208	2.50	3.10	- 0.60				
	2500 - 2400	0.555	2.40	3.30	- 0.90				
	2400 - 2300	0.676	2.25	3.40	- 1.25				
	2300 - 2200	0.630	1.85	3.60	- 1.75				
	2200 - 2100	0.606	1.60	3.80	- 2.20				
	2100 - 2000	0.671	1.50	4.00	- 2.50				
	< 2000	0.500	1.30	5.00	- 3.70				
	Total	3.869	1.87	3.77	- 1.90				

Table 9.2.3. Canada - Mass balance study results.

Part 4 of 8 Mass balance versus altitude

Peyto Glacier, Alberta 51° 40' N, 116° 33' W

Year	Altitude interval m a.sl.	Area km²	Winter balance m	Summer balance m	Net balance m	Balance period Start	Winter to	Summer to	Equilibrium line m a.sl.
	> 3100	0.028	2.20	0.50	+ 1.70				
	3100 - 3000	0.181	2.19	0.50	+ 1.55				
	3000 - 2900	0.800	2.05	0.70	+ 1.35				
	2900 - 2800	1.837	1.90	0.80	+ 1.10				
	2800 - 2700	2.347	1.75	0.90	+ 0.85				
1965-1966	2700 - 2600	2.687	1.57	1.21	+ 0.36	.09.65.	7.06.66.	.09.66.	2610
	2600 - 2500	2.884	1.21	1.89	- 0.68				
	2500 - 2400	1.293	1.00	2.38	- 1.38				
	2400 - 2300	0.944	0.75	2.78	- 2.03				
	2300 - 2200	0.750	0.71	3.35	- 2.64				
	2200 - 2100	0.079	0.44	3.40	- 2.96				
	< 2100	0.006	0.75	3.25	- 2.50				
	Total	13.836	1.44	1.55	- 0.11				
	> 3100	0.028	2.50	1.30	+ 1.20				
	3100 - 3000	0.181	2.50	1.40	+ 1.10				
	3000 - 2900	0.800	2.40	1,50	+ 0.90				
	2900 - 2800	1.837	2.25	1.70	+ 0.65				
	2800 - 2700	2.347	2.05	1.85	+ 0.20				
1966-1967	2700 - 2600	2.687	1.69	2.20	- 0.51	.09.66.	3.06.67.	28.09.67.	2720
	2600 - 2500	2.884	1.47	2.57	- 1.10				
	2500 - 2400	1.293	1.32	3.46	- 2.14				
	2400 - 2300	0.944	0.89	4.18	- 3.29				
	2300 - 2200	0.750	0.75	4.51	- 3.76				
	2200 - 2100	0.079	0.65	4.51	- 3.86				
	< 2100	0.006	0.67	4.17	- 3.50				
	Total	13.936	1.69	2.49	- 0.80				
	> 3100	0.028	1.75	0.86	+ 0.89				
	3100 - 3000	0.181	1.75	0.85	+ 0.90				
	3000 - 2900	0.800	1.75	0.85	+ 0.90				
	2900 - 2800	1.837	1.75	0.85	+ 0.89				
	2800 - 2700	2.347	1.67	0.98	+ 0.68				
1967-1968	2700 - 2600	2.687	1.49	1.27	+ 0.22	28.09.67.	8.05.68.	6.10.68.	2620
	2600 - 2500	2.884	1.28	1.79	- 0.51				
	2500 - 2400	1.293	1.16	2.43	- 1.27				
	2400 - 2300	0.944	0.68	3.14	- 2.46				
	2300 - 2200	0.750	0.77	3.29	- 2.52				
	2200 - 2100	0.079	0.65	3.34	- 2.69				
	< 2100	0.006	0.75	3.25	- 2.50				
	Total	13.836	1.40	1.60	- 0.20				
	> 3100	0.028	1.95	0.60	+ 1.35				
	3100 - 3000	0.181	1.90	0.70	+ 1.20				
	3000 - 2900	0.800	1.75	0.80	+ 0.95				
	2900 - 2800	1.837	1.60	0.90	+ 0.70				
	2800 - 2700	2.347	1.45	1,10	+ 0.30				
1968-1969	2700 - 2600	2.687	1.25	1.55	- 0.30	6.10.68.	12.05.69.	30.09.69.	2700
	2600 - 2500	2.884	1.22	2.28	- 1.06				
	2500 - 2400	1.293	0.94	3.00	- 2.06				
	2400 - 2300	0.944	0.69	3.36	- 2.67				
	2300 - 2200	0.750	0.75	3.84	- 3.09				
	2200 - 2100	0.079	0.30	3.76	- 3.49				
	< 2100	0.006	0.25	3.67	- 3.52				
	Total	13.836	1.25	1.90	- 0.65				

Table 9.2.3. Canada - Mass balance study results.

Part 5 of 8 Mass balance versus altitude

Peyto Glacier, Alberta 51° 40' N, 116° 33' W

Year	Altitude interval m a.sl.	Area km^2	Winter balance m	Summer balance m	Net balance m	Balance period Start	Winter to	Summer to	Equilibrium line m a.sl.
1969-1970	> 3100	0.028	1.50	1.25	+ 0.25	28.09.69.	13.05.70	7.09.70.	2780
	3100 - 3000	0.181	1.50	1.25	+ 0.25				
	3000 - 2900	0.800	1.50	1.25	+ 0.25				
	2900 - 2800	1.837	1.39	1.25	+ 0.13				
	2800 - 2700	2.347	1.25	1.70	- 0.45				
	2700 - 2600	2.687	1.00	2.25	- 1.25				
	2600 - 2500	2.884	0.95	2.60	- 1.65				
	2500 - 2400	1.293	0.85	3.52	- 2.67				
	2400 - 2300	0.944	0.65	4.01	- 3.36				
	2300 - 2200	0.750	0.60	4.27	- 3.67				
	2200 - 2100	0.079	0.40	4.40	- 4.00				
	< 2100	0.006	0.25	5.00	- 4.75				
	Total	13.836	1.05	2.45	- 1.40				

Per Ardua Glacier, N.W.T. 81° 27' N, 76° 35' W

Year	Altitude interval m a.sl.	Aerea km^2	Winter balance m	Summer balance m	Net balance m	Balance period Start	Winter to	Summer to	Equilibrium line m a.sl.
1967-1968	1700 - 1600	0.148	0.15	0.08	+ 0.07	1.09.67.	24.05.68.	1.09.68.	1350
	1600 - 1500	0.243	0.15	0.08	+ 0.07				
	1500 - 1400	0.524	0.14	0.08	+ 0.06				
	1400 - 1300	0.576	0.15	0.15	0.00				
	1300 - 1200	0.596	0.19	0.20	- 0.01				
	1200 - 1100	0.658	0.19	0.30	- 0.11				
	1100 - 1000	0.839	0.21	0.50	- 0.29				
	1000 - 900	0.273	0.15	0.75	- 0.60				
	900 - 800	0.376	0.15	1.04	- 0.89				
	800 - 700	0.201	0.15	1.45	- 1.30				
	700 - 600	0.071	0.09	2.10	- 2.01				
	600 - 500	0.073	0.06	2.06	- 2.00				
	500 - 400	0.072	0.06	2.13	- 2.07				
	400 - 300	0.032	0.05	2.25	- 2.20				
	Total	4.682	0.17	0.49	- 0.32				

Table 9.2.3. Canada - Mass balance study results.

Part 6 of 8 Mass balance versus altitude

Ram River Glacier, Alberta 51° 51' N, 116° 11' W

Year	Altitude interval m a.sl.	Area km^2	Winter balance m	Summer balance m	Net balance m	Balance period Start	Winter to	Summer to	Equilibrium line m a.sl.
1965-1966	> 3000	0.004	1.25	0.25	+ 1.00	.10.65.	11.06.66.	4.09.66.	2760
	3000 - 2900	0.113	1.36	0.13	+ 1.23				
	2900 - 2800	0.469	1.28	0.70	+ 0.58				
	2800 - 2700	0.751	0.97	1.05	- 0.08				
	2700 - 2600	0.419	0.72	1.45	- 0.73				
	< 2600	0.094	0.68	2.45	- 1.77				
	Total	1.850	1.00	1.06	- 0.06				
1966-1967	> 3000	0.004	1.00	1.75	- 0.75	4.09.66.	6.06.67.	24.09.67.	3000
	3000 - 2900	0.113	1.12	1.58	- 0.46				
	2900 - 2800	0.469	1.08	1.53	- 0.47				
	2800 - 2700	0.751	0.96	1.78	- 0.82				
	2700 - 2600	0.419	0.75	2.80	- 2.05				
	< 2600	0.094	0.23	3.49	- 3.26				
	Total	1.850	0.93	2.08	- 1.15				
1967-1968	> 3000	0.004	1.05	0.44	+ 0.61	24.09.67.	9.06.68.	17.09.68.	2760
	3000 - 2900	0.113	1.05	0.44	+ 0.61				
	2900 - 2800	0.469	1.05	0.44	+ 0.56				
	2800 - 2700	0.751	0.92	0.50	+ 0.42				
	2700 - 2600	0.419	0.78	0.89	- 0.11				
	< 2600	0.094	0.70	1.42	- 0.71				
	Total	1.850	0.91	0.62	+ 0.29				
1968-1969	> 3000	0.004	1.25	0.75	+ 0.50	17.09.68.	11.05.69.	5.10.69.	2760
	3000 - 2900	0.113	1.25	0.75	+ 0.50				
	2900 - 2800	0.469	1.02	0.75	+ 0.27				
	2800 - 2700	0.751	0.79	0.87	- 0.08				
	2700 - 2600	0.419	0.60	1.71	- 1.11				
	< 2600	0.094	0.25	2.21	- 1.96				
	Total	1.850	0.80	1.09	- 0.29				
1969-1970	> 3000	0.004	1.25	1.75	- 0.50	5.10.69.	21.05.70.	.09.70.	> 3000
	3000 - 2900	0.113	1.25	1.75	- 0.50				
	2900 - 2800	0.469	1.00	1.75	- 0.75				
	2800 - 2700	0.751	0.75	1.75	- 1.00				
	2700 - 2600	0.419	0.61	2.70	- 2.09				
	< 2600	0.094	0.25	3.10	- 2.85				
	Total	1.850	0.79	2.03	- 1.24				

Table 9.2.3. Canada - Mass balance study results.

Part 7 of 8 Mass balance versus altitude

Decade Glacier, N.W.T. 69° 39' N, 69° 55' W

Year	Altitude interval m a.sl.	Area km^2	Winter balance m	Summer balance m	Net balance m	Balance period Start	Winter to	Summer to	Equilibrium line m a.sl.
	1500 - 1400	0.461	0.13	0.52	- 0.39				
	1400 - 1300	0.770	0.26	0.51	- 0.25				
	1300 - 1200	1.645	0.24	0.69	- 0.45				
	1200 - 1100	1.519	0.29	0.84	- 0.55				
	1100 - 1000	1.141	0.32	0.76	- 0.44				
1965-1966	1000 - 900	1.304	0.27	1.19	- 0.92	30.09.65.	28.06.66.	30.09.67.	above glacier
	900 - 800	0.870	0.32	1.44	- 1.12				
	800 - 700	0.496	0.22	1.77	- 1.55				
	700 - 600	0.327	0.23	1.73	- 1.50				
	600 - 500	0.096	0.05	1.62	- 1.57				
	500 - 400	0.025	0.00	1.62	- 1.62				
	Total	8.654	0.26	0.97	- 0.71				
	1500 - 1400	0.461	0.15	0.04	+ 0.11				
	1400 - 1300	0.770	0.29	0.15	+ 0.14				
	1300 - 1200	1.645	0.22	0.09	+ 0.13				
	1200 - 1100	1.519	0.31	0.10	+ 0.21				
	1100 - 1000	1.141	0.36	0.19	+ 0.71				
1966-1967	1000 - 900	1.304	0.29	0.37	+ 0.08	30.09.66.	29.06.67.	30.09.67.	930
	900 - 800	0.870	0.31	0.59	- 0.28				
	800 - 700	0.496	0.13	1.09	- 0.96				
	700 - 600	0.327	0.24	0.77	- 0.53				
	600 - 500	0.096	0.03	0.65	- 0.62				
	500 - 400	0.025	0.00	0.65	- 0.65				
	Total	8.654	0.27	0.29	- 0.02				
	1500 - 1400	0.461	0.26	0.00	+ 0.26				
	1400 - 1300	0.770	0.38	0.02	+ 0.36				
	1300 - 1200	1.645	0.38	0.07	+ 0.31				
	1200 - 1100	1.519	0.41	0.06	+ 0.35				
	1100 - 1000	1.141	0.35	0.04	+ 0.31				
1967-1968	1000 - 900	1.304	0.25	0.10	+ 0.15	30.09.67.	23.07.68.	30.09.68.	770
	900 - 800	0.870	0.40	0.18	+ 0.22				
	800 - 700	0.496	0.16	0.23	- 0.07				
	700 - 600	0.327	0.13	0.29	- 0.16				
	600 - 500	0.096	0.01	0.35	- 0.34				
	500 - 400	0.025	0.00	0.35	- 0.35				
	Total	8.654	0.33	0.09	+ 0.24				
	1500 - 1400	0.461	0.32	0.57	- 0.25				
	1400 - 1300	0.770	0.32	0.75	- 0.43				
	1300 - 1200	1.645	0.32	0.81	- 0.49				
	1200 - 1100	1.519	0.24	1.01	- 0.77				
	1100 - 1000	1.141	0.31	1.10	- 0.79				
1968-1969	1000 - 900	1.304	0.26	1.24	- 0.98	30.09.68.	15.04.69.	30.09.69.	above glacier
	900 - 800	0.870	0.26	1.39	- 1.13				
	800 - 700	0.496	0.20	0.57	- 0.38				
	700 - 600	0.327	0.03	1.60	- 1.57				
	600 - 500	0.096	0.03	1.92	- 1.89				
	500 - 400	0.025	0.00	2.10	- 2.10				
	Total	8.654	0.27	1.02	- 0.75				

Table 9.2.3. Canada - Mass balance study results.

Part 8 of 8 Mass balance versus altitude

Decade Glacier, N.W.T. 69° 39' N, 69° 55' W

Year	Altitude interval m a.sl.	Area km^2	Winter balance m	Summer balance m	Net balance m	Balance period Start	Winter to	Summer to	Equilibrium line m a.sl.
1969-1970	1500 - 1400	0.461	0.23	0.01	+ 0.22	30.09.69	17.07.70.	30.09.70.	865
	1400 - 1300	0.770	0.34	0.01	+ 0.33				
	1300 - 1200	1.645	0.28	0.01	+ 0.27				
	1200 - 1100	1.519	0.23	0.02	+ 0.21				
	1100 - 1000	1.141	0.27	0.06	+ 0.21				
	1000 - 900	1.304	0.27	0.15	+ 0.12				
	900 - 800	0.870	0.20	0.24	- 0.04				
	800 - 700	0.496	0.16	0.57	- 0.41				
	700 - 600	0.327	0.10	1.09	- 0.99				
	600 - 500	0.096	0.00	1.87	- 1.87				
	500 - 400	0.025	0.00	2.00	- 2.00				
	Total	8.654	0.24	0.16	+ 0.08				

Table 9.2.4. USA - Mass balance study results.

Part 1 of 4 Description of the glaciers (see also chapter 2)

For references see chapter 4, for abbreviations of the sponsoring agencies see chapter 2.

No.	Glacier	Area (km^2)	Altitudes High (m)	Altitudes Low (m)	References	Principal Investigator Sponsor
1	Gulkana	23.7	2500	1136	Reger (1968)	L.R. Mayo; USGS (F)
6	Wolverine	18.0	1650	400		L.R. Mayo; USGS (F)
43	Sherman	54.1	725	112	Bull and Marangunic (1967)	C. Bull; IPS
114	Blue	4.2	2350	1275	La Chapelle (1965)	E. La Chapelle; UW (AS)
121	South Cascade	2.8	2190	1610	Meier and Tangborn (1965)	M.F. Meier; USGS (T)
129	Nisqually	6.5	4300	1410	Meier (1966, 1968) Veatsch (1970)	S. Hodge, UW (G)
132	Eliot	1.8	3290	1920	Handewith (1959), Matthes and Philips (1943)	N.A. Dodge; MAZ
133	Maclure	0.17	3760	3590		D. Scully; USGS (S)
136	Grasshopper	0.41	3350	3145	Alford and Clark (1968)	D.L. Alford; UC (INSTAAR)
140	McCall	6.2	2440	1520	Wendler (1970)	G. Wendler, C. Benson, UC
141	Isabelle	0.12	3920	3660	Lloyd (1970)	D. Lloyd, UC

Table 9.2.4. USA - Mass balance study results.

Part 2 of 4 Mass balances 1964/65 and 1965/66

Ex...

$\bar{b}_n/\bar{b}_a$ net or annual balance averaged over the glacier, in m. If 2 numbers, upper is $\bar{b}_n$ lower is $\bar{b}_a$

Error standard error of $\bar{b}_n$ or $\bar{b}_a$, in m.

$\bar{e}_t/\bar{e}_a$ total or annual exchange, averaged over the glacier, in m.

$\bar{b}_w$ winter balance, averaged over the glacier, in m.

ELA altitude of the equilibrium line, in m.

AAR accumulation area ratio (accumulation area divided by glacier area).

ΔL change in glacier length, in m.

year	No.	Glacier	Balance Year Begin	End	$\bar{b}_n/\bar{b}_a$	Error	$\bar{e}_t/\bar{e}_a$	$\bar{b}_w$	ELA	AAR	ΔL	Remarks
1964 - 65	1	Gulkana										
	6	Wolverine										
	43	Sherman		1. 8.	+ 0.4	0.2			550		- 25	
	114	Blue			- 0.32	0.03			1900			
	121	South Cascade	30. 9.	1.11.	- 0.1	0.1	7.1	3.5	1880	0.53	- 12	
	129	Nisqually										
	132	Eliot		1. 9.	+ 0.7				2330	0.67	+ 8	b_n estimated from AAR
	133	Maclure										
	136	Gr...										
1965 - 66	1	Gulkana	1.10.	30. 9.	0 + 0.1	0.19	1.94	1.00	1770	0.59	- 44	$\bar{b}_n$(fi)=-0.2
	6	Wolverine	1.10.	30. 9.	- 0.26	0.34	3.92	1.83	1250	0.56	- 4	
	43	Sherman		1. 8.	+ 0.2	0.2			560		- 8	
	114	Blue			+ 0.58	0.06			1655			
	121	South Cascade	2.11.	16.10.	- 1.0 - 0.9	0.1	6.0	2.5	2140	0.13	- 16	$\bar{b}_n$(fi)=-1.1
	129	Nisqually										
	132	Eliot		22. 9.	- 0.5				2380	0.53	- 16	b_n estimated from AAR
	133	Maclure										
	136	Grasshopper										

Table 9.2.4. USA - Mass balance study results.

Part 3 of 4 Mass balances 1966/67 and 1967/68

Explanation

$\bar{b}_n / \bar{b}_a$ net or annual balance, averaged over the glacier, in m. If 2 numbers, upper is $\bar{b}_n$, lower is $\bar{b}_a$

Error standard error of $\bar{b}_n$ or $\bar{b}_a$, in m.

$\bar{e}_t / \bar{e}_a$ total or annual exchange, averaged over the glacier, in m.

b_w winter balance, averaged over the glacier, in m.

ELA altitude of the equilibrium line, in m.

AAR accumulation area ratio (accumulation area divided by glacier area).

ΔL change in glacier length, in m.

year	No.	Glacier	Balance Year Begin	End	$\bar{b}_n/\bar{b}_a$	Error	$\bar{e}_t/\bar{e}_a$	b_w	ELA	AAR	ΔL	Remarks
1966 - 67	1	Gulkana			- 0.5	0.2	3.3	1.2	1800	0.55	- 80	preliminary values
	6	Wolverine			- 2.0	0.5	4.7	1.3	1460	0.10	- 5	preliminary values
	43	Sherman	1. 8.	1. 8.	- 0.8	0.20			600		+ 10	
	114	Blue			+ 0.59	0.06			1675			
	121	South Cascade	17.10.	1.10.	- 0.6 - 0.8	0.1	7.2	3.3	1870	0.55	- 17	$\bar{b}_a$ value is preliminary
	129	Nisqually										
	132	Eliot		19. 9.	- 1.6				2670	0.31	0	b_n estimated from AAR
	133	Maclure	20.10.	15.10.	+ 1.21	0.1	5.7	3.5	3640	0.86	+ 0.5	ΔL estimated
	136	Grashopper	9.	9. 9.	+ 0.81	0.2	3.4	2.1	3180	0.60	0	
1967 - 68	1	Gulkana			- 0.8	0.2	4.0	1.5	1850	0.48	-30	preliminary value
	6	Wolverine			- 1.5	0.3	4.9	1.6	1350	0.35	- 5	preliminary value
	43	Sherman			0	0.2			575		+10	
	114	Blue			+ 0.27	0.1						
	121	South Cascade	2.10.	3.10.	0 0	0.1	6.0	3.0	1850	0.73	-12	$\bar{b}_a$ value is preliminary
	129	Nisqually		5.10.					2080	0.62		
	132	Eliot		28. 9.	- 2.8				2710	0.15		b_n estimated from AAR
	133	Maclure	16.10.	12.10.	- 0.76	0.2	3.22	1.23	3740	0.18		
	136	Grasshopper	9. 9.	27. 8.	+ 0.53	0.2	3.97	2.25	3160	0.63	0	

Table 9.2.4. USA - Mass balance study results.

Part 4 of 4 Mass balances 1968/69 and 1969/70

Explanation

$\bar{b}_n/\bar{b}_a$ net or annual balance, averaged over the glacier, in m. If 2 numbers, upper is $\bar{b}_n$ lower is $\bar{b}_a$

Error standard error of $\bar{b}_n$ or $\bar{b}_a$ in m.

$\bar{e}_t/\bar{e}_a$ total or annual exchange, averaged over the glacier, in m.

$\bar{b}_w$ winter balance, averaged over the glacier, in m.

ELA altitude of the equilibrium line, in m.

AAR accumulation area ratio (accumulation area divided by glacier area).

ΔL change in galcier length, in m.

year	No.	Glacier	Balance Year Begin	End	$\bar{b}_n/\bar{b}_a$	Error	$\bar{e}_t/\bar{e}_a$	$\bar{b}_w$	ELA	AAR	ΔL	Remarks
1968 - 69	1	Gulkana			- 1.2	0.2	2.6	0.6	2000	0.05	- 30	preliminary values
	6	Wolverine			- 1.0	0.3	5.2	1.9	1250	0.55	- 5	preliminary values
	43	Sherman										
	114	Blue										
	121	South Cascade	4.10.	6.10	- 0.7 - 0.7	0.1	7.1	3.2	1910	0.47	- 17	$\bar{b}_a$ value is preliminary
	129	Nisqually										
	132	Eliot			+ 0.7				2320	0.64		
	133	Maclure			+ 0.7		4.3	2.5	3670	0.80		
	136	Grasshopper										
	140	McCall			- 0.28			0.13	2210	0.40		
	141	Isabelle			+ 1.2	0.4		2.8	3700			ELA approximate
1969 - 70	1	Gulkana			+ 0.3	0.2	1.9	0.9	1700	0.65	- 20	preliminary values
	6	Wolverine			+ 1.0	0.3	5.2	2.9	1000	0.85	- 2	preliminary values
	43	Sherman										
	114	Blue										
	121	South Cascade	7.10	18.10	- 1.3 - 1.2	0.1	6.0	2.4	2050	0.06	- 11	$\bar{b}_n$ value is preliminary
	129	Nisqually										
	132	Eliot			- 0.3				2380	0.56		
	133	Maclure										
	136	Grasshopper										

Table 9.2.5. USA - Mass balance study results.

Part 1 of 2 Mass balance versus altitude 1964/65 and 1965/66

Explanation :

ΔS area of glacier within each altitude interval, in km^2.
b_n net balance, in m.
b_w winter balance, in m.
b_a annual balance, in m.

Period	Altitude	No. 114 Blue-			No. 121 South Cascade								
		ΔS	b_n	b_w	ΔS	b_n	b_w	ΔS	b_n	b_w	ΔS	b_n	b_w
1964 - 65	1400	0.06	- 3.8										
	1500	0.45	- 3.3										
	1600	0.68	- 2.4										
	1700	0.72	- 1.3		0.12	- 4.5	1.5						
	1800	0.43	- 0.5		0.43	- 2.5	2.1						
	1900	0.28	- 0.1		0.74	- 0.2	3.7						
	2000	0.20	+ 0.4		0.65	+ 0.8	3.9						
	2100	0.40	+ 0.9		0.66	+ 1.3	4.2						
	2200	0.59	+ 1.6		0.23	+ 0.6	3.7						
	2300	0.34	+ 2.2										
	2400	0.11											

Period	Altitude	No. 1 Gulkana			No. 6 Wolverine			No. 114 Blue			No. 121 South Cascade		
		ΔS	b_a	b_w	ΔS	b_a	b_w	ΔS	b_n	b_w	ΔS	b_n	b_w
1965 - 66	500				0.1	- 6.4	0.3						
	600				0.2	- 6.5	0.6						
	700				0.7	- 4.9	1.0						
	800				0.7	- 4.3	1.0						
	900				0.3	- 4.0	1.0						
	1000				0.9	- 3.6	1.2						
	1100				2.6	- 1.3	1.5						
	1200	0.7	- 2.0	0.2	1.8	- 0.5	1.8						
	1300	0.8	- 3.3	0.4	1.7	- 0.1	1.6						
	1400	1.6	- 3.2	0.4	5.5	+ 0.9	2.0	0.06	- 2.1				
	1500	1.8	- 2.8	0.5	2.5	+ 1.5	2.5	0.45	- 1.8				
	1600	1.7	- 1.9	0.6	0.9	+ 2.0	2.8	0.68	- 1.1				
	1700	2.7	- 0.9	0.8	0.2	+ 2.2	3.0	0.72	0		0.12	- 4.4	0.8
	1800	2.8	- 0.2	1.0				0.43	+ 0.4		0.42	- 2.7	1.9
	1900	3.4	+ 0.7	1.2				0.28	+ 0.7		0.74	- 1.1	2.6
	2000	3.3	+ 0.7	1.2				0.20	+ 0.9		0.65	- 0.4	2.5
	2100	2.2	+ 1.1	1.3				0.40	+ 1.3		0.66	- 0.4	3.1
	2200	1.5	+ 1.2	1.2				0.59	+ 1.9		0.23	- 0.1	2.6
	2300	1.0	+ 1.5	1.4				0.34	+ 2.4				
	2400	0.2	+ 1.9	1.0				0.11					
	2500	0.02	+ 2.0	0.5									

Table 9.2.5. USA - Mass balance study results.

Part 2 of 2 Mass balance versus altitude 1966/67 and 1967/68

Explanation:

ΔS area of glacier within each altitude interval, in km^2.
b_n net balance, in m.
b_w winter balance, in m.

	Altitude	No. 114 Blue			No. 121 South Cascade			No. 133 Maclure			No. 136 Grasshopper		
		ΔS	b_n	b_w	ΔS	b_n	b_w	ΔS	b_n	b_w	ΔS	b_n	b_w
1966 - 67	1400	0.06	- 2.9										
	1500	0.45	- 2.4										
	1600	0.68	- 1.5										
	1700	0.72	- 0.3		0.11	- 4.8	1.8						
	1800	0.43	+ 0.2		0.42	- 3.3	2.5						
	1900	0.28	+ 0.4		0.73	- 0.5	3.5						
	2000	0.20	+ 0.7		0.65	+ 0.5	3.6						
	2100	0.40	+ 1.4		0.66	+ 0.6	3.7						
	2200	0.59	+ 2.5		0.23	+ 0.6	3.4						
	2300	0.34	+ 3.5										
	2400	0.11											
	3100										0.21	0	1.5
	3200										0.20	+ 1.6	2.4
	3700							0.077	+ 0.3	2.4			
	3800							0.086	+ 2.1	4.4			

	Altitude	No. 121 South Cascade			No. 129 Nisqually			No. 133 Maclure			No. 136 Grasshopper		
		ΔS	b_n	b_w	ΔS	b_n	b_w	ΔS	b_n	b_w	ΔS	b_n	b_w
1967 - 68	1600				0.12	- 8.1	0.8						
	1700	0.11	- 4.8	1.4	0.13	- 7.3	1.0						
	1800	0.42	- 1.8	2.4	0.18	- 6.1	1.3						
	1900	0.73	+ 0.2	3.4	0.36	- 4.5	2.0						
	2000	0.65	+ 0.7	3.3	0.32	- 2.6	2.9						
	2100	0.66	+ 0.7	3.2	0.23	- 0.8	4.3						
	2200	0.23	+ 0.3	2.1	0.21	+ 1.2	5.8						
	3000-3100										0.21	0	1.85
	3200										0.20	+ 1.0	2.50
	3700							0.077	- 1.43	0.74			
	3800							0.086	- 0.11	1.69			

	Altitude	1968 - 69			1969 - 70		
		ΔS	b_n	b_w	ΔS	b_n	b_w
South Cascade	1700	0,08	- 5,6	1,4	0,07	- 4,39	1,2
	1800	0,42	- 2,8	2,7	0,41	- 2,82	1,96
	1900	0,73	- 0,45	3,5	0,73	- 1,06	2,71
	2000	0,65	+ 0,21	3,5	0,65	- 0,46	2,53
	2100	0,66	+ 0,53	3,4	0,66	- 0,22	2,60
	2200	0,23	+ 0,2	2,6	0,23	- 0,32	2,18

Table 9.2.6. Norway - Mass balance study results.

Glacier	Area	Accumulation	Ablation	Balance	Equilibrium line	Area	Accumulation	Ablation	Balance	Equilibrium line
	km^2	g/cm^2	g/cm^2	g/cm^2	m a.sl.	km^2	g/cm^2	g/cm^2	g/cm^2	m a.sl.
			1964 / 65					1965 / 66		
Southern Norway										
Alfotbreen		364	316	+ 47			247	408	- 161	
Tunsbergdalsbreen							157	266	- 109	
Nigardsbreen		229	138	+ 91			176	268	- 92	
Midtre Folgefonni		237	233	+ 4			168	308	- 140	
Hardangerjøkulen		205	154	+ 51			160	224	- 64	
Storbreen		154	120	+ 34			125	186	- 61	
Hellstugubreen		129	77	+ 52			95	162	- 67	
Grasubreen		77	36	+ 41			72	101	- 29	
Northern Norway										
Blaisen		200	146	+ 54			112	239	- 127	
Storsteinfjellbreen		169	125	+ 44			105	188	- 83	
Cainhavarre		141	120	+ 21			112	207	- 95	
	km^2	m	m	m	m a.sl.	km^2	m	m	m	m a.sl.
			1966 / 67					1967 / 68		
Southern Norway										
Alfotbreen	4.76	4.46	3.18	+1.28	950	4.76	4.55	3.60	+0.95	1075
Vesledalsbreen	4.22	2.06	1.71	+0.35	1400	4.22	3.14	2.50	+0.64	1320
Tunsbergdalsbreen	43.77	3.31	1.52	+1.79	1160	43.77	2.74	2.70	+0.04	1270
Nigardsbreen	40.86	3.40	1.24	+2.16	1310	47.03	2.72	2.50	+0.22	1550
Midtre Folgefonni	19.74	3.59	2.23	+1.36	1350	19.52	3.36	2.72	+0.64	1365
Hardangerjøkulen	17.55	2.44	1.25	+1.19	1540	17.55	2.68	2.15	+0.53	1600
Omnsbreen						1.52	2.20	2.38	-0.18	1530
Storbreen	5.45	1.89	1.17	+0.72	1570	5.45	1.59	1.54	+0.05	1700
Hellstugubreen	3.38	1.48	0.93	+0.55	1800	3.33	1.38	1.49	-0.11	1875
Vestre Memurubre						9.06	1.70	1.46	+0.24	1820
Austre Memurubre						8.86	1.77	1.76	+0.01	1960
Grasubreen	2.39	1.45	0.74	+0.71	-	2.53	1.03	1.11	-0.08	2140
Store Supphellebre	11.99	2.72	1.50	+1.22	1190					
Northern Norway										
Blaisen	2.18	1.38	2.35	-0.97	1175	2.18	1.62	1.36	+0.26	1010
Storsteinsfjellbreen	6.12	1.37	1.77	-0.40	1450	6.12	1.44	0.99	+0.45	1275
Cainhavarre	0.68	1.63	1.79	-0.16	1450	0.68	1.31	1.05	+0.26	1290
Spitsbergen										
Midre Lovenbreen						6.03	0.48	0.51	-0.03	295
Austre Brøggerbreen		0.77	1.42	0.65	400	6.08	0.57	0.67	-0.10	295
Southern Norway			1968 / 69					1969/70		
Alfotbreen	4.82	2.66	4.83	-2.17	(1550)		2.60	3.83	-1.23	
Vesledalsbreen	4.22	1.26	3.44	-2.18	(1850)		1.52	2.66	-1.14	
Tunsbergdalsbreen	50.11	1.53	3.22	-1.69	1700		1.54	2.38	-0.84	
Nigardsbreen	47.03	1.95	3.26	-1.31	1850		1.73	2.29	-0.56	
Midtre Folgefonni							2.07	2.69	-0.62	
Hardangerjøkulen	17.53	1.07	2.97	-1.90	(1950)		1.29	1.89	-0.60	
Omnsbreen	1.50	1.09	3.68	-2.59	?					
Storbreen	5.36	1.22	2.64	-1.42	(2020)		0.97	1.69	-0.72	
Hellstugubreen	3.33	0.95	2.23	-1.28	2130		0.69	1.73	-1.01	
Vestre Memurubre	9.05	1.05	2.11	-1.06	2170		0.84	1.63	-0.79	
Austre Memurubre	8.86	0.99	2.45	-1.46	2130		0.81	1.71	-0.90	
Grasubreen	2.53	0.74	2.04	-1.37	(2350)		0.57	1.23	-0.66	
Spitsbergen										
Midre Lovénbreen	6.03	0.41	1.25	-0.84	(650)		0.36	0.89	-0.53	
Austre Brøggerbreen	6.08	0.40	1.33	-0.93	(650)		0.37	0.91	-0.54	

( ) Theoretical line above the highest point of the glacier.

Table 9.2.7. Norway - Mass balance study results.

Part 1 of 22 Mass balance versus altitude Tunsbergdalsbreen

Altitude interval	Area	Winter balance			Summer balance			Net balance		
		Total	Specific		Total	Specific		Total	Specific	
m a sl	km^2	10^6 m^3	m	l/s km^2	10^6 m^3	m	l/s km^2	10^6 m^3	m	l/s km^2
1950 - 1900	0.384	1.027	2.67	85	0.432	1.12	35	+ 0.595	+ 1.55	+ 50
1900 - 1800	2.808	7.511	2.67	85	3.941	1.40	44	+ 3.570	+ 1.27	+ 41
1800 - 1700	8.788	20.783	2.36	75	15.480	1.76	55	+ 5.303	+ 0.60	+ 20
1700 - 1600	7.008	14.575	2.08	66	14.068	2.00	63	+ 0.507	+ 0.08	+ 3
1600 - 1500	4.088	7.375	1.84	58	9.309	2.27	72	- 1.934	- 0.43	- 14
1500 - 1400	2.908	4.397	1.51	48	7.530	2.58	82	- 3.133	- 1.07	- 34
1400 - 1300	1.202	1.509	1.25	39	3.408	2.83	90	- 1.899	- 1.58	- 51
1300 - 1200	0.992	0.978	0.98	31	3.054	3.07	97	- 2.076	- 2.09	- 66
1200 - 1100	3.344	2.956	0.88	28	11.379	3.40	107	- 8.423	- 2.52	- 79
1100 - 1000	3.824	2.944	0.77	24	14.053	3.67	116	- 11.109	- 2.90	- 92
1000 - 900	3.336	2.252	0.67	21	12.811	3.84	122	- 10.559	- 3.17	- 101
900 - 800	2.451	1.138	0.46	15	10.044	4.09	129	- 8.906	- 3.63	- 114
800 - 700	2.040	1.019	0.50	16	8.415	4.12	130	- 7.396	- 3.62	- 114
700 - 600	0.416	0.156	0.37	12	1.820	4.37	138	- 1.664	- 4.00	- 126
600 - 540	0.184	0.069	0.37	12	0.805	4.37	138	- 0.736	- 4.00	- 126
1950 - 540	43.773	68.689	1.57	50	116.549	2.66	84	- 47.860	- 1.09	- 34

1966/67

1930 - 1900	0.384	1.523	3.97	125	0.134	0.35	11	+ 1.389	+ 3.62	+ 114
1900 - 1800	2.808	12.964	4.62	146	1.053	0.37	12	+ 11.911	+ 4.25	+ 134
1800 - 1700	8.788	40.941	4.67	147	4.710	0.54	17	+ 36.231	+ 4.13	+ 130
1700 - 1600	7.008	29.903	4.27	135	5.452	0.78	25	+ 24.451	+ 3.49	+ 110
1600 - 1500	4.088	15.966	3.90	123	4.243	1.04	33	+ 11.723	+ 2.86	+ 90
1500 - 1400	2.908	9.962	3.39	197	3.841	1.32	42	+ 6.021	+ 2.07	+ 65
1400 - 1300	1.202	3.304	2.75	87	1.953	1.62	51	+ 1.351	+ 1.13	+ 36
1300 - 1200	0.992	2.464	2.49	79	1.926	1.95	62	+ 0.538	+ 0.54	+ 17
1200 - 1100	3.344	7.094	2.12	67	7.595	2.27	72	- 0.501	- 0.15	- 5
1100 - 1000	3.824	7.370	1.93	61	9.520	2.49	79	- 2.150	- 0.56	- 18
1000 - 900	3.336	6.255	1.86	59	9.302	2.79	88	- 3.047	- 0.93	- 29
900 - 800	2.451	3.936	1.61	51	7.610	3.10	98	- 3.674	- 1.49	- 47
800 - 700	2.040	2.601	1.28	40	6.806	3.34	105	- 4.205	- 2.06	- 65
700 - 600	0.416	0.468	1.13	36	1.508	3.62	114	- 1.040	- 2.49	- 78
600 - 540	0.184	0.207	1.13	36	0.713	3.87	122	- 0.506	- 2.74	- 86
1930 - 540	43.773	144.858	3.31	104	66.366	1.51	48	+ 78.492	+ 1.79	+ 56

1967/68

1930 - 1900	0.384	1.440	3.75	118	0.472	1.23	39	+ 0.968	+ 2.52	+ 79
1900 - 1800	2.808	10.530	3.75	118	3.861	1.38	44	+ 6.669	+ 2.37	+ 74
1800 - 1700	8.788	32.593	3.71	117	13.452	1.53	48	+ 19.141	+ 2.18	+ 69
1700 - 1600	7.008	22.274	3.18	100	11.698	1.67	53	+ 10.576	+ 1.51	+ 47
1600 - 1500	4.088	11.485	2.80	88	7.665	1.88	59	+ 3.820	+ 0.92	+ 29
1500 - 1400	2.908	7.997	2.75	87	6.326	2.18	69	+ 1.671	+ 0.57	+ 18
1400 - 1300	1.202	3.288	2.74	87	3.100	2.58	81	+ 0.188	+ 0.16	+ 6
1300 - 1200	0.992	2.772	2.79	88	2.920	2.94	94	- 0.148	- 0.15	- 6
1200 - 1100	3.344	7.598	2.27	72	11.750	3.51	111	- 4.152	- 1.24	- 39
1100 - 1000	3.824	7.850	2.05	65	14.980	3.92	124	- 7.130	- 1.87	- 59
1000 - 900	3.336	5.538	1.66	52	14.845	4.45	140	- 9.307	- 2.79	- 88
900 - 800	2.451	3.370	1.37	43	12.209	4.98	157	- 8.839	- 3.61	- 114
800 - 700	2.040	2.632	1.29	41	11.177	5.48	173	- 8.545	- 4.19	- 132
700 - 600	0.416	0.468	1.13	36	2.558	6.15	194	- 2.090	- 5.02	- 158
600 - 540	0.184	0.207	1.13	36	1.242	6.75	213	- 1.035	- 5.62	- 177
1930 - 540	43.773	120.042	2.74	86	118.255	2.70	85	+ 1.787	+ 0.04	+ 1

Table 9.2.7. Norway - Mass balance study results.

Part 2 of 22 Mass balance versus altitude Tunsbergdalsbreen 1968/69

Altitude interval	Area	Winter balance			Summer balance			Net balance		
		Total	Specific		Total	Specific		Total	Specific	
m a sl	km^2	10^6 m^3	m	l/s km^2	10^6 m^3	m	l/s km^2	10^6 m^3	m	l/s km^2
1930 - 1900	0.38	0.82	2.13	67	0.69	1.80	57	+ 0.13	+ 0.33	+ 10
1900 - 1800	2.81	5.97	2.13	67	5.05	1.80	57	+ 0.92	+ 0.33	+ 10
1800 - 1700	8.79	18.59	2.12	67	17.14	1.95	62	+ 1.45	+ 0.17	+ 5
1700 - 1600	8.44	16.41	1.94	61	17.94	2.13	67	- 1.53	- 0.19	- 6
1600 - 1500	6.95	12.38	1.78	56	17.17	2.47	78	- 4.79	- 0.69	- 22
1500 - 1400	4.28	6.80	1.59	50	11.78	2.75	87	- 4.98	- 1.16	- 37
1400 - 1300	1.62	2.44	1.51	48	5.26	3.25	103	- 2.82	- 1.74	- 55
1300 - 1200	1.24	1.51	1.22	39	4.51	3.64	115	- 3.00	- 2.42	- 76
1200 - 1100	3.34	3.68	1.10	35	14.21	4.25	134	- 10.53	- 3.15	- 99
1100 - 1000	3.82	3.40	0.89	28	18.35	4.80	152	- 14.95	- 3.91	- 123
1000 - 900	3.34	2.25	0.67	21	17.51	5.25	166	- 15.26	- 4.58	- 145
900 - 800	2.45	1.53	0.63	20	14.09	5.75	182	- 12.56	- 5.12	- 162
800 - 700	2.04	0.87	0.43	14	13.26	6.50	205	- 12.39	- 6.07	- 191
700 - 600	0.42	0.16	0.38	12	3.12	7.50	237	- 2.96	- 7.12	- 225
600 - 540	0.18	0.07	0.38	12	1.56	8.50	268	- 1.49	- 8.12	- 256
1930 - 540	50.11	76.86	1.53	48	161.65	3.22	102	- 84.79	- 1.69	- 54

1969/70

Altitude interval	Area	Winter balance			Summer balance			Net balance		
		Total	Specific		Total	Specific		Total	Specific	
m a sl	km^2	10^6 m^3	m	l/s km^2	10^6 m^3	m	l/s km^2	10^6 m^3	m	l/s km^2
1930 - 1900	0.38	0.75	1.97	62	0.33	0.87	28	+ 0.42	+ 1.11	+ 32
1900 - 1800	2.81	6.48	2.31	73	2.46	0.88	28	+ 4.02	+ 1.43	+ 45
1800 - 1700	8.79	19.23	2.19	69	10.20	1.16	37	+ 9.03	+ 1.03	+ 33
1700 - 1600	8.44	15.23	1.80	54	12.34	1.46	46	+ 2.89	+ 0.34	+ 11
1600 - 1500	6.95	11.39	1.64	52	12.16	1.75	55	- 0.77	- 0.11	- 4
1500 - 1400	4.28	6.54	1.53	48	9.63	2.25	71	- 3.09	- 0.72	- 23
1400 - 1300	1.62	2.33	1.44	45	4.04	2.49	78	- 1.71	- 1.06	- 34
1300 - 1200	1.24	1.74	1.40	44	3.69	2.98	94	- 1.95	- 1.57	- 50
1200 - 1100	3.34	3.85	1.15	36	11.41	3.42	108	- 7.56	- 2.26	- 71
1100 - 1000	3.82	3.87	1.01	32	14.33	3.75	118	- 10.46	- 2.74	- 87
1000 - 900	3.34	2.54	0.76	24	14.20	4.25	134	- 11.66	- 3.49	- 110
900 - 800	2.45	1.56	0.64	20	11.54	4.71	149	- 9.98	- 4.07	- 128
800 - 700	2.04	1.28	0.63	20	9.77	4.79	151	- 8.49	- 4.16	- 131
700 - 600	0.42	0.26	0.63	20	2.21	5.25	166	- 1.95	- 4.64	- 146
600 - 540	0.18	0.12	0.63	20	0.95	5.25	166	- 0.83	- 4.61	- 146
1930 - 540	50.11	77.17	1.54	49	119.26	2.38	75	- 42.09	- 0.84	- 27

Table 9.2.7. Norway - Mass balance study results.

Part 3 of 22 Mass balance versus altitude Nigardsbreen

1964/65

Altitude interval m a sl	Area km²	Winter balance Total 10^6 m³	Winter balance Specific m	Winter balance Specific l/s km²	Summer balance Total 10^6 m³	Summer balance Specific m	Summer balance Specific l/s km²	Net balance Total 10^6 m³	Net balance Specific m	Net balance Specific l/s km²
2000 - 1900	0.16	0.44	2.75	87	0.06	0.38	12	+ 0.38	+ 2.37	+ 75
1900 - 1800	3.76	9.82	2.61	83	1.43	0.38	12	+ 8.39	+ 2.23	+ 71
1800 - 1700	9.76	24.48	2.51	80	5.76	0.59	19	+ 18.72	+ 1.92	+ 61
1700 - 1600	11.84	29.87	2.52	80	9.96	0.84	27	+ 19.91	+ 1.68	+ 53
1600 - 1500	5.68	12.45	2.19	69	7.10	1.25	40	+ 5.35	+ 0.94	+ 30
1500 - 1400	3.70	7.71	2.08	66	5.81	1.57	50	+ 1.90	+ 0.51	+ 16
1400 - 1300	1.90	3.65	1.92	61	4.41	2.32	74	- 0.76	- 0.40	- 13
1300 - 1200	0.72	1.26	1.75	56	1.80	2.50	79	- 0.54	- 0.75	- 24
1200 - 1100	0.32	0.48	1.50	48	1.04	3.26	103	- 0.56	- 1.76	- 56
1100 - 1000	0.54	0.68	1.25	40	1.96	3.60	114	- 1.28	- 2.35	- 74
1000 - 900	0.42	0.53	1.25	40	1.89	4.50	143	- 1.36	- 3.25	- 103
900 - 800	0.48	0.60	1.25	40	2.58	5.42	172	- 1.98	- 4.17	- 132
800 - 700	0.40	0.50	1.25	40	2.53	6.40	203	- 2.03	- 5.15	- 163
700 - 600	0.44	0.44	1.00	32	3.18	7.29	231	- 2.74	- 6.29	- 199
600 - 500	0.28	0.25	0.88	28	2.29	8.28	263	- 2.04	- 7.40	- 235
500 - 400	0.20	0.14	0.68	22	1.90	9.48	301	- 1.76	- 8.80	- 279
400 - 300	0.26	0.18	0.68	22	2.69	10.50	333	- 2.51	- 9.82	- 311
2000 - 300	40.86	93.48	2.29	73	56.39	1.38	44	+ 37.09	+ 0.91	+ 29

1965/66

Altitude interval m a sl	Area km²	Winter balance Total 10^6 m³	Winter balance Specific m	Winter balance Specific l/s km²	Summer balance Total 10^6 m³	Summer balance Specific m	Summer balance Specific l/s km²	Net balance Total 10^6 m³	Net balance Specific m	Net balance Specific l/s km²
1950 - 1900	0.16	0.38	2.38	75	0.20	1.25	40	+ 0.18	+ 1.13	+ 36
1900 - 1800	3.76	8.93	2.38	75	5.68	1.51	48	+ 3.25	+ 0.87	+ 27
1800 - 1700	9.76	20.80	2.13	68	17.55	1.80	57	+ 3.25	+ 0.33	+ 11
1700 - 1600	11.84	22.63	1.91	60	25.62	2.16	68	- 2.99	- 0.25	- 8
1600 - 1500	5.68	9.46	1.67	53	15.02	2.64	84	- 5.55	- 0.98	- 31
1500 - 1400	3.70	5.13	1.39	44	11.50	3.11	99	- 6.38	- 1.72	- 55
1400 - 1300	1.90	1.93	1.02	32	7.12	3.75	119	- 5.19	- 2.73	- 87
1300 - 1200	0.72	0.65	0.90	29	3.24	4.50	143	- 2.59	- 3.60	- 114
1200 - 1100	0.32	0.28	0.88	28	1.51	4.72	150	- 1.23	- 3.84	- 122
1100 - 1000	0.54	0.42	0.78	25	2.97	5.51	175	- 2.55	- 4.72	- 150
1000 - 900	0.42	0.28	0.68	21	2.53	6.02	191	- 2.25	- 5.35	- 170
900 - 800	0.48	0.32	0.68	21	3.12	6.50	206	- 2.80	- 5.83	- 185
800 - 700	0.40	0.22	0.56	18	3.00	7.50	238	- 2.78	- 6.95	- 220
700 - 600	0.44	0.16	0.38	12	3.41	7.77	246	- 3.24	- 7.38	- 234
600 - 500	0.28	0.10	0.36	11	2.42	8.64	274	- 2.32	- 8.28	- 263
500 - 400	0.20	0.06	0.33	10	1.96	9.80	311	- 1.90	- 9.48	- 301
400 - 300	0.26	0.10	0.38	12	2.73	10.50	333	- 2.63	- 10.12	- 321
1950 - 300	40.86	71.87	1.76	56	109.57	2.68	85	- 37.70	- 0.92	- 29

1966/67

Altitude interval m a sl	Area km²	Winter balance Total 10^6 m³	Winter balance Specific m	Winter balance Specific l/s km²	Summer balance Total 10^6 m³	Summer balance Specific m	Summer balance Specific l/s km²	Net balance Total 10^6 m³	Net balance Specific m	Net balance Specific l/s km²
1950 - 1900	0.16	0.92	5.75	182	0.06	0.37	12	+ 0.86	+ 5.38	+ 170
1900 - 1800	3.76	17.35	4.62	146	1.41	0.37	12	+ 15.94	+ 4.25	+ 134
1800 - 1700	9.76	39.83	4.06	128	3.83	0.39	13	+ 36.00	+ 3.67	+ 115
1700 - 1600	11.84	41.15	3.47	110	8.07	0.68	22	+ 33.08	+ 2.79	+ 88
1600 - 1500	5.68	16.89	2.97	95	6.23	1.10	35	+ 10.66	+ 1.87	+ 60
1500 - 1400	3.70	10.77	2.91	92	6.47	1.75	55	+ 4.30	+ 1.16	+ 37
1400 - 1300	1.90	4.92	2.59	82	4.28	2.25	71	+ 0.64	+ 0.34	+ 11
1300 - 1200	0.72	1.66	2.31	73	1.98	2.75	87	- 0.32	- 0.44	- 14
1200 - 1100	0.32	0.72	2.25	71	1.12	3.50	111	- 0.40	- 1.25	- 40
1100 - 1000	0.54	1.21	2.25	71	2.13	4.14	131	- 0.92	- 1.89	- 60
1000 - 900	0.42	0.79	1.88	60	1.89	4.50	142	- 1.10	- 2.62	- 82
900 - 800	0.48	0.84	1.75	56	2.32	4.83	153	- 1.48	- 3.08	- 97
800 - 700	0.40	0.70	1.75	56	2.30	5.75	182	- 1.60	- 4.00	- 126
700 - 600	0.44	0.67	1.52	48	2.86	6.50	206	- 2.19	- 4.98	- 158
600 - 500	0.28	0.35	1.25	40	1.92	6.85	217	- 1.57	- 5.60	- 177
500 - 400	0.20	0.19	0.95	30	1.58	7.90	250	- 1.39	- 6.95	- 220
400 - 300	0.26	0.23	0.87	27	2.30	8.85	280	- 2.07	- 7.98	- 253
1950 - 300	40.86	139.19	3.40	107	50.75	1.24	39	+ 88.44	+ 2.16	+ 68

Table 9.2.7. Norway - Mass balance study results.

Part 4 of 22 Mass balance versus altitude Nigardsbreen 1967/68

Altitude interval	Area	Winter balance			Summer balance			Net balance		
		Total	Specific		Total	Specific		Total	Specific	
m a sl	km^2	$10^6\ m^3$	m	l/s km^2	$10^6\ m^3$	m	l/s km^2	$10^6\ m^3$	m	l/s km^2
1950 - 1900	0.16	0.60	3.75	118	0.18	1.13	36	+ 0.42	+ 2.62	+ 82
1900 - 1800	3.76	14.10	3.75	118	5.08	1.35	43	+ 9.02	+ 2.40	+ 75
1800 - 1700	9.76	34.50	3.54	112	15.40	1.58	50	+ 19.10	+ 1.96	+ 62
1700 - 1600	12.76	39.34	3.08	97	24.90	1.95	61	+ 14.44	+ 1.13	+ 36
1600 - 1500	9.28	21.63	2.33	74	21.67	2.34	74	- 0.04	- 0.01	0
1500 - 1400	5.26	10.20	1.94	61	15.78	3.00	95	- 5.58	- 1.06	- 34
1400 - 1300	2.06	3.35	1.63	51	7.42	3.60	114	- 4.07	- 1.97	- 63
1300 - 1200	0.72	1.13	1.56	49	3.06	4.25	135	- 1.93	- 2.69	- 86
1200 - 1100	0.32	0.44	1.38	44	1.57	4.90	155	- 1.13	- 3.52	- 111
1100 - 1000	0.54	0.61	1.13	36	3.05	5.65	178	- 2.44	- 4.52	- 142
1000 - 900	0.42	0.47	1.13	36	2.67	6.35	200	- 2.20	- 5.22	- 164
900 - 800	0.48	0.54	1.13	36	3.36	7.00	221	- 2.82	- 5.87	- 185
800 - 700	0.40	0.39	0.97	31	3.12	7.80	246	- 2.73	- 6.83	- 215
700 - 600	0.44	0.33	0.75	24	3.74	8.50	268	- 3.41	- 7.75	- 244
600 - 500	0.28	0.21	0.75	24	2.58	9.20	284	- 2.37	- 8.45	- 260
500 - 400	0.20	0.11	0.53	17	2.00	10.00	316	- 1.89	- 9.47	- 299
400 - 300	0.19	0.12	0.45	14	2.03	10.70	322	- 1.91	- 10.25	- 308
1950 - 300	47.03	128.07	2.72	86	117.61	2.50	79	+ 10.46	+ 0.22	+ 6

1968/69

1950 - 1900	0.16	0.38	2.38	75	0.34	2.12	67	+ 0.04	+ 0.26	+ 8
1900 - 1800	3.76	8.44	2.24	71	8.64	2.30	73	- 0.20	- 0.06	- 2
1800 - 1700	9.76	22.21	2.28	72	24.72	2.53	80	- 2.51	- 0.25	- 8
1700 - 1600	12.76	28.13	2.20	69	34.77	2.72	86	- 6.64	- 0.52	- 17
1600 - 1500	9.28	16.84	1.81	57	28.38	3.06	97	- 11.54	- 1.25	- 40
1500 - 1400	5.26	9.05	1.72	54	18.93	3.60	114	- 9.88	- 1.88	- 60
1400 - 1300	2.06	3.31	1.61	51	8.52	4.14	131	- 5.21	- 2.53	- 80
1300 - 1200	0.72	1.03	1.43	45	3.30	4.58	144	- 2.27	- 3.15	- 99
1200 - 1100	0.32	0.40	1.25	39	1.76	5.50	174	- 1.36	- 4.25	- 135
1100 - 1000	0.54	0.61	1.13	36	3.20	5.93	187	- 2.59	- 4.80	- 151
1000 - 900	0.42	0.37	0.88	28	2.73	6.50	205	- 2.36	- 5.62	- 177
900 - 800	0.48	0.39	0.81	26	3.60	7.50	237	- 3.21	- 6.69	- 211
800 - 700	0.40	0.23	0.58	18	3.30	8.25	260	- 3.07	- 7.67	- 242
700 - 600	0.44	0.17	0.38	12	3.96	9.00	284	- 3.79	- 8.62	- 272
600 - 500	0.28	0.11	0.38	12	2.82	10.07	318	- 2.71	- 9.69	- 306
500 - 400	0.20	0.04	0.20	6	2.28	11.40	360	- 2.24	- 11.20	- 354
400 - 300	0.19	0.02	0.13	4	2.20	12.94	408	- 2.18	- 12.85	- 404
1950 - 300	47.03	91.73	1.95	62	153.45	3.26	103	- 61.72	- 1.31	- 41

1969/70

1950 - 1900	0.16	0.50	3.13	99	0.14	0.88	28	+ 0.36	+ 2.25	+ 71
1900 - 1800	3.76	8.90	2.37	75	4.13	1.10	35	+ 4.77	+ 1.27	+ 40
1800 - 1700	9.76	19.26	1.97	62	13.21	1.36	43	+ 6.05	+ 0.61	+ 19
1700 - 1600	12.76	23.54	1.84	58	23.06	1.81	57	+ 0.48	+ 0.03	+ 1
1600 - 1500	9.28	14.71	1.58	50	20.91	2.25	71	- 6.20	- 0.67	- 21
1500 - 1400	5.26	7.68	1.46	46	14.34	2.73	86	- 6.66	- 1.27	- 40
1400 - 1300	2.06	2.85	1.38	44	6.53	3.17	100	- 3.68	- 1.79	- 57
1300 - 1200	0.72	0.89	1.24	41	2.80	3.89	123	- 1.91	- 2.65	- 84
1200 - 1100	0.32	0.36	1.13	36	1.44	4.50	142	- 1.08	- 3.37	- 106
1100 - 1000	0.54	0.61	1.13	36	2.75	5.09	161	- 2.14	- 3.96	- 125
1000 - 900	0.42	0.47	1.13	36	2.31	5.50	174	- 1.84	- 4.37	- 138
900 - 800	0.48	0.44	0.91	29	3.12	6.50	205	- 2.68	- 5.59	- 176
800 - 700	0.40	0.35	0.88	28	2.90	7.25	229	- 2.55	- 6.37	- 201
700 - 600	0.44	0.37	0.84	27	3.40	7.73	244	- 3.03	- 6.89	- 217
600 - 500	0.28	0.18	0.64	20	2.41	8.60	271	- 2.23	- 7.96	- 251
500 - 400	0.20	0.13	0.65	20	1.96	9.80	309	- 1.83	- 9.15	- 289
400 - 300	0.19	0.12	0.63	20	2.04	10.73	339	- 1.92	- 10.10	- 319
1950 - 300	47.03	81.36	1.73	55	107.45	2.29	72	- 26.09	- 0.56	- 17

Table 9.2.7. Norway - Mass balance study results.

Part 5 of 22 Mass balance versus altitude Storsteinfjellbreen 1964/65

Altitude interval	Area	Winter balance			Summer balance			Net balance		
		Total	Specific		Total	Specific		Total	Specific	
m a sl	km^2	$10^6\ m^3$	m	l/s km^2	$10^6\ m^3$	m	l/s km^2	$10^6\ m^3$	m	l/s km^2
1850 - 1800	0.022	0.039	1.77	56	0.003	0.13	4	+ 0.036	+ 1.64	+ 52
1800 - 1750	0.041	0.072	1.76	56	0.010	0.24	8	+ 0.062	+ 1.52	+ 48
1750 - 1700	0.134	0.245	1.83	58	0.053	0.40	13	+ 0.192	+ 1.43	+ 45
1700 - 1650	0.173	0.355	2.05	65	0.098	0.57	18	+ 0.257	+ 1.48	+ 47
1650 - 1600	0.174	0.438	2.52	80	0.108	0.63	20	+ 0.330	+ 1.89	+ 60
1600 - 1550	0.207	0.542	2.62	83	0.173	0.84	27	+ 0.369	+ 1.78	+ 56
1550 - 1500	0.357	0.726	2.03	64	0.363	1.02	32	+ 0.363	+ 1.01	+ 32
1500 - 1450	0.484	1.016	2.10	67	0.545	1.13	36	+ 0.471	+ 0.97	+ 31
1450 - 1400	0.898	1.539	1.72	55	1.010	1.13	36	+ 0.529	+ 0.59	+ 19
1400 - 1350	1.032	1.772	1.72	55	1.161	1.13	36	+ 0.611	+ 0.59	+ 19
1350 - 1300	0.675	1.160	1.72	55	0.783	1.16	37	+ 0.377	+ 0.56	+ 18
1300 - 1250	0.743	1.055	1.42	45	1.025	1.38	44	+ 0.030	+ 0.04	+ 1
1250 - 1200	0.379	0.499	1.32	42	0.609	1.61	51	- 0.110	- 0.29	- 9
1200 - 1150	0.300	0.372	1.24	39	0.525	1.75	55	- 0.153	- 0.51	- 16
1150 - 1100	0.215	0.239	1.11	35	0.437	2.03	64	- 0.198	- 0.92	- 29
1100 - 1050	0.120	0.124	1.03	33	0.270	2.25	71	- 0.146	- 1.22	- 39
1050 - 1000	0.080	0.080	1.00	32	0.220	2.75	87	- 0.140	- 1.75	- 55
1000 - 950	0.061	0.068	1.11	35	0.198	3.25	103	- 0.130	- 2.14	- 68
950 - 920	0.025	0.031	1.25	40	0.081	3.25	103	- 0.041	- 2.00	- 63
1850 - 920	6.120	10.372	1.69	54	7.672	1.25	40	+ 2.700	+ 0.44	+ 14

1965/66

Altitude interval	Area	Winter balance			Summer balance			Net balance		
		Total	Specific		Total	Specific		Total	Specific	
m a sl	km^2	$10^6\ m^3$	m	l/s km^2	$10^6\ m^3$	m	l/s km^2	$10^6\ m^3$	m	l/s km^2
1850 - 1800	0.022	0.037	1.68	53	0.027	1.23	39	+ 0.010	+ 0.45	+ 14
1800 - 1750	0.041	0.069	1.68	53	0.051	1.24	39	+ 0.018	+ 0.44	+ 14
1750 - 1700	0.134	0.214	1.60	51	0.167	1.24	39	+ 0.047	+ 0.36	+ 11
1700 - 1650	0.173	0.272	1.57	50	0.216	1.24	39	+ 0.056	+ 0.33	+ 10
1650 - 1600	0.174	0.265	1.52	48	0.217	1.24	39	+ 0.048	+ 0.28	+ 9
1600 - 1550	0.208	0.280	1.40	44	0.258	1.24	39	+ 0.022	+ 0.16	+ 5
1550 - 1500	0.357	0.474	1.33	42	0.446	1.25	40	+ 0.028	+ 0.08	+ 2
1500 - 1450	0.484	0.601	1.24	39	0.630	1.30	41	- 0.029	- 0.06	- 2
1450 - 1400	0.898	1.047	1.17	37	1.272	1.42	45	- 0.225	- 0.25	- 8
1400 - 1350	1.032	1.111	1.08	34	1.712	1.61	51	- 0.601	- 0.53	- 17
1350 - 1300	0.675	0.671	0.99	31	1.218	1.80	57	- 0.547	- 0.81	- 26
1300 - 1250	0.743	0.644	0.87	28	1.583	2.13	68	- 0.939	- 1.26	- 40
1250 - 1200	0.379	0.298	0.79	25	0.990	2.61	83	- 0.692	- 1.82	- 58
1200 - 1150	0.300	0.223	0.74	24	0.926	3.09	98	- 0.703	- 2.35	- 74
1150 - 1100	0.215	0.135	0.63	20	0.698	3.25	103	- 0.563	- 2.62	- 83
1100 - 1050	0.120	0.071	0.59	19	0.431	3.59	114	- 0.360	- 3.00	- 95
1050 - 1000	0.080	0.037	0.46	15	0.316	3.96	126	- 0.279	- 3.50	- 111
1000 - 950	0.061	0.034	0.56	18	0.259	4.25	135	- 0.224	- 3.69	- 117
950 - 920	0.025	0.017	0.68	22	0.106	4.25	135	- 0.089	- 3.57	- 113
1850 - 920	6.120	6.447	1.05	34	11.527	1.88	60	- 5.080	- 0.83	- 26

Table 9.2.7. Norway - Mass balance study results.

Part 6 of 22 Mass balance versus altitude Storsteinfjellbreen 1966/67

Altitude interval	Area	Winter balance			Summer balance			Net balance		
		Total	Specific		Total	Specific		Total	Specific	
m a sl	km^2	$10^6\ m^3$	m	$l/s\ km^2$	$10^6\ m^3$	m	$l/s\ km^2$	$10^6\ m^3$	m	$l/s\ km^2$
1850 - 1800	0.022	0.041	1.87	59	0.019	0.87	27	+ 0.022	+ 1.00	+ 32
1800 - 1750	0.041	0.077	1.87	59	0.036	0.87	27	+ 0.041	+ 1.00	+ 32
1750 - 1700	0.134	0.244	1.82	57	0.118	0.87	27	+ 0.126	+ 0.95	+ 30
1700 - 1650	0.173	0.307	1.78	56	0.151	0.87	27	+ 0.156	+ 0.91	+ 29
1650 - 1600	0.174	0.291	1.67	53	0.196	1.12	35	+ 0.095	+ 0.55	+ 18
1600 - 1550	0.208	0.361	1.74	55	0.234	1.12	35	+ 0.127	+ 0.62	+ 20
1550 - 1500	0.357	0.536	1.50	47	0.490	1.37	43	+ 0.046	+ 0.13	+ 4
1500 - 1450	0.484	0.733	1.51	48	0.664	1.37	43	+ 0.069	+ 0.14	+ 5
1450 - 1400	0.898	1.279	1.42	45	1.459	1.62	51	- 0.180	- 0.20	- 6
1400 - 1350	1.032	1.461	1.41	44	1.677	1.62	51	- 0.216	- 0.21	- 7
1350 - 1300	0.674	0.961	1.43	45	1.140	1.69	53	- 0.179	- 0.26	- 8
1300 - 1250	0.743	0.854	1.15	36	1.551	2.09	66	- 0.697	- 0.94	- 30
1250 - 1200	0.379	0.411	1.08	34	0.852	2.25	71	- 0.441	- 1.17	- 37
1200 - 1150	0.300	0.303	1.01	32	0.700	2.33	74	- 0.400	- 1.32	- 42
1150 - 1100	0.215	0.219	1.02	32	0.592	2.75	87	- 0.373	- 1.73	- 54
1100 - 1050	0.120	0.134	1.12	35	0.370	3.08	97	- 0.236	- 1.96	- 62
1050 - 1000	0.080	0.083	1.04	33	0.278	3.48	110	- 0.195	- 2.44	- 77
1000 - 950	0.061	0.057	0.93	29	0.229	3.75	118	- 0.172	- 2.82	- 89
950 - 920	0.025	0.022	0.88	28	0.093	3.72	117	- 0.071	- 2.84	- 89
1850 - 920	6.120	8.374	1.37	43	10.849	1.77	56	- 2.475	- 0.40	- 13

1967/68

Altitude interval	Area	Winter balance			Summer balance			Net balance		
		Total	Specific		Total	Specific		Total	Specific	
m a sl	km^2	$10^6\ m^3$	m	$l/s\ km^2$	$10^6\ m^3$	m	$l/s\ km^2$	$10^6\ m^3$	m	$l/s\ km^2$
1850 - 1800	0.022	0.041	1.86	58	0.009	0.41	13	+ 0.032	+ 1.45	+ 45
1800 - 1750	0.041	0.077	1.88	59	0.018	0.44	14	+ 0.059	+ 1.44	+ 45
1750 - 1700	0.134	0.251	1.87	59	0.067	0.50	16	+ 0.184	+ 1.37	+ 43
1700 - 1650	0.173	0.321	1.86	58	0.090	0.52	16	+ 0.231	+ 1.34	+ 42
1650 - 1600	0.174	0.316	1.82	57	0.099	0.57	18	+ 0.217	+ 1.25	+ 39
1600 - 1550	0.208	0.362	1.74	55	0.129	0.62	19	+ 0.233	+ 1.12	+ 36
1550 - 1500	0.357	0.554	1.55	49	0.239	0.67	21	+ 0.315	+ 0.88	+ 28
1500 - 1450	0.484	0.750	1.55	49	0.353	0.73	23	+ 0.397	+ 0.82	+ 26
1450 - 1400	0.898	1.262	1.41	44	0.727	0.81	26	+ 0.535	+ 0.60	+ 18
1400 - 1350	1.032	1.508	1.46	46	0.949	0.92	29	+ 0.559	+ 0.54	+ 17
1350 - 1300	0.674	0.972	1.44	45	0.721	1.07	34	+ 0.251	+ 0.37	+ 11
1300 - 1250	0.743	0.968	1.30	41	0.921	1.24	39	+ 0.047	+ 0.06	+ 2
1250 - 1200	0.379	0.454	1.20	38	0.531	1.40	44	- 0.077	- 0.20	- 6
1200 - 1150	0.300	0.386	1.29	41	0.432	1.44	45	- 0.046	- 0.15	- 4
1150 - 1100	0.215	0.264	1.23	39	0.316	1.47	46	- 0.052	- 0.24	- 7
1100 - 1050	0.120	0.165	1.38	43	0.181	1.51	48	- 0.016	- 0.13	- 5
1050 - 1000	0.080	0.096	1.20	38	0.126	1.58	50	- 0.030	- 0.38	- 12
1000 - 950	0.061	0.062	1.02	32	0.100	1.64	51	- 0.038	- 0.62	- 19
950 - 920	0.025	0.026	1.04	33	0.043	1.72	54	- 0.017	- 0.68	- 21
1850 - 920	6.120	8.835	1.44	45	6.051	0.99	31	+ 2.784	+ 0.45	+ 14

Table 9.2.7. Norway - Mass balance study results.

Part 7 of 22 Mass balance versus altitude Blaisen

1964/65

Altitude interval	Area	Winter balance			Summer balance			Net balance		
		Total	Specific		Total	Specific		Total	Specific	
m a sl	km^2	10^6 m^3	m	1/s km^2	10^6 m^3	m	1/s km^2	10^6 m^3	m	1/s km^2
1200 - 1150	0.04	0.11	2.75	87	0.06	1.38	44	+ 0.05	+ 1.37	+ 43
1150 - 1100	0.24	0.63	2.63	83	0.33	1.38	44	+ 0.30	+ 1.25	+ 40
1100 - 1050	0.66	1.46	2.21	70	0.91	1.38	44	+ 0.55	+ 0.83	+ 26
1050 - 1000	0.65	1.18	1.82	58	0.91	1.40	44	+ 0.27	+ 0.42	+ 13
1000 - 950	0.42	0.67	1.60	51	0.66	1.57	50	+ 0.01	+ 0.03	+ 1
950 - 900	0.15	0.26	1.73	55	0.27	1.80	57	- 0.01	- 0.07	- 2
900 - 850	0.02	0.04	2.00	63	0.04	2.13	67	0.00	- 0.13	- 4
1200 - 850	2.18	4.35	2.00	63	3.18	1.46	45	+ 1.17	+ 0.54	+ 17

1965/66

Altitude interval	Area	Winter balance			Summer balance			Net balance		
		Total	Specific		Total	Specific		Total	Specific	
m a sl	km^2	10^6 m^3	m	1/s km^2	10^6 m^3	m	1/s km^2	10^6 m^3	m	1/s km^2
1200 - 1150	0.040	0.061	1.53	48	0.085	2.12	67	- 0.024	- 0.60	- 19
1150 - 1100	0.237	0.340	1.43	45	0.504	2.12	67	- 0.164	- 0.69	- 22
1100 - 1050	0.661	0.842	1.27	40	1.443	2.18	69	- 0.601	- 0.91	- 29
1050 - 1000	0.653	0.707	1.08	34	1.584	2.43	77	- 0.877	- 1.35	- 43
1000 - 950	0.421	0.363	0.86	27	1.102	2.62	83	- 0.739	- 1.76	- 56
950 - 900	0.152	0.114	0.75	24	0.442	2.91	92	- 0.328	- 2.16	- 68
900 - 850	0.020	0.014	0.70	22	0.065	3.25	103	- 0.051	- 2.55	- 81
1200 - 850	2.184	2.441	1.12	36	5.225	2.39	76	- 2.784	- 1.27	- 40

1966/67

Altitude interval	Area	Winter balance			Summer balance			Net balance		
		Total	Specific		Total	Specific		Total	Specific	
m a sl	km^2	10^6 m^3	m	1/s km^2	10^6 m^3	m	1/s km^2	10^6 m^3	m	1/s km^2
1200 - 1150	0.040	0.075	1.88	60	0.075	1.88	60	0.000	0.00	- 0
1150 - 1100	0.237	0.406	1.71	54	0.475	2.00	63	- 0.069	- 0.29	- 9
1100 - 1050	0.661	0.992	1.50	48	1.459	2.21	70	- 0.467	- 0.71	- 22
1050 - 1000	0.653	0.857	1.31	42	1.597	2.45	78	- 0.740	- 1.14	- 36
1000 - 950	0.421	0.486	1.15	37	1.032	2.45	78	- 0.546	- 1.30	- 41
950 - 900	0.152	0.173	1.14	36	0.430	2.83	90	- 0.259	- 1.69	- 54
900 - 850	0.020	0.022	1.15	36	0.065	3.25	108	- 0.042	- 2.10	- 66
1200 - 850	2.184	3.011	1.38	44	5.133	2.35	75	- 2.121	- 0.97	- 31

1967/68

Altitude interval	Area	Winter balance			Summer balance			Net balance		
		Total	Specific		Total	Specific		Total	Specific	
m a sl	km^2	10^6 m^3	m	1/s km^2	10^6 m^3	m	1/s km^2	10^6 m^3	m	1/s km^2
1200 - 1150	0.040	0.075	1.88	59	0.035	0.88	28	+ 0.040	+ 1.00	+ 32
1150 - 1100	0.237	0.441	1.86	59	0.211	0.89	28	+ 0.230	+ 0.97	+ 31
1100 - 1050	0.661	1.153	1.74	55	0.809	1.22	39	+ 0.344	+ 0.52	+ 16
1050 - 1000	0.653	1.034	1.58	50	0.932	1.43	45	+ 0.102	+ 0.16	+ 5
1000 - 950	0.421	0.592	1.41	45	0.678	1.61	51	- 0.086	- 0.20	- 6
950 - 900	0.152	0.214	1.41	45	0.275	1.81	57	- 0.061	- 0.40	- 13
900 - 850	0.020	0.033	1.65	52	0.038	1.90	60	- 0.005	- 0.25	- 8
1200 - 850	2.184	3.542	1.62	51	2.978	1.36	43	+ 0.564	+ 0.26	+ 8

Table 9.2.7. Norway - Mass balance study results.

Part 8 of 22 Mass balance versus altitude Hardangerjøkulen 1964/65

Altitude interval	Area	Winter balance			Summer balance			Net balance		
		Total	Specific		Total	Specific		Total	Specific	
m a sl	km^2	$10^6\ m^3$	m	l/s km^2	$10^6\ m^3$	m	l/s km^2	$10^6\ m^3$	m	l/s km^2
1964/65										
1900 - 1850	0.064		1.96	62		0.70	22		+ 1.26	+ ,40
1850 - 1800	3.638		2.20	70		0.86	27		+ 1.34	+ 43
1800 - 1750	3.862		2.35	75		1.03	33		+ 1.32	+ 42
1750 - 1700	3.968		2.28	72		1.25	40		+ 1.03	+ 33
1700 - 1650	2.102		2.01	64		1.48	47		+ 0.53	+ 17
1650 - 1600	0.952		1.81	57		1.75	56		+ 0.06	+ 2
1600 - 1550	0.660		1.62	51		2.07	66		- 0.45	- 14
1550 - 1500	0.557		1.50	48		2.41	76		- 0.91	- 29
1500 - 1450	0.321		1.24	39		2.70	86		- 1.46	- 46
1450 - 1400	0.191		1.06	34		3.04	96		- 1.98	- 63
1400 - 1350	0.114		1.07	34		3.40	108		- 2.33	- 74
1350 - 1300	0.080		1.10	35		3.75	119		- 2.65	- 84
1300 - 1250	0.272		1.17	37		4.10	130		- 2.93	- 93
1250 - 1200	0.325		1.08	34		4.50	143		- 3.47	- 110
1200 - 1150	0.324		1.00	32		4.87	154		- 3.80	- 121
1150 - 1100	0.112		1.01	32		5.32	169		- 4.31	- 137
1100 - 1050	0.056		0.98	31		5.80	184		- 4.82	- 153
1050 - 1000	0.048		0.82	26		6.25	198		- 5.43	- 172
1900 - 1000	17.65		2.05	65		1.54	49		+ 0.51	+ 16
1965/66										
1900 - 1850	0.065	0.094	1.45	46	0.098	1.50	48	- 0.003	- 0.05	- 2
1850 - 1800	3.638	6.330	1.74	55	5.457	1.50	48	+ 0.873	+ 0.24	+ 8
1800 - 1750	3.862	6.952	1.80	57	6.527	1.69	54	+ 0.386	+ 0.10	+ 3
1750 - 1700	3.968	6.865	1.73	55	7.539	1.90	60	- 0.675	- 0.17	- 5
1700 - 1650	2.102	3.447	1.64	52	4.603	2.19	69	- 1.156	- 0.55	- 17
1650 - 1600	0.952	1.447	1.52	48	2.323	2.44	77	- 0.876	- 0.92	- 29
1600 - 1550	0.660	0.911	1.38	44	1.782	2.70	86	- 0.871	- 1.32	- 42
1550 - 1500	0.557	0.696	1.25	40	1.682	3.02	96	- 0.986	- 1.77	- 56
1500 - 1450	0.321	0.337	1.05	33	1.075	3.35	106	- 0.738	- 2.30	- 73
1450 - 1400	0.191	0.166	0.87	28	0.707	3.70	117	- 0.540	- 2.83	- 90
1400 - 1350	0.114	0.093	0.82	26	0.475	4.17	133	- 0.382	- 3.35	- 106
1350 - 1300	0.080	0.065	0.81	26	0.382	4.78	152	- 0.317	- 3.97	- 126
1300 - 1250	0.272	0.231	0.85	27	1.447	5.32	169	- 1.216	- 4.47	- 142
1250 - 1200	0.325	0.253	0.78	25	1.895	5.83	185	- 1.641	- 5.05	- 160
1200 - 1150	0.324	0.253	0.78	25	2.106	6.50	206	- 1.853	- 5.72	- 181
1150 - 1100	0.112	0.095	0.85	27	0.801	7.15	227	- 0.706	- 6.30	- 200
1100 - 1050	0.045	0.041	0.92	29	0.353	7.87	260	- 0.322	- 6.95	- 220
1050 - 1000	0.028	0.024	0.87	28	0.202	8.37	265	- 0.180	- 7.50	- 238
1900 - 1000	17.62	28.30	1.60	51	39.45	2.24	71	- 11.20	- 0.64	- 20
1966/67										
1900 - 1850	0.075	0.17	2.25	71	0.01	0.61	17	+ 0.13	+ 1.71	+ 54
1850 - 1800	3.638	9.28	2.55	81	2.18	0.60	19	+ 7.10	+ 1.95	+ 62
1800 - 1750	3.862	10.74	2.78	88	2.90	0.75	24	+ 7.84	+ 2.03	+ 64
1750 - 1700	3.940	10.48	2.66	84	3.59	0.91	29	+ 6.89	+ 1.75	+ 56
1700 - 1650	2.102	5.26	2.50	79	2.52	1.20	38	+ 2.74	+ 1.30	+ 41
1650 - 1600	0.952	2.19	2.30	73	1.39	1.46	46	+ 0.80	+ 0.84	+ 27
1600 - 1550	0.660	1.39	2.10	67	1.17	1.78	56	+ 0.22	+ 0.32	+ 10
1550 - 1500	0.557	1.07	1.92	61	1.18	2.12	67	- 0.11	- 0.20	- 6
1500 - 1450	0.321	0.56	1.75	56	0.77	2.45	77	- 0.21	- 0.65	- 20
1450 - 1400	0.191	0.30	1.57	50	0.54	2.84	90	- 0.24	- 1.27	- 40
1400 - 1350	0.110	0.16	1.42	45	0.35	3.24	103	- 0.19	- 1.72	- 54
1350 - 1300	0.082	0.10	1.27	40	0.30	3.65	116	- 0.20	- 2.38	- 73
1300 - 1250	0.270	0.31	1.15	36	1.11	4.12	130	- 0.80	- 2.97	- 94
1250 - 1200	0.325	0.34	1.05	33	1.49	4.60	145	- 1.15	- 3.55	- 113
1200 - 1150	0.324	0.32	1.00	32	1.65	5.10	161	- 1.33	- 4.10	- 130
1150 - 1100	0.110	0.10	0.95	30	0.62	5.62	178	- 0.52	- 4.67	- 148
1100 - 1050	0.030	0.03	1.00	32	0.18	6.15	194	- 0.15	- 5.15	- 163
1900 - 1050	17.55	42.80	2.44	77	21.98	1.25	40	+ 20.82	+ 1.19	+ 38

Table 9.2.7. Norway - Mass balance study results.

Part 9 of 22 Mass balance versus altitude Hardangerjøkulen 1967/68

Altitude interval	Area	Winter balance			Summer balance			Net balance		
		Total	Specific		Total	Specific		Total	Specific	
m a sl	km^2	10^6 m^3	m	l/s km^2	10^6 m^3	m	l/s km^2	10^6 m^3	m	l/s km^2
1900 - 1850	0.080	0.208	2.60	82	0.132	1.65	52	+ 0.076	+ 0.95	+ 30
1850 - 1800	3.742	11.525	3.08	98	6.286	1.68	53	+ 5.239	+ 1.40	+ 44
1800 - 1750	3.870	12.074	3.12	99	6.850	1.77	56	+ 5.224	+ 1.35	+, 43
1750 - 1700	3.910	12.082	3.09	98	7.312	1.87	59	+ 4.770	+ 1.22	+ 39
1700 - 1650	2.082	5.372	2.58	82	4.185	2.01	64	+ 1.187	+ 0.57	+ 18
1650 - 1600	0.931	1.694	1.82	58	2.002	2.15	68	- 0.308	- 0.33	- 10
1600 - 1550	0.640	1.299	2.03	64	1.536	2.40	76	- 0.237	- 0.37	- 12
1550 - 1500	0.545	0.932	1.71	54	1.395	2.56	81	- 0.464	- 0.85	- 27
1500 - 1450	0.321	0.449	1.40	44	1.027	3.20	101	- 0.578	- 1.80	- 57
1450 - 1400	0.191	0.177	0.93	25	0.678	3.55	112	- 0.501	- 2.62	- 83
1400 - 1350	0.110	0.114	1.04	33	0.438	3.98	126	- 0.324	- 2.94	- 93
1350 - 1300	0.082	0.080	0.98	31	0.353	4.30	136	- 0.273	- 3.32	- 105
1300 - 1250	0.270	0.310	1.15	36	1.269	4.70	148	- 0.959	- 3.55	- 112
1250 - 1200	0.325	0.302	0.93	29	1.664	5.12	163	- 1.362	- 4.19	- 133
1200 - 1150	0.324	0.275	0.85	27	1.798	5.55	176	- 1.523	- 4.70	- 149
1150 - 1100	0.105	0.090	0.90	29	0.651	6.20	197	- 0.561	- 5.30	- 169
1100 - 1050	0.025	0.024	0.95	30	0.167	6.68	215	- 0.141	- 5.73	- 172
1900 - 1050	17.55	47.01	2.68	85	37.74	2.15	68	+ 9.27	+ 0.53	+ 17

1968/69

1900 - 1850	0.070	0.075	1.07	34	0.154	2.21	70	- 0.079	- 1.14	- 36
1850 - 1800	3.375	4.781	1.28	41	8.441	2.26	72	- 3.660	- 0.98	- 31
1800 - 1750	3.866	5.258	1.36	43	9.549	2.47	78	- 4.291	- 1.11	- 35
1750 - 1700	3.910	4.770	1.22	39	10.401	2.66	84	- 5.631	- 1.44	- 46
1700 - 1650	2.084	2.000	0.96	30	6.002	2.88	91	- 4.002	- 1.92	- 61
1650 - 1600	0.936	0.571	0.61	19	2.930	3.13	99	- 2.359	- 2.52	- 80
1600 - 1550	0.640	0.333	0.52	16	2.189	3.42	108	- 1.856	- 2.90	- 92
1550 - 1500	0.542	0.287	0.53	17	2.060	3.80	120	- 1.772	- 3.27	- 103
1500 - 1450	0.319	0.140	0.44	14	1.360	4.26	134	- 1.218	- 3.82	- 121
1450 - 1400	0.196	0.073	0.37	12	0.921	4.70	148	- 0.849	- 4.33	- 137
1400 - 1350	0.112	0.034	0.31	10	0.577	5.15	163	- 0.542	- 4.84	- 153
1350 - 1300	0.084	0.032	0.38	12	0.473	5.63	177	- 0.441	- 5.25	- 166
1300 - 1250	0.270	0.108	0.40	13	1.647	6.10	192	- 1.539	- 5.70	- 181
1250 - 1200	0.315	0.107	0.34	11	2.076	6.59	208	- 1.969	- 6.25	- 198
1200 - 1150	0.321	0.093	0.29	9	2.285	7.12	226	- 2.192	- 6.83	- 216
1150 - 1100	0.115	0.038	0.33	10	0.879	7.65	242	- 0.842	- 7.32	- 233
1100 - 1050	0.022	0.007	0.34	11	0.180	8.20	260	- 0.173	- 7.86	- 251
1900 - 1050	17.53	18.71	1.07	34	52.12	2.97	94	- 33.41	- 1.90	- 60

1969/70

1900 - 1850	0.074	0.101	1.36	43	0.104	1.40	44	- 0.003	- 0.06	- 2
1850 - 1800	3.358	4.965	1.48	47	4.735	1.41	45	+ 0.230	+ 0.07	+ 2
1800 - 1750	3.763	5.454	1.45	46	5.454	1.45	46	0.000	0.00	0
1750 - 1700	4.033	5.536	1.37	43	6.251	1.55	49	- 0.715	- 0.18	- 6
1700 - 1650	2.219	2.680	1.21	38	3.861	1.74	55	- 1.180	- 0.53	- 17
1650 - 1600	0.971	1.108	1.14	36	1.893	1.95	62	- 0.785	- 0.81	- 25
1600 - 1550	0.624	0.757	1.22	39	1.385	2.22	70	- 0.628	- 1.00	- 32
1550 - 1500	0.569	0.662	1.16	37	1.451	2.55	80	- 0.789	- 1.39	- 44
1500 - 1450	0.371	0.389	1.05	33	1.076	2.90	92	- 0.687	- 1.85	- 57
1450 - 1400	0.176	0.167	0.95	30	0.577	3.28	104	- 0.410	- 2.33	- 73
1400 - 1350	0.109	0.093	0.85	27	0.403	3.70	117	- 0.310	- 2.85	- 90
1350 - 1300	0.078	0.059	0.75	24	0.321	4.12	130	- 0.262	- 3.37	- 106
1300 - 1250	0.265	0.186	0.70	22	1.208	4.56	144	- 1.022	- 3.86	- 122
1250 - 1200	0.308	0.200	0.65	20	1.540	5.00	158	- 1.340	- 4.35	- 137
1200 - 1150	0.312	0.199	0.64	20	1.716	5.50	174	- 1.517	- 4.86	- 153
1150 - 1100	0.109	0.065	0.60	20	0.651	5.97	188	- 0.586	- 5.37	- 170
1100 - 1050	0.059	0.035	0.60	20	0.381	6.45	203	- 0.346	- 5.85	- 185
1900 - 1050	17.44	22.65	1.29	41	33.00	1.89	60	- 10.35	- 0.60	- 19

Table 0.2.7. Norway - Mass balance study results.

Part 10 of 22 Mass balance versus altitude Gråsubreen 1964/65

Altitude interval m a sl	Area km^2	Winter balance Total 10^6 m^3	Winter balance Specific m	Winter balance Specific l/s km^2	Summer balance Total 10^6 m^3	Summer balance Specific m	Summer balance Specific l/s km^2	Net balance Total 10^6 m^3	Net balance Specific m	Net balance Specific l/s km^2
2260 - 2250	0.007	0.008	0.87	28	0.001	0.12	4	+ 0.007	+ 0.75	+ 24
2250 - 2200	0.129	0.127	0.98	31	0.016	0.12	4	+ 0.111	+ 0.86	+ 27
2200 - 2150	0.274	0.256	0.93	30	0.034	0.12	4	+ 0.222	+ 0.81	+ 26
2150 - 2100	0.362	0.231	0.64	20	0.052	0.14	4	+ 0.179	+ 0.49	+ 16
2100 - 2050	0.385	0.259	0.67	21	0.130	0.34	11	+ 0.129	+ 0.34	+ 11
2050 - 2000	0.456	0.319	0.68	22	0.171	0.38	12	+ 0.148	+ 0.30	+ 9
2000 - 1950	0.478	0.383	0.80	25	0.221	0.46	15	+ 0.162	+ 0.34	+ 11
1950 - 1900	0.209	0.183	0.87	28	0.147	0.70	22	+ 0.036	+ 0.17	+ 5
1900 - 1850	0.086	0.075	0.87	28	0.083	0.97	31	- 0.008	- 0.10	- 3
2260 - 1850	2.386	1.841	0.77	24	0.855	0.36	11	+ 0.986	+ 0.41	+ 13

1965/66

Altitude interval m a sl	Area km^2	Winter balance Total 10^6 m^3	Winter balance Specific m	Winter balance Specific l/s km^2	Summer balance Total 10^6 m^3	Summer balance Specific m	Summer balance Specific l/s km^2	Net balance Total 10^6 m^3	Net balance Specific m	Net balance Specific l/s km^2
2260 - 2250	0.007	0.006	0.87	28	0.002	0.38	12	+ 0.004	+ 0.49	+ 16
2250 - 2200	0.129	0.094	0.73	23	0.048	0.38	12	+ 0.046	+ 0.35	+ 11
2200 - 2150	0.274	0.200	0.73	23	0.184	0.68	22	+ 0.016	+ 0.05	+ 2
2150 - 2100	0.362	0.251	0.69	22	0.300	0.83	26	- 0.049	- 0.14	- 4
2100 - 2050	0.385	0.249	0.65	21	0.349	0.91	29	- 0.100	- 0.26	- 8
2050 - 2000	0.456	0.299	0.66	21	0.464	1.02	32	- 0.165	- 0.36	- 11
2000 - 1950	0.478	0.373	0.78	25	0.596	1.25	40	- 0.223	- 0.47	- 15
1950 - 1900	0.209	0.177	0.85	27	0.303	1.45	46	- 0.126	- 0.60	- 19
1900 - 1850	0.086	0.075	0.87	28	0.152	1.77	56	- 0.077	- 0.90	- 28
2260 - 1850	2.386	1.724	0.72	23	2.398	1.01	32	- 0.674	- 0.29	- 9

1966/67

Altitude interval m a sl	Area km^2	Winter balance Total 10^6 m^3	Winter balance Specific m	Winter balance Specific l/s km^2	Summer balance Total 10^6 m^3	Summer balance Specific m	Summer balance Specific l/s km^2	Net balance Total 10^6 m^3	Net balance Specific m	Net balance Specific l/s km^2
2260 - 2250	0.007	0.013	1.86	59	0.004	0.62	20	+ 0.009	+ 1.24	+ 39
2250 - 2200	0.129	0.200	1.55	47	0.081	0.62	20	+ 0.119	+ 0.93	+ 27
2200 - 2150	0.274	0.408	1.49	47	0.171	0.62	20	+ 0.237	+ 0.87	+ 27
2150 - 2100	0.362	0.439	1.22	38	0.226	0.62	20	+ 0.213	+ 0.60	+ 18
2100 - 2050	0.385	0.563	1.46	46	0.241	0.62	20	+ 0.322	+ 0.84	+ 26
2050 - 2000	0.456	0.613	1.34	42	0.352	0.77	24	+ 0.261	+ 0.57	+ 18
2000 - 1950	0.478	0.728	1.52	48	0.419	0.87	27	+ 0.309	+ 0.65	+ 21
1950 - 1900	0.209	0.350	1.67	53	0.193	0.92	29	+ 0.157	+ 0.75	+ 24
1900 - 1850	0.086	0.157	1.83	58	0.097	1.12	35	+ 0.060	+ 0.71	+ 23
2260 - 1850	2.386	3.471	1.45	46	1.784	0.74	23	+ 1.687	+ 0.71	+ 23

Table 9.2.7. Norway - Mass balance study results.

Part 11 of 22 Mass balance versus altitude Gråsubreen

1967/68

Altitude interval	Area	Winter balance			Summer balance			Net balance		
		Total	Specific		Total	Specific		Total	Specific	
m a sl	km^2	10^6 m^3	m	l/s km^2	10^6 m^3	m	l/s km^2	10^6 m^3	m	l/s km^2
2260 - 2250	0.031	0.043	1.36	43	0.021	0.67	21	+ 0.021	+ 0.68	+ 22
2250 - 2200	0.178	0.210	1.18	37	0.120	0.67	21	+ 0.090	+ 0.50	+ 16
2200 - 2150	0.309	0.331	1.07	34	0.270	0.88	28	+ 0.060	+ 0.20	+ 6
2150 - 2100	0.386	0.366	0.95	30	0.381	0.99	31	- 0.015	- 0.04	- 1
2100 - 2050	0.417	0.388	0.93	29	0.442	1.06	34	- 0.055	- 0.13	- 4
2050 - 2000	0.461	0.466	1.01	32	0.546	1.18	38	- 0.081	- 0.17	- 6
2000 - 1950	0.459	0.490	1.07	34	0.601	1.31	42	- 0.111	- 0.24	- 8
1950 - 1900	0.218	0.229	1.05	33	0.313	1.43	46	- 0.084	- 0.39	- 12
1900 - 1870	0.068	0.072	1.06	34	0.103	1.52	48	- 0.032	- 0.47	- 15
2260 - 1870	2.528	2.593	1.03	33	2.799	1.11	35	- 0.206	- 0.08	- 3

1968/69

Altitude interval	Area	Winter balance			Summer balance			Net balance		
		Total	Specific		Total	Specific		Total	Specific	
m a sl	km^2	10^6 m^3	m	l/s km^2	10^6 m^3	m	l/s km^2	10^6 m^3	m	l/s km^2
2260 - 2250	0.031	0.026	0.84	26	0.039	1.25	40	- 0.013	- 0.41	- 14
2250 - 2200	0.178	0.134	0.75	24	0.222	1.25	40	- 0.088	- 0.49	- 16
2200 - 2150	0.309	0.221	0.72	23	0.449	1.45	46	- 0.228	- 0.74	- 23
2150 - 2100	0.386	0.216	0.56	17	0.760	1.97	62	- 0.544	- 1.41	- 45
2100 - 2050	0.417	0.268	0.64	20	0.876	2.10	66	- 0.608	- 1.46	- 46
2050 - 2000	0.461	0.318	0.69	22	0.979	2.12	67	- 0.661	- 1.43	- 45
2000 - 1950	0.459	0.344	0.75	24	1.051	2.29	72	- 0.707	- 1.54	- 48
1950 - 1900	0.218	0.171	0.78	25	0.602	2.76	87	- 0.431	- 1.98	- 62
1900 - 1870	0.068	0.059	0.87	27	0.213	3.13	99	- 0.154	- 2.26	- 72
2260 - 1870	2.528	1.852	0.74	23	5.191	2.04	65	- 3.459	- 1.37	- 43

1969/70

Altitude interval	Area	Winter balance			Summer balance			Net balance		
		Total	Specific		Total	Specific		Total	Specific	
m a sl	km^2	10^6 m^3	m	l/s km^2	10^6 m^3	m	l/s km^2	10^6 m^3	m	l/s km^2
2280 - 2250	0.031	0.024	0.77	24	0.019	0.63	20	+ 0.005	+ 0.14	+ 4
2250 - 2200	0.178	0.122	0.69	22	0.111	0.63	20	+ 0.011	+ 0.06	+ 2
2200 - 2150	0.309	0.186	0.60	19	0.224	0.73	23	- 0.038	- 0.13	- 4
2150 - 2100	0.386	0.153	0.40	12	0.382	0.99	31	- 0.229	- 0.59	- 19
2100 - 2050	0.417	0.189	0.45	14	0.490	1.18	37	- 0.301	- 0.73	- 23
2050 - 2000	0.461	0.234	0.51	16	0.669	1.45	46	- 0.435	- 0.94	- 30
2000 - 1950	0.459	0.309	0.67	21	0.745	1.62	51	- 0.436	- 0.95	- 30
1950 - 1900	0.218	0.169	0.78	25	0.354	1.63	51	- 0.185	- 0.85	- 27
1900 - 1870	0.068	0.059	0.87	28	0.124	1.83	58	- 0.065	- 0.96	- 30
2280 - 1870	2.528	1.445	0.57	18	3.118	1.23	39	- 1.673	- 0.66	- 21

Table 9.2.7. Norway - Mass balance study results.

Part 12 of 22 Mass balance versus altitude Cainhavarre 1964/65

Altitude interval	Area	Winter balance			Summer balance			Net balance		
		Total	Specific		Total	Specific		Total	Specific	
m a sl	km^2	10^6 m^3	m	l/s km^2	10^6 m^3	m	l/s km^2	10^6 m^3	m	l/s km^2
1540 - 1500	0.087	0.121	1.39	44	0.064	0.74	24	+ 0.057	+ 0.65	+ 20
1500 - 1450	0.143	0.190	1.33	42	0.143	1.00	32	+ 0.047	+ 0.33	+ 10
1450 - 1400	0.135	0.182	1.35	43	0.177	1.31	42	+ 0.005	+ 0.04	+ 1
1400 - 1350	0.125	0.189	1.51	48	0.172	1.38	44	+ 0.017	+ 0.13	+ 4
1350 - 1300	0.088	0.140	1.59	50	0.121	1.38	44	+ 0.019	+ 0.21	+ 7
1300 - 1250	0.074	0.098	1.32	42	0.102	1.38	44	- 0.004	- 0.06	- 2
1250 - 1210	0.031	0.041	1.32	42	0.043	1.38	44	- 0.002	- 0.06	- 2
1540 - 1210	0.683	0.961	1.41	45	0.822	1.20	38	+ 0.139	+ 0.21	+ 7

1965/66

Altitude interval	Area	Winter balance			Summer balance			Net balance		
		Total	Specific		Total	Specific		Total	Specific	
m a sl	km^2	10^6 m^3	m	l/s km^2	10^6 m^3	m	l/s km^2	10^6 m^3	m	l/s km^2
1540 - 1500	0.087	0.102	1.17	37	0.141	1.62	51	- 0.039	- 0.45	- 14
1500 - 1450	0.143	0.163	1.14	36	0.252	1.76	56	- 0.089	- 0.62	- 20
1450 - 1400	0.135	0.145	1.07	34	0.286	2.12	67	- 0.141	- 1.05	- 33
1400 - 1350	0.125	0.136	1.08	34	0.279	2.23	71	- 0.143	- 1.15	- 37
1350 - 1300	0.088	0.099	1.12	36	0.209	2.38	75	- 0.110	- 1.25	- 40
1300 - 1250	0.074	0.086	1.16	37	0.176	2.38	75	- 0.090	- 1.22	- 39
1250 - 1210	0.031	0.036	1.16	37	0.073	2.36	75	- 0.037	- 1.20	- 38
1540 - 1210	0.683	0.767	1.12	36	1.416	2.07	66	- 0.649	- 0.95	- 30

1966/67

Altitude interval	Area	Winter balance			Summer balance			Net balance		
		Total	Specific		Total	Specific		Total	Specific	
m a sl	km^2	10^6 m^3	m	l/s km^2	10^6 m^3	m	l/s km^2	10^6 m^3	m	l/s km^2
1540 - 1500	0.087	0.141	1.62	51	0.098	1.13	36	+ 0.043	+ 0.49	+ 15
1500 - 1450	0.143	0.231	1.62	51	0.211	1.48	47	+ 0.020	+ 0.14	+ 4
1450 - 1400	0.135	0.228	1.69	53	0.253	1.87	59	- 0.025	- 0.20	- 6
1400 - 1350	0.125	0.184	1.47	46	0.260	2.08	66	- 0.076	- 0.61	- 20
1350 - 1300	0.088	0.151	1.72	54	0.183	2.08	66	- 0.032	- 0.36	- 12
1300 - 1250	0.074	0.123	1.66	52	0.154	2.08	66	- 0.031	- 0.42	- 14
1250 - 1210	0.031	0.052	1.68	53	0.064	2.06	65	- 0.012	- 0.38	- 12
1540 - 1210	0.683	1.110	1.63	51	1.223	1.79	57	- 0.113	- 0.16	- 5

1967/68

Altitude interval	Area	Winter balance			Summer balance			Net balance		
		Total	Specific		Total	Specific		Total	Specific	
m a sl	km^2	10^6 m^3	m	l/s km^2	10^6 m^3	m	l/s km^2	10^6 m^3	m	l/s km^2
1540 - 1500	0.087	0.118	1.35	43	0.065	0.75	24	+ 0.053	+ 0.60	+ 19
1500 - 1450	0.143	0.187	1.31	42	0.124	0.87	27	+ 0.063	+ 0.44	+ 15
1450 - 1400	0.135	0.169	1.25	39	0.128	0.95	30	+ 0.041	+ 0.30	+ 9
1400 - 1350	0.125	0.157	1.26	40	0.135	1.08	34	+ 0.022	+ 0.18	+ 6
1350 - 1300	0.088	0.118	1.34	42	0.110	1.25	39	+ 0.008	+ 0.09	+ 3
1300 - 1250	0.074	0.101	1.36	43	0.105	1.41	44	- 0.004	- 0.05	- 1
1250 - 1210	0.031	0.043	1.39	44	0.050	1.61	51	- 0.007	- 0.22	- 7
1540 - 1210	0.683	0.893	1.31	41	0.717	1.05	33	+ 0.176	+ 0.26	+ 8

Table 9.2.7. Norway - Mass balance study results.

Part 13 of 22 Mass balance versus altitude Hellstugubreen 1964/65

Altitude interval	Area	Winter balance			Summer balance			Net balance		
		Total	Specific		Total	Specific		Total	Specific	
m a sl	km^2	10^6 m^3	m	l/s km^2	10^6 km^2	m	l/s km^2	10^6 m^3	m	l/s km^2
2200 - 2100	0.075	0.149	1.99	63	0.009	0.12	4	+ 0.140	+ 1.87	+ 59
2100 - 2050	0.291	0.494	1.70	54	0.036	0.12	4	+ 0.458	+ 1.57	+ 50
2050 - 2000	0.213	0.338	1.59	50	0.027	0.13	4	+ 0.311	+ 1.46	+ 46
2000 - 1950	0.392	0.529	1.35	43	0.124	0.32	10	+ 0.405	+ 1.03	+ 32
1950 - 1900	0.610	0.795	1.30	41	0.229	0.38	12	+ 0.566	+ 0.93	+ 30
1900 - 1850	0.380	0.467	1.23	39	0.213	0.56	18	+ 0.254	+ 0.67	+ 21
1850 - 1800	0.347	0.432	1.25	40	0.252	0.73	23	+ 0.180	+ 0.52	+ 17
1800 - 1750	0.144	0.176	1.22	39	0.148	1.03	33	+ 0.028	+ 0.20	+ 6
1750 - 1700	0.141	0.199	1.41	45	0.165	1.17	37	+ 0.034	+ 0.24	+ 8
1700 - 1650	0.207	0.259	1.25	40	0.285	1.38	44	- 0.026	- 0.13	- 4
1650 - 1600	0.195	0.231	1.18	37	0.324	1.66	53	- 0.093	- 0.48	- 15
1600 - 1550	0.198	0.178	0.90	29	0.377	1.90	60	- 0.199	- 1.00	- 32
1550 - 1500	0.125	0.077	0.62	20	0.268	2.14	68	- 0.191	- 1.53	- 49
1500 - 1450	0.062	0.022	0.35	11	0.147	2.37	75	- 0.125	- 2.02	- 64
2200 - 1450	3.380	4.346	1.29	41	2.604	0.77	24	+ 1.742	+ 0.52	+ 17

1965/66

Altitude interval	Area	Winter balance			Summer balance			Net balance		
		Total	Specific		Total	Specific		Total	Specific	
m a sl	km^2	10^6 m^3	m	l/s km^2	10^6 m^3	m	l/s km^2	10^6 m^3	m	l/s km^2
2200 - 2100	0.075	0.116	1.55	49	0.028	0.37	12	+ 0.088	+ 1.17	+ 37
2100 - 2050	0.291	0.446	1.53	48	0.151	0.52	16	+ 0.295	+ 1.01	+ 32
2050 - 2000	0.213	0.281	1.32	42	0.133	0.62	20	+ 0.148	+ 0.70	+ 22
2000 - 1950	0.392	0.442	1.13	36	0.357	0.91	29	+ 0.085	+ 0.22	+ 7
1950 - 1900	0.610	0.624	1.02	32	0.684	1.12	36	- 0.060	- 0.09	- 3
1900 - 1850	0.380	0.339	0.89	28	0.573	1.51	48	- 0.234	- 0.61	- 19
1850 - 1800	0.347	0.295	0.85	27	0.639	1.84	58	- 0.344	- 0.99	- 31
1800 - 1750	0.144	0.120	0.83	26	0.324	2.25	71	- 0.204	- 1.41	- 45
1750 - 1700	0.141	0.107	0.75	24	0.322	2.29	73	- 0.215	- 1.53	- 49
1700 - 1650	0.207	0.145	0.70	22	0.543	2.62	83	- 0.398	- 1.92	- 61
1650 - 1600	0.195	0.119	0.61	19	0.523	2.68	85	- 0.404	- 2.06	- 65
1600 - 1550	0.198	0.099	0.50	16	0.576	2.90	92	- 0.477	- 2.40	- 76
1550 - 1500	0.125	0.055	0.44	14	0.403	3.22	102	- 0.348	- 2.78	- 88
1500 - 1450	0.062	0.024	0.38	12	0.217	3.50	111	- 0.193	- 3.11	- 99
2200 - 1450	3.380	3.212	0.95	30	5.473	1.62	51	- 2.261	- 0.67	- 21

1966/67

Altitude interval	Area	Winter balance			Summer balance			Net balance		
		Total	Specific		Total	Specific		Total	Specific	
m a sl	km^2	10^6 m^3	m	l/s km^2	10^6 m^3	m	l/s km^2	10^6 m^3	m	l/s km^2
2200 - 2100	0.075	0.178	2.37	75	0.009	0.12	4	+ 0.169	+ 2.25	+ 71
2100 - 2050	0.291	0.678	2.33	74	0.036	0.12	4	+ 0.642	+ 2.20	+ 69
2050 - 2000	0.213	0.404	1.90	60	0.080	0.37	12	+ 0.324	+ 1.53	+ 48
2000 - 1950	0.392	0.717	1.83	58	0.147	0.37	12	+ 0.570	+ 1.46	+ 46
1950 - 1900	0.610	0.985	1.61	51	0.381	0.62	19	+ 0.604	+ 0.99	+ 31
1900 - 1850	0.380	0.602	1.59	50	0.332	0.87	24	+ 0.270	+ 0.72	+ 23
1850 - 1800	0.347	0.485	1.40	44	0.383	1.10	34	+ 0.102	+ 0.30	+ 9
1800 - 1750	0.144	0.159	1.10	35	0.189	1.31	41	- 0.030	- 0.21	- 7
1750 - 1700	0.141	0.146	1.04	33	0.194	1.37	43	- 0.048	- 0.33	- 10
1700 - 1650	0.207	0.227	1.10	35	0.285	1.37	43	- 0.058	- 0.27	- 8
1650 - 1600	0.195	0.201	1.03	33	0.317	1.62	51	- 0.116	- 0.59	- 18
1600 - 1550	0.198	0.153	0.77	24	0.365	1.84	58	- 0.212	- 1.07	- 34
1550 - 1500	0.125	0.054	0.43	14	0.266	2.13	67	- 0.212	- 1.70	- 54
1500 - 1450	0.062	0.017	0.27	8	0.147	2.37	75	- 0.130	- 2.10	- 66
2200 - 1450	3.380	5.006	1.48	47	3.131	0.93	29	+ 1.875	+ 0.55	+ 18

Table 9.2.7. Norway - Mass balance study results.

Part 14 of 22 Mass balance versus altitude Hellstugubreen 1967/68

Altitude interval	Area	Winter balance			Summer balance			Net balance		
		Total	Specific		Total	Specific		Total	Specific	
m a sl	km^2	$10^6\ m^3$	m	l/s km^2	$10^6\ m^3$	m	l/s km^2	$10^6\ m^3$	m	l/s km^2
2200 - 2100	0.078	0.143	1.83	58	0.083	1.06	34	+ 0.060	+ 0.77	+ 24
2100 - 2050	0.261	0.479	1.83	58	0.277	1.06	34	+ 0.202	+ 0.77	+ 25
2050 - 2000	0.202	0.346	1.71	54	0.214	1.06	34	+ 0.132	+ 0.65	+ 21
2000 - 1950	0.386	0.622	1.61	51	0.457	1.18	38	+ 0.164	+ 0.43	+ 14
1950 - 1900	0.610	0.915	1.50	48	0.722	1.18	38	+ 0.193	+ 0.32	+ 10
1900 - 1850	0.377	0.494	1.31	42	0.519	1.37	44	- 0.025	- 0.07	- 2
1850 - 1800	0.350	0.464	1.33	42	0.505	1.44	46	- 0.040	- 0.12	- 4
1800 - 1750	0.141	0.172	1.22	39	0.229	1.62	52	- 0.057	- 0.41	- 13
1750 - 1700	0.142	0.195	1.37	44	0.245	1.73	55	- 0.050	- 0.35	- 11
1700 - 1650	0.207	0.263	1.27	40	0.388	1.87	59	- 0.125	- 0.60	- 19
1650 - 1600	0.192	0.222	1.15	37	0.409	2.12	67	- 0.187	- 0.97	- 31
1600 - 1550	0.195	0.176	0.90	29	0.429	2.20	70	- 0.253	- 1.30	- 41
1550 - 1500	0.124	0.079	0.64	20	0.314	2.52	80	- 0.235	- 1.89	- 60
1500 - 1450	0.060	0.033	0.55	17	0.181	3.01	95	- 0.148	- 2.46	- 78
2200 - 1450	3.326	4.604	1.38	44	4.972	1.49	47	- 0.368	- 0.11	- 4

1968/69

Altitude interval	Area	Winter balance			Summer balance			Net balance		
		Total	Specific		Total	Specific		Total	Specific	
m a sl	km^2	$10^6\ m^3$	m	l/s km^2	$10^6\ m^3$	m	l/s km^2	$10^6\ m^3$	m	l/s km^2
2200 - 2100	0.078	0.105	1.35	43	0.107	1.37	43	- 0.002	- 0.02	0
2100 - 2050	0.261	0.345	1.32	42	0.361	1.39	44	- 0.016	- 0.07	- 2
2050 - 2000	0.202	0.250	1.24	39	0.328	1.62	51	- 0.078	- 0.38	- 11
2000 - 1950	0.386	0.394	1.02	32	0.627	1.62	51	- 0.233	- 0.60	- 19
1950 - 1900	0.610	0.604	0.99	31	1.050	1.72	54	- 0.446	- 0.73	- 23
1900 - 1850	0.377	0.404	1.07	34	0.747	1.98	63	- 0.343	- 0.91	- 29
1850 - 1800	0.350	0.331	0.95	30	0.873	2.49	79	- 0.542	- 1.54	- 49
1800 - 1750	0.141	0.112	0.79	25	0.370	2.62	83	- 0.258	- 1.83	- 58
1750 - 1700	0.142	0.122	0.86	27	0.408	2.87	91	- 0.286	- 2.01	- 64
1700 - 1650	0.207	0.136	0.66	21	0.603	2.91	92	- 0.467	- 2.25	- 71
1650 - 1600	0.192	0.128	0.67	21	0.600	3.12	99	- 0.427	- 2.45	- 78
1600 - 1550	0.195	0.110	0.56	18	0.658	3.37	106	- 0.548	- 2.81	- 88
1550 - 1500	0.124	0.072	0.58	18	0.450	3.62	114	- 0.378	- 3.04	- 96
1500 - 1450	0.060	0.040	0.67	21	0.232	3.87	122	- 0.192	- 3.20	- 101
2200 - 1450	3.325	3.153	0.95	30	7.414	2.23	70	- 4.261	- 1.28	- 40

1969/70

Altitude interval	Area	Winter balance			Summer balance			Net balance		
		Total	Specific		Total	Specific		Total	Specific	
m a sl	km^2	$10^6\ m^3$	m	l/s km^2	$10^6\ m^3$	m	l/s km^2	$10^6\ m^3$	m	l/s km^2
2200 - 2100	0.078	0.089	1.14	36	0.049	0.63	20	+ 0.040	+ 0.51	+ 16
2100 - 2050	0.261	0.333	1.28	40	0.228	0.87	27	+ 0.105	+ 0.41	+ 13
2050 - 2000	0.202	0.230	1.14	36	0.195	0.97	31	+ 0.035	+ 0.17	+ 5
2000 - 1950	0.386	0.341	0.88	28	0.434	1.12	35	- 0.093	- 0.24	- 7
1950 - 1900	0.610	0.429	0.70	22	0.766	1.26	40	- 0.337	- 0.56	- 18
1900 - 1850	0.377	0.272	0.72	23	0.588	1.56	49	- 0.316	- 0.84	- 27
1850 - 1800	0.350	0.161	0.46	15	0.613	1.75	55	- 0.452	- 1.29	- 41
1800 - 1750	0.141	0.058	0.41	13	0.273	1.94	61	- 0.215	- 1.53	- 48
1750 - 1700	0.142	0.076	0.54	17	0.319	2.25	97	- 0.243	- 1.71	- 54
1700 - 1650	0.207	0.091	0.44	14	0.466	2.25	71	- 0.375	- 1.81	- 57
1650 - 1600	0.192	0.085	0.44	14	0.528	2.75	87	- 0.443	- 2.31	- 73
1600 - 1550	0.195	0.082	0.42	13	0.564	2.89	91	- 0.482	- 2.47	- 78
1550 - 1500	0.124	0.045	0.36	11	0.403	3.25	102	- 0.358	- 2.89	- 91
1500 - 1450	0.060	0.013	0.22	7	0.225	3.75	118	- 0.212	- 3.53	- 111
2200 - 1450	3.326	2.305	0.70	22	5.651	1.70	54	- 3.346	- 1.00	- 32

Table 9.2.7. Norway - Mass balance study results.

Part 15 of 22 Mass balance versus altitude Alfotbreen 1964/65

Altitude interval	Area	Winter balance			Summer balance			Net balance		
		Total	Specific		Total	Specific		Total	Specific	
m a sl	km^2	10^6 m^3	m	l/s km^2	10^6 m^3	m	l/s km^2	10^6 m^3	m	l/s km^2
1400 - 1350	0.26	0.97	3.75	119	0.62	2.37	75	+ 0.35	+ 1.35	+ 43
1350 - 1300	1.02	3.82	3.75	119	2.61	2.56	81	+ 1.22	+ 1.20	+ 38
1300 - 1250	0.84	3.15	3.75	119	2.39	2.84	90	+ 0.76	+ 0.90	+ 29
1250 - 1200	0.64	2.37	3.69	117	1.97	3.07	97	+ 0.40	+ 0.63	+ 20
1200 - 1150	0.51	1.78	3.49	110	1.71	3.35	106	+ 0.07	+ 0.14	+ 4
1150 - 1100	0.50	1.74	3.50	111	1.78	3.58	114	- 0.04	- 0.08	- 3
1100 - 1050	0.40	1.40	3.49	110	1.52	3.77	120	- 0.12	- 0.30	- 10
1050 - 1000	0.31	1.05	3.40	108	1.24	4.03	128	- 0.21	- 0.65	- 20
1000 - 950	0.16	0.57	3.53	112	0.70	4.32	137	- 0.13	- 0.81	- 26
950 - 900	0.09	0.35	3.75	119	0.42	4.53	144	- 0.07	- 0.78	- 25
900 - 850	0.02	0.08	3.75	119	0.09	4.65	147	- 0.02	- 1.10	- 35
1400 - 850	4.75	17.28	3.64	115	15.05	3.16	100	+ 2.23	+ 0.47	+ 15

1965/66

Altitude interval	Area	Winter balance			Summer balance			Net balance		
		Total	Specific		Total	Specific		Total	Specific	
m a sl	km^2	10^6 m^3	m	l/s km^2	10^6 m^3	m	l/s km^2	10^6 m^3	m	l/s km^2
1400 - 1350	0.259	0.693	2.68	85	0.842	3.25	103	- 0.149	- 0.57	- 18
1350 - 1300	1.020	2.760	2.71	86	3.315	3.25	103	- 0.555	- 0.54	- 17
1300 - 1250	0.841	2.278	2.71	86	3.093	3.68	117	- 0.815	- 0.97	- 31
1250 - 1200	0.642	1.676	2.61	83	2.562	3.99	127	- 0.886	- 1.38	- 44
1200 - 1150	0.510	1.233	2.42	77	2.168	4.25	135	- 0.935	- 1.83	- 58
1150 - 1100	0.496	1.105	2.23	71	2.238	4.51	143	- 1.133	- 2.28	- 72
1100 - 1050	0.403	0.875	2.17	69	1.924	4.77	151	- 1.049	- 2.60	- 82
1050 - 1000	0.309	0.602	1.95	62	1.598	5.17	164	- 0.996	- 3.22	- 102
1000 - 950	0.162	0.309	1.91	60	0.947	5.85	185	- 0.638	- 3.94	- 125
950 - 900	0.093	0.193	2.08	66	0.581	6.25	198	- 0.388	- 4.17	- 132
900 - 850	0.020	0.043	2.15	68	0.135	6.75	214	- 0.092	- 4.60	- 146
1400 - 850	4.755	11.767	2.47	78	19.403	4.08	129	- 7.636	- 1.61	- 51

1966/67

Altitude interval	Area	Winter balance			Summer balance			Net balance		
		Total	Specific		Total	Specific		Total	Specific	
m a sl	km^2	10^6 m^3	m	l/s km^2	10^6 m^3	m	l/s km^2	10^6 m^3	m	l/s km^2
1400 - 1350	0.259	1.125	4.34	137	0.615	2.37	75	+ 0.510	+ 1.97	+ 62
1350 - 1300	1.020	4.635	4.54	143	2.677	2.62	83	+ 1.958	+ 1.92	+ 60
1300 - 1250	0.841	3.768	4.48	141	2.246	2.68	85	+ 1.522	+ 1.80	+ 56
1250 - 1200	0.642	2.817	4.39	139	1.881	2.93	93	+ 0.936	+ 1.46	+ 46
1200 - 1150	0.510	2.237	4.39	139	1.655	3.24	102	+ 0.582	+ 1.15	+ 37
1150 - 1100	0.496	2.219	4.47	141	1.817	3.66	116	+ 0.402	+ 0.81	+ 25
1100 - 1050	0.403	1.789	4.44	140	1.620	4.02	127	+ 0.169	+ 0.42	+ 13
1050 - 1000	0.309	1.332	4.31	136	1.320	4.27	135	+ 0.012	+ 0.04	+ 1
1000 - 950	0.162	0.749	4.62	146	0.721	4.45	140	+ 0.028	+ 0.17	+ 6
950 - 900	0.093	0.432	4.64	146	0.442	4.75	150	- 0.010	- 0.11	- 4
900 - 850	0.020	0.095	4.75	150	0.097	4.87	154	- 0.003	- 0.12	- 4
1400 - 850	4.755	21.198	4.46	141	15.091	3.18	100	+ 6.107	+ 1.28	+ 41

Table 9.2.7. Norway - Mass balance study results.

Part 16 of 22 Mass balance versus altitude Alfotbreen 1967/68

Altitude interval	Area	Winter balance			Summer balance			Net balance		
		Total	Specific		Total	Specific		Total	Specific	
m a sl	km^2	10^6 m^3	m	l/s km^2	10^6 m^3	m	l/s km^2	10^6 m^3	m	l/s km^2
1400 - 1350	0.259	1.234	4.76	150	0.809	3.12	99	+ 0.425	+ 1.64	+ 51
1350 - 1300	1.020	5.063	4.96	157	3.240	3.18	100	+ 1.823	+ 1.78	+ 57
1300 - 1250	0.841	4.000	4.76	150	2.838	3.38	107	+ 1.162	+ 1.38	+ 43
1250 - 1200	0.642	2.943	4.58	145	2.217	3.45	109	+ 0.726	+ 1.13	+ 36
1200 - 1150	0.510	2.373	4.65	147	1.850	3.63	115	+ 0.523	+ 1.02	+ 32
1150 - 1100	0.496	2.278	4.60	145	1.925	3.88	123	+ 0.353	+ 0.72	+ 22
1100 - 1050	0.403	1.669	4.14	130	1.653	4.10	129	+ 0.016	+ 0.01	0
1050 - 1000	0.309	1.132	3.66	116	1.325	4.29	135	- 0.193	- 0.63	- 19
1000 - 950	0.162	0.555	3.43	108	0.730	4.50	142	- 0.175	- 1.07	- 34
950 - 900	0.093	0.321	3.45	109	0.430	4.62	146	- 0.109	- 1.17	- 37
900 - 890	0.020	0.067	3.35	106	0.092	4.63	146	- 0.025	- 1.28	- 40
1400 - 890	4.755	21.635	4.55	144	17.109	3.60	114	+ 4.526	+ 0.95	+ 30

1968/69

Altitude interval	Area	Winter balance			Summer balance			Net balance		
		Total	Specific		Total	Specific		Total	Specific	
m a sl	km^2	10^6 m^3	m	l/s km^2	10^6 m^3	m	l/s km^2	10^6 m^3	m	l/s km^2
1378 - 1350	0.274	0.782	2.83	89	1.055	3.85	121	- 0.273	- 1.02	- 32
1350 - 1300	1.014	2.842	2.80	88	4.113	4.08	129	- 1.271	- 1.28	- 41
1300 - 1250	0.808	2.288	2.85	90	3.434	4.25	134	- 1.146	- 1.40	- 44
1250 - 1200	0.754	2.021	2.67	84	3.582	4.75	150	- 1.561	- 2.08	- 66
1200 - 1150	0.672	1.708	2.54	80	3.528	5.25	166	- 1.820	- 2.71	- 86
1150 - 1100	0.544	1.356	2.51	79	3.040	5.60	177	- 1.684	- 3.09	- 98
1100 - 1050	0.356	0.889	2.51	79	2.081	5.85	185	- 1.192	- 3.34	- 105
1050 - 1000	0.224	0.515	2.32	73	1.372	6.12	193	- 0.857	- 3.80	- 120
1000 - 950	0.128	0.300	2.34	74	0.784	6.12	193	- 0.484	- 3.78	- 119
950 - 900	0.046	0.119	2.59	82	0.282	6.13	193	- 0.163	- 3.54	- 112
900 - 870	0.004	0.016	2.63	83	0.025	6.13	193	- 0.009	- 3.50	- 111
1378 - 870	4.824	12.836	2.66	84	23.296	4.83	153	- 10.460	- 2.17	- 69

1969/70

Altitude interval	Area	Winter balance			Summer balance			Net balance		
		Total	Specific		Total	Specific		Total	Specific	
m a sl	km^2	10^6 m^3	m	l/s km^2	10^6 m^3	m	l/s km^2	10^6 m^3	m	l/s km^2
1378 - 1350	0.274	0.734	2.68	85	0.799	2.88	91	- 0.065	- 0.20	- 6
1350 - 1300	1.014	2.677	2.64	83	3.043	3.00	95	- 0.366	- 0.36	- 12
1300 - 1250	0.808	2.163	2.67	84	2.724	3.87	106	- 0.561	- 0.70	- 22
1250 - 1200	0.754	1.996	2.63	83	2.986	3.96	125	- 0.990	- 1.33	- 42
1200 - 1150	0.672	1.673	2.50	79	2.830	4.21	133	- 1.157	- 1.71	- 54
1150 - 1100	0.544	1.351	2.48	78	2.423	4.45	141	- 1.072	- 1.97	- 63
1100 - 1050	0.356	0.870	2.44	77	1.685	4.73	149	- 0.815	- 2.29	- 72
1050 - 1000	0.224	0.538	2.40	76	1.105	4.93	156	- 0.567	- 2.53	- 80
1000 - 950	0.128	0.369	2.88	91	0.632	4.94	156	- 0.263	- 2.06	- 65
950 - 900	0.046	0.144	3.13	99	0.224	4.88	154	- 0.080	- 1.75	- 55
900 - 870	0.004	0.014	3.50	11	0.020	4.88	154	- 0.006	- 1.38	- 143
1378 - 870	4.824	12.529	2.60	82	18.462	3.83	121	- 5.933	- 1.23	- 39

Table 9.2.7. Norway - Mass balance study results.

Part 17 of 22 Mass balance versus altitude Folgefonni - West

Altitude interval	Area	Winter balance			Summer balance			Net balance		
		Total	Specific		Total	Specific		Total	Specific	
m a sl	km^2	$10^6 m^3$	m	l/s km^2	$10^6 m^3$	m	l/s km^2	$10^6 m^3$	m	l/s km^2
1964/65										
1660 - 1650	0.387	1.072	2.77	88	0.484	1.25	40	+ 0.588	+ 1.52	+ 48
1650 - 1600	1.460	4.182	2.86	91	2.099	1.44	46	+ 2.083	+ 1.42	+ 45
1600 - 1550	3.701	9.490	2.56	81	6.670	1.80	57	+ 2.820	+ 0.76	+ 24
1550 - 1500	2.842	6.763	2.38	75	5.401	1.90	60	+ 1.362	+ 0.48	+ 15
1500 - 1450	1.848	3.960	2.14	68	4.101	2.22	70	- 0.141	- 0.08	- 3
1450 - 1400	1.270	2.744	2.16	69	3.019	2.38	75	- 0.275	- 0.22	- 7
1400 - 1350	1.181	2.445	2.07	65	3.237	2.74	87	- 0.792	- 0.67	- 21
1350 - 1300	0.985	1.754	1.78	56	3.069	3.12	99	- 1.315	- 1.34	- 43
1300 - 1250	0.651	1.259	1.93	61	2.264	3.48	110	- 1.005	- 1.55	- 49
1250 - 1200	0.474	0.694	1.46	46	1.940	4.09	130	- 1.246	- 2.63	- 83
1200 - 1150	0.301	0.407	1.35	43	1.337	4.44	141	- 0.930	- 3.09	- 98
1150 - 1100	0.130	0.208	1.60	51	0.645	4.96	157	- 0.437	- 3.36	- 107
1100 - 1050	0.044	0.077	1.75	56	0.231	5.25	166	- 0.154	- 3.50	- 111
1660 - 1050	15.274	35.055	2.30	73	34.497	2.26	72	+ 0.558	+ 0.04	+ 1
1965/66										
1660 - 1650	0.387	0.738	1.91	60	0.871	2.25	71	- 0.133	- 0.34	- 11
1650 - 1600	1.460	2.764	1.89	60	3.285	2.25	71	- 0.521	- 0.36	- 11
1600 - 1550	3.701	6.424	1.74	55	9.024	2.44	77	- 2.600	- 0.70	- 22
1550 - 1500	2.842	4.836	1.70	54	7.702	2.71	86	- 2.866	- 1.01	- 32
1500 - 1450	1.848	3.006	1.63	52	5.728	3.10	98	- 2.722	- 1.47	- 47
1450 - 1400	1.270	2.163	1.70	54	4.504	3.55	113	- 2.341	- 1.85	- 59
1400 - 1350	1.181	2.032	1.72	55	4.288	3.63	115	- 2.256	- 1.91	- 60
1350 - 1300	0.985	1.399	1.42	45	3.694	3.75	119	- 2.295	- 2.33	- 74
1300 - 1250	0.651	0.985	1.51	48	2.767	4.25	135	- 1.782	- 2.74	- 87
1250 - 1200	0.474	0.498	1.05	33	2.144	4.52	143	- 1.646	- 3.47	- 110
1200 - 1150	0.301	0.223	0.74	24	1.430	4.75	151	- 1.207	- 4.01	- 127
1150 - 1100	0.130	0.153	1.18	37	0.663	5.10	162	- 0.510	- 3.92	- 124
1100 - 1030	0.044	0.061	1.39	44	0.231	5.25	166	- 0.170	- 3.86	- 122
1660 - 1030	15.274	25.282	1.66	53	46.331	3.03	96	- 21.050	- 1.37	- 43
1966/67										
1660 - 1650	0.387	1.657	4.28	135	0.387	1.00	32	+ 1.270	+ 3.28	+ 103
1650 - 1600	1.460	6.004	4.14	131	1.734	1.19	38	+ 4.270	+ 2.95	+ 93
1600 - 1550	3.701	13.970	3.78	119	5.126	1.38	44	+ 8.844	+ 2.40	+ 75
1550 - 1500	2.842	10.227	3.60	114	4.667	1.64	52	+ 5.560	+ 1.96	+ 62
1500 - 1450	1.848	6.217	3.36	106	3.720	2.02	64	+ 2.497	+ 1.34	+ 42
1450 - 1400	1.270	4.286	3.38	107	3.240	2.55	80	+ 1.046	+ 0.83	+ 27
1400 - 1350	1.181	3.715	3.15	99	3.442	2.92	92	+ 0.273	+ 0.23	+ 7
1350 - 1300	0.985	2.966	3.02	95	3.520	3.58	113	- 0.554	- 0.56	- 18
1300 - 1250	0.651	1.971	3.03	96	2.603	4.00	126	- 0.632	- 0.97	- 30
1250 - 1200	0.474	1.218	2.57	81	2.032	4.28	135	- 0.814	- 1.71	- 54
1200 - 1150	0.301	0.776	2.58	81	1.329	4.42	140	- 0.553	- 1.84	- 59
1150 - 1100	0.130	0.440	3.39	107	0.618	4.75	150	- 0.178	- 1.36	- 43
1100 - 1030	0.044	0.165	3.75	118	0.209	4.75	150	- 0.044	- 1.00	- 32
1660 - 1030	15.274	53.612	3.51	111	32.627	2.14	68	+ 20.985	+ 1.37	+ 43
1967/68										
1660 - 1650	0.377	1.414	3.75	118	0.829	2.20	70	+ 0.584	+ 1.55	+ 49
1650 - 1600	1.461	5.625	3.85	122	3.265	2.23	71	+ 2.361	+ 1.62	+ 51
1600 - 1550	3.681	13.914	3.77	119	8.399	2.28	72	+ 5.515	+ 1.50	+ 47
1550 - 1500	2.805	10.324	3.68	116	6.712	2.39	75	+ 3.612	+ 1.29	+ 41
1500 - 1450	1.821	6.490	3.56	112	4.745	2.61	82	+ 1.745	+ 0.96	+ 30
1450 - 1400	1.260	4.149	3.29	104	3.528	2.80	88	+ 0.620	+ 0.49	+ 15
1400 - 1350	1.175	3.718	3.16	100	3.437	2.93	92	+ 0.281	+ 0.24	+ 8
1350 - 1300	0.994	2.587	2.60	83	3.299	3.32	105	- 0.712	- 0.72	- 23
1300 - 1250	0.642	1.661	2.59	82	2.375	3.70	117	- 0.713	- 1.11	- 35
1250 - 1200	0.467	0.984	2.11	67	1.994	4.27	136	- 1.011	- 2.17	- 69
1200 - 1150	0.293	0.561	1.91	61	1.405	4.79	152	- 0.844	- 2.88	- 91
1150 - 1100	0.144	0.301	2.09	66	0.719	4.99	158	- 0.419	- 2.91	- 92
1100 - 1050	0.039	0.088	2.25	71	0.201	5.12	163	- 0.113	- 2.87	- 91
1050 - 1030	0.004	0.009	2.25	71	0.020	5.12	163	- 0.011	- 2.87	- 91
1660 - 1030	15.163	51.825	3.42	107	40.928	2.69	85	+ 10.895	+ 0.72	+ 23

Table 9.2.7. Norway - Mass balance study results.

Part 18 of 22 Mass balance versus altitude Folgefonni - East

1964/65

Altitude interval m a sl	Area km^2	Winter balance Total 10^6 m^3	Winter balance Specific m	Winter balance Specific l/s km^2	Summer balance Total 10^6 m^3	Summer balance Specific m	Summer balance Specific l/s km^2	Net balance Total 10^6 m^3	Net balance Specific m	Net balance Specific l/s km^2
1610 - 1600	0.084	0.232	2.76	88	0.139	1.61	51	+ 0.093	+ 1.15	+ 36
1600 - 1550	1.296	3.395	2.62	83	2.280	1.76	56	+ 1.115	+ 0.86	+ 27
1550 - 1500	1.001	2.650	2.65	84	2.001	2.00	63	+ 0.649	+ 0.65	+ 20
1500 - 1450	0.445	1.427	3.21	102	1.001	2.25	71	+ 0.426	+ 0.96	+ 30
1450 - 1400	0.303	0.872	2.88	91	0.682	2.25	71	+ 0.190	+ 0.63	+ 20
1400 - 1350	0.294	0.650	2.21	70	0.773	2.63	83	- 0.123	- 0.42	- 13
1350 - 1300	0.223	0.464	2.08	66	0.645	2.89	92	- 0.181	- 0.81	- 26
1300 - 1250	0.234	0.537	2.29	73	0.822	3.51	111	- 0.285	- 1.22	- 39
1250 - 1200	0.199	0.341	1.71	54	0.835	4.20	133	- 0.494	- 2.49	- 79
1200 - 1150	0.224	0.422	1.88	60	1.046	4.67	148	- 0.624	- 2.79	- 88
1150 - 1100	0.132	0.233	1.76	56	0.693	5.25	166	- 0.460	- 3.49	- 111
1100 - 1050	0.028	0.053	1.88	60	0.147	5.25	166	- 0.094	- 3.37	- 107
1610 - 1050	4.463	11.276	2.53	80	11.064	2.48	79	+ 0.212	+ 0.05	+ 2

1965/66

Altitude interval m a sl	Area km^2	Winter balance Total 10^6 m^3	Winter balance Specific m	Winter balance Specific l/s km^2	Summer balance Total 10^6 m^3	Summer balance Specific m	Summer balance Specific l/s km^2	Net balance Total 10^6 m^3	Net balance Specific m	Net balance Specific l/s km^2
1610 - 1600	0.084	0.140	1.67	53	0.232	2.75	87	- 0.092	- 1.08	- 34
1600 - 1550	1.296	2.280	1.76	56	3.564	2.75	87	- 1.284	- 0.99	- 31
1550 - 1500	1.001	0.979	1.98	63	2.605	2.75	87	- 0.626	- 0.77	- 24
1500 - 1450	0.445	1.005	2.26	72	1.446	3.25	103	- 0.441	- 0.99	- 31
1450 - 1400	0.303	0.622	2.05	65	0.985	3.25	103	- 0.363	- 1.20	- 38
1400 - 1350	0.294	0.469	1.60	51	1.102	3.75	119	- 0.633	- 2.15	- 68
1350 - 1300	0.223	0.361	1.62	51	0.837	3.75	119	- 0.476	- 2.13	- 68
1300 - 1250	0.234	0.414	1.77	56	0.995	4.25	135	- 0.581	- 2.48	- 79
1250 - 1200	0.199	0.227	1.14	36	0.887	4.46	141	- 0.660	- 3.32	- 105
1200 - 1150	0.224	0.207	0.92	29	1.064	4.75	151	- 0.857	- 3.83	- 121
1150 - 1100	0.132	0.116	0.88	28	0.693	5.25	166	- 0.577	- 4.37	- 138
1100 - 1050	0.028	0.024	0.86	27	0.147	5.25	166	- 0.123	- 4.39	- 139
1610 - 1050	4.463	7.844	1.76	56	14.557	3.26	103	- 6.713	- 1.50	- 48

1966/67

Altitude interval m a sl	Area km^2	Winter balance Total 10^6 m^3	Winter balance Specific m	Winter balance Specific l/s km^2	Summer balance Total 10^6 m^3	Summer balance Specific m	Summer balance Specific l/s km^2	Net balance Total 10^6 m^3	Net balance Specific m	Net balance Specific l/s km^2
1610 - 1600	0.084	0.315	3.75	118	0.126	1.50	47	+ 0.189	+ 2.25	+ 71
1600 - 1550	1.296	4.914	3.79	120	2.160	1.67	53	+ 2.754	+ 2.12	+ 67
1550 - 1500	1.001	4.094	4.09	128	1.976	1.97	62	+ 2.118	+ 2.12	+ 66
1500 - 1450	0.445	1.932	4.35	137	1.001	2.25	71	+ 0.931	+ 2.10	+ 66
1450 - 1400	0.303	1.234	4.08	129	0.846	2.79	88	+ 0.388	+ 1.29	+ 41
1400 - 1350	0.294	1.132	3.85	121	0.940	3.20	101	+ 0.192	+ 0.60	+ 20
1350 - 1300	0.223	0.868	3.89	123	0.859	3.85	121	+ 0.009	+ 0.04	+ 2
1300 - 1250	0.234	0.901	3.85	121	0.902	3.85	121	- 0.001	- 0.00	- 0
1250 - 1200	0.199	0.671	3.37	106	0.846	4.25	134	- 0.175	- 0.88	- 28
1200 - 1150	0.224	0.700	3.12	98	0.998	4.46	141	- 0.298	- 1.34	- 43
1150 - 1100	0.132	0.395	3.00	95	0.627	4.75	150	- 0.232	- 1.75	- 55
1100 - 1050	0.028	0.079	2.82	89	0.133	4.75	150	- 0.054	- 1.93	- 61
1610 - 1050	4.463	17.235	3.86	122	11.414	2.56	81	+ 5.821	+ 1.30	+ 41

1967/68

Altitude interval m a sl	Area km^2	Winter balance Total 10^6 m^3	Winter balance Specific m	Winter balance Specific l/s km^2	Summer balance Total 10^6 m^3	Summer balance Specific m	Summer balance Specific l/s km^2	Net balance Total 10^6 m^3	Net balance Specific m	Net balance Specific l/s km^2
1610 - 1600	0.074	0.252	3.39	108	0.164	2.20	70	+ 0.089	+ 1.19	+ 38
1600 - 1550	1.283	4.694	3.66	116	2.961	2.31	73	+ 1.733	+ 1.35	+ 43
1550 - 1500	1.002	3.470	3.46	110	2.489	2.48	79	+ 0.981	+ 0.98	+ 31
1500 - 1450	0.413	1.435	3.47	110	1.085	2.62	83	+ 0.350	+ 0.85	+ 27
1450 - 1400	0.300	0.970	3.24	103	0.829	2.77	88	+ 0.141	+ 0.47	+ 15
1400 - 1350	0.278	0.724	2.60	83	0.800	2.87	91	- 0.076	- 0.27	- 0
1350 - 1300	0.215	0.542	2.52	80	0.670	3.11	99	- 0.128	- 0.60	- 19
1300 - 1250	0.232	0.559	2.42	77	0.881	3.80	121	- 0.321	- 1.39	- 44
1250 - 1200	0.185	0.407	2.20	70	0.762	4.12	131	- 0.355	- 1.92	- 61
1200 - 1150	0.229	0.498	2.18	69	0.944	4.12	131	- 0.446	- 1.95	- 62
1150 - 1100	0.118	0.236	2.00	63	0.515	4.37	139	- 0.280	- 2.37	- 75
1100 - 1065	0.025	0.052	2.09	66	0.108	4.37	139	- 0.056	- 2.28	- 72
1610 - 1065	4.354	13.840	3.18	101	12.209	2.80	89	+ 1.631	+ 0.37	+ 12

Table 9.2.7. Norway - Mass balance study results.

Part 19 of 22 Mass balance versus altitude Vesledalsbreen 1966/67

Altitude interval m a sl	Area km^2	Winter balance Total 10^6 m^3	Winter balance Specific m	Winter balance Specific l/s km^2	Summer balance Total 10^6 m^3	Summer balance Specific m	Summer balance Specific l/s km^2	Net balance Total 10^6 m^3	Net balance Specific m	Net balance Specific l/s km^2
1730 - 1700	0.014	0.024	1.71	54	0.012	0.86	27	+ 0.012	+ 0.85	+ 27
1700 - 1650	0.115	0.205	1.78	56	0.101	0.88	28	+ 0.104	+ 0.90	+ 27
1650 - 1600	0.186	0.310	1.67	53	0.163	0.78	25	+ 0.147	+ 0.89	+ 28
1600 - 1550	0.704	1.464	2.04	65	0.754	1.07	34	+ 0.710	+ 0.97	+ 31
1550 - 1500	0.521	1.069	2.10	66	0.626	1.20	38	+ 0.443	+ 0.90	+ 28
1500 - 1450	0.468	1.128	2.32	73	0.668	1.38	44	+ 0.460	+ 0.94	+ 29
1450 - 1400	0.724	1.625	2.26	71	1.217	1.68	53	+ 0.418	+ 0.58	+ 18
1400 - 1350	0.719	1.433	2.00	63	1.511	2.10	66	- 0.078	- 0.10	- 3
1350 - 1300	0.420	0.762	1.82	57	1.051	2.51	79	- 0.289	- 0.69	- 22
1300 - 1250	0.120	0.270	2.25	71	0.338	2.82	89	- 0.068	- 0.57	- 18
1250 - 1200	0.147	0.316	2.15	66	0.514	3.50	111	- 0.198	- 1.35	- 45
1200 - 1150	0.050	0.074	1.47	46	0.212	4.24	134	- 0.138	- 2.77	- 88
1150 - 1125	0.015	0.024	1.60	51	0.071	4.73	149	- 0.047	- 3.13	- 98
1730 - 1125	4.221	8.714	2.06	65	7.238	1.71	54	+ 1.476	+ 0.35	+ 11

1967/68

1730 - 1700	0.014	0.045	3.25	103	0.029	2.06	65	+ 0.016	+ 1.19	+ 38
1700 - 1650	0.115	0.374	3.25	103	0.237	2.06	65	+ 0.137	+ 1.19	+ 38
1650 - 1600	0.186	0.605	3.25	103	0.383	2.06	65	+ 0.222	+ 1.19	+ 38
1600 - 1550	0.704	2.268	3.22	102	1.450	2.06	65	+ 0.818	+ 1.16	+ 37
1550 - 1500	0.521	1.692	3.25	103	1.107	2.13	67	+ 0.585	+ 1.12	+ 36
1500 - 1450	0.486	1.597	3.29	104	1.033	2.13	67	+ 0.564	+ 1.16	+ 37
1450 - 1400	0.724	2.436	3.36	106	1.814	2.51	78	+ 0.622	+ 0.85	+ 28
1400 - 1350	0.719	2.118	2.95	94	2.003	2.79	88	+ 0.115	+ 0.16	+ 6
1350 - 1300	0.420	1.274	3.03	96	1.289	3.07	97	- 0.025	- 0.04	- 1
1300 - 1250	0.120	0.353	2.94	93	0.396	3.30	104	- 0.043	- 0.36	- 11
1250 - 1200	0.147	0.347	2.36	75	0.549	3.73	118	- 0.202	- 1.37	- 43
1200 - 1150	0.050	0.111	2.22	70	0.200	4.00	126	- 0.089	- 1.78	- 56
1150 - 1125	0.015	0.033	2.20	69	0.062	4.13	130	- 0.029	- 1.93	- 61
1730 - 1125	4.221	13.253	3.14	99	10.552	2.50	79	+ 2.701	+ 0.64	+ 20

Table 9.2.7. Norway - Mass balance study results.

Part 20 of 22 Mass balance versus altitude Vestre Memurubre 1967/68

Altitude interval	Area	Winter balance			Summer balance			Net balance		
		Total	Specific		Total	Specific		Total	Specific	
m a sl	km^2	10^6 m^3	m	l/s km^2	10^6 m^3	m	l/s km^2	10^6 m^3	m	l/s km^2
2230 - 2200	0.017	0.032	1.88	59	0.014	0.82	26	+ 0.018	+ 1.06	+ 33
2200 - 2150	0.089	0.169	1.89	60	0.078	0.88	28	+ 0.091	+ 1.01	+ 32
2150 - 2100	0.148	0.279	1.89	60	0.142	0.95	30	+ 0.137	+ 0.94	+ 30
2100 - 2050	0.218	0.384	1.76	55	0.225	1.03	33	+ 0.159	+ 0.73	+ 22
2050 - 2000	0.260	0.503	1.93	61	0.286	1.10	35	+ 0.217	+ 0.83	+ 26
2000 - 1950	0.664	1.255	1.89	60	0.777	1.17	37	+ 0.478	+ 0.72	+ 23
1950 - 1900	1.469	2.763	1.88	59	1.851	1.26	40	+ 0.912	+ 0.62	+ 19
1900 - 1850	3.185	5.564	1.75	55	4.363	1.37	43	+ 1.201	+ 0.38	+ 12
1850 - 1800	1.324	2.117	1.60	50	2.092	1.58	50	+ 0.025	+ 0.02	0
1800 - 1750	0.796	1.132	1.42	45	1.465	1.84	58	- 0.333	- 0.42	- 13
1750 - 1700	0.510	0.720	1.41	44	1.071	2.10	66	- 0.351	- 0.69	- 22
1700 - 1650	0.262	0.345	1.33	42	0.608	2.32	73	- 0.263	- 0.99	- 31
1650 - 1600	0.086	0.126	1.46	46	0.218	2.53	80	- 0.092	- 1.07	- 34
1600 - 1570	0.022	0.032	1.45	46	0.060	2.73	86	- 0.028	- 1.28	- 40
2230 - 1570	9.058	15.421	1.70	54	13.250	1.46	46	+ 2.171	+ 0.24	+ 8

1968/69

Altitude interval	Area	Winter balance			Summer balance			Net balance		
		Total	Specific		Total	Specific		Total	Specific	
m a sl	km^2	10^6 m^3	m	l/s km^2	10^6 m^3	m	l/s km^2	10^6 m^3	m	l/s km^2
2230 - 2200	0.017	0.022	1.30	41	0.023	1.37	43	- 0.001	- 0.07	- 2
2200 - 2150	0.089	0.116	1.30	41	0.122	1.38	44	- 0.006	- 0.08	- 3
2150 - 2100	0.148	0.189	1.28	40	0.204	1.38	44	- 0.015	- 0.10	- 3
2100 - 2050	0.218	0.271	1.24	39	0.309	1.42	45	- 0.038	- 0.18	- 6
2050 - 2000	0.260	0.317	1.22	39	0.403	1.55	49	- 0.086	- 0.33	- 10
2000 - 1950	0.664	0.751	1.13	36	1.079	1.63	51	- 0.328	- 0.50	- 16
1950 - 1900	1.469	1.585	1.08	34	2.567	1.75	55	- 0.982	- 0.67	- 21
1900 - 1850	3.185	3.318	1.04	33	6.192	1.94	61	- 2.874	- 0.90	- 28
1850 - 1800	1.324	1.304	0.98	31	3.258	2.46	78	- 1.954	- 1.48	- 47
1800 - 1750	0.796	0.733	0.92	29	2.202	2.77	87	- 1.469	- 1.85	- 58
1750 - 1700	0.510	0.474	0.93	29	1.517	2.98	94	- 1.043	- 2.05	- 65
1700 - 1650	0.262	0.266	1.01	32	0.819	3.12	99	- 0.553	- 2.11	- 67
1650 - 1600	0.086	0.110	1.28	40	0.291	3.38	107	- 0.181	- 2.10	- 66
1600 - 1570	0.022	0.025	1.14	36	0.077	3.50	111	- 0.052	- 2.36	- 75
2230 - 1570	9.050	9.481	1.05	33	19.063	2.11	67	- 9.582	- 1.06	- 34

1969/70

Altitude interval	Area	Winter balance			Summer balance			Net balance		
		Total	Specific		Total	Specific		Total	Specific	
m a sl	km^2	10^6 m^3	m	l/s km^2	10^6 m^3	m	l/s km^2	10^6 m^3	m	l/s km^2
2230 - 2200	0.017	0.023	1.38	43	0.011	0.63	20	+ 0.012	+ 0.75	+ 24
2200 - 2150	0.085	0.117	1.38	44	0.053	0.63	20	+ 0.064	+ 0.75	+ 24
2150 - 2100	0.144	0.198	1.38	44	0.090	0.63	20	+ 0.108	+ 0.75	+ 24
2100 - 2050	0.212	0.292	1.38	44	0.169	0.80	25	+ 0.123	+ 0.58	+ 18
2050 - 2000	0.242	0.308	1.27	40	0.212	0.88	28	+ 0.096	+ 0.39	+ 13
2000 - 1950	0.656	0.655	1.00	32	0.738	1.25	40	- 0.083	- 0.25	- 4
1950 - 1900	1.469	1.347	0.92	29	1.933	1.32	42	- 0.586	- 0.40	- 13
1900 - 1850	3.185	2.466	0.77	24	4.724	1.48	47	- 2.258	- 0.71	- 22
1850 - 1800	1.324	0.925	0.70	22	2.421	1.83	58	- 1.496	- 1.13	- 36
1800 - 1750	0.791	0.584	0.74	23	1.843	2.33	74	- 1.259	- 1.59	- 50
1750 - 1700	0.510	0.365	0.72	23	1.403	2.75	87	- 1.038	- 2.03	- 65
1700 - 1650	0.262	0.193	0.74	23	0.720	2.75	87	- 0.527	- 2.01	- 63
1650 - 1600	0.086	0.091	1.06	34	0.280	3.26	103	- 0.189	- 2.20	- 69
1600 - 1570	0.022	0.025	1.14	36	0.072	3.27	103	- 0.047	- 2.13	- 68
2230 - 1570	9.005	7.589	0.84	27	14.669	1.63	51	- 7.080	- 0.79	- 25

Table 9.2.7. Norway - Mass balance study results.

Part 21 of 22 Mass balance versus altitude Austre Memurubre

1967/68

Altitude interval	Area	Winter balance			Summer balance			Net balance		
		Total	Specific		Total	Specific		Total	Specific	
m a sl	km^2	10^6 m^3	m	l/s km^2	10^6 m^3	m	l/s km^2	10^6 m^3	m	l/s km^2
2250 - 2200	0.034	0.075	2.22	70	0.038	1.12	36	+ 0.037	+ 1.10	+ 35
2200 - 2150	0.224	0.548	2.45	78	0.252	1.13	36	+ 0.296	+ 1.32	+ 42
2150 - 2100	0.605	1.276	2.11	67	0.681	1.13	36	+ 0.595	+ 0.98	+ 31
2100 - 2050	0.864	1.518	1.76	56	0.989	1.14	36	+ 0.529	+ 0.61	+ 19
2050 - 2000	1.068	1.967	1.84	58	1.485	1.39	44	+ 0.482	+ 0.45	+ 14
2000 - 1950	1.289	2.380	1.85	59	2.000	1.55	49	+ 0.380	+ 0.29	+ 9
1950 - 1900	1.412	2.309	1.64	52	2.421	1.72	54	- 0.112	- 0.08	- 3
1900 - 1850	0.859	1.487	1.73	55	1.644	1.91	61	- 0.157	- 0.18	- 6
1850 - 1800	0.789	1.269	1.61	51	1.665	2.11	67	- 0.396	- 0.50	- 16
1800 - 1750	0.921	1.538	1.67	53	2.117	2.30	73	- 0.580	- 0.63	- 20
1750 - 1700	0.522	0.860	1.65	52	1.389	2.66	84	- 0.529	- 1.01	- 32
1700 - 1650	0.241	0.394	1.64	52	0.776	3.22	102	- 0.383	- 1.59	- 50
1650 - 1630	0.032	0.052	1.62	52	0.115	3.62	115	- 0.063	- 2.00	- 63
2250 - 1630	8.858	15.672	1.77	56	15.574	1.76	56	+ 0.098	+ 0.01	0

1968/69

Altitude interval	Area	Winter balance			Summer balance			Net balance		
		Total	Specific		Total	Specific		Total	Specific	
m a sl	km^2	10^6 m^3	m	l/s km^2	10^6 m^3	m	l/s km^2	10^6 m^3	m	l/s km^2
2250 - 2200	0.034	0.039	1.15	35	0.030	0.88	28	+ 0.009	+ 0.27	+ 7
2200 - 2150	0.224	0.277	1.24	38	0.221	0.99	31	+ 0.056	+ 0.25	+ 7
2150 - 2100	0.605	0.676	1.12	35	0.714	1.18	37	- 0.038	- 0.06	- 2
2100 - 2050	0.864	0.904	1.05	32	1.325	1.53	48	- 0.421	- 0.48	- 16
2050 - 2000	1.068	1.111	1.04	32	1.908	1.79	56	- 0.797	- 0.75	- 24
2000 - 1950	1.289	1.357	1.05	32	2.677	2.08	66	- 1.320	- 1.03	- 34
1950 - 1900	1.410	1.281	0.91	28	3.552	2.52	80	- 2.271	- 1.61	- 52
1900 - 1850	0.859	0.847	0.99	31	2.362	2.75	87	- 1.515	- 1.76	- 56
1850 - 1800	0.789	0.751	0.95	32	2.422	3.07	97	- 1.671	- 2.12	- 65
1800 - 1750	0.921	0.840	0.91	28	3.166	3.44	109	- 2.326	- 2.53	- 81
1750 - 1700	0.522	0.482	0.92	28	2.040	3.91	123	- 1.558	- 2.99	- 95
1700 - 1650	0.241	0.218	0.90	28	1.142	4.74	150	- 0.924	- 3.84	- 122
1650 - 1630	0.032	0.028	0.87	28	0.176	5.50	174	- 0.148	- 4.63	- 146
2250 - 1630	8.858	8.811	0.99	31	21.735	2.45	77	- 12.924	- 1.46	- 46

1969/70

Altitude interval	Area	Winter balance			Summer balance			Net balance		
		Total	Specific		Total	Specific		Total	Specific	
m a sl	km^2	10^6 m^3	m	l/s km^2	10^6 m^3	m	l/s km^2	10^6 m^3	m	l/s km^2
2275 - 2250	0.005	0.007	1.40	44	0.002	0.63	20	+ 0.005	+ 0.77	+ 24
2250 - 2200	0.036	0.040	1.11	35	0.021	0.63	20	+ 0.019	+ 0.48	+ 15
2200 - 2150	0.208	0.251	1.21	38	0.144	0.69	22	+ 0.144	+ 0.52	+ 16
2150 - 2100	0.586	0.670	1.14	36	0.444	0.76	24	+ 0.226	+ 0.38	+ 12
2100 - 2050	0.831	0.749	0.90	28	0.847	1.02	32	- 0.098	- 0.12	- 12
2050 - 2000	1.045	0.870	0.83	26	1.260	1.21	38	- 0.390	- 0.38	- 4
2000 - 1950	1.287	1.053	0.81	25	1.753	1.36	43	- 0.700	- 0.55	- 18
1950 - 1900	1.401	0.991	0.71	22	2.330	1.66	52	- 1.339	- 0.95	- 30
1900 - 1850	0.825	0.582	0.71	22	1.640	1.99	63	- 1.058	- 1.28	- 41
1850 - 1800	0.773	0.593	0.77	24	1.740	2.25	71	- 1.147	- 1.48	- 47
1800 - 1750	0.921	0.682	0.74	23	2.178	2.36	75	- 1.496	- 1.62	- 52
1750 - 1700	0.522	0.371	0.71	22	1.568	3.00	95	- 1.197	- 2.29	- 73
1700 - 1650	0.241	0.166	0.69	22	0.827	3.43	108	- 0.661	- 2.74	- 86
1650 - 1630	0.032	0.020	0.63	22	0.120	3.75	118	- 0.100	- 3.12	- 96
2275 - 1630	8.713	7.045	0.81	26	14.871	1.71	54	- 7.826	- 0.90	- 28

Table 9.2.7. Norway - Mass balance study results.

Part 22 of 22 Mass balance versus altitude Trollbergdalsbreen 1969/70

Altitude interval	Area	Winter balance			Summer balance			Net balance		
		Total	Specific		Total	Specific		Total	Specific	
m a sl	km^2	10^6 m^3	m	l/s km^2	10^6 m^3	m	l/s km^2	10^6 m^3	m	l/s km^2
1370 - 1350	0.003	0.006	2.00	63	0.011	3.66	116	- 0.005	- 1.66	- 53
1350 - 1300	0.050	0.106	2.12	67	0.187	3.74	119	- 0.081	- 1.62	- 52
1300 - 1250	0.045	0.096	2.13	67	0.167	3.71	118	- 0.071	- 1.58	- 51
1250 - 1200	0.190	0.404	2.12	67	0.701	3.69	117	- 0.297	- 1.57	- 50
1200 - 1150	0.191	0.361	1.90	60	0.701	3.68	117	- 0.340	- 1.78	- 57
1150 - 1100	0.137	0.292	2.13	67	0.496	3.62	115	- 0.204	- 1.49	- 48
1100 - 1050	0.556	1.025	1.84	58	2.216	3.98	126	- 1.191	- 2.14	- 68
1050 - 1000	0.607	0.907	1.50	48	2.702	4.45	141	- 1.795	- 2.95	- 93
1000 - 950	0.202	0.290	1.43	45	1.041	5.14	163	- 0.751	- 3.71	- 118
950 - 900	0.074	0.096	1.30	41	0.425	5.74	182	- 0.329	- 4.44	- 141
1370 - 900	2.055	3.583	1.74	55	8.647	4.21	133	- 5.064	- 2.47	- 78

Engabreen 1969/70

Altitude interval	Area	Winter balance			Summer balance			Net balance		
		Total	Specific		Total	Specific		Total	Specific	
m a sl	km^2	10^6 m^3	m	l/s km^2	10^6 m^3	m	l/s km^2	10^6 m^3	m	l/s km^2
1594 - 1500	0.124	0.274	2.21	70	0.217	1.75	55	+ 0.057	+ 0.46	+ 15
1500 - 1450	0.276	0.651	2.36	75	0.482	1.75	55	+ 0.169	+ 0.61	+ 20
1450 - 1400	2.232	5.593	2.51	80	4.042	1.81	57	+ 1.551	+ 0.70	+ 23
1400 - 1350	5.036	12.600	2.50	79	11.171	2.22	70	+ 1.429	+ 0.28	+ 9
1350 - 1300	4.312	10.768	2.49	79	9.700	2.25	71	+ 1.068	+ 0.24	+ 8
1300 - 1250	3.088	7.597	2.46	78	8.130	2.63	83	- 0.533	- 0.17	- 5
1250 - 1200	5.252	11.423	2.17	69	14.585	2.78	88	- 3.162	- 0.61	- 19
1200 - 1150	5.178	10.413	2.01	64	16.260	3.14	99	- 5.847	- 1.13	- 35
1150 - 1100	2.630	4.658	1.77	56	8.550	3.25	103	- 3.892	- 1.48	- 47
1100 - 1050	2.792	4.681	1.68	53	9.870	3.55	112	- 5.189	- 1.87	- 59
1050 - 1000	1.992	3.256	1.63	52	7.460	3.75	119	- 4.204	- 2.12	- 67
1000 - 900	2.460	3.826	1.56	49	9.595	3.90	124	- 5.769	- 2.34	- 75
900 - 800	1.128	1.532	1.36	43	4.930	4.37	138	- 3.398	- 3.01	- 95
800 - 700	0.552	0.621	1.13	36	2.725	4.94	157	- 2.104	- 3.81	- 121
700 - 600	0.442	0.409	0.93	29	2.320	5.25	166	- 1.911	- 4.32	- 137
600 - 80	1.107	0.832	0.71	22	7.560	6.48	205	- 6.728	- 5.77	- 183
1594 - 80	38.661	79.134	2.05	65	117.697	3.04	96	- 38.563	- 0.99	- 31

Table 9.2.8. Sweden - Mass balance study results.

Mass balance of Størglaciären 1966 - 1970

Year		1966	1967	1968	1969	1970
End of accumulation season		27.5.	22.5.	21.5.	20.5.	24.5.
Beginning of ablation season		1.6.	1.6.	3.6.	4.6.	26.5.
End of ablation season		7.9.	21.9.	10.9.	23.9.	21.9.
Total winter balance	10^6 m^3	3.73	4.19	3.93	3.13	3.08
	g/cm^2	121	136	130	102	101
Total summer balance	10^6 m^3	5.36	4.89	4.25	6.27	7.78
	g/cm^2	174	159	141	205	254
Net balance	10^6 m^3	- 1.63	- 0.69	- 0.33	- 3.14	- 4.70
	g/cm^2	- 53	- 23	- 11	- 103	- 153
Equilibrium line	m a.sl.	1500	1500	1480	1570	1610

Table 9.2.9. Sweden - Mass balance study results.

Spot observations of mass balance 1965 - 1967
on 6 glaciers south of Kebnekaise

Glacier	Altitude of stake m a.sl.	Net balance cm of water	Period
Mikkajökeln	1125	- 332	4.8.65 - 18.7.67
	1075	- 442	
Ruotesglaciären	1180	- 293	3.8.65 - 20.7.67
	1100	- 372	
	1090	- 463	
Suotasglaciären	1450	- 163	30.7.65 - 21.7.67
	1330	- 332	
Vartasglaciären	1330	- 202	31.7.65 - 22.7.67
Hyllglaciären	1420	- 120	29.7.65 - 28.7.67
Salajekna	1090	- 253	2.8.65 - 3.8.67
	1020	- 414	
	1010	- 446	
Storglaciären	1400	- 115	1.8.65 - 23.7.67
	1350	- 264	
	1300	- 275	
	1220	- 386	
	1150	- 346	

Table 9.2.10. Austria - Mass balance study results.

Part 1 of 2

(for abbreviations see chapter 4)

Glacier	Year	S_c km^2	B_c 10^6 m^3	$\bar{b}_c$ g/cm^2	S_a km^2	B_a 10^6 m^3	$\bar{b}_a$ g/cm^2	S km^2	B 10^6 m^3	$\bar{b}$ g/cm^2	S_c/S	S_c/S_a	E m a.sl.
Hintereis-ferner	1965/66	6.83	+6.97	+102.0	2.22	-3.86	-173.9	9.05	+3.11	+ 34.5	0.76	3.08	2850
	1966/67	6.20	+5.04	+ 81.2	2.83	-4.86	-172.0	9.03	+0.18	+ 2.0	0.69	2.20	2920
	1967/68	6.63	+6.73	+101.5	2.40	-3.68	-153.0	9.03	+3.05	+ 34.0	0.73	2.77	2850
	1968/69	5.06	+2.48	+ 49.0	3.95	-6.36	-161.0	9.01	-3.88	- 43.1	0.56	1.28	2960
	1969/70	4.41	+1.92	+ 43.6	4.60	-6.90	-150.0	9.01	-4.98	- 55.2	0.49	0.96	3030
Kessel-wand-ferner	1965/66	3.31	+4.20	+127	0.75	-1.79	-239	4.06	+2.41	+ 59	0.82	4.42	3040
	1966/67	3.28	+3.03	+ 92	0.66	-1.86	-282	3.94	+1.18	+ 30	0.83	4.97	3070
	1967/68	3.30	+3.28	+ 99	0.64	-1.50	-234	3.94	+1.78	+ 45	0.84	5.16	3060
	1968/69	3.06	+1.68	+ 55	0.88	-2.28	-259	3.94	-0.60	- 15	0.77	3.48	3090
	1969/70	3.10	+2.07	+ 67	0.84	-2.05	-244	3.94	+0.02	0	0.79	3.69	3080
Vernagt-ferner	1965/66	8.968	+9.650	+107.6	0.585	-0.740	-126.6	9.553	+8.909	+ 93.3	0.94	15.34	2935
	1966/67	6.717	+2.747	+ 40.9	2.814	-1.958	- 69.6	9.522	+0.789	+ 8.3	0.71	2.39	3015
	1967/68	8.148	+4.028	+ 49.4	1.374	-1.158	- 84.2	9.522	+2.870	+ 30.1	0.86	5.93	2995
Langtaler-ferner	1962/63	1.302	+0.610	+ 46.8	1.629	-2.522	-154.2	2.931	-1.912	- 65.2	0.44	0.80	2945
	1963/64	1.081	+0.410	+ 37.9	1.850	-3.495	-188.9	2.931	-3.085	-105.2	0.37	0.58	2990
	1964/65	2.057	+2.897	+140.9	0.846	-0.660	- 78.1	2.903	+2.237	+ 77.0	0.71	2.43	2770
	1965/66	1.880	+2.524	+134.2	1.022	-1.137	-111.3	2.902	+1.387	+ 47.8	0.65	1.84	2815
	1966/67	1.856	+1.235	+ 66.5	1.045	-1.418	-135.8	2.901	-0.183	- 6.3	0.64	1.78	2820
	1967/68	1.912	+1.577	+ 82.5	0.988	-1.026	-103.4	2.900	+0.551	+ 19.0	0.66	1.94	2795
	1968/69	1.355	+0.337	+ 24.9	1.695	-1.908	-112.6	3.050	-1.571	- 51.5	0.44	0.80	2917
	1969/70	1.084	+0.265	+ 24.4	1.965	-2.419	-123.1	3.049	-2.154	- 70.6	0.36	0.55	2975
Stubacher Sonn-blickkees	1963/64	0.325	+0.070	+ 21.5	1.290	-1.408	-109.1	1.615	-1.339	- 82.9	0.20	0.25	*)
	1964/65	1.747	+3.528	+201.9	0.025	-0.028	-112.0	1.772	+3.500	+197.6	0.99	70.73	
	1965/66	1.571	+1.490	+ 94.8	0.200	-0.186	- 93.0	1.772	+1.304	+ 73.6	0.89	7.88	
	1966/67	1.282	+0.610	+ 47.6	0.490	-0.326	- 66.5	1.772	+0.284	+ 16.0	0.72	2.62	
	1967/68	1.340	+0.718	+ 53.6	0.431	-0.300	- 69.6	1.772	+0.418	+ 23.6	0.76	3.12	
	1968/69	0.740	+0.263	+ 35.5	1.032	-0.701	- 67.9	1.772	-0.438	- 24.7	0.42	0.72	
	1969/70	1.211	+0.618	+ 51.0	0.561	-0.363	- 64.7	1.772	+0.255	+ 14.4	0.68	2.16	

*) For the year 1963/64 preliminary values.

Table 9.2.10. Austria - Mass balance study results.

Part 2 of 2 Hintereisferner

budget year	net accumulation		net ablation		mass balance			mean altitude of the equilibrium line	ratio of areas	
1.X.-30.IX.	S_c	B_c	S_a	B_a	S	B	$\bar{b}$		S_c/S	S_c/S_a
	km^2	$10^3\ m^3$	km^2	$10^6\ m^3$	km^2	$10^6\ m^3$	$10^{-1}\ g/cm^2$	m a.sl.		
1952/53	5.44	+ 1.66	4.80	- 7.19	10.24	- 5.53	- 540	3020	0.53	1.13
1953/54	7.04	+ 3.03	3.16	- 5.95	10.20	- 2.92	- 286	2970	0.69	2.23
1954/55	7.57	+ 5.20	2.58	- 4.43	10.15	+ 0.77	+ 76	2850	0.75	2.93
1955/56	7.01	+ 3.12	3.10	- 5.97	10.11	- 2.78	- 275	2920	0.69	2.26
1966/57	6.51	+ 3.74	3.55	- 5.64	10.06	- 1.90	- 189	2930	0.65	1.83
1957/58	3.49	+ 1.49	6.53	-11.32	10.02	- 9.83	- 981	3100	0.35	0.53
1958/59	3.42	+ 1.26	6.55	- 8.87	9.97	- 7.61	- 763	3060	0.34	0.52
1959/60	7.15	+ 4.32	2.77	- 4.94	9.92	- 0.62	- 62	2880	0.72	2.58
1960/61	6.27	+ 4.11	3.61	- 6.14	9.88	- 2.03	- 205	2940	0.63	1.74
1961/62	3.57	+ 1.27	5.64	- 7.66	9.21*)	- 6.41	- 696	3080	0.39	0.63
1962/63	4.83	+ 3.20	4.33	- 8.72	9.16	- 5.52	- 603	3010	0.53	1.12
1963/64	2.29	+ 0.81	6.77	-12.08	9.06	-11.27	-1244	3180	0.25	0.34
1964/65	7.36	+10.67	1.69	- 2.30	9.05	+ 8.37	+ 925	2770	0.81	4.36
1965/66	6.83	+ 6.97	2.22	- 3.86	9.05	+ 3.11	+ 344	2850	0.76	3.08
1966/67	6.20	+ 5.04	2.83	- 4.86	9.03	+ 0.18	+ 20	2920	0.69	2.20
1967/68	6.63	+ 6.73	2.40	- 3.68	9.03	+ 3.05	+ 338	2850	0.73	2.76
1968/69	5.06	+ 2.48	3.95	- 6.36	9.01	- 3.88	- 431	2960	0.56	1.28
1969/70	4.41	+ 1.92	4.60	- 6.90	9.01	- 4.98	- 552	3030	0.49	0.96

*) Two small, apron shaped glaciers on the right slope of the valley are separated from Hintereisferner since 1962, Their mass balance is not taken into account any more.

Remark: In order to determine the mean altitude of the equilibrium line, two methods have been used:

1. (formerly used): the arithmetic mean of the altitude of equidistant points on the equilibrium line analysed on a map.

2. (now used): ordinate for b = 0 in the graph net balance versus altitude. Till 1961 the graph was constructed for countour line intervals of 100 m, since 1962 for those of 50 m.

Table 9.2.11. **Austria - Mass balance study results.**

Part 1 of 3 Mass balance versus altitude

Hintereisferner and Kesselwandferner 1966/67 - 1969/70

(for abbreviations see chapter 4)

Altitude interval	Area	1966/67		1967/68		1968/69		1969/70	
		B	$\overline{b}$	B	$\overline{b}$	B	$\overline{b}$	B	$\overline{b}$
m a.sl.	km^2	10^3 m^3	g/cm^2	10^3 m^3	g/cm^2	10^3 m^3	g/cm^2	10^3 m^3	g/cm^2
				H i n t e r e i s f e r n e r					
3700 - 3650	0.024	+ 22.63	+ 94	+ 20.12	+ 84	+ 0.63	+ 3	+ 2.50	+ 13
3650 - 3600	0.026	+ 24.38	+ 94	+ 21.00	+ 81	+ 1.00	+ 4	+ 5.87	+ 20
3600 - 3550	0.029	+ 26.49	+ 91	+ 31.37	+108	+ 5.88	+ 20	+ 6.25	+ 26
3550 - 3500	0.028	+ 28.48	+102	+ 27.36	+ 98	+ 3.88	+ 14	+ 6.13	+ 20
3500 - 3450	0.091	+ 88.81	+ 98	+ 75.50	+ 83	+ 36.63	+ 40	+ 37.50	+ 44
3450 - 3400	0.150	+ 125.25	+ 84	+ 158.01	+105	+ 63.76	+ 42	+ 80.62	+ 51
3400 - 3350	0.283	+ 282.34	+100	+ 359.12	+127	+ 126.25	+ 45	+ 135.38	+ 46
3350 - 3300	0.415	+ 512.93	+124	+ 587.49	+142	+ 263.87	+ 64	+ 261.62	+ 62
3300 - 3250	0.456	+ 517.78	+114	+ 616.00	+135	+ 203.52	+ 45	+ 228.25	+ 51
3250 - 3200	0.511	+ 538.68	+105	+ 675.37	+132	+ 186.36	+ 36	+ 183.13	+ 36
3200 - 3150	0.699	+ 704.35	+101	+ 847.12	+121	+ 303.74	+ 43	+ 245.62	+ 36
3150 - 3100	0.815	+ 756.09	+ 93	+ 954.64	+117	+ 417.13	+ 51	+ 284.63	+ 34
3100 - 3050	0.816	+ 646.89	+ 79	+ 854.78	+105	+ 301.76	+ 37	+ 155.00	+ 19
3050 - 3000	0.628	+ 275.37	+ 44	+ 496.50	+ 79	+ 126.75	+ 20	- 20.00	- 3
3000 - 2950	0.638	+ 183.46	+ 29	+ 392.24	+ 61	+ 45.87	+ 7	- 194.75	- 30
2950 - 2900	0.583	+ 11.07	+ 2	+ 203.76	+ 35	- 72.62	- 13	- 378.38	- 65
2900 - 2850	0.501	- 87.56	- 17	+ 77.50	+ 15	- 207.12	- 54	- 471.88	- 96
2850 - 2800	0.426	- 213.43	- 50	- 53.13	- 12	- 442.50	-104	- 573.00	-136
2800 - 2750	0.581	- 645.02	-111	- 297.00	- 51	- 891.75	-154	-1033.00	-172
2750 - 2700	0.355	- 652.48	-184	- 412.00	-116	- 795.75	-224	- 829.25	-235
2700 - 2650	0.379	- 894.27	-236	- 665.00	-175	-1111.75	-293	-1012.00	-278
2650 - 2600	0.233	- 651.74	-280	- 575.25	-247	- 832.75	-357	- 703.25	-313
2600 - 2550	0.156	- 548.01	-351	- 496.25	-318	- 659.00	-422	- 557.25	-374
2550 - 2500	0.118	- 471.64	-400	- 419.05	-355	- 542.00	-475	- 471.50	-429
2500 - 2450	0.075	- 322.14	-430	- 342.25	-456	- 348.00	-527	- 313.50	-490
2450 - 2400	0.016	- 81.09	-507	- 89.50	-559	- 65.00	-650	- 48.50	-485
Total	9.032	+ 177.62	+ 2	+3048.45	+ 34	-3881.21	- 43	-4973.76	- 55
				K e s s e l w a n d f e r n e r					
3500 - 3450	0.026	+ 18.4	+ 71.6	+ 22.6	+ 86.9	+ 11.6	+ 44.6	+ 10.0	+ 38.2
3450 - 3400	0.029	+ 18.4	+ 64.1	+ 25.1	+ 86.5	+ 11.5	+ 37.4	+ 8.8	+ 28.7
3400 - 3350	0.050	+ 36.4	+ 73.5	+ 45.8	+ 91.6	+ 16.3	+ 32.0	+ 8.6	+ 16.9
3350 - 3300	0.267	+ 275.0	+102.9	+ 316.3	+118.5	+ 147.0	+ 55.0	+ 169.5	+ 63.4
3300 - 3250	0.525	+ 570.4	+108.7	+ 688.9	+131.2	+ 447.9	+ 85.3	+ 577.8	+110.0
3250 - 3200	0.769	+ 864.9	+112.5	+ 933.5	+121.4	+ 548.1	+ 71.2	+ 635.4	+ 82.5
3200 - 3150	0.729	+ [illegible]	+ [illegible]	+ 730.7	+100.2	+ 344.5	+ 47.5	+ 131.3	+ 50.5
3150 - 3100	0.519	+ 378.0	+ 72.8	+ 399.7	+ 77.0	+ 117.2	+ 22.6	+ 182.7	+ 35.1
3100 - 3050	0.444	+ 81.4	+ 18.3	+ 43.6	+ 9.8	- 275.7	- 62.0	- 304.2	- 68.4
3050 - 3000	0.181	- 300.0	-165.7	- 210.3	-116.0	- 411.2	-227.2	- 387.9	-214.3
3000 - 2950	0.122	- 365.7	-300.5	- 303.0	-248.4	- 370.8	-304.1	- 383.0	-314.2
2950 - 2900	0.062	- 249.6	-400.6	- 211.0	-340.3	- 239.6	-376.2	- 228.3	-358.4
2900 - 2850	0.055	- 211.2	-381.9	- 189.5	-344.5	- 278.4	-500.8	- 174.7	-314.2
2850 - 2800	0.098	- 382.3	-390.5	- 333.0	-339.8	- 419.2	-429.5	- 306.9	-314.4
2800 - 2750	0.053	- 223.3	-418.9	- 167.5	-316.0	- 221.8	-431.8	- 200.7	-374.4
2750 - 2700	0.006	- 18.0	-300.0	- 12.0	-200.0	- 23.4	-543.0	- 19.8	-450.0
Total	3.935	+1175.1	+ 29.9	+1779.9	+ 45.2	- 596.0	- 15.1	+ 18.6	+ 0.5

Table 9.2.11. Austria - Mass balance study results.

Part 2 of 3 Mass balance versus altitude

Langtalerferner 1962/63 - 1969/70

(for abbreviations see chapter 4)

Altitude interval	S 10^3 m^2	B 10^3 m^3	$\bar{b}$ cm	S 10^3 m^2	B 10^3 m^3	$\bar{b}$ cm	S 10^3 m^2	B 10^3 m^3	$\bar{b}$ cm
m a.sl.	1962/63			1963/64			1964/65		
> 3300	53.0	+ 51.8	+ 97.7	53.0	+ 40.7	+ 76.8	17.0	+ 42.4	+249.4
3300 - 3250	122.0	+ 126.5	+103.7	121.5	+ 105.3	+ 86.7	108.0	+ 279.2	+258.5
3250 - 3200	144.0	+ 137.9	+ 95.8	144.0	+ 90.4	+ 62.8	127.0	+ 316.5	+249.2
3200 - 3150	193.0	+ 110.7	+ 57.3	193.0	+ 81.1	+ 42.0	174.0	+ 420.6	+241.7
3150 - 3100	165.0	+ 77.7	+ 47.1	164.5	+ 36.8	+ 22.4	158.0	+ 339.3	+214.7
3100 - 3050	146.0	+ 42.9	+ 29.4	146.0	+ 20.9	+ 14.3	151.0	+ 293.5	+194.4
3050 - 3000	162.0	+ 27.1	+ 16.7	162.0	+ 7.7	+ 4.7	169.0	+ 293.1	+173.4
3000 - 2950	171.0	+ 7.9	+ 4.6	171.0	- 1.4	- 0.8	181.9	+ 268.8	+147.8
2950 - 2900	212.0	- 2.8	- 1.3	212.0	- 63.0	- 29.7	194.0	+ 240.7	+124.1
2900 - 2850	294.0	- 94.0	- 32.0	294.0	- 221.4	- 75.3	307.0	+ 277.8	+ 90.5
2850 - 2800	256.5	- 232.0	- 90.4	256.5	- 376.4	-146.7	273.9	+ 102.7	+ 37.5
2800 - 2750	237.5	- 314.4	-132.4	237.5	- 459.9	-193.6	230.0	- 8.0	- 3.5
2750 - 2700	204.0	- 369.5	-181.1	204.0	- 538.5	-264.0	232.0	- 98.1	- 42.3
2700 - 2650	191.0	- 411.2	-215.3	191.0	- 532.2	-278.6	198.0	- 124.4	- 62.8
2650 - 2600	138.5	- 356.1	-257.1	138.6	- 420.9	-303.7	148.0	- 106.5	- 71.9
2600 - 2550	120.0	- 339.7	-283.1	120.0	- 412.6	-343.8	116.9	- 132.9	-113.7
2550 - 2500	83.0	- 239.2	-288.2	83.0	- 292.5	-352.4	81.9	- 104.1	-127.1
2500 >	39.0	- 135.5	-347.4	39.0	- 149.4	-383.1	35.0	- 63.7	-182.0
Total	2931.5	-1911.9	- 65.2	2930.6	-3085.3	-105.3	2902.6	+2236.9	+ 77.1
	1965/66			1966/67			1967/68		
> 3300	17.0	+ 39.1	+230.0	16.5	+ 22.1	+133.9	16.5	+ 22.1	+133.9
3300 - 3250	108.0	+ 261.2	+241.9	107.7	+ 150.6	+139.8	107.7	+ 149.2	+138.5
3250 - 3200	127.0	+ 299.6	+235.9	126.8	+ 158.7	+125.1	126.8	+ 148.1	+116.8
3200 - 3150	174.0	+ 388.7	+223.4	173.6	+ 187.0	+107.7	173.6	+ 196.4	+113.1
3150 - 3100	158.0	+ 305.0	+193.0	157.7	+ 147.8	+ 93.7	157.7	+ 172.6	+109.4
3100 - 3050	151.0	+ 264.1	+174.9	150.7	+ 132.7	+ 88.0	150.7	+ 151.5	+100.5
3050 - 3000	169.0	+ 270.0	+159.8	169.2	+ 123.3	+ 72.9	169.2	+ 155.1	+ 91.7
3000 - 2950	181.8	+ 253.0	+139.1	181.8	+ 103.8	+ 57.1	181.8	+ 161.6	+ 88.9
2950 - 2900	194.0	+ 210.9	+108.7	194.4	+ 82.4	+ 42.4	194.4	+ 163.2	+ 84.0
2900 - 2850	307.0	+ 190.6	+ 62.1	307.0	+ 94.1	+ 30.7	307.0	+ 182.7	+ 59.5
2850 - 2800	273.9	+ 20.8	+ 7.6	274.1	+ 15.4	+ 5.6	274.1	+ 66.0	+ 24.1
2800 - 2750	230.0	- 72.0	- 31.3	229.6	- 124.7	- 54.3	229.6	- 49.0	- 21.3
2750 - 2700	232.0	- 109.0	- 47.0	231.7	- 247.7	-106.9	231.7	- 135.5	- 58.5
2700 - 2650	197.9	- 147.2	- 74.4	197.9	- 289.0	-146.0	197.9	- 192.5	- 97.3
2650 - 2600	148.0	- 239.7	-161.9	148.0	- 222.5	-150.3	148.0	- 164.5	-111.1
2600 - 2550	116.9	- 247.3	-211.5	117.2	- 214.8	-183.3	117.2	- 211.8	-180.7
2550 - 2500	81.9	- 201.3	-245.8	81.9	- 194.0	-236.8	81.9	- 175.9	-214.8
2500 >	35.0	- 99.6	-284.6	34.8	- 108.7	-312.5	34.8	- 88.2	-253.4
Total	2902.4	+1386.9	+ 47.8	2900.6	- 183.5	- 6.3	2900.6	+ 551.1	+ 19.0
	1968/69			1969/70					
> 3300	17.6	+ 5.3	+ 30.1	17.6	+ 5.3	+ 30.1			
3300 - 3250	111.4	+ 57.9	+ 52.0	111.4	+ 47.6	+ 42.7			
3250 - 3200	129.8	+ 63.7	+ 49.1	129.8	+ 59.9	+ 46.1			
3200 - 3150	178.6	+ 64.6	+ 36.2	178.6	+ 68.8	+ 38.5			
3150 - 3100	160.4	+ 48.1	+ 30.0	160.4	+ 29.6	+ 18.4			
3100 - 3050	159.6	+ 37.1	+ 23.2	159.6	+ 20.8	+ 13.0			
3050 - 3000	149.6	+ 15.9	+ 10.6	149.6	+ 14.7	+ 9.8			
3000 - 2950	178.7	+ 17.0	+ 9.5	178.6	- 2.7	- 1.5			
2950 - 2900	214.3	+ 3.8	+ 1.8	214.3	- 25.2	- 11.7			
2900 - 2850	320.5	- 42.5	- 13.3	320.5	- 61.3	- 19.1			
2850 - 2800	245.9	- 62.5	- 25.4	245.9	- 89.6	- 36.4			
2800 - 2750	300.0	- 237.5	- 79.2	300.0	- 343.2	-114.4			
2750 - 2700	235.4	- 290.4	-123.4	235.4	- 346.3	-147.1			
2700 - 2650	216.1	- 320.5	-148.3	216.1	- 391.6	-181.2			
2650 - 2600	167.5	- 277.5	-165.7	167.5	- 376.0	-224.5			
2600 - 2550	121.3	- 252.4	-208.1	121.3	- 307.9	-253.8			
2550 - 2500	90.8	- 239.4	-263.6	90.8	- 270.1	-297.5			
2500 >	51.8	- 161.0	-310.8	51.8	- 186.8	-360.6			
Total	3049.3	-1570.3	- 51.5	3049.2	-2154.0	- 70.6			

Table 9.2.11. Austria - Mass balance study results.

Part 3 of 3 Mass balance versus altitude Vernagtferner 1965/66 - 1967/68

for abbreviations see chapter 4

Altitude interval	1965/66			1966/67			1967/68		
	S	B	$\bar{b}$	S	B	$\bar{b}$	S	B	$\bar{b}$
m a.sl.	10^3 m^2	10^3 m^3	g/cm^2	10^3 m^2	10^3 m^3	g/cm^2	10^3 m^2	10^3 m^3	g/cm^2
> 3600	3.3	+ 3.6	+109.1	5.5	- 0.1	- 1.8	5.5	+ 1.7	+ 30.9
3600 - 3550	7.3	+ 11.1	+152.0	6.7	+ 0.1	+ 1.5	6.7	+ 2.0	+ 29.9
3550 - 3500	35.2	+ 68.1	+193.5	40.8	+ 9.7	+ 23.8	40.8	+ 16.3	+ 40.0
3500 - 3450	169.7	+ 361.2	+212.8	155.4	+ 87.0	+ 56.0	155.4	+ 93.0	+ 59.8
3450 - 3400	237.7	+ 376.2	+158.3	236.5	+ 58.0	+ 24.5	236.5	+ 104.6	+ 44.2
3400 - 3350	289.1	+ 448.7	+155.2	300.8	+ 42.5	+ 14.1	300.8	+ 97.7	+ 32.5
3350 - 3300	515.8	+ 663.4	+128.6	515.6	+158.6	+ 30.8	515.6	+ 232.6	+ 45.1
3300 - 3250	979.8	+1248.3	+127.4	1017.6	+473.7	+ 46.5	1017.6	+ 651.6	+ 64.0
3250 - 3200	1102.7	+1327.7	+120.4	1146.9	+500.1	+ 43.6	1146.9	+ 669.0	+ 58.3
3200 - 3150	1306.7	+1513.2	+115.8	1214.9	+397.5	+ 32.7	1214.9	+ 670.5	+ 55.2
3150 - 3100	1291.8	+1450.7	+112.3	1323.7	+414.6	+ 31.3	1323.7	+ 644.7	+ 48.7
3100 - 3050	1194.9	+1143.5	+ 95.7	1169.8	+167.7	+ 14.3	1169.8	+ 483.5	+ 41.3
3050 - 3000	919.0	+ 681.9	+ 74.2	922.0	+ 66.6	+ 7.2	922.0	+ 253.4	+ 27.5
3000 - 2950	648.6	+ 239.4	+ 36.9	653.8	-254.3	- 38.9	653.8	- 107.4	- 16.4
2950 - 2900	417.7	- 44.5	- 10.7	404.7	-484.8	-119.8	404.7	- 290.6	- 71.8
2900 - 2850	240.5	- 172.6	- 71.8	230.8	-392.5	-170.1	230.8	- 258.5	-112.0
2850 - 2800	112.3	- 180.8	-161.0	106.5	-244.2	-229.3	106.5	- 216.5	-203.3
2800 - 2750	57.6	- 153.8	-267.0	54.0	-154.4	-285.9	54.0	- 129.5	-239.8
< 2750	23.5	- 76.0	-323.4	16.0	- 56.7	-354.4	16.0	- 47.6	-297.5
Total	9553.2	+8909.3	+ 93.3	9522.0	+789.1	+ 8.3	9522.0	+2870.5	+ 30.1

Table 9.2.12. France - Mass balance study results.

Part 1 of 2 Glacier de Sarennes [1] 1965/66 - 1968/69

Year	Date of measurement	Number of days	Ablation (m of water equivalent) Snow	Ice	Total	Daily mean	Balance since 1.9. (beginning of budget year) m
1965 -1966	15.06.						+ 2.14
		16	0.20		0.20	0.0125	
	1.07.						+ 1.94
		21	0.10		0.10	0.0048	
	22.07.						+ 1.84
		20	0.73		0.73	0.0365	
	11.08.						+ 1.11
		22	0.08		0.08	0.0031	
	2.09.						+ 1.03
		20	0.33		0.33	0.0165	
	22.09.						+ 0.70
		22	0.28		0.28	0.0127	
	14.10.						+ 0.42
	Total	121	1.72		1.27	0.0141	+ 0.42
1966 -1967	13.06.						+ 1.63
		20	0.16		0.16	0.008	
	3.07.						+ 1.47
		23	0.61		0.61	0.0265	
	26.07.						+ 0.86
		14	0.42		0.42	0.030	
	8.08.						+ 0.44
		22	0.25	0.35	0.60	0.0273	
	29.08.						- 0.16
		31	0.19	0.06	0.25	0.008	
	29.09.						- 0.41
	Total	110	1.63	0.41	2.04		- 0.41
1967 -1968	12.06.						+ 1.13
		19	0.06		0.06	0.00316	
	1.07.						+ 1.07
		18	0.48		0.48	0.02666	
	19.07.						+ 0.59
		19	0.15		0.15	0.00789	
	7.08.						+ 0.44
		22	0.12		0.12	0.00545	
	29.08.						+ 0.32
		29	-0.008	-0.012	-0.02	-0.00069	
	27.09.						+ 0.34
	Total	107	0.802	-0.012	0.79		+ 0.34
1968 -1969	3.06.						+ 1.59
		29	0.10		0.10	0.0034	
	2.07.						+ 1.49
		27	0.61		0.61	0.0226	
	29.07.						+ 0.88
		16	0.29		0.29	0.0181	
	14.08.						+ 0.59
		20	0.288	0.012	0.30	0.015	
	3.09.						+ 0.29
		28	0.075	0.085	0.16	0.0057	
	1.10.						+ 0.13
		29	0.127	0.363	0.49	0.169	
	30.10.						- 0.36
	Total	149	1.490	0.460	1.95		

(1 cf. index map Fig. 2 Fluctuations of Glaciers 1959-1965, Glacier Nr. 29.

Table 9.2.12. France - Mass balance study results.

Part 2 of 2 Glacier de Sarennes 1969/70

Year	Date of measurement	Number of days	Ablation (m of water equivalent)				Blance since 1.9. (beginning of budget year) m
			Snow	Ice	Total	Daily mean	
	19.06.						+ 1.82
		14	0.13		0.13	0.0093	
	3.07.						+ 1.69
		25	0.70		0.70	0.0280	
	28.07.						+ 0.99
1969 -1970		23	0.48		0.48	0.0209	
	20.08.						+ 0.51
		25	0.46	0.02	0.48	0.0192	
	14.09.						+ 0.03
		16	0.04	0.40	0.44	0.0275	
	1.10.						- 0.41
	Total	103	1.81	0.42	2.23		

Table 9.2.13. Germany - Mass balance study results.

Nördlicher Schneeferner 1964/65 - 1968/69

B u d g e t y e a r			1964/65	1965/66	1966/67	1967/68
Accumulation area	S_c	(km^2)	0.325	0.325	0.307	0.335
Net accumulation	B_c	(10^6 m^3)	+ 0.610	+ 0.324	+ 0.190	+ 0.203
Specific net accumulation	b_c	(cm)	+187.7	+99.7	+61.9	+62.3
Ablation area	S_a	(km^2)	0.0	0.0	0.018	0.0
Net ablation	B_a	(10^6 m^3)	0.0	0.0	- 0.003	0.0
Specific net ablation	b_a	(cm)	-	-	-16.7	-
Total area glacier	S	(km^2)	0.325	0.325	0.325	0.335
Net balance	B	(10^6 m^3)	+ 0.610	+ 0.324	+ 0.187	+ 0.208
Specific net balance	b	(cm)	+187.7	+99.7	+57.6	+62.3
S_c / S			1.0	1.0	0.94	1.0
S_c / S_a			-	-	17.06	-
Mean altitude of equilibrium line		(m)	(<2560)	(<2560)	(<2560)	(<2560)

Table 9.2.14. Germany - Mass balance study results.

Mass balance versus altitude Nördlicher Schneeferner 1964/65 - 1968/69

Altitude interval m a. sl.	1966/67 S 10^3 m^2	B 10^3 m^2	b cm	1967/68 S 10^3 m^2	B 10^3 m^2	b cm
< 2600	47.2	+ 9.1	+ 19.3	47.2	+ 10.6	+ 22.5
2600 - 2650	118.3	+ 44.2	+ 37.4	125.3	+ 72.0	+ 57.5
2650 - 2700	105.2	+ 83.4	+ 79.3	109.1	+ 94.1	+ 86.2
2700 - 2750	35.5	+ 35.2	+ 99.2	35.2	+ 25.1	+ 71.4
2750 - 2800	16.6	+ 13.6	+ 81.9	15.8	+ 6.0	+ 37.8
> 2800	2.2	+ 1.6	+ 72.7	2.1	+ 0.6	+ 30.0
Total	325.0	+187.1	+ 57.6	334.8	+208.4	+ 62.3

Table 9.2.15. Italy - Mass balance study results.

Mass balance versus altitude

Marmolada

Year	Altitude interval	Area	Accumulation		Ablation		Net balance	
	m a.sl.	m^2	m^3	m	m^3	m	m^3	m
1964-1965	2600 - 2700	28.500	34.000	1.19	51.500	1.80	- 17.500	- 0.61
	2700 - 2800	130.500	156.000	1.19	111.000	0.85	+ 45,000	+ 0.34
	2800 - 2900	114.000	137.000	1.20	35.500	0.31	+ 101.500	+ 0.89
	2900 - 3000	106.000	113.500	1.07	-	-	+ 113.500	+ 1.07
	3000 - 3100	84.000	81.500	0.97	-	-	+ 81.500	+ 0.97
	3100 - 3200	90.000	73.000	0.81	-	-	+ 73.000	+ 0.81
	3200 - 3300	72.500	54.000	0.74	-	-	+ 54.000	+ 0.74
	2600 - 3300	625.500	649.000	1.04	198.000	0.32	+ 451.000	+ 0.72
1965-1966	2600 - 2700	226.500	181.000	0.80	498.500	2.20	- 317.500	- 1.40
	2700 - 2800	365.500	326.000	0.89	329.000	0.90	- 3.000	- 0.01
	2800 - 2900	309.500	302.500	0.98	108.500	0.35	+ 194.000	+ 0.63
	2900 - 3000	257.500	266.000	1.03	23.000	0.09	+ 243.000	+ 0.94
	3000 - 3100	200.500	227.000	1.13	-	-	+ 227.000	+ 1.13
	3100 - 3200	125.500	149.000	1.19	-	-	+ 149.000	+ 1.19
	3200 - 3300	56.500	61.500	1.09	-	-	+ 61.500	+ 1.09
	2600 - 3300	1 541.500	1 513.000	0.98	959.000	0.62	+ 554.000	+ 0.36

Ghiacciaio del Caresèr

Year	Altitude interval	Area	Accumulation		Ablation		Net balance	
	m a.sl.	km^2	10^6 m^3	mm	10^6 m^3	mm	10^6 m^3	mm
1966-1967	2850 - 2900	0.0560	0.0600	1070	0.1190	2120	- 0.0590	- 1050
	2900 - 2950	0.2110	0.2250	1060	0.4135	1950	- 0.1885	- 890
	2950 - 3000	0.3500	0.3640	1040	0.6335	1810	- 0.2695	- 770
	3000 - 3050	0.9010	0.8530	950	1.4690	1630	- 0.6160	- 680
	3050 - 3100	1.0685	1.0325	970	1.5495	1450	- 0.5170	- 480
	3100 - 3150	1.3045	1.3140	1010	1.5915	1220	- 0.2775	- 210
	3150 - 3200	0.4605	0.5145	1120	0.4700	1020	+ 0.0445	+ 100
	> 3200	0.3660	0.4205	1150	0.3730	1020	+ 0.0470	+ 130
	2850 - 3350	4.7175	4.7835	1010	6.6195	1400	- 1.8360	- 390
1967-1968	2850 - 2900	0.0560	0.0460	820	0.0680	1210	- 0.0220	- 390
	2900 - 2950	0.2090	0.1540	740	0.2320	1110	- 0.0780	- 370
	2950 - 3000	0.3480	0.2515	720	0.3725	1070	- 0.1210	- 350
	3000 - 3050	0.9020	0.5790	640	0.7945	880	- 0.2150	- 240
	3050 - 3100	1.0655	0.7885	740	0.6075	570	+ 0.1810	+ 170
	3100 - 3150	1.2975	1.1200	860	0.3890	300	+ 0.7310	+ 560
	3150 - 3200	0.4630	0.4130	890	0.0455	100	+ 0.3665	+ 790
	> 3200	0.3630	0.3720	1020	- 0.0365	- 100	+ 0.4085	+ 1120
	2850 - 3350	4.7040	3.7240	790	2.4730	530	+ 1.2510	+ 260
1968-1969	2855 - 2900	0.0550	0.0525	950	0.1140	2070	- 0.0615	- 1120
	2900 - 2950	0.2115	0.1950	920	0.3975	1880	- 0.2025	- 960
	2950 - 3000	0.3525	0.3305	940	0.5710	1620	- 0.2405	- 680
	3000 - 3050	0.9035	0.8435	930	1.2290	1360	- 0.3855	- 430
	3050 - 3100	1.0675	1.0500	980	1.1420	1070	- 0.0920	- 90
	3100 - 3150	1.3070	1.3485	1030	0.8625	660	+ 0.4860	+ 370
	3150 - 3200	0.4605	0.4745	1030	0.2395	520	+ 0.2350	+ 510
	3200 - 3250	0.2510	0.2635	1050	0.1055	420	+ 0.1580	+ 630
	> 3250	0.1150	0.1210	1050	0.0320	280	+ 0.0890	+ 770
	2855 - 3350	4.7235	4.6790	990	4.6930	990	- 0.0140	∿ 0
1969-1970	2850 - 2900	0.0530	0.0555	1050	0.1440	2720	- 0.0885	- 1670
	2900 - 2950	0.2100	0.2115	1010	0.5480	2610	- 0.3365	- 1600
	2950 - 3000	0.3535	0.3455	980	0.8660	2450	- 0.5205	- 1470
	3000 - 3050	0.8900	0.8270	930	1.8690	2100	- 1.0420	- 1170
	3050 - 3100	1.0570	0.9950	940	1.9660	1860	- 0.9710	- 920
	3100 - 3150	1.3025	1.2635	970	1.5890	1220	- 0.3255	- 250
	3150 - 3200	0.4550	0.5305	1160	0.4505	990	+ 0.0800	+ 170
	3200 - 3250	0.2470	0.2935	1190	0.1455	590	+ 0.1480	+ 600
	> 3250	0.1140	0.1375	1210	0.0340	300	+ 0.1035	+ 910
	2850 - 3350	4.6820	4.6595	990	7.6120	1620	- 2.9525	- 630

Table 9.2.16. Switzerland - Mass balance study results.

Griesgletscher, Aletschgletscher, Limmerngletscher and Silvrettagletscher 1965/66 - 1969/70

Glacier	Balance year	Area km²	Mean specific mass balance kg/m²	Equilibrium line m a.sl.
Gries	3.10.65 - '3.10.66	6.69	- 279	2780
	3.10.66 - 13.10.67	6.54	+ 260	2800
	13.10.67 - 11.10.68	6.38[1]	+ 332	2710
	11.10.68 - 7.10.69	6.38[1]	+ 269	2740
	7.10.69 - 12.10.70	6.38[1]	- 519	3040
Aletsch	1.10.65 - 30.09.66	124.80	+ 591	
	1.10.66 - 30.09.67	124.15	+ 307	
	1.10.67 - 30.09.68	124.00[2]	+ 705	
	1.10.68 - 30.09.69	123.71[3]	+ 321	
	1.10.69 - 30.09.70	123.26[4]	- 129	
Limmern	16.09.65 - 17.09.66	3.29[5]	+ 505	2420
	17.09.66 - 17.09.67	3.29[5]	- 363	2860
	17.09.67 - 9.09.68	3.29[5]	+ 397	2530
	9.09.68 - 5.09.69	3.29[5]	- 140	2740
	5.09.69 - 6.09.70	3.29[5]	- 158	2820
Silvretta	29.09.65 - 23.09.66	3.33[6]	+1096	2510
	23.09.66 - 29.09.67	3.33[6]	+ 262	2715
	29.09.67 - 24.09.68	3.33[6]	+ 456	2645
	24.09.68 - 25.09.69	3.33[6]	- 294	2800
	25.09.69 - 26.09.70	3.33[6]	+ 72	2730

[1] area of 1.9.1967

[2] area estimated for 15.9.1968

[3] area estimated for 13.9.1969

[4] area estimated for 11.9.1970

[5] area of 11.9.1959

[6] area of 2.10.1956

Table 9.2.17. Switzerland - Mass balance study results.

Griesgletscher

Mass balance versus altitude 1969/70

Altitude interval m a.sl.	Area km2	Mass balance $10^6 m^3$ of water	Mean specific mass balance kg/m2
3400 - 3300	0.077	+ 0.056	+ 727
3300 - 3200	0.139	+ 0.076	+ 547
3200 - 3100	0.497	+ 0.175	+ 352
3100 - 3000	1.552	+ 0.497	+ 320
3000 - 2900	0.976	- 0.030	- 31
2900 - 2800	0.707	- 0.175	- 248
2800 - 2700	0.689	- 0.202	- 293
2700 - 2600	0.968	- 1.440	- 1 487
2600 - 2500	0.503	- 1.178	- 2 342
2500 - 2400	0.261	- 0.834	- 3 195
2400 - 2300	0.011	- 0.257 (1	-23 363 (1
3400 - 2300	6.380	- 3.312	- 519

(1 Included the part of the snout melted in contact with the storage lake.

Hydrological balances 1961/62 - 1969/70

Balance year	Runoff 1) kg/m2	Evaporation 2) kg/m2	Storage variation 3) kg/m2	Mean precipitation of the catchment area kg/m2 4)	Precipitation measured with storage gauge kg/m2 5)
3.10.1961 - 2.10.1962	2335	200	- 702	1833	-
2.10.1962 - 3.10.1963	2103	200	+ 21	2324	-
3.10.1963 - 2.10.1964	2466	200	- 567	2099	-
2.10.1964 - 5.10.1965	1449	200	+ 608	2257	1956
5.10.1965 - 3.10.1966	1626	200	- 183	1643	1343
3.10.1966 - 13.10.1967	1964	200	+ 167	2331	1874
13.10.1967 - 11.10.1968	1557	200	+ 208	1962	1887
11.10.1968 - 7.10.1969	1631	200	+ 168	1999	1642
7.10.1969 - 12.10.1970	1858	200	- 325	1733	1331

(1 The runoff values refer to the balance year October 1 to September 30. They differ by ± 2% from the values for the indicated balance year.

Runoff data for the preceding years:

1956/57	1957/58	1958/59	1959/60	1960/61	
2113	2577	2411	2173	2360	kg/m2

(2 Taken invariable

(3 Values for the total catchment area (10.18 km2) calculated from the values for the glaciated area (6.38 km2).

(4 Calculated from runoff, evaporation and storage variation.

(5 Position of the storage gauge: coordinates: 672.340 km / km 145.800; 2510 m a.sl., see fig.9.2.1.

Fig. 9.2.1.

GRIES GLACIER

mass balance 1969/70

Drainage basin of the storage lake Gries $S = 10.18 km^2$
Glaciated area $G = 6.38 km^2$ (1.9.1967) $\frac{G}{S} = 62.7\ \%$
Mean specific mass balance 7.10.1969 – 12.10.1970 = $-519\ kg\ m^{-2}$
Altitude of equilibrium line 12.10.1970 = 3040 m a.s.l.
Variation in the position of the glacier front 17.10.1969 – 18.10.1970 = −25.5 m

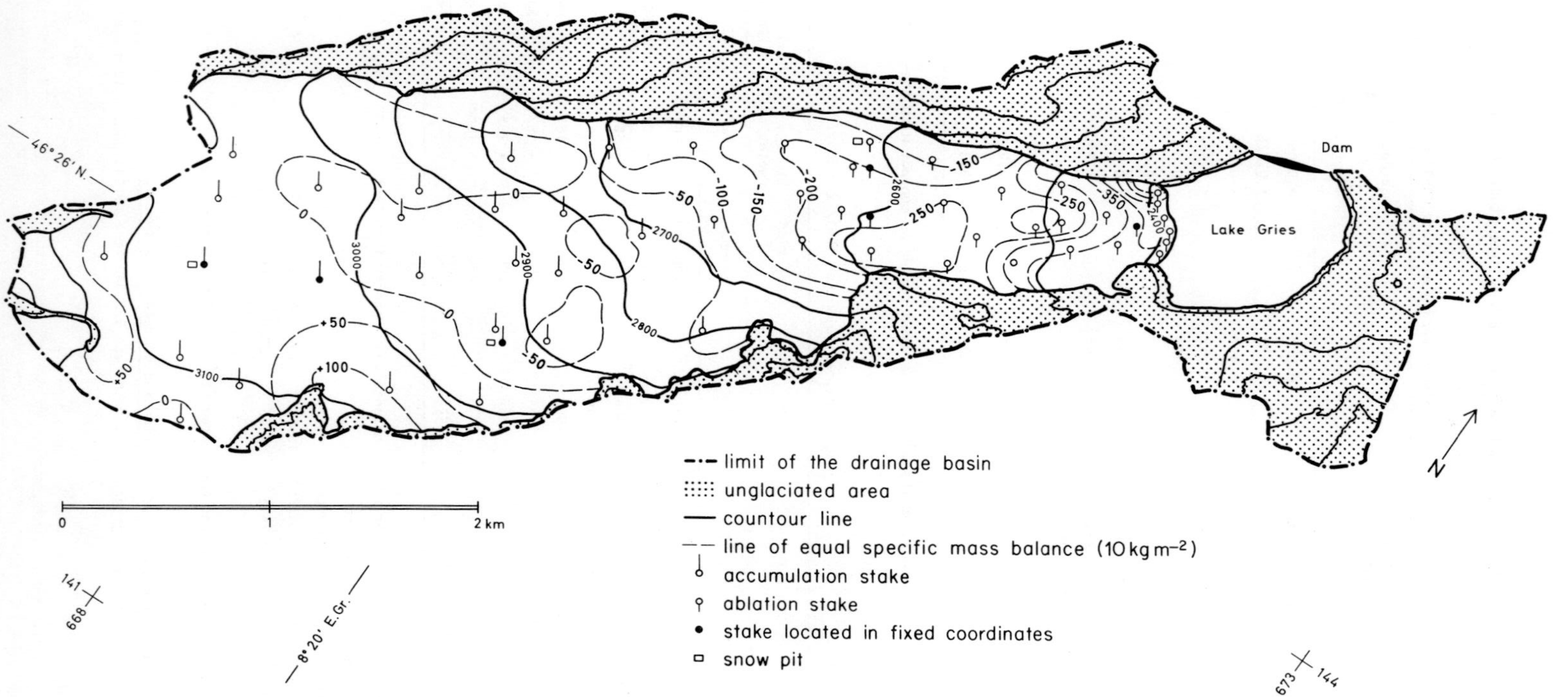

Table 9.2.18. Switzerland - Mass balance study results.

Part 1 of 2 Aletschgletscher: hydrological balances 1964/65 - 1969/70

Catchment area till end of 1964:

Limnigraph station: Massa-Massaboden 686.93 m a.sl., coordinates: 643.9 / 131.5

total area: S_M = 202.00 km^2

mean altitude: 2899.9 m a.sl. (1957)

glacier area: S_G = 129.76 km^2 (1957)

mean altitude of glacier surface: 3037.9 m a.sl. (1957)

Catchment area since 1965:

Limnigraph station: Massa-Blatten bei Naters 1445.96 m a.sl., coordinates: 643.700 / 137.290

total area: S_B = 194.70 km^2

mean altitude: 2950.3 m a.sl. (1957)

P_M, P_B = precipitation
E_M, E_B = evaporation
R_M, R_B = runoff
M_M, M_B = storage variation
} referred to catchment area S_M (Massaboden) and S_B (Blatten bei Naters) respectively.

B = mass balance — referred to glacier area S_G

For catchment area "Massaboden": (1 (2

Balance year 1.10.-30.9.	P_M kg/m^2	E_M kg/m^2	R_M kg/m^2	M_M kg/m^2	S_G (4 km^2	B kg/m^2
1964/65	2601	210	1606	+ 785	126.14	+ 1257

For catchment area "Blatten bei Naters":

Balance year 1.10.-30.9.	P_B kg/m^2	E_B kg/m^2	R_B kg/m^2	M_B kg/m^2	S_G (4 km^2	B kg/m^2
1964/65	2652	210	1627	+ 815	126.14	+ 1257
1965/66	2391	210	1802	+ 379	124.80	+ 591
1966/67	2426	210	2020	+ 196	124.15	+ 307
1967/68	2358	210	1699	+ 449	124.00	+ 705
1968/69	2241	210	1827	+ 204	123.71	+ 321
1969/70	2270	210	2142	- 82	123.26	- 129

(1 For data of the years 1922/23 - 1964/65 see the publication: P. Kasser (1967): Fluctuations of Glaciers, Vol. 1, Table 11, UNESCO and IAHS.

(2 The precipitation is not measured in the catchment area "Massaboden", but it is calculated from the precipitation of the three neighbouring stations Fiesch, Kippel and Grindelwald by the relation:

$P_M = 2/3(P_{Fiesch} + P_{Kippel} + P_{Grindelwald})$

This equation was found using the balance values for the period October 1, 1927 to September 30, 1957 as follows:

B determined from photogrammetric surveys in the years 1927 and 1957 (5

R_M measured at the limnigraph station Massaboden

E_M estimated as 210 kg/m^2·year

$M_M = B \cdot S_G/S_M$ it follows: $P_M = R_M + E_M + M_M$

Table 9.2.18. **Switzerland - Mass balance study results.**

Part 2 of 2 Aletschgletscher: hydrological balances 1964/65 - 1969/70

(3 Determination of P_B:

$$P_B \cdot S_B = P_M \cdot S_M - P_D (S_M - S_B)$$

where $$\frac{P_D}{P_{D(1901-40)}} = \frac{(N_{Fiesch} + N_{Kippel})}{(N_{Fiesch} + N_{Kippel})_{(1901-40)}}$$

P_D for the period 1901-1940 was estimated being 1200 kg/m^2.

(4 The glacier area S_G was calculated by assuming that the variation of the position of the glacier snout is proportional to the variation of the glacier area S_G. For the period 1927-1957 the variation of S_G amounted to - 8.14 km^2, that one of the position of the snout to - 451.6 m.

(5 The variation of the glacier volume for the period 1927-1957 is to be found in the publication mentioned under no. 1. The mass balance B was calculated from the variation of the volume and the density γ = 0.9 kg/dm^3.

Table 9.2.19. Switzerland - Mass balance study results.

Mass balance versus altitude 1969/70

Limmerngletscher — Balance year: 5.9.1969 - 6.9.1970

Altitude interval m a.sl.	Area km^2	Mass balance $10^6 m^3$ of water	Mean specific mass balance kg/m^2
3400 - 3300	0.053	+ 0.040	+ 755
3300 - 3200	0.034	+ 0.012	+ 353
3200 - 3100	0.014	+ 0.003	+ 214
3100 - 3000	0.079	+ 0.014	+ 177
3000 - 2900	0.264	+ 0.063	+ 239
2900 - 2800	0.572	+ 0.125	+ 219
2800 - 2700	0.980	- 0.091	- 93
2700 - 2600	0.529	- 0.253	- 478
2600 - 2500	0.257	- 0.063	- 245
2500 - 2400	0.235	- 0.107	- 455
2400 - 2300	0.101	- 0.104	- 1030
2300 - 2200	0.124	- 0.114	- 919
2200 - 2100	0.048	- 0.045	- 938
2100 - 3400	3.290	- 0.520	- 158

Silvrettagletscher — Balance year: 25.9.1969 - 26.9.1970

Altitude interval m a.sl.	Area km^2	Mass balance $10^6 m^3$ of water	Mean specific mass balance kg/m^2
3100 - 3000	0.337	+ 0.356	+ 1056
3000 - 2900	0.658	+ 0.483	+ 734
2900 - 2800	0.638	+ 0.327	+ 512
2800 - 2700	0.848	+ 0.077	+ 91
2700 - 2600	0.379	- 0.164	- 433
2600 - 2500	0.363	- 0.551	- 1518
2500 - 2400	0.107	- 0.288	- 2692
2400 - 3100	3.330	+ 0.240	+ 72

Fig. 9.2.2.

LIMMERN and PLATTALVA GLACIER

mass balance 1969/70

Glaciated area G = 3.29 km^2 (11.9.1959)

Mean specific mass balance 5.9.1969 – 6.9.1970 = –158 kg m^{-2}

Altitude of equilibrium line 6.9.1970 = 2820 m a.s.l.

Variation in the position of the glacier front 7.9.1969 – 2.9.1970 = –0.6 m

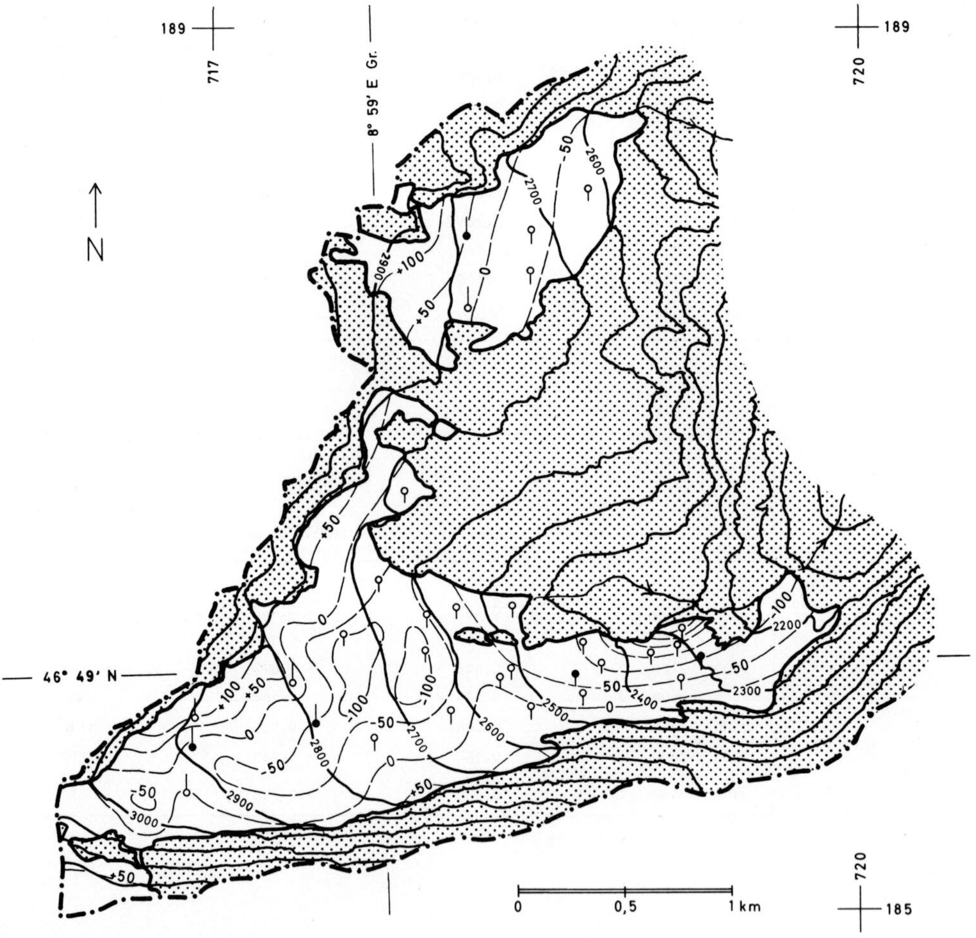

–·– limit of the drainage basin

:::::: unglaciated area

—— countour line

– – line of equal specific mass balance (10 kg m^{-2})

accumulation stake

ablation stake

• stake located in fixed coordinates

▫ snow pit

Fig. 9.2.3.

SILVRETTA GLACIER

mass balance 1969/70

Glaciated area G = 3.33 km² (2.10.1956)

Mean specific mass balance 25.9.1969 – 26.9.1970 = +72 kg m⁻²

Altitude of equilibrium line 26.9.1970 = 2730 m a.s.l.

Variation in the position of the glacier front 27.9.1969 – 24.9.1970 = –4.0 m

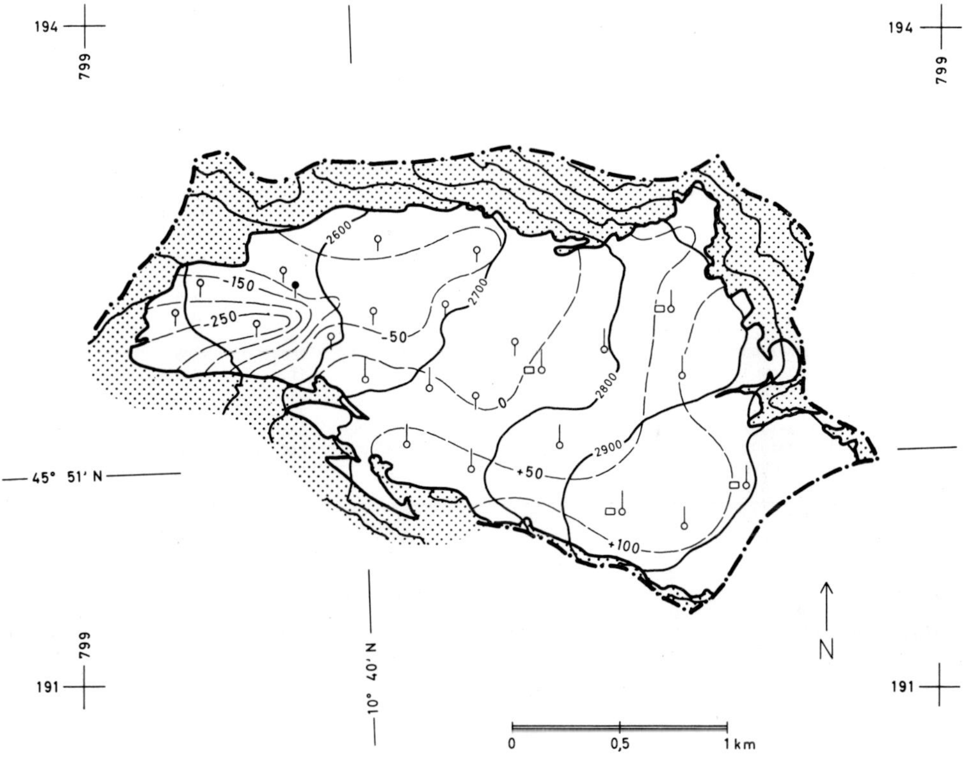

—·— limit of the drainage basin

∷∷ unglaciated area

—— countour line

— — line of equal specific mass balance (10 kg m⁻²)

accumulation stake

ablation stake

• stake located in fixed coordinates

▫ snow pit

Table 9.2.20. USSR - Mass balance study results.

Part 1 of 3

Tsentralny Tuyuksu Glacier; Tien - Shan 1967 - 1968

Altitude interval	m	3400--3500	3500--3600	3600--3700	3700--3800	3800--3900	3900--4000	4000--4100	4100--4219	3400--4219
Area	km^2	0.35	0.37	0.32	0.91	0.41	0.30	0.34	0.15	3.15
Total accumulation	mm	480	580	750	930	1123	920	650	570	810
Total ablation	mm	3080	2630	2380	1620	1170	940	700	500	1700
Inner alimentation	mm					220	410	300	250	113
Mass balance	mm	-2600	-2050	-1630	-690	+ 173	+ 390	+ 250	+ 320	- 780
	10^3 m^3	- 910	- 758	- 522	- 628	+ 71	+ 117	+ 85	+ 48	-2457

P o l a r U r a l

Glacier	Budget year	Firn basin area 10^3 m^2	Net balance of accumulation area 10^3 t	Net balance of ablation area 10^3 t	The winter balance g/cm^2	The summer balance g/cm^2	The total mass balance g/cm^2
Obrucheva's Glacier	1959-1960	130	60	225	175	225	- 50
Glacier area: 0.30 km^2	1960-1961	160	100	170	280	301	- 21
	1961-1962	150	170	220	350	365	- 15
	1962-1963	50	30	270	250	323	- 73
	1963-1964	5	5	405	110	240	-130
	1964-1965	120	55	170	242	278	- 36
	1965-1966	200	170	82	203	175	+ 28
	1966-1967	199	174	115	376	356	+ 20
	1967-1968	272	538	10	241	160	+181
	1968-1969	31	23	200	148	203	- 55
	1969-1970	205	176	75	220	186	+ 34
Glacier Igan	1958-1959	250	195	480	265	299	- 34
Glacier area: 0.81 km^2	1959-1960	160	45	620	140	208	- 68
	1960-1961	250	185	400	220	246	- 26
	1961-1962	330	315	320	320	321	- 1
	1962-1963	60	15	1080	210	337	-127
	1963-1964	100	60	910	100	200	-100
	1964-1965	125	65	620	250	316	- 66
	1965-1966	305	129	287	175	195	- 20
	1966-1967	454	418	221	314	283	+ 31
	1967-1968	808	1852	0	341	92	+249
	1968-1969	197	122	575	142	200	- 58
	1969-1970	282	285	277	178	177	+ 1

C a u c a s u s

Glacier	Budget year	Total accumulation 10^{-1} gr/cm^2	Total ablation 10^{-1} gr/cm^2	Mass balance 10^{-1} gr/cm^2	The altitude of equilibrium line, m
Dzhankuat Glacier	1967 - 1968	199 ± 13	199 ± 9	0 ± 16	3090 ± 20
	1968 - 1969	188 ± 15	293 ± 12	-105 ± 19	3280 ± 40
	1969 - 1970	251 ± 13	199 ± 50	52 ± 14	3050 ± 10
	1970 - 1971	258 ± 11	258 ± 4	0 ± 13	3170 ± 10

Table 9.2.20. USSR - Mass balance study results.

Part 2 of 3

C A U C A S U S

Glacier	Budget year	Total accumulation 10^6 m^3	Total accumulation 10^{-1} gr/cm^2	Total ablation 10^{-6} m^3	Total ablation 10^{-1} gr/cm^2	Mass balance 10^6 m^3	Mass balance 10^{-1} gr/cm^2
	1959 - 60	91.8	254	75.0	207	+ 16.8	+ 47
	1960 - 61	69.6	192	92.1	254	- 22.5	- 62
	1961 - 62	73.7	204	108.6	300	- 34.9	- 96
	1962 - 63	94.0	260	85.1	235	+ 8.9	+ 25
	1963 - 64	83.4	230	82.2	227	+ 11.2	+ 3
Bezingi glacier	1964 - 65	71.0	196	92.6	256	- 21.6	- 60
	1965 - 66	104.3	288	103.0	284	+ 1.3	+ 4
	1966 - 67	85.4	236	75.0	207	+ 10.4	+ 29
	1967 - 68	68.3	189	101.3	280	- 33.0	- 91
	1968 - 69	71.0	196	90.2	249	- 19.2	- 53
	1969 - 70	87.6	242	82.8	229	+ 4.8	+ 13
	1965 - 66	2.7	168	2.8	174	- 0.1	- 6
	1966 - 67	2.5	155	1.0	62	+ 1.5	+ 94
Kelbashi glacier	1967 - 68	2.1	132	2.0	125	+ 0.1	+ 7
	1968 - 69	2.2	137	1.7	108	+ 0.4	+ 29
	1969 - 70	2.7	169	1.5	95	+ 1.2	+ 74
	1959 - 60	19.5	201	18.9	195	+ 0.6	+ 6
	1960 - 61	18.4	190	24.5	253	- 6.1	- 63
	1961 - 62	15.4	159	19.2	198	- 3.8	- 39
	1962 - 63	27.9	288	16.8	173	+ 11.1	+115
	1963 - 64	19.0	196	19.4	200	- 0.4	- 4
Zeiskiy glacier	1963 - 65	17.0	175	19.4	200	- 2.4	- 25
	1965 - 66	21.0	223	21.2	219	+ 0.4	+ 4
	1966 - 67	30.3	312	16.6	171	+ 13.7	+141
	1967 - 68	19.3	199	21.2	219	- 1.9	- 20
	1968 - 69	16.4	169	21.2	219	- 4.8	- 50
	1969 - 70	22.1	217	17.6	181	+ 3.5	+ 36

T I E N - S H A N

Glacier	Budget year	Accumulation 10^6 m^3	Ablation 10^6 m^3	Regimen 10^6 m^3	Specific annual mass balance m	Begin of ablation period, date	Glacier area km^2	Glacier area year
	10.09.1958-10.10.1959	3.36	4.93	8.29	-1.57	12.7		
	10.10.1959-20.09.1960	3.04	4.37	7.41	-1.33	7.7		
	20.09.1960-20.09.1961	2.16	5.85	8.01	-3.69	10.6		
	20.09.1961-08.09.1962	2.99	3.38	6.37	-0.39	28.6		
	08.09.1962-20.09.1963	3.04	3.21	6.25	-0.17	2.7		
Karabatkak glacier	20.09.1963-01.09.1964	4.07	3.41	7.48	+0.66	18.7	4.58	1964
	01.09.1964-28.08.1965	3.24	3.43	6.67	-0.10	20.6		
	28.08.1965-15.09.1966	4.00	4.70	8.70	-0.70	18.6		
	15.09.1966-28.08.1967	3.63	3.58	7.21	+0.05	3.7		
	28.08.1967-03.09.1968	2.79	5.58	8.37	-2.79	19.6		
	03.09.1968-28.08.1969	3.03	2.04	5.07	+0.99	28.6		
	28.08.1969-10.09.1970	2.98	1.78	4.76	+1.20	23.6		

Table 9.2.20. USSR - Mass balance study results.

Part 3 of 3

T i e n - S h a n

G l a c i e r	Budget year	Net accumulation		Net ablation		Mass balance			The altitude snow line
		km^2	10^6 m^3	km^2	10^6 m^3	km^2	10^6 m^3	10^{-1} g/cm^2	m
Tsentralny Tuyuksu glacier	1956 - 1957	1.57	+ 0.86	1.66	- 1.31	3.23	- 0.45	- 15	3760
	1957 - 1958	2.22	+ 1.52	1.01	- 0.47	3.23	+ 1.05	+ 33	3670
	1958 - 1959	1.19	+ 0.64	2.03	- 2.03	3.22	- 1.39	- 42	3800
	1959 - 1960	1.75	+ 0.84	1.47	- 1.17	3.22	- 0.33	- 10	3740
	1960 - 1961	1.10	+ 0.66	2.11	- 2.45	3.21	- 1.79	- 56	3820
	1961 - 1962	1.00	+ 0.59	2.21	- 2.78	3.21	- 2.19	- 69	3840
	1962 - 1963	2.31	+ 2.08	0.90	- 0.64	3.21	+ 1.44	+ 44	3640
	1963 - 1964	2.43	+ 2.28	0.88	- 0.56	3.21	+ 1.72	+ 52	3600
Igly Tuyuksu glacier	1956 - 1957	0.95	+ 0.56	1.04	- 0.50	1.99	+ 0.06	+ 3	3680
	1957 - 1958	1.26	+ 1.04	0.73	- 0.14	1.99	+ 0.90	+ 45	3600
	1958 - 1959	0.83	+ 0.46	1.16	- 0.80	1.99	- 0.34	- 17	3750
	1959 - 1960	0.95	+ 0.94	1.04	- 0.52	1.99	+ 0.42	+ 21	3680
	1960 - 1961	0.83	+ 0.57	1.16	- 1.02	1.99	- 0.45	- 23	3760
	1961 - 1962	0.65	+ 0.52	1.34	- 1.25	1.99	- 0.73	- 37	3780
	1962 - 1963	1.19	+ 1.02	0.80	- 0.22	1.99	+ 0.80	+ 40	3620
	1963 - 1964	1.26	+ 1.10	0.73	- 0.19	1.99	+ 0.91	+ 46	3600
Molodezhnyy glacier	1956 - 1957	0.78	+ 0.60	0.90	- 0.44	1.68	+ 0.16	+ 9	3650
	1957 - 1958	0.95	+ 0.82	0.73	- 0.14	1.68	+ 0.68	+ 41	3600
	1958 - 1959	0.66	+ 0.54	1.02	- 0.72	1.68	- 0.18	- 11	3700
	1959 - 1960	0.77	+ 0.68	0.90	- 0.42	1.68	+ 0.26	+ 16	3650
	1960 - 1961	0.66	+ 0.64	1.02	- 0.74	1.68	- 0.10	- 6	3700
	1961 - 1962	0.57	+ 0.50	1.11	- 1.06	1.68	- 0.56	- 33	3750
	1962 - 1963	0.84	+ 1.05	0.84	- 0.44	1.68	+ 0.61	+ 36	3630
	1963 - 1964	0.91	+ 1.10	0.77	- 0.32	1.68	+ 0.78	+ 46	3610
Shokalskiy glacier *	1962 - 1963	0.54	+ 0.54	0.26	- 0.07	0.80	+ 0.47	+ 60	3950
	1963 - 1964	0.76	+ 0.70	0.04	- 0.01	0.80	+ 0.69	+ 87	3850
Mametova's glacier	1956 - 1957	0.20	+ 0.14	0.31	- 0.09	0.51	+ 0.05	+ 10	3800
	1957 - 1958	0.26	+ 0.22	0.25	- 0.03	0.51	+ 0.19	+ 38	3750
	1958 - 1959	0.15	+ 0.11	0.36	- 0.17	0.51	- 0.06	- 11	3825
	1959 - 1960	0.20	+ 0.17	0.31	- 0.08	0.51	+ 0.09	+ 18	3790
	1960 - 1961	0.15	+ 0.13	0.36	- 0.19	0.51	- 0.06	- 11	3825
	1961 - 1962	0.15	+ 0.12	0.36	- 0.19	0.51	- 0.07	- 15	3825
	1962 - 1963	0.23	+ 0.23	0.28	- 0.13	0.51	+ 0.10	+ 20	3770
	1963 - 1964	0.27	+ 0.28	0.24	- 0.07	0.51	+ 0.21	+ 41	3730
Teu-Northern glacier	1962 - 1963	1.44	+ 1.41	0.46	- 0.21	1.90	+ 1.20	+ 63	3795
	1963 - 1964	1.58	+ 1.45	0.32	- 0.11	1.90	+ 1.34	+ 71	3770
Teu-Southern glacier	1962 - 1963	1.38	+ 1.35	0.50	- 0.26	1.88	+ 1.09	+ 58	3820
	1963 - 1964	1.47	+ 1.35	0.41	- 0.07	1.88	+ 1.28	+ 68	3800
Korzhenevsky glacier	1964 - 1966	25.70	60.00	12.30	48.60	38.00	7.30	+ 19	3960-4000

A l t a i

Bolshoy Berel glacier	1962 - 1967	5.33	50.8	4.12	44.1	9.45	+ 6.7	+0.7	3000
Malyi Berel glacier	1962 - 1967	3.39	36.1	3.45	35.5	6.84	+ 0.6	+0.1	2800

* The right component of the right branch

Table 9.2.21. **New Zealand - Mass balance study results.**

Tasman Glacier 43.5° S, 170.3° E Green.

Specific mass balances 1966/67 - 1970/71

(see also Fig. 9.2.4.) mass balance in water equivalent

Budget year	Site	Altitude m a. sl.	Winter balance m	Summer balance m	Net balance m
1966 - 1967	1	2340	+ 3.38	- 0.98	+ 2.40
	2	2055	+ 2.02	?	?
	3	1970	+ 1.39	- 1.39 (approx.)	0 (approx.)
	4	1820	+ 1.36	?	Negative
	5	1731	+ 1.24	?	Negative
	6	1545	+ 0.55	?	Negative
1967 - 1968	1	2340	+ 7.26	- 1.72	+ 5.54
	2	2055	+ 7.09	- 1.90	+ 5.19
	3	1970	+ 5.28	- 1.46	- 3.82
	4	1820	+ 4.35	?	Positive
	5	1731	- 3.19	- 2.39	+ 0.80
	6	1545	+ 1.30	?	Negative
1968 - 1969	1	2340	+ 5.96	- 1.38	+ 4.58
	2	2060	+ 4.94	- 1.12	+ 3.82
	3	1974	+ 4.27	- 1.16	+ 3.11
	4	1822	+ 3.21	- 2.04	- 1.17
	5	1735	+ 2.45	- 2.03	+ 0.42
	6	1549	+ 1.32	?	Negative
1969 - 1970	1	2340	+ 4.30	- 3.14	+ 1.25
	2	2060	+ 3.25	- 3.50	- 0.25
	3	1974	+ 2.66	?	Negative
	4	1822	+ 2.33	?	Negative
	5	1735	+ 2.64	?	Negative
	6	1545	+ 0.45	?	Negative
1970 - 1971	1	2340	+ 4.05	- 2.09	+ 1.96
	2	2060	+ 3.61	- 1.87	+ 1.74
	3	1974	+ 2.60	- 2.80	- 0.20
	4	1822	+ 2.34	- 4.09	- 1.75
	5	1735	+ 3.19	- 5.09	- 1.90
	6	1549	+ 1.15	- 6.35	- 5.20

Budget year	Net balance in m of water equivalent on site 1 (2340 m)	2 (2055 m)	3 (1970 m)	4 (1820 m)	5 (1731 m)	6 (1545 m)
1966 - 1967	+ 2.40	?	ca 0	Neg.	Neg.	Neg.
1967 - 1968	+ 5.54	+ 5.19	- 3.82	Pos.	+ 0.80	Neg.
1968 - 1969	+ 4.58	+ 3.82	+ 3.11	- 1.17	+ 0.42	Neg.
1969 - 1970	+ 1.25	- 0.25	Neg.	Neg.	Neg.	Neg.
1970 - 1971	+ 1.96	+ 1.74	- 0.20	- 1.75	- 1.90	- 5.20

Fig. 9.2.4. New Zealand

UPPER TASMAN GLACIER

Location of the stakes

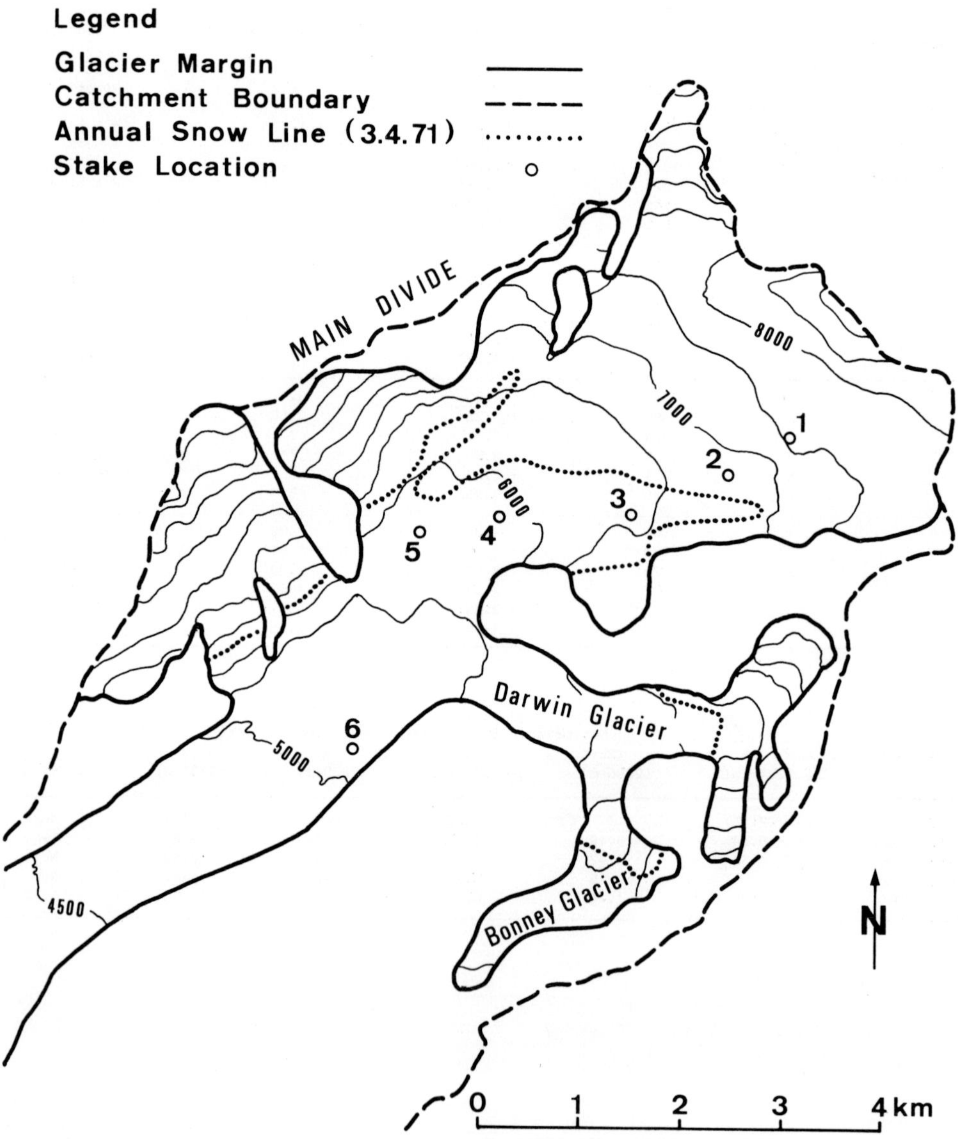

Contour interval 500 ft. (Glacier surface)

Table 9.2.22. Deception Island - Mass balance study results.

Glacier G1, 63.0 °S, 60.6 °W Green. Mass balance versus altitude 1968/69, 1969/70 and 1970/71

Elevation m a.sl.	Area m^2	1968 - 1969 B_n m^3	b_n m	1969 - 1970 B_w m^3	b_w m	B_s m^3	b_s m	B_n m^3	b_n m	1970 - 1971 B_w m^3	b_w m	B_s m^3	b_s m	B_n m^3	b_n m
> 400	1 600	1 000	0.62	1 600	0.99	- 500	-0.29	1 100	0.70	700	0.41	- 500	-0.29	200	0.12
375 - 400	78 900	41 800	0.53	48 800	0.62	- 24 400	-0.31	24 400	0.31	28 400	0.36	- 25 200	-0.32	3 200	0.04
350 - 375	45 700	24 700	0.54	38 400	0.84	- 20 600	-0.45	17 800	0.39	19 200	0.42	- 16 500	-0.36	2 700	0.06
325 - 350	40 100	20 400	0.51	44 100	1.10	- 23 300	-0.58	20 800	0.52	20 500	0.51	- 20 100	-0.50	400	0.01
300 - 325	40 300	13 700	0.34	29 400	0.73	- 27 000	-0.67	2 400	0.06	23 400	0.58	- 30 600	-0.76	- 7 200	-0.18
275 - 300	31 100	300	0.01	11 500	0.37	- 24 600	-0.79	- 13 100	-0.42	16 500	0.53	- 30 500	-0.98	- 14 000	-0.45
250 - 275	31 400	1 900	0.06	26 700	0.85	- 23 800	-0.76	2 900	0.09	15 100	0.48	- 33 000	-1.05	- 17 900	-0.57
225 - 250	52 800	- 9 000	-0.17	37 000	0.70	- 63 400	-1.20	- 26 400	-0.50	29 100	0.55	- 69 700	-1.32	- 40 600	-0.77
200 - 225	26 300	-19.500	-0.74	15 800	0.60	- 45 500	-1.73	- 29 700	-1.13	7 600	0.29	- 45 000	-1.71	- 37 400	-1.42
175 - 200	33 000	-32 700	-0.99	18 800	0.57	- 68 600	-2.08	- 49 800	-1.51	10 300	0.31	- 61 400	-1.86	- 51 100	-1.55
150 - 175	17 900	-20 200	-1.13	7 200	0.40	- 47 500	-2.65	- 40 300	-2.25	3 900	0.22	- 37 900	-2.12	- 34 000	-1.90
125 - 150	15 100	-20 300	-1.34	9 800	0.65	- 40 800	-2.70	- 31 000	-2.05	3 300	0.22	- 35 000	-2.32	- 31 700	-2.10
< 125	3 600	- 5 800	-1.62	3 800	1.05	- 9 800	-2.73	- 6 000	-1.68	700	0.20	- 9 500	-2.64	- 8 800	-2.44
Total	417 800	- 3 700	-0.01	292 900	0.70	-419 800	-1.00	-126 900	-0.30	178 700	0.43	-414 900	-0.99	-236 200	-0.56

Budget year	Area 10^3 m^2	B_w 10^3 m^3	$\bar{b}_w$ m	B_s 10^3 m^3	$\bar{b}_s$ m	B_n 10^3 m^3	$\bar{b}_n$ m
1968 - 1969	418					- 4	- 0.01 ± 0.15
1969 - 1970	418	+ 293	+ 0.70 ± 0.15	- 420	- 1.00 ± 0.15	- 127	- 0.30 ± 0.10
1970 - 1971	418	+ 179	+ 0.43 ± 0.15	- 415	- 0.99 ± 0.20	- 236	- 0.56 ± 0.15

B_N (b_N) total (specific) net balance

B_W (b_W) total (specific) winter balance

B_S (b_S) total (specific) summer balance

Fig. 9. 2. 5. Deception Island - Location map and map of the Glacier G 1

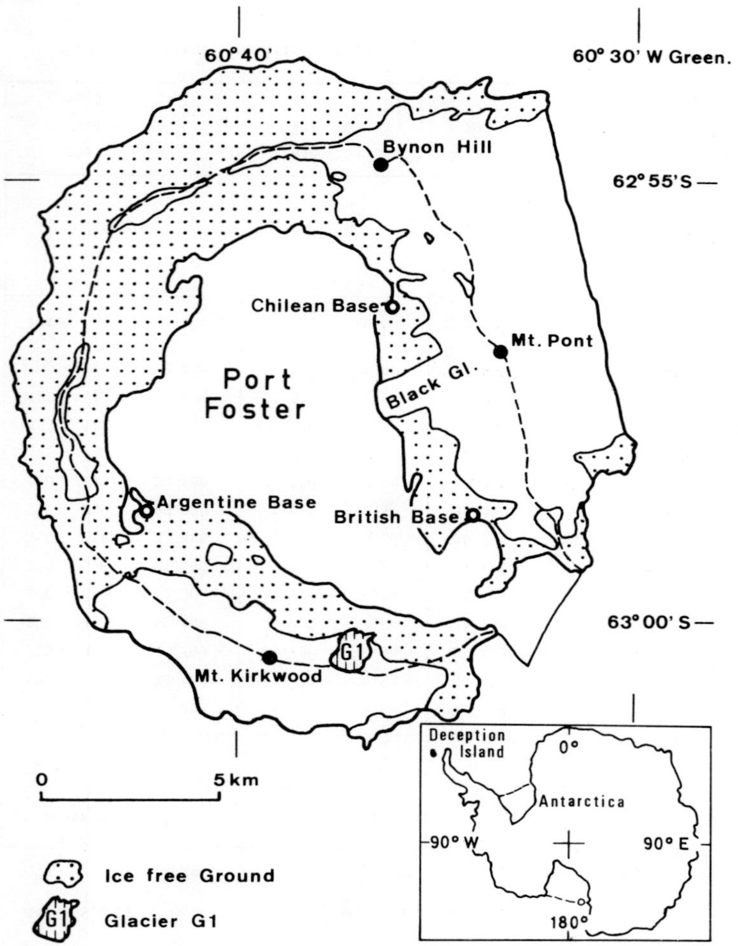

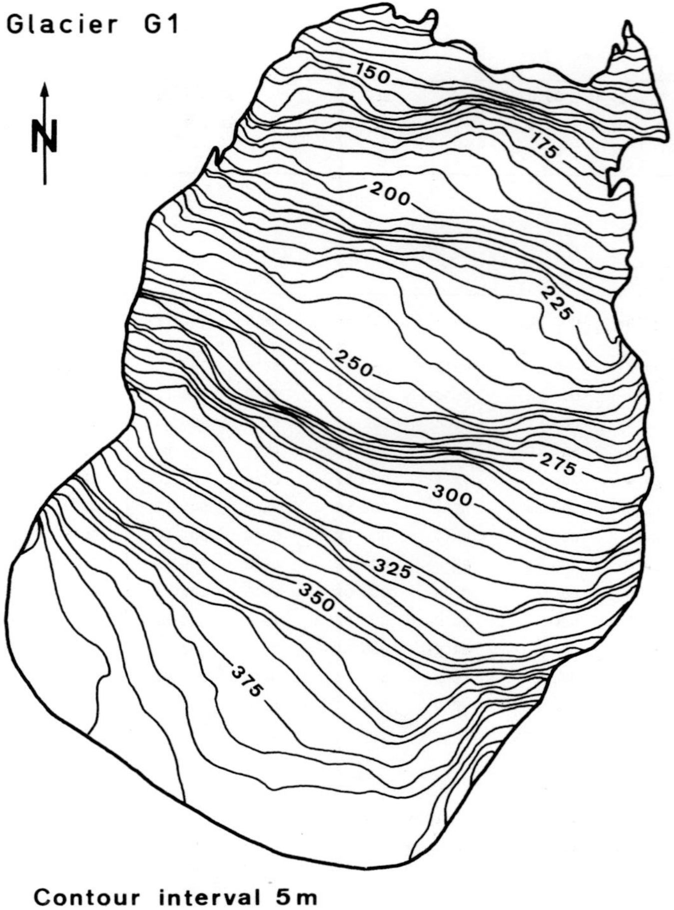

Table 9.3.1. USA - Changes in thickness and area.

Part 1 of 2

for references see chapter 5, for abbreviations of the sponsoring agencies see chapter 2

No. Name	Aerea km²	Altitude of glacier m	standard[1)] error (m)	method of measurement	Altitude range m	Aerea (Date) $m^2 \cdot 10^3$	Aerea (Date) $m^2 \cdot 10^3$	Aerea (Change) $m^2 \cdot 10^3$	Altitude Changes and Dates m	Investigator (References)
43 Sherman	54.1	725[2)]– 112	not computed est: > 2	Triangulation survey to stakes					3.7.66–23.8.67	C.Bull, IPS (Ref. Bull and Marangunic, 1967)
					100 – 200				+ 4.5	
					200 – 300				0.0	
					300 – 400				0.0	
					400 – 500				0.0	
					500 – 600				– 1.5	
					600 – 700				– 1.5	
51 Slide	17.2 (1959)	1219– 152	0.5	Theodolite triangulation to points on glacier, and direct levelling across and up the glacier		(1958)			15.6.65–25.6.68	J.R.Reid UND (Ref. Reid 1969)
					200				+ 10	
					200 – 300	950			+ 11	
					300 – 400	900			– 2	
115 Coleman	–	3200–1250	1 (est)	Terrestrial photogrammetry		(1964)		(1965)	19.9.64–19.9.65	A.E.Harrison UW (EE) (Ref. Bengtson 1956)
					1200–1300	7		+ 17.0	+ 25.1	
					1300–1400	79		+ 9.8	+ 6.2	
					1400–1500	117		+ 12.5	+ 4.0	
					1500–1600	261		+ 10.2	+ 2.9	
121 South Cascade	2.8	2190–1610	0.5	Photogrammetry (1961) triangulation surveys to stakes on glacier (1967)		(1961)	(1967)	1961–1967	12.9.61–22.9.67	M.F.Meier (Ref. Meier and Tangborn 1965)
					1600–1700	133	111	– 22	– 15.1	
					1700–1800	450	420	– 30	– 9.1	
					1800–1900	755	734	– 21	– 3.1	
					1900–2000	664	650	– 14	– 1.2	
					2000–2100	672	655	– 17	– 1.0	
					2100–2200	220	233	– 13	+ 0.1	

Table 9.3.1. USA - Changes in thickness and area.

Part 2 of 2

for references see chapter 5, for abbreviations of the sponsoring agencies see chapter 2.

No. Name	Aerea km²	Altitude of glacier m	standard[1] error (m)	method of measurement	Altitude range m	Aerea (Date) $m^2 \cdot 10^3$	Aerea (Date) $m^2 \cdot 10^3$	Aerea (Change) $m^2 \cdot 10^3$	Altitude Changes and Dates m					Investigator (References)
129 Nisqually	6.5 (1966)	4300-1410 (1966)	1.0 (est)	Aerial photogrammetry		(1966)			27.8.63-12.9.64	12.9.64-31.8.65	31.8.65-31.8.66	31.8.66-28.8.67	28.8.67-24.9.68	A.Johnson G.C.Giles D.Richardson USGS (Ref. Meier 1968)
					1600-1700	140			+ 4	+ 1	- 2	- 2	- 4	
					1800-1900	364			- 1	- 1	- 3	- 2	- 5	
					2000-2100	238			0	+ 1	- 6	0	- 5	
134 Grinnell	1.27 (1960)	2255-1945 (1960)	ca.0.8	2 profiles across glacier measured since 1950 and a third profile measured since 1957		(1960)		1964-1968	1964-1968					A.Johnson USGS/NPS (Ref. Alden, 1914)
					below2000	280		ca.100[3]	below 2100 m decrease of 1 - 2 m; changes above 2100 m not determined					
					2000-2100	620								
					2100-2200	250								
					above2200	120								
135 Sperry	1.16	2680-2250	ca.0.7-1.0	transverse profile across entire glacier. One longitudinal profile extending up glacier about 670m from 1969 terminus, one longitudinal profile extending up glacier about 200m from 1969 terminus		(1960)		(1964-1968)	1964-1968					A.Johnson USGS/NPS (Ref.Dyson, 1948)
					below2300	220		ca.100-150[3]	Very little change. It appears that above 2350m the surface elevation has remained much the same with a possible indication of an increase of 2 to 3m. Below 2350m there was a slight decrease, about 1 to 2m. The terminus continues to recede on a very irregular pattern with consequent slight decrease in surface altitude of the aerea immediately above the terminus					
					2300-2400	430								
					2400-2500	300								
					2500-2600	170								
					above2600	40								

1) Standard error of altitude determinations averaged over the aerea of one 100 m increment

2) Exclusive of tributaries which reach 1676 m

3) Due primarily to terminal recession

Table 9.3.2. Austria - Changes in thickness, area and volume.

Part 1 of 2 Vernagtferner and Guslarferner 1889 - 1912, 1912 - 1938, 1938 - 1969

see also annexed maps.

Altitude interval	Area 10³ m²				Change in volume 10³ m³			Altitude change of the glacier surface related to mean areas of the periods m and m/year					
m a.sl.	1889	1912	1938	1969	1889 1912	1912 1938	1938 1969	1889 - 1912		1912 - 1938		1938 - 1969	
Vernagtferner													
3650 - 3600	4	5	4	4	+ 7	- 14	0	+ 1.6	+0.07	- 3.0	-0.12	0.0	0.00
3600 - 3550	12	12	10	8	+ 40	- 88	- 38	+ 3.4	+0.15	- 8.1	-0.31	- 4.3	-0.14
3550 - 3500	56	55	35	34	+ 42	- 480	- 82	+ 0.8	+0.03	- 10.6	-0.41	- 2.4	-0.08
3500 - 3450	215	204	187	182	- 100	- 960	- 150	- 0.5	-0.02	- 4.9	-0.19	- 0.8	-0.03
3450 - 3400	314	292	252	251	- 435	- 1335	- 152	- 1.4	-0.06	- 4.9	-0.19	- 0.6	-0.02
3400 - 3350	366	348	304	298	- 590	- 1715	- 120	- 1.7	-0.07	- 5.3	-0.20	- 0.4	-0.01
3350 - 3300	660	600	522	535	- 1138	- 2800	+ 72	- 1.8	-0.08	- 5.0	-0.19	+ 0.1	0.00
3300 - 3250	1138	1101	1039	1039	- 2132	- 4150	+ 472	- 1.9	-0.08	- 3.9	-0.15	+ 0.5	+0.02
3250 - 3200	1280	1199	1144	1082	- 4460	- 4878	- 435	- 3.6	-0.16	- 4.2	-0.16	- 0.4	-0.01
3200 - 3150	1422	1392	1368	1345	- 7200	- 5110	- 1568	- 5.1	-0.22	- 3.7	-0.14	- 1.2	-0.04
3150 - 3100	1440	1404	1354	1259	- 8935	- 5408	- 3480	- 6.3	-0.27	- 3.9	-0.15	- 2.7	-0.09
3100 - 3050	1376	1369	1304	1192	-10015	- 5552	- 7378	- 7.3	-0.32	- 4.2	-0.16	- 5.9	-0.19
3050 - 3000	1011	1041	1009	880	- 9252	- 5062	-10595	- 9.0	-0.39	- 4.9	-0.19	- 11.2	-0.36
3000 - 2950	670	746	776	638	- 6498	- 3472	-13975	- 9.2	-0.40	- 4.6	-0.18	- 19.7	-0.64
2950 - 2900	428	451	424	408	- 4108	- 2632	-14635	- 9.4	-0.41	- 6.0	-0.23	- 35.1	-1.13
2900 - 2850	293	343	303	230	- 2540	- 3778	-12320	- 8.0	-0.35	- 11.7	-0.45	- 46.2	-1.49
2850 - 2800	243	276	208	109	- 1000	- 5252	-10882	- 3.9	-0.17	- 21.7	-0.83	- 68.6	-2.21
2800 - 2750	192	237	135	55	+ 5	- 7470	- 9605	0.0	0.00	- 40.1	-1.54	-101.0	-3.26
2750 - 2700	110	142	82	14	+ 758	- 7362	- 6275	+ 6.0	+0.26	- 65.9	-2.53	}-140.0	}-4.52
2700 - 2650	132	143	20		+ 882	- 6518	- 1788	+ 6.4	+0.28	- 79.9	-3.07		
2650 - 2600	111	120			+ 575	- 6098		+ 5.0	+0.22				
2600 - 2550	81	58			- 390	- 3078		- 5.6	-0.24	}-100.6	}-3.87		
2550 - 2500	22	10			- 450	- 280		-28.0	-1.22				
Total	11576	11548	10480	9563	-56934	-83492	-92924	- 4.9	-0.21	- 7.6	-0.29	- 9.3	-0.30
Guslarferner													
3500 - 3450	4	5	4	4	0	0	0	0.0	0.00	0.0	0.00	0.0	0.00
3450 - 3400	32	27	23	22	- 125	- 78	0	- 4.2	-0.18	- 3.1	-0.12	0.0	0.00
3400 - 3350	70	71	53	51	- 232	- 342	+ 60	- 3.3	-0.14	- 5.5	-0.21	+ 1.2	+0.04
3350 - 3300	169	155	140	136	- 592	- 498	+ 102	- 3.7	-0.16	- 3.4	-0.13	+ 0.7	+0.02
3300 - 3250	246	207	203	200	- 1695	- 518	+ 92	- 7.5	-0.33	- 2.5	-0.10	+ 0.5	+0.02
3250 - 3200	242	263	237	218	- 1842	- 865	- 122	- 7.3	-0.32	- 3.5	-0.13	- 0.5	-0.02
3200 - 3150	342	318	315	285	- 1632	- 1178	- 860	- 4.9	-0.21	- 3.7	-0.14	- 2.9	-0.09
3150 - 3100	369	335	315	298	- 2558	- 1178	- 1642	- 7.3	-0.32	- 3.6	-0.14	- 5.4	-0.17
3100 - 3050	501	519	432	378	- 2710	- 2045	- 2565	- 5.3	-0.23	- 4.3	-0.16	- 6.3	-0.20
3050 - 3000	538	543	511	422	- 2265	- 3290	- 4240	- 4.2	-0.18	- 6.2	-0.24	- 9.1	-0.29
3000 - 2950	464	487	443	367	- 2062	- 3442	- 5950	- 4.3	-0.19	- 7.4	-0.28	- 14.7	-0.47
2950 - 2900	455	444	425	295	- 2168	- 3518	- 8788	- 4.8	-0.21	- 8.1	-0.31	- 24.4	-0.79
2900 - 2850	366	332	318	233	- 3050	- 3210	- 9870	- 8.7	-0.38	- 9.9	-0.38	- 35.9	-1.16
2850 - 2800	172	192	123	67	- 2800	- 2355	- 6998	-15.4	-0.67	- 14.9	-0.57	- 73.5	-2.37
2800 - 2750	101	116	93	10	- 1278	- 1995	- 4358	-11.8	-0.51	- 19.1	-0.73		
2750 - 2700	68	56	35		- 760	- 1602	- 1815	-12.3	-0.54	- 35.0	-1.35	}- 88.7	}-2.81
2700 - 2650	50	38	1		- 792	- 878	- 19	-18.1	-0.79	- 43.4	-1.67		
2650 - 2600	31	5			- 408	- 81		}-22.3	}-0.97				
2600 - 2550	7				- 65								
Total	4227	4113	3671	2986	-27034	-27073	-46973	- 6.5	-0.28	- 7.0	-0.27	- 14.1	-0.46

Table 9.3.2. Austria - Changes in thickness, area and volume.

Part 2 of 2 Hintereisferner 1894 - 1920, 1920 - 1940 and 1953 - 1962

Altitude interval m a.sl.	Area 10^3 m^2		Change in volume 10^3 m^3	Altitude change of the glacier surface			
					m/year	m/year	m/year
	1953	1962	1953-1962	1953-1962 (2)	1953-1962 (2)	1894-1920 (3)	1920-1940 (3)
> 3600	39	37	± 0	± 0	± 0	-	-
3600 - 3500	53	59	- 5	- 0.84	-0.09	-	-
3500 - 3400	213	223	+ 250	+ 1.14	+0.13	-0.02	+0.02
3400 - 3300	675	699	+ 1450	+ 2.10	+0.23	-0.11	+0.01
3300 - 3200	933	975	+ 2750	+ 2.83	+0.31	-0.29	+0.01
3200 - 3100	1577	1697	+ 3300	+ 1.95	+0.22	-0.36	-0.02
3100 - 3000	1601	1612	+ 2950	+ 1.82	+0.20	-0.37	-0.15
3000 - 2900	1445	1394	- 250	- 0.18	-0.02	-0.42	-0.14
2900 - 2800	1155	1030	- 7950	- 6.98	-0.78	-0.39	-0.42
2800 - 2700	1043	966	-12750	- 12.83	-1.43	-0.42	-1.22
2700 - 2600	677	655	-11150	- 17.17	-1.91	-0.44	-2.22
2600 - 2500	369	305	-10150	- 28.24	-3.14	-0.46	-2.96
2500 - 2450	86	86	- 4200	- 46.12	-5.12	} -0.59	} -3.70
2450 - 2400	51	35	- 2575	- 56.98	-6.33		
2400 - 2383.5	8	11	- 413	-(17.93)	-		
2383.5-2370.5(1)	2	-	- 13	-(6.50)	-		
Total	9927	9784	-39156	- 3.97	-0.44	-0.38	-0.63

(1) 2383.5 m: altitude of snout in 1962
2370.5 m: altitude of snout in 1953

(2) Change in altitude detremined for only area of 1962

(3) See: R. Finsterwalder 1953, Zeitschrift für Gletscherkunde II, 2: 189-239.

Table 9.3.3. France - Changes in thickness.

Glacier d'Argentière 1956 to 1971

Mean thickness changes in profiles

Profile	Mean surface heigt in metres above sea-level in the year 1958	Mean thickness changes in metres in the years						
		1956/58	1958/59	1959/60	1959/69	1960/69	1969/70	1970/71
ancien	1817.1	+ 4.6	- 7.6	+ 0.5		+10.5	+ 1.5	+ 1.7
1	1833.1		- 2.3	- 1.8				
2	1832.0		- 2.0	- 2.4				
3	2399.8		- 0.2		+ 7.2		+ 0.2	- 2.0
6	2554.1		- 2.4		+ 6.4		0.0	- 1.5
7A	2711.3		- 1.9	+ 1.8		+ 7.5	0.0	- 1.2
7B	2754.1		- 1.3	+ 2.4		+ 4.6	+ 0.1	- 1.1
8	2808.7		- 0.7	+ 2.2				

Fig. 9.3.1. France - Glacier d'Argentière, location of the profiles.

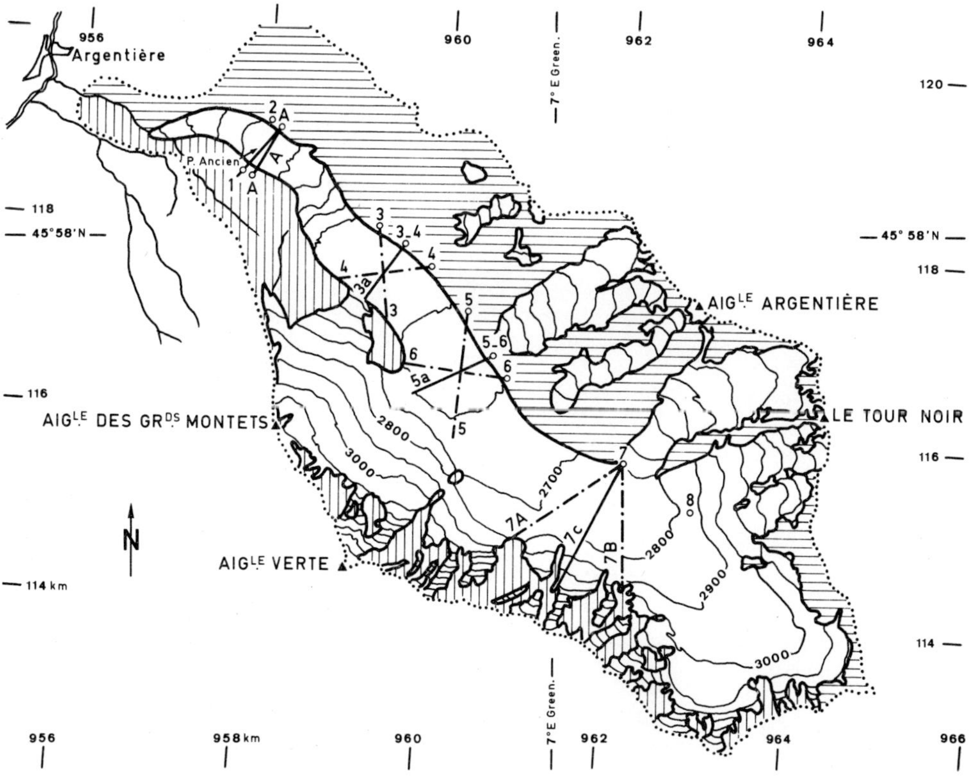

Table 9.3.4. Germany - Changes in area.

Nördlicher Schneeferner 1856 - 1969

Year	Number of years	Area km2	Change of area km2	Terrestrial photogrammetrical survey carried out by	Reference
1856		0.68		—	R. Finsterwalder, 1951 Die Gletscher der Bayerischen Alpen Jahrbuch des Deutschen Alpenvereins, p. 60-66
	36		- 0.07		
1892		0.61		S. Finsterwalder	
	57		- 0.22		
1949		0.39		Institut für Photogrammetrie der TH München	
	12		- 0.053		
1961		0.337		Kommision für Glaziologie	
	4		- 0.012		
1965		0.3250		Kommision für Glaziologie	
	4		+ 0.0729		
1969		0.3979		Kommision für Glaziologie	

Table 9.3.5. Switzerland - Changes in Thickness, area and volume

Part 1 of 6 Mattmarkgletscher 1956/57 - 1966/67

Glacier area 1956 in 10^3 m^2.

Altitude interval m a. sl.	1 Kessjen	2 Hohlaub	3 Allalin	4 Hangend	5 Fluchthorn Schwarzberg-grat	6 Schwarzberg	7 Seewinen	8 Monte Moro	9 Tälliboden	10 Ofental	Total
4200 - 4150			12								12
4150 - 4100			59								59
4100 - 4050			89								89
4050 - 4000		4	94								98
4000 - 3950		7	112		2						121
3950 - 3900		11	180		9						200
3900 - 3850		17	245		7						269
3850 - 3800		45	185		6						236
3800 - 3750		38	199		14						251
3750 - 3700		26	255		40						321
3700 - 3650		32	354		47						433
3650 - 3600		36	365		54	11					466
3600 - 3550		44	408		73	21					546
3550 - 3500		103	428		35	82					648
3500 - 3450		107	549		16	139					811
3450 - 3400		121	548		22	188					879
3400 - 3350		144	456		26	306					932
3350 - 3300		147	715	26	1	361					1250
3300 - 3250	2	140	806	81	28	465	2				1524
3250 - 3200	6	188	706	68	12	549	15				1544
3200 - 3150	15	245	356	27	7	495	65				1210
3150 - 3100	26	197	405	25	9	475	141				1278
3100 - 3050	46	292	465	18	1	472	334			2	1630
3050 - 3000	126	168	307	15		507	325			15	1463
3000 - 2950	168	147	334	26		733	229			44	1681
2950 - 2900	159	150	275	5		430	264	13	7	69	1372
2900 - 2850	76	37	380			487	153	87	33	94	1347
2850 - 2800	2	20	237			191	131	85	48	102	816
2800 - 2750		22	106			258	75	32	69	102	664
2750 - 2700		14	77			177	44	1	105	73	491
2700 - 2650		17	53			36	40		36	81	263
2650 - 2600			43			0	66		6	26	141
2600 - 2550			28			0	36				64
2550 - 2500			28			3	0				31
2500 - 2450			24			52					76
2450 - 2400			24			20					44
2400 - 2350			18			2					20
2350 - 2300			9								9
2300 - 2250											
Total	626	2519	9934	291	409	6460	1920	218	304	608	23289

Table 9.3.5. Switzerland - Changes in thickness, area and volume.

Part 2 of 6 Mattmarkgletscher 1956/57 - 1966/67

Glacier area 1967 in 10^3 m^2

Altitude interval m a. sl.	1 Kessjen	2 Hohlaub	3 Allalin	4 Hangend	5 Fluchthorn Schwarzberg-grat	6 Schwarzberg	7 Seewinen	8 Monte Moro	9 Tälliboden	10 Ofental	Total
4200 - 4150			12								12
4150 - 4100			61								61
4100 - 4050			92								92
4050 - 4000		4	94								98
4000 - 3950		7	113		2						122
3950 - 3900		12	186		9						207
3900 - 3850		18	238		8						264
3850 - 3800		44	190		8						242
3800 - 3750		40	205		19						264
3750 - 3700		32	268		34						334
3700 - 3650		32	352		46						430
3650 - 3600		35	364		51	8					458
3600 - 3550		54	407		69	15					545
3550 - 3500		104	437		31	78					650
3500 - 3450		108	537		17	133					795
3450 - 3400		135	553		23	192					903
3400 - 3350		150	467	1	14	309					941
3350 - 3300	0	162	748	26	0	357					1293
3300 - 3250	3	140	812	82	21	450	1				1509
3250 - 3200	8	201	689	63	7	547	15				1530
3200 - 3150	20	238	357	25	7	485	60				1192
3150 - 3100	28	198	419	10	9	487	152				1303
3100 - 3050	46	295	480	9	1	480	296			0	1607
3050 - 3000	123	163	302	16		621	359			6	1590
3000 - 2950	157	134	337	30		652	238			35	1583
2950 - 2900	129	133	285	6		409	249	14	4	61	1290
2900 - 2850	64	34	385			451	152	68	20	92	1266
2850 - 2800		17	225			163	124	76	36	83	724
2800 - 2750		20	107			212	69	33	50	77	568
2750 - 2700		0	74			162	28	0	83	60	407
2700 - 2650		0	55			48	16		45	70	234
2650 - 2600			39			0	27		7	27	100
2600 - 2550			26			0	23				49
2550 - 2500			24			0	0				24
2500 - 2450			8			29					37
2450 - 2400			0			8					8
2400 - 2350			0			0					0
2350 - 2300			0								0
2300 - 2250											
Total	578	2510	9948	268	376	6296	1809	191	245	511	22732

Table 9.3.5. Switzerland - Changes in thickness, area and volume.

Part 3 of 6 Mattmarkgletscher 1956/57 - 1966/67

Change in area 1956 - 1967 in $10^3 m^2$

Altitude interval m a. sl.	1 Kessjen	2 Hohlaub	3 Allalin	4 Hangend	5 Fluchthorn Schwarzberg-grat	6 Schwarzberg	7 Seewinen	8 Monte Moro	9 Tälliboden	10 Ofental	Total
4200 - 4150			0								0
4150 - 4100			+ 2								+ 2
4100 - 4050			+ 3								+ 3
4050 - 4000		0	0								0
4000 - 3950		0	+ 1		0						+ 1
3950 - 3900		+ 1	+ 6		0						+ 7
3900 - 3850		+ 1	- 7		+ 1						- 5
3850 - 3800		- 1	+ 5		+ 2						+ 6
3800 - 3750		+ 2	+ 6		+ 5						+ 13
3750 - 3700		+ 6	+13		- 6						+ 13
3700 - 3650		- 0	- 2		- 1						- 3
3650 - 3600		- 1	- 1		- 3	- 3					- 8
3600 - 3550		+10	- 1		- 4	- 6					- 1
3550 - 3500		+ 1	+ 9		- 4	- 4					+ 2
3500 - 3450		+ 1	-12		+ 1	- 6					- 16
3450 - 3400		+14	+ 5		+ 1	+ 4					+ 24
3400 - 3350		+ 6	+11	+ 1	-12	+ 3					+ 9
3350 - 3300		+15	+33	0	- 1	- 4					+ 43
3300 - 3250	+ 1	0	+ 6	+ 1	- 7	- 15	- 1				- 15
3250 - 3200	+ 2	+13	-17	- 5	- 5	- 2	0				- 14
3200 - 3150	+ 5	- 7	+ 1	- 2	0	- 10	- 5				- 18
3150 - 3100	+ 2	+ 1	+14	-15	0	+ 12	+ 11				+ 25
3100 - 3050	0	+ 3	+15	- 9	0	+ 8	- 38			- 2	- 23
3050 - 3000	- 3	- 5	- 5	+ 1		+114	+ 34			- 9	+127
3000 - 2950	-11	-13	+ 3	+ 4		- 81	+ 9			- 9	- 98
2950 - 2900	-30	-17	+10	+ 1		- 21	- 15	+ 1	- 3	- 8	- 82
2900 - 2850	-12	- 3	+ 5			- 36	- 1	-19	-13	- 2	- 81
2850 - 2800	- 2	- 3	-12			- 28	- 7	- 9	-12	-19	- 92
2800 - 2750		- 2	+ 1			- 46	- 6	+ 1	-19	-25	- 96
2750 - 2700		-14	- 3			- 15	- 16	- 1	-22	-13	- 84
2700 - 2650		-17	+ 2			+ 12	- 24		+ 9	-11	- 29
2650 - 2600			- 4			0	- 39		+ 1	+ 1	- 41
2600 - 2550			- 2			0	- 13				- 15
2550 - 2500			- 4			- 3	0				- 7
2500 - 2450			-16			- 23					- 39
2450 - 2400			-24			- 12					- 36
2400 - 2350			-18			- 2					- 20
2350 - 2300			- 9								- 9
2300 - 2250											
Total	-48	- 9	+14	-23	-33	-164	-111	-27	-59	-97	-557

Table 9.3.5. Switzerland - Changes in thickness, area and volume.

Part 4 of 6 Mattmarkgletscher 1956/57 - 1966/67

Change in volume 1956 - 1967 in 10^3 m^3

Altitude interval m a. sl.	1 Kessjen	2 Hohlaub	3 Allalin	4 Hangend	5 Fluchthorn Schwarzberg-grat	6 Schwarzberg	7 Seewinen	8 Monte Moro	9 Tälliboden	10 Ofental	Total
4200 - 4150			+ 23								+ 23
4150 - 4100			+ 105								+ 105
4100 - 4050			+ 196								+ 196
4050 - 4000		- 3	+ 245								+ 242
4000 - 3950		- 15	+ 264		- 4						+ 245
3950 - 3900		- 38	+ 359		- 15						+ 336
3900 - 3850		- 42	+ 260		- 8						+ 210
3850 - 3800		- 70	+ 144		+ 15						+ 59
3800 - 3750		-122	+ 334		+ 20						+ 232
3750 - 3700		- 2	+ 572		+ 50						+ 620
3700 - 3650		+ 2	+ 838		+ 68						+ 908
3650 - 3600		-210	+ 768		+ 55	+ 5					+ 618
3600 - 3550		-330	+ 510		+ 65	- 2					+ 243
3550 - 3500		-425	+ 648		+ 30	- 108					+ 145
3500 - 3450		-548	+ 370		+ 0	- 382					- 560
3450 - 3400		-415	- 8		+ 2	- 372					- 793
3400 - 3350		-155	+ 105	0	+ 2	- 8					- 56
3350 - 3300	0	+282	+ 905	+ 40	+ 0	+ 115					+1342
3300 - 3250	0	+605	+ 1582	+158	+ 18	- 70	0				+2293
3250 - 3200	+ 25	+778	+ 1198	+148	+ 38	- 125	- 5				+2057
3200 - 3150	+ 95	+760	+ 732	+ 30	+ 15	- 382	- 98				+1152
3150 - 3100	+ 172	+390	+ 725	- 40	- 15	- 432	+ 102				+ 902
3100 - 3050	+ 108	+335	+ 1108	- 2	- 2	+ 18	- 520			0	+1045
3050 - 3000	- 10	+295	+ 1058	+ 62		+2945	- 558			+ 15	+3807
3000 - 2950	- 112	+ 15	+ 670	+ 75		+3425	+ 458			+ 10	+4541
2950 - 2900	- 760	-115	+ 652	+ 5		+ 488	+ 342	- 28	- 58	- 140	+ 386
2900 - 2850	- 708	- 32	+ 1018			- 418	+ 80	- 215	- 158	- 232	- 665
2850 - 2800	- 9	- 88	+ 1028			- 882	- 25	- 468	- 218	- 372	-1034
2800 - 2750		-205	+ 780			-1930	- 185	- 292	- 418	- 732	-2982
2750 - 2700		-225	+ 642			-2352	- 305	- 1	- 792	- 850	-3883
2700 - 2650		-115	+ 455			- 965	- 548		- 610	- 790	-2573
2650 - 2600			+ 340			0	-1088		- 30	- 175	- 953
2600 - 2550			+ 152			0	- 725				- 573
2550 - 2500			- 132			- 12	- 0				- 144
2500 - 2450			- 295			- 165					- 460
2450 - 2400			- 220			- 238					- 458
2400 - 2350			- 108			- 32					- 140
2350 - 2300			- 18								- 18
2300 - 2250											
Total	-1199	+307	+18005	+476	+334	-1879	-3075	-1004	-2284	-3266	+6415

Table 9.3.5. Switzerland - Changes in thickness, area and volume.

Part 5 of 6 Mattmarkgletscher 1956/57 - 1966/67

Mean change in altitude of the glacier surface 1956 - 1967, in m

Altitude interval m a. sl.	1 Kessjen	2 Hohlaub	3 Allalin	4 Hangend	5 Fluchthorn Schwarzberg-grat	6 Schwarzberg	7 Seewinen	8 Monte Moro	9 Tälliboden	10 Ofental	Total
4200 - 4150			+ 1.9								+ 1.9
4150 - 4100			+ 1.8								+ 1.8
4100 - 4050			+ 2.2								+ 2.2
4050 - 4000		- 0.8	+ 2.6								+ 2.5
4000 - 3950		- 2.1	+ 2.3		-2.0						+ 2.0
3950 - 3900		- 3.3	+ 2.0		-1.7						+ 1.7
3900 - 3850		- 2.4	+ 1.1		-1.1						+ 0.8
3850 - 3800		- 1.6	+ 0.8		+2.1						+ 0.2
3800 - 3750		- 3.1	+ 1.7		+1.2						+ 0.9
3750 - 3700		- 0.1	+ 2.2		+1.4						+ 1.9
3700 - 3650		+ 0.1	+ 2.4		+1.5						+ 2.1
3650 - 3600		- 5.9	+ 2.1		+1.0	+ 0.5					+ 1.3
3600 - 3550		- 6.7	+ 1.2		+0.9	- 0.1					+ 0.4
3550 - 3500		- 4.1	+ 1.5		+0.9	- 1.4					+ 0.2
3500 - 3450		- 5.1	+ 0.7		+0.0	- 2.8					- 0.7
3450 - 3400		- 3.2	- 0.0		+0.1	- 2.0					- 0.9
3400 - 3350		- 1.1	+ 0.2	0.0	+0.1	- 0.0					- 0.6
3350 - 3300	0.0	+ 1.8	+ 1.2	+1.5	+0.0	+ 0.3					+ 1.1
3300 - 3250	0.0	+ 4.3	+ 2.0	+1.9	+0.7	- 0.2	0.0				+ 1.5
3250 - 3200	+3.6	+ 2.5	+ 1.7	+2.3	+4.0	- 0.2	- 0.3				+ 1.3
3200 - 3150	+1.8	+ 3.2	+ 2.1	+1.2	+2.1	- 0.8	- 1.6				+ 1.0
3150 - 3100	+6.4	+ 2.0	+ 1.8	-2.3	-1.7	- 0.9	+ 0.7				+ 0.7
3100 - 3050	+2.3	+ 1.1	+ 2.3	-0.1	-2.0	+ 0.0	- 1.6			0.0	+ 0.6
3050 - 3000	-0.1	+ 1.8	+ 3.5	+4.0		+ 5.2	- 1.6			+ 1.4	+ 2.5
3000 - 2950	-0.7	+ 0.1	+ 2.0	+2.7		+ 5.0	+ 2.0			+ 0.3	+ 2.8
2950 - 2900	-1.9	- 0.8	+ 2.3	+0.9		+ 1.2	+ 1.3	-2.1	-10.5	+ 2.2	+ 0.3
2900 - 2850	-1.0	- 0.9	+ 2.7			- 0.9	+ 0.5	-2.8	- 6.0	- 2.5	- 0.5
2850 - 2800	-9.0	- 4.8	+ 4.4			- 5.0	- 0.2	-5.8	- 5.2	- 4.0	- 1.3
2800 - 2750		- 9.8	+ 7.3			- 8.2	- 2.6	-9.0	- 7.0	- 8.2	- 4.8
2750 - 2700		-32.2	+ 8.5			-13.9	- 8.5	-2.0	- 8.4	-12.8	- 8.6
2700 - 2650		-13.5	+ 8.4			-23.0	- 2.0		-15.0	-10.5	-10.3
2650 - 2600			+ 8.3			0.0	-23.4		- 4.6	- 6.6	- 7.9
2600 - 2550			+ 5.6			0.0	-24.6				-10.1
2550 - 2500			- 5.1			- 8.0	- 0.0				- 5.2
2500 - 2450			-18.4			- 4.1					- 8.2
2450 - 2400			-18.3			-17.0					-17.6
2400 - 2350			-12.0			-32.0					-14.0
2350 - 2300			- 4.0								- 4.0
2300 - 2250											
Total	-2.0	+ 0.1	+ 1.8	+1.7	+0.8	- 0.3	- 1.6	-4.9	- 8.3	- 5.8	+ 0.3

Table 9.3.5. Switzerland - Changes in thickness, area and volume.

Part 6 of 6 Mattmarkgletscher 1956/57 - 1966/67

Mean change in altitude of the glacier surface 1956 - 1967 in cm/year

Altitude interval m a. sl.	1 Kessjen	2 Hohlaub	3 Allalin	4 Hangend	5 Fluchthorn Schwarzberg-grat	6 Schwarzberg	7 Seewinen	8 Monte Moro	9 Tälliboden	10 Ofental	Total
4200 - 4150			+ 17								+ 17
4150 - 4100			+ 16								+ 16
4100 - 4050			+ 20								+ 20
4050 - 4000		- 7	+ 24								+ 23
4000 - 3950		- 19	+ 21		-18						+ 18
3950 - 3900		- 30	+ 18		-15						+ 15
3900 - 3850		- 22	+ 10		-10						+ 7
3850 - 3800		- 15	+ 7		+19						+ 2
3800 - 3750		- 28	+ 15		+11						+ 8
3750 - 3700		- 1	+ 20		+13						+ 17
3700 - 3650		+ 1	+ 22		+14						+ 19
3650 - 3600		- 54	+ 19		+ 9	+ 5					+ 12
3600 - 3550		- 61	+ 11		+ 8	- 1					+ 4
3550 - 3500		- 37	+ 14		+ 8	- 13					+ 2
3500 - 3450		- 46	+ 6		+ 0	- 25					- 6
3450 - 3400		- 29	- 0		+ 1	- 18					- 8
3400 - 3350		- 10	+ 2	0	+ 1	- 0					- 5
3350 - 3300	0	+ 16	+ 11	+14	+ 0	+ 3					+ 10
3300 - 3250	0	+ 39	+ 18	+17	+ 6	- 2	0				+ 14
3250 - 3200	+33	+ 23	+ 15	+21	+36	- 2	- 3				+ 12
3200 - 3150	+16	+ 29	+ 19	+11	+19	- 7	- 15				+ 9
3150 - 3100	+58	+ 18	+ 16	-21	-15	- 8	+ 6				+ 6
3100 - 3050	+21	+ 10	+ 21	- 1	-18	+ 0	- 15			0	+ 5
3050 - 3000	- 1	+ 16	+ 32	+36		+ 47	- 15			+ 13	+ 23
3000 - 2950	- 6	+ 1	+ 18	+25		+ 45	+ 18			+ 3	+ 25
2950 - 2900	-17	- 7	+ 21	+ 8		+ 11	+ 12	-19	- 95	+ 20	+ 3
2900 - 2850	- 9	- 8	+ 25			- 8	+ 5	-25	- 55	- 23	- 5
2850 - 2800	-82	- 44	+ 40			- 45	- 2	-53	- 47	- 36	- 12
2800 - 2750		- 89	+ 66			- 74	- 24	-82	- 64	- 75	- 44
2750 - 2700		-293	+ 77			-126	- 77	-18	- 76	-116	- 78
2700 - 2650		-123	+ 76			-209	- 18		-136	- 95	- 94
2650 - 2600			+ 75			0	-213		- 42	- 60	- 72
2600 - 2550			+ 51			0	-224				- 92
2550 - 2500			- 46			- 73	- 0				- 47
2500 - 2450			-167			- 37					- 74
2450 - 2400			-166			-155					-160
2400 - 2350			-109			-291					-127
2350 - 2300			- 36								- 36
2300 - 2250											
Total	-18	+ 1	+ 16	+15	+ 7	- 3	- 15	-45	- 75	- 53	+ 3

Table 9.3.6. USSR - Changes in thickness.

Part 1 of 54 Variations in the surface elevation and horizontal component of the displacement of marked points on the glacier surface.

Caucasus.

H - altitude in m.a.sl.
dH - difference in elevation of glacier surface in m
D_h - displacement of point on the glacier surface in m
$\bar{D}_h$ - mean of S for n points
n - number of points
L - distance from point of support, m.

Glacier Mayli. Profile 1.

Point L	A 100 H	dH	D_h	B 130 H	dH	D_h	C 160 H	dH	D_h	D 190 H	dH	D_h	E 220 H	dH	D_h	$\bar{D}_h$	n	days
25.08.1961	2516.9			2519.5			2519.5			2513.8			2520.2					
		- 6.2			- 2.3			+ 0.4			+ 6.1			- 5.1				
01.09.1962	2510.7			2517.2			2519.9			2519.9			2515.3					
		+ 2.7	57.7		+ 2.5	55.5		+ 1.8	56.6		+ 2.4	51.5		+ 3.9	50.1	59.7	7	394
28.09.1963	2513.4			2519.7			2521.7			2522.3			2519.2					
		+ 3.7	41.9		+ 1.3	43.9		+ 3.5	46.2		+ 3.6	50.3			44.9	43.6	7	306
30.08.1964	2517.4			2521.0			2525.2			2525.9								
			51.9			55.4			59.6			67.2			61.8	56.8	7	407
10.09.1965	2517.1																	
			46.4			50.9			52.5			64.5			55.9	54.3	7	349
25.08.1966	2520.8			2532.2			2530.2			2529.7			2528.2					
		+ 2.3	62.9		- 2.3	68.2		+ 1.7	75.4		+ 1.3	75.6		- 2.8	77.3	71.2	7	421
20.10.1967	2523.1			2529.9			2531.9			2531.0			2525.4					
		0.0	62.8		- 6.5	66.6		- 4.6	73.9		- 2.0	69.9		+ 5.9	67.2	67.2	7	327
10.09.1968	2523.1			2523.4			2527.3			2529.0			2531.3					
		- 1.7	49.4		+ 2.3	53.9		+ 3.1	59.1		+ 4.4	57.0		- 0.9	54.6	54.8	7	344
20.08.1969	2521.4			2525.7			2530.4			2533.4			2530.4					

Table 9.3.6. USSR - Changes in thickness.

Part 2 of 54 Variations in the surface elevation and horizontal component of the displacement of marked points on the glacier surface

Caucasus Glacier Gergeti. Profile 1.

Point	A			B			C			D			E					
L		145			175			205			235			265				
	H	dH	D_h	H	dH	D_h	H	dH	D_h	H	dH	D_h	H	dH	D_h	$\bar{D}_h$	n	days
1960	3080.2			3079.8			3078.1			3076.2			3071.6					
		+ 1.9			+ 0.4			- 1.9			+ 1.4			+ 0.9				
1961	3082.1			3080.2			3076.2			3077.6			3072.5					
		- 2.3			+ 0.1			+ 3.3			+ 0.5			+ 1.7				
1962	3079.8			3080.3			3079.5			3078.1			3074.2					
		+ 2.5	32.4		+ 2.0	32.7		+ 2.7	34.0		+ 1.6	35.2		+ 0.5	35.1	33.1	7	355
1963	3082.3			3082.3			3082.2			3078.7			3074.7					
1966	3095.3			3095.6			3098.1			3093.5			3088.6					
		- 3.0	48.7		0.0	52.5		- 3.5	54.9		- 2.8	59.8		- 2.6	60.4	57.1	7	406
1967	3092.3			3095.6			3094.6			3090.7			3086.0					
		+ 0.9	65.9		- 0.2	66.2		+ 3.4	66.0		+ 4.3	68.9		+ 3.8	68.9	68.8	7	355
1968	3093.2			3095.4			3098.0			3095.0			3089.8					
		+ 4.4			+ 2.6			+ 1.0			- 1.3			- 0.1			7	
1969	3097.6			3098.0			3099.0			3093.7			3089.7					
		- 1.5			- 0.3			- 1.7			+ 3.3			+ 5.5			7	
1970	3096.1			3097.7			3097.3			3097.0			3095.2					

Glacier Lazg - Tsiti. Profile 1.

Point	A			B			C			D			E					
L		47			67			87			107			127				
	H	dH	D_h	H	dH	D_h	H	dH	D_h	H	dH	D_h	H	dH	D_h	$\bar{D}_h$	n	days
17.09.1966	3190.0			3190.9			3190.9			3191.3			3193.3					
		+ 2.7			+ 3.6	341		+ 2.2	276		+ 3.1	253		+ 5.1	379	259	11	337
19.09.1967	3192.7			3194.5			3193.1			3194.4			3196.4					
		- 3.3	414		- 2.2	498		- 2.3	476		- 2.2	483		- 2.4	406	395	11	345
29.08.1968	3189.4			3192.3			3190.8			3192.2			3194.0					
		+ 2.3	355		+ 2.0	388		+ 1.1	359		+ 1.0	382		+ 1.7	380	373	11	362
25.08.1969	3191.7			3194.3			3191.9			3193.2			3195.7					
		+ 0.5	344		- 0.3	358		+ 0.4	426		+ 0.7	383		+ 0.1	402	383	11	364
25.08.1970	3192.2			3194.0			3192.3			3193.9			3195.8					

Table 9.3.6. USSR - Changes in thickness.

Part 3 of 54 Variations in the surface elevation and horizontal component of the displacement of marked points on the glacier surface

Caucasus Glacier Gergeti. Profile 2.

Point L	A 1270			B 1220			C 1170			D 1120			E 1070					
	H	dH	D_h	H	dH	D_h	H	dH	D_h	H	dH	D_h	H	dH	D_h	D_h	n	days
1959																		
			85.3			79.6			79.5			79.4			69.1			339
1960	3613.1			3614.0			3609.4			3597.4			3581.5					
		+ 2.2	115.2		- 1.4	120.1		- 4.0	119.2		+95.8	113.9		+ 1.5			15	333
1961	3615.3			3612.6			3605.4			3693.2			3583.0					
		- 0.8	112.8		- 0.3	112.2		+ 0.4	105.6		+ 1.4	102.7		- 1.4	95.5	80.6	15	374
1962	3614.5			3612.3			3605.8			3594.6			3581.6					
		- 1.4	109.3		- 0.1	108.0		- 0.5	107.9		- 5.2	99.3		+ 0.2	97.0	73.1	15	364
1963	3613.1			3612.2			3605.3			3589.4			3581.4					
			101.0			109.6			104.2			102.2			99.4	(74.4)	15	397
1964																		
			109.8			109.0			107.2			103.8			98.7	77.7	15	396
1965																		
			110.4			107.9			106.1			108.1			104.3	73.8	15	342
1966	3610.9			3611.6			3609.8			3602.7			3590.7					
		+ 0.3	117.4		+ 0.8	109.3		- 0.1	105.4		- 2.2	101.7		- 2.5	98.6	77.6	15	373
1967	3611.2			3612.4			3609.7			3600.5			3588.2					
		+ 0.7	120.4		- 0.5	119.5		- 0.3	125.1		+ 1.3	118.7		+ 1.9	112.1	85.5	15	347
1968	3611.9			3611.9			3609.4			3601.8			3590.1					
		0.0	70.2		- 0.2	69.9		0.0	67.9		+ 1.3	69.4		+ 0.2	67.6	47.6	15	312
1969	3611.9			3611.7			3609.4			3603.1			3590.3					
		+ 2.2	61.6		+ 2.3	70.3		+ 1.4	65.5		- 0.2	63.5		+ 0.6	61.8	45.5	15	331
1970	3614.1			3614.0			3610.8			3602.9			3590.9					

Table 9.3.6. USSR - Changes in thickness.

Part 4 of 54 Variations in the surface elevation and horizontal component of the displacement of marked points on the glacier surface

Caucasus Glacier Suatisi middle. Profile 1.

Point L	A 130 H	dH	D_h	B 160 H	dH	D_h	C 190 H	dH	D_h	D 220 H	dH	D_h	E 250 H	dH	D_h	$\bar{D}_h$	n	days
1960	3080.1			3070.7			3067.6			3064.3			3065.4					
					+ 9.8												11	
10.09.1962	3077.9			3080.5			3072.2			3071.8			3063.4					
		- 0.4	29.2		- 0.9	29.9		+ 2.3	30.5		- 3.1	33.6		+ 0.6	35.8	31.2	11	367
11.09.1963	3077.5			3079.6			3074.5			3068.7			3064.0					
		0.0	13.1		- 0.3	19.5		+ 4.8	28.0		+10.8	29.3		+15.2	31.4	26.5	11	369
14.09.1964	3077.5			3079.3			3079.3			3079.5			3079.2					
			32.4			33.5			39.0			43.7			47.7	41.1	11	370
1965																		
			36.5			34.9			39.6			45.9			45.1	44.5	11	357
11.09.1966	3065.6			3069.1			3071.0			3074.0			3082.0					
		-11.1	16.5		- 7.2	22.3		+ 9.8	12.2		-13.9	31.9		-27.4	32.0	27.3	11	371
17.09.1967	3054.5			3061.9			3060.8			3060.1			3054.6					
		+ 1.7	19.6		- 1.7	27.9		- 0.5	28.5		+ 4.0	39.5		+12.5	43.5	29.6	11	339
22.08.1968	3056.2			3060.2			3060.3			3064.1			3067.1					
		+ 1.3	19.9		+ 2.1	32.5		+ 1.7	36.8		+ 1.9	43.8		+ 1.9	46.8	36.9	11	369
25.08.1969	3057.5			3062.3			3062.0			3066.0			3069.0					
		+ 0.2	22.8		0.0	30.0		+ 1.2	39.0		- 0.2	45.0		- 0.2	48.6	37.6	11	364
24.08.1970	3057.7			3062.3			3063.2			3065.8			3068.8					

Glacier Koruldash. Profile 1.

Point L	A 77 H	dH	D_h	B 95 H	dH	D_h	C 114 H	dH	D_h	D 133 H	dH	D_h	H	dH	D_h	$\bar{D}_h$	n	days
23.08.1965	2334.2			2334.0			2335.8			2335.8								
		- 0.5	3.7		- 1.2	5.5		- 0.4	4.1			3.4				4.2	6	386
12.09.1966	2333.7			2332.8			2335.4											
		+ 0.6	3.9		+ 0.3	5.0		- 1.4	4.0			2.9				3.9	6	368
16.09.1967	2334.3			2333.1			2334.0			2335.3								
		0.0	2.5		+ 0.1	3.3		+ 0.3	3.6		+ 0.3	2.9				2.9	6	372
23.09.1968	2334.3			2333.2			2334.3			2335.6								
		+ 0.4	2.2		+ 0.4	2.9		+ 0.3	2.9		+ 0.8	2.9				2.5	6	337
24.08.1969	2334.7			2333.6			2334.6			2336.4								
		+ 0.7	3.6		+ 0.8	3.7		+ 1.0	3.7		+ 0.7	3.3				3.6	6	368
26.09.1970	2335.4			2334.4			2335.6			2337.1								

Table 9.3.6. USSR - Changes in thickness.

Part 5 of 54 Variations in the surface elevation and horizontal component of the displacement of marked points on the glacier surface

Caucasus Glacier Devdoraki. Profile 1.

Point L		A 185			B 215			C 245			D 275			E 305				
	H	dH	D_h	H	dH	D_h	H	dH	D_h	H	dH	D_h	H	dH	D_h	$\bar{D}_h$	n	days
1960																		
			54.0			68.3			69.3			61.9			62.7	61.3	7	369
31.08.1961	2387.9			2383.4			2385.5			2384.1			2374.2					
		+ 2.7	74.3		+ 2.5	67.9		+ 0.9	72.1		+ 3.3	74.2		+ 6.3	83.7	73.2	7	371
6.09.1962	2390.6			2385.9			2386.4			2387.4			2380.5					
		+ 7.2	53.1		+ 1.5	49.7		- 1.3	51.6		- 2.8	51.8		- 4.9	51.2	48.2	7	384
28.09.1963	2397.8			2387.4			2385.1			2384.6			2375.6					
			68.0			66.6			90.3			88.5			88.1	78.7	7	341
1964																		
			50.6			57.1			62.4			68.3			72.7	62.2	7	370
1965																		
			50.2			49.9			58.9			59.5			67.1	62.8	7	359
1966																		
			61.7			61.8			61.2			117.5			92.3	65.3	7	412
14.10.1967	2387.2			2387.4			2386.0			2383.4			2374.5					
		- 0.6	54.5		- 3.0	60.6		- 1.9	62.0		- 2.4	63.8		+ 1.6	63.7	59.7	7	338
16.09.1968	2386.4			2384.4			2384.1			2381.0			2376.1					
		+ 0.9	52.8		0.0	58.4		- 1.0	59.7		+ 1.7	59.6		- 1.2	61.4	57.5	7	341
23.08.1969	2387.3			2384.4			2384.0			2382.7			2374.3					
		- 0.2	53.8		+ 0.3	60.4		- 0.1	59.4		+ 0.1	61.2		+ 2.0	61.9	58.9	7	364
22.07.1970	2387.1			2384.7			2383.9			2382.8			2376.9					

Glacier Chalati. Profile 1.

Point L		A 176			B 201			C 225			D 252			E 277				
	H	dH	D_h	H	dH	D_h	H	dH	D_h	H	dH	D_h	H	dH	D_h	$\bar{D}_h$	n	days
			86.5			87.5			89.3			93.2			93.9	89.2	9	368
09.09.1967	2160.6			2163.0			2166.5			2170.5			2174.5					
		- 0.2	85.7		+ 1.8	93.2		- 0.7	95.0		+ 0.5	97.7		+ 0.3	100.3	94.5	9	386
16.09.1968	2160.4			2164.8			2165.8			2171.0			2174.8					
		+ 2.1	79.3		- 0.4	83.7		+ 2.0	85.9		+ 0.8	86.1		+ 0.9	91.3	84.7	9	320
16.08.1969	2162.5			2164.4			2167.8			2171.8			2175.7					
		+ 1.3	82.3		+ 1.3	86.7		+ 1.6	87.8		+ 3.3	104.3		+ 1.5	101.5	92.5	9	360
11.08.1970	2163.8			2165.7			2169.4			2175.1			2177.2					

Table 9.3.6. USSR - Changes in thickness.

Part 6 of 54 Variations in the surface elevation and horizontal component of the displacement of marked points on the glacier surface

Caucasus Glacier Kirtisho. Profile 1.

Point L	A 98			B 129			C 162			D 191			E 221					
	H	dH	D_h	H	dH	D_h	H	dH	D_h	H	dH	D_h	H	dH	D_h	$\bar{D}_h$	n	days
05.09.1965	2733.5			2733.3			2738.8			2739.3			2738.4					
		- 3.4			- 0.2			- 0.1			- 4.9			-11.1				
21.09.1966	2730.1			2733.1			2738.7			2734.4			2727.3					
		+11.5			+13.2			+ 3.4			- 2.0			+15.7				
25.09.1967	2741.6			2746.3			2742.1			2742.4			2743.0					
		+ 3.1			- 3.4			+ 3.4			+ 4.0			+ 3.1				
16.09.1968	2744.7			2742.9			2745.5			2746.4			2746.1					
		- 2.4	28.9		- 2.4	30.1		- 6.1	29.9		- 5.7	33.6		- 4.9	36.7	31.0	9	350
01.09.1969	2742.3			2740.5			2739.4			2740.7			2741.2					
		- 0.4			+ 0.2	30.8		- 1.0	34.6		+ 0.4	35.7		0.0	38.6	33.8	9	352
19.08.1970	2741.9			2740.7			2738.4			2741.4			2741.2					

Glacier Abano. Profile 1.

Point L	A 113			B 183			C 233			D 283			E 333					
	H	dH	D_h	H	dH	D_h	H	dH	D_h	H	dH	D_h	H	dH	D_h	$\bar{D}_h$	n	days
			50.6			72.9			70.6			71.6			55.2	49.8		376
04.09.1961	3383.1			3387.8			3387.5			3384.8			3384.5					
		- 2.4	45.7		- 4.6	51.0		- 1.5	56.0		+ 0.4	53.9		+ 0.4	51.2	(46.4)	10	382
22.09.1962	3380.7			3383.2			3386.0			3385.2			3384.9					
		+ 0.5	42.3		+ 4.3	43.8		+ 2.1	51.9		+ 0.5	50.5		+ 0.3	48.3	46.5	10	361
14.09.1963	3381.2			3387.5			3388.1			3385.7			3385.2					
			45.8			46.3			50.6			47.0			47.8	48.1	10	361
1964																		
			37.6			41.0			49.5			58.4			52.3	45.1	10	354
1965																		
																	10	
05.09.1966	3386.0			3381.8			3378.6			3378.5			3378.2					
		+ 1.4			- 2.2			- 0.2			- 5.0			- 0.6	-	-	10	
14.09.1967	3387.4			3379.6			3378.4			3373.5			3377.6					
		- 7.8	38.0		0.0	44.9		+ 2.9	60.9		+ 3.3	47.4		+ 2.0	48.2	48.0	10	358
06.09.1968	3379.6			3379.6			3375.5			3376.8			3379.6					
		+ 7.8	37.7		- 0.2	51.2		- 0.5	52.5		- 1.4	59.0		+ 1.2	60.6	53.4	10	346
18.08.1969	3387.4			3379.4			3375.0			3375.4			3378.4					
		+ 0.4	46.7		+ 2.0	47.1		+ 2.0	59.0		+ 0.7	67.9		+ 0.6	70.0	65.8	10	367
20.08.1970	3387.8			3381.4			3377.0			3376.1			3379.0					

Table 9.3.6. USSR - Changes in thickness.

Part 7 of 54 Variations in the surface elevation and horizontal component of the displacement of marked points on the glacier surface

Tien-Shan Tsentralny Tuyuksu Glacier Profile 1

point no.	distance from fixed point	date	1959 11.07.		1961 14.09.		1963 26.09.		1964 12.08.		1965 28.09.
		days		793		739		320		412	
1	179	H			3402.7		3398.8		3395.1		3391.2
		ΔH				-3.9		-3.7		-3.9	
2	194	H	3407.9		3402.3		3398.8		3394.8		3391.3
		ΔH		-5.6		-3.5		-4.0		-3.5	
3	209	H	3407.8		3401.9		3398.4		3394.4		3391.4
		ΔH		-5.9		-3.5		-4.0		-3.0	
4	219	H	3407.6		3401.7		3398.3		3394.3		3391.3
		ΔH		-5.9		-3.4		-4.0		-3.0	
5	248	H	3407.4		3401.5		3397.4		3393.4		3390.4
		ΔH		-5.9		-4.1		-4.0		-3.0	
6	276	H	3406.1		3400.8		3396.4		3392.3		3390.2
		ΔH		-5.3		-4.4		-4.1		-2.1	
7	299	H	3403.8		3399.5		3395.2		3391.2		3390.5
		ΔH		-4.3		-4.3		-4.0		-0.7	
8	328	H	3406.8		3397.9		3393.4		3390.6		3387.9
		ΔH		-8.9		-4.5		-2.8		-2.7	
9	358	H	3408.3		3400.9		3396.0		3391.5		3388.6
		ΔH		-7.4		-4.9		-4.5		-2.9	
10	389	H	3405.6		3401.9		3397.1		3393.0		3389.6
		ΔH		-3.7		-4.8		-4.1		-3.4	

Remark: The horizontal component of the displacement was not measured.

Table 9.3.6. USSR - Changes in thickness.

Part 8 of 54 Variations in the surface elevation and horizontal component of the displacement of marked points on the glacier surface

Tien-Shan Tsentralny Tuyuksu Glacier Profile 2

point no.	distance from fixed point	date / days	1959 05.10.	709	1961 14.09.	383	1963 02.10.	356	1964 23.09.	370	1965 28.09.	360	1966 23.09.	358	1967 16.09.	308	1968 21.07.	393	1969 18.08.	361	1970 22.08.
1	174	H	3434.0		3431.2		3425.0		3425.8		3421.5		3423.1		3419.8		3419.4		3418.7		3413.2
		ΔH		-2.8		-6.2		+0.8		-4.3		+1.6		-3 3		-0.4		-0.7		-5.5	
		D_h								0.2		←		0.4				0.0		0.0	
2	193	H	3435.9		3433.0		3426.4		3427.2		3421.9		3425.3		3421.2		3420.8		3420.0		3415.5
		ΔH		-2.9		-6.6		+0.8		-5.3		+3.4		-4.1		-0.4		-0.8		-4.5	
		D_h		3.5						0.4		←		0.7				0.2		0.0	
3	214	H	3435.7		3434.9		3427.8		3428.8		3422.4		3427.8		3423.9		3422.7		3421.3		3418.4
		ΔH		-0.8		-7.1		+1.0		-6.4		+5.4		-3.9		-1.2		-1.4		-2.9	
		D_h		3.5						0.7		←		1.1				0.3		0.3	
4	238	H	3437.1		3435.5		3430.9		3430.8		3423.1		3430.6		3427.1		3424.2		3422.8		3420.5
		ΔH		-1.6		-4.6		-0.1		-7.7		+7.5		-3.5		-2.9		-1.4		-2.3	
		D_h		3.0						1.1		←		1.6				0.5		0.5	
5	276	H	3437.4		3436.3		3431.1		3431.6		3426.4		3430.9		3426.9		3424.2		3423.0		3421.2
		ΔH		-1.1		-5.2		+0.5		-5.2		+4.5		-4.0		-2.7		-1.2		-2.8	
		D_h		2.8						1.2		←		3.2				0.8		0.6	
6	316	H	3440.4		3436.7		3430.8		3430.4		3428.8		3429.0		3425.0		3424.1		3420.8		3419.8
		ΔH		-3.7		-5.9		-0.4		-1.6		+0.2		-4.0		-0.9		-3.3		-1.0	
		D_h		3.0						1.3		←		3.2				1.2		1.0	
7	356	H	3441.2		3438.0		3432.3		3431.8		3429.4		3432.7		3427.5		3425.7		3423.9		3422.0
		ΔH		-3.2		-5.7		-0.5		-2.4		+3.3		-5.2		-2.0		-1.8		-1.9	
		D_h		3.1						1.4		←		3.9				1.2		1.3	
8	386	H	3441.4		3440.6		3436.0		3435.0		3431.1		3435.5		3429.4		3428.5		3426.9		3425.3
		ΔH		-0.8		-4.6		-1.0		-3.9		+4.4		-6.1		-0.9		-0.6		-1.6	
		D_h		3.3						1.2		←		4.4				1.1		1.1	
9	411	H	3441.2		3441.1		3436.1		3435.1		3432.9		3437.1		3430.5		3430.4		3427.5		3426.3
		ΔH		-0.1		-5.0		-1.0		-2.2		+4.2		-6.6		-0.1		-2.9		-1.2	
		D_h		4.0						0.9		←		4.7				1.2		1.0	
10	460	H	3435.6		3437.8		3433.2		3432.2		3430.4		3434.7		3428.0		3432.4		3425.0		3424.4
		ΔH		+2.2		-4.6				-1.8		+4.3		-6.7		-0.6		-2.4		-0.6	
		D_h		5.7						0.7		←		4.4				0.2			

Table 9.3.6. USSR - Changes in thickness.
Part 9 of 54 Variations in the surface elevation and horizontal component of the displacement of marked points on the glacier surface
Tien-Shan Tsentralny Tuyuksu Glacier Profile 3

point no.	distance from fixed point	date	1959 05.10.		1961 14.09.		1963 02.10.		1964 23.09.		1965 28.09.		1966 23.09.		1967 16.09.		1968 21.07.		1969 18.08.		1970 21.08.
		days		710		748		357		370		360		358		309		393		368	
1	123	H	3476.4		3476.2		3472.8		3473.3		3472.7		3474.4		3474.7		3474.7		3471.8		3469.5
		ΔH		-0.2		-3.4		+0.5		-0.6		+1.7		+0.3		0.0		-2.9		-2.3	
		D_h		10.0						4.2		4.8		5.0				3.8		3.4	
2	167	H	3473.5		3474.1		3470.4		3470.8		3470.2		3469.9		3469.9		3470.2		3468.4		3468.3
		ΔH		+0.6		-3.7		+0.4		-0.6		-0.3		-0.1		+0.3		-1.8		-0.1	
		D_h		10.0						4.5		5.0		5.0				4.9		4.8	
3	204	H	3473.1		3472.5		3469.1		3469.7		3469.0		3469.3		3468.7		3469.3		3466.3		3466.9
		ΔH		-0.6		-3.4		+0.6		-0.7		+0.3		-0.6		+0.6		-3.0		+0.6	
		D_h		12.0						5.1		4.7		6.7				5.8		5.6	
4	233	H	3473.1		3471.9		3468.7		3470.0		3469.2		3469.8		3469.0		3469.7		3466.9		3467.4
		ΔH		-1.2		-3.2		+1.3		-0.8		+0.6		-0.8		+0.7		-2.8		+0.5	
		D_h		10.5						5.5		4.8		5.2				6.1		5.9	
5	269	H	3474.1		3473.5		3470.5		3470.9		3469.6		3469.8		3469.6		3469.7		3467.5		3466.9
		ΔH		-0.6		-3.0		+0.4		-1.3		+0.2		0.0		+0.1		-2.1		-0.6	
		D_h		9.0						5.7		5.7		4.1				6.1		5.8	
6	301	H	3474.0		3473.7		3470.7		3471.5		3469.9		3469.5		3469.8		3470.2		3467.2		3466.1
		ΔH		-0.3		-3.0		+0.8		-0.6		-0.4		+0.3		+0.4		-3.0		-1.1	
		D_h		8.2						5.8		5.9		4.0				6.0		6.0	
7	341	H	3474.7		3474.5		3470.9		3471.9		3470.2		3470.3		3470.7		3471.2		3468.6		3468.4
		ΔH		-0.2		-3.6		+1.0		-1.7		+0.1		+0.4		+0.5		-2.6		-0.2	
		D_h		7.4						5.5		5.5		4.0				5.7		5.6	
8	374	H	3475.0		3475.7		3472.4		3472.4		3471.6		3471.1		3471.5		3471.7		3468.8		3469.3
		ΔH		+0.7		-3.3		0.0		-0.8		-0.5		+0.4		+0.2		-2.9		+0.5	
		D_h		7.4						5.4		4.6		4.5				5.5		5.5	
9	408	H	3475.4		3475.2		3472.1		3473.1		3471.2		3471.4		3471.1		3471.5		3469.2		3468.4
		ΔH		-0.2		-3.1		+1.0		-1.9		+0.2		-0.4		+0.4		-2.3		-0.8	
		D_h		6.5						5.4		4.4		3.7				4.7		4.5	
10	442	H	3474.0		3474.7		3471.5		3472.8		3470.5		3470.7		3470.1		3470.6		3468.2		3466.8
		ΔH		+0.7		-3.2		+1.3		-2.3		+0.2		-0.6		+0.5		-2.4		-1.4	
		D_h		7.0						3.8		4.7		4.5				4.0		3.3	
11	477	H	3472.1		3471.4		3467.9		3468.7		3467.1		3470.0		3469.1		3469.5		3467.2		3465.6
		ΔH		-0.7		-3.5		+0.8		-1.6		+2.9		-0.8		+0.4		-2.3		-1.6	
		D_h		7.5						2.0		5.5		5.0				3.2		1.9	

Table 9.3.6. USSR - Changes in thickness.

Part 10 of 54 Variations in the surface elevation and horizontal component of the displacement of marked points on the glacier surface
Tien-Shan Tsentralny Tuyuksu Glacier Profile 4

point no.	distance from fixed point	date	1959 03.09.		1960 16.09.		1961 14.09.	distance from fixed point	1963 02.07.		1963 03.10.		1964 22.09.		1965 05.10.		1966 22.09.		1967 14.09.		1968 24.07.		1969 19.08.		1970 23.08.
		days		378		363				93		354		378		352		357		313		391		261	
		H	3547.0		3546.1		3546.2		3544.4		3545.1		3545.7		3546.4		3546.0		3546.4		3543.5		3544.8		3545.2
1	70	ΔH		-0.9		+0.1		139		+0.7		+0.6		+0.7		-0.4		+0.4		-2.9		+1.3		+0.4	
		D_h		11.1		12.2				2.7		8.7		7.0		8.5		8.5				10.0		9.6	
		H	3547.4		3546.4		3546.6		3545.7		3545.7		3546.3		3546.2		3545.9		3546.7		3545.9		3544.1		3545.3
2	94	ΔH		-1.0		+0.2		180		0.0		+0.6		-0.1		-0.3		+0.8		-0.8		-1.8		+1.2	
		D_h		12.0		14.1				3.4		14.3		7.1		11.1		8.6				16.1		12.4	
		H	3548.0		3547.2		3547.2		3545.0		3544.7		3545.7		3544.9		3545.6		3545.5		3545.1		3543.4		3544.0
3	141	ΔH		-0.8		-0.0		219		-0.3		+1.0		-0.8		+0.7		-0.1		-0.4		-1.7		+0.6	
		D_h		12.7		14.5				4.2		12.3		9.7		11.7		9.0				13.0		13.3	
		H	3547.6		3546.5		3546.7		3544.9		3544.3		3544.6		3544.4		3545.1		3544.7		3545.3		3543.4		3543.7
4	190	ΔH		-1.1		+0.2		258		-0.6		+0.3		-0.2		+0.7		-0.4		+0.6		-1.9		+0.3	
		D_h		12.8		13.9				3.8		12.2		12.3		12.5		10.6				14.4		14.0	
		H	3548.3		3546.0		3545.7		3544.4		3544.0		3544.8		3545.0		3545.2		3545.8		3545.9		3543.9		3544.4
5	242	ΔH		-2.3		-0.3		303		-0.4		+0.8		+0.2		+0.2		+0.6		+0.1		-2.0		+0.5	
		D_h		14.3		12.9				3.7		12.1		14.4		11.0		11.0				16.2		14.7	
		H	3548.5		3547.2		3546.2		3544.5		3544.3		3544.9		3545.5		3545.3		3545.5		3545.8		3544.2		3544.9
6	294	ΔH		-1.3		-1.0		343		-0.2		+0.6		+0.6		-0.2		+0.2		+0.3		-1.6		+0.7	
		D_h		13.2		14.3				3.9		12.3		14.4		11.3		12.1				14.6		15.3	
		H	3548.6		3547.2		3546.5		3544.9		3544.4		3545.4		3546.2		3546.2		3546.1		3546.6		3544.7		3545.5
7	331	ΔH		-1.4		-0.7		383		-0.5		+1.0		+0.8		0.0		-0.1		+0.5		-1.9		+0.8	
		D_h		13.4		14.0				3.8		12.8		14.6		12.0		11.4				16.6		15.9	
		H	3548.0		3547.6		3546.4		3546.0		3545.6		3546.3		3546.9		3546.9		3546.9		3547.3		3545.6		3546.0
8	366	ΔH		0.4		-1.2		423		-0.4		+0.7		+0.6		0.0		0.0		+0.4		-1.7		+0.4	
		D_h		12.6		13.0				4.0		12.5		14.8		11.5		11.1				15.4		15.7	
		H	3548.2		3548.2		3547.1		3545.2		3545.1		3545.3		3546.3		3545.9		3546.2		3546.8		3545.1		3545.6
9	400	ΔH		0.0		-1.1		463		-0.1		+0.2		+1.0		-0.4		+0.3		+0.6		-1.7		+0.5	
		D_h		12.7		13.1				3.9		10.4		16.5		9.4		10.6				15.9		15.5	
		H	3547.3		3548.0		3547.4		3544.6		3544.4		3544.2		3545.9		3546.7		3546.0		3546.0		3544.3		3545.0
10	434	ΔH		+0.7		-0.6		503		-0.2		-0.2		+1.7		+0.8		-0.7		0.0		-1.7		+0.7	
		D_h		12.5		12.8				3.9		10.5		17.3		5.0		10.0				14.8		14.8	
		H							3545.2		3544.8		3545.4		3545.8		3545.4		3545.5		3545.9		3544.2		3545.2
11		ΔH						543		-0.4		+0.6		+0.4		-0.4		+0.1		+0.4		-1.7		+1.0	
		D_h								3.3		11.1		19.1		10.0		8.8				14.2		13.8	
		H							3542.6		3542.4		3544.0		3545.3		3541.3		3542.9		3544.5		3542.6		3545.7
12		ΔH						583		-0.2		+1.6		+1.3		-4.0		+1.6		+1.6		-1.9		+3.1	
		D_h								4.2		9.4		11.2		10.5		9.5				12.8		12.5	

Table 9.3.6. USSR - Changes in thickness.

Part 11 of 54 Variations in the surface elevation and horizontal component of the displacement of marked points on the glacier surface

Tien-Shan Tsentralny Tuyuksu Glacier Profile 5

point no.	distance from fixed point	date / days	1959 05.10.	291	1960 23.07.	412	1960 16.09.	365	1961 15.09.	365	1962 15.09.
1	180	H	3618.5		3619.3		3619.2		3618.7		3618.5
		ΔH		+0.8		-0.1		-0.5		-0.2	
		D_h		12.5		2.8		14.2			
2	238	H	3619.5		3620.1		3620.1		3618.5		3618.3
		ΔH		+0.6		0.0		-1.6		-0.2	
		D_h		13.0		3.0		15.0			
3	269	H	3620.6		3621.6		3621.2		3620.9		3620.6
		ΔH		+1.0		-0.4		-0.3		-0.3	
		D_h		13.0		3.0		18.4			
4	303	H	3623.3		3624.1		3623.4		3622.6		3621.9
		ΔH		+0.8		-0.7		-0.8		-0.7	
		D_h		14.0		3.1		18.5			
5	337	H	3627.0		3627.8		3626.4		3625.3		3625.0
		ΔH		+0.8		-1.4		-1.1		-0.3	
		D_h		15.0		3.2		18.6			
6	382	H	3629.0		3629.5		3629.3		3627.9		3627.0
		ΔH		+0.5		-0.2		-1.4		-0.9	
		D_h		13.6		5.0		18.6			
7	430	H	3629.7		3630.5		3629.4		3628.7		3627.9
		ΔH		+0.8		-1.1		-0.7		-0.8	
		D_h		14.5		3.2		20.0		18.0	
8	477	H	3628.2		3629.1		3628.8		3627.4		3626.7
		ΔH		+0.9		-0.3		-1.4		-0.7	
		D_h		18.0		2.9		20.1		20.0	
9	520	H	3626.9		3627.6		3627.2		3625.9		3625.3
		ΔH		+0.7		-0.4		-1.3		-0.6	
		D_h		13.9		3.5		17.8			

point no.	distance from fixed point		1963 12.07.	52	1963 02.10.	355	1964 22.09.	378	1965 05.10.	352	1966 22.09.	357	1967 14.09.	303	1968 23.07.	398	1969 25.08.	363	1970 23.08.
1	186	H	3619.2		3619.3		3618.9		3618.7		3620.6		3621.7		3621.9		3618.4		3619.0
		ΔH		+0.1		-0.2		-0.2		+1.9		+1.1		+0.2		-3.5		+0.6	
		D_h		5.1		16.7		18.1		17.0		16.4				21.3		18.4	
2	232	H	3618.3		3618.4		3619.9		3617.7		3621.8		3620.2		3620.4		3619.8		3620.6
		ΔH		+0.1				-2.2		+4.1		-1.5		+0.2		-0.6		+0.8	
		D_h		5.0		17.5		18.3		20.0		14.1				21.9		19.8	
3	270	H	3621.2		3621.4		3623.2		3622.4		3621.8		3623.0		3623.3		3620.6		3621.9
		ΔH		+0.2				-0.8		-0.6		+1.2		+0.3		-2.7		+1.3	
		D_h		5.2		18.8		20.4		16.0		17.6				22.4		20.8	
4	313	H	3622.8		3623.1		3624.7		3625.5		3624.8		3625.2		3626.1		3625.1		3625.2
		ΔH		+0.3				+0.1		-0.7		+0.4		+0.9		-1.0		+0.1	
		D_h		5.5		19.1		21.4		18.0		18.7				22.4		20.4	
5	349	H	3627.4		3627.1		3627.8		3627.2		3627.2		3628.1		3630.7		3627.0		3629.0
		ΔH		-0.3				-1.2		0.0		+0.9		+2.6		-3.7		+2.0	
		D_h		5.5		19.3		22.4		19.0		17.6				23.9		21.7	
6	391	H	3628.0		3628.2		3630.4		3629.3		3629.1		3630.1		3632.0		3629.0		3629.5
		ΔH		+0.2				-1.6		-0.2		+1.0		+1.9		-4.0		+0.5	
		D_h		5.5		19.7		26.1		13.0		18.0				23.6		21.6	
7	432	H	3628.4		3628.5		3630.6		3630.3		3630.3		3631.1		3632.3		3629.8		3629.9
		ΔH		+0.1				-0.3		0.0		+0.8		+1.2		-2.5		+0.1	
		D_h		5.5		18.7		29.0		10.0		16.0				26.1		21.8	
8	465	H	3628.7		3628.4		3630.1		3630.2		3630.0		3629.6		3630.8		3628.8		3629.0
		ΔH		0.3				-1.0		-0.2		-0.4		+1.2		-2.0		+0.2	
		D_h		5.5		18.6		29.3		8.0		14.9				29.0		20.8	
9	529	H							3624.7		3624.6		3626.0		3624.6		3622.1		3622.6
		ΔH								-0.1		+1.4		-1.4		-2.8		+0.5	
		D_h														23.4		20.3	

Table 9.3.6. USSR - Changes in thickness.

Part 12 of 54 Variations in the surface elevation and horizontal component of the displacement of marked points on the glacier surface

Tien-Shan Tsentralny Tuyuksu Glacier Profile 6

point no.	distance from fixed point	date	1959 05.10.		1961 15.09.	distance from fixed point	1963 01.10.		1964 22.09.		1965 29.09.		1966 22.09.		1967 14.09.		1968 21.07.		1969 24.08.		1970 23.08.
		days		710				357		372		357		357		310		399		364	
		H	3676.3		3674.8		3671.7		3672.2		3675.3		3679.4		3678.8		3680.0		3677.2		3678.8
1	518	ΔH		-1.5		467		+0.5		+2.1		+4.1		-0.6		+1.2		-2.8		+1.1	
		D_h		27.0				17.8		10.3		9.0		12.5				16.8		15.0	
		H	3672.6		3671.1		3669.7		3670.8		3671.1		3673.7		3674.6		3675.4		3673.0		3673.7
2	467	ΔH		-1.5		429		+1.1		+0.3		+2.6		+0.9		+0.8		-2.4		+0.7	
		D_h		29.3				17.9		16.4		12.0		14.0				19.9		18.3	
		H	3670.5		3669.5		3669.6		3669.9		3669.3		3671.0		3672.4		3673.4		3671.2		3671.9
3	423	ΔH		-1.0		387		+0.3		-0.6		+1.7		+1.4		+1.0		-2.2		+0.7	
		D_h		33.4				18.2		19.6		18.0		14.1				22.2		19.9	
		H	3671.9		3670.9		3669.9		3671.0		3669.7		3671.2		3672.8		3673.3		3671.5		3672.0
4	364	ΔH		-1.0		350		+1.1		-1.3		+1.5		+1.6		+0.5		-1.8		+0.5	
		D_h		36.6				18.6		19.0		19.0		16.8				20.8		19.0	
		H	3676.6		3674.2		3671.8		3673.2		3673.5		3675.6		3676.0		3676.2		3674.5		3675.0
5	293	ΔH		-2.4		310		+1.4		+0.3		+2.1		+0.4		+0.2		-1.7		+0.5	
		D_h		38.2				20.1		19.4		19.5		19.0				19.6		18.3	
		H	3673.4		3671.2		3674.9		3676.6		3672.1		3675.3		3674.6		3677.0		3673.7		3673.0
6	201	ΔH		-2.2		268		+1.7		-4.5		+3.2		-0.7		+2.4		-3.3		-0.7	
		D_h		35.7				19.1		20.4		18.0		21.0				23.1		21.1	
		H	3672.2		3671.2		3673.8		3676.0		3672.6		3675.5		3674.4		3676.7		3673.5		3672.7
7	145	ΔH		-1.0		228		+2.2		-3.4		+2.9		-1.1		+2.3		-3.2		-0.8	
		D_h		28.9				19.1		19.3		20.0		19.0				22.8		20.0	
		H					3669.3		3672.5												
8	88	ΔH		22.7		191		+3.2													
		D_h						17.7		20.2		22.0		19.0				21.5		19.3	
		H					3673.0		3673.0												
9		ΔH				144		0.0													
		D_h						14.0		23.6		24.1		19.0				19.9		18.1	

Table 9.3.6. USSR - Changes in thickness.

Part 13 of 54 Variations in the surface elevation and horizontal component of the displacement of marked points on the glacier surface

Tien-Shan Tsentralny Tuyuksu Glacier Profile 7

point no.	distance from fixed point	date	1959 05.10.		1961 15.09.		1962 12.09.	distance from fixed point	1963 04.07.		1963 01.10.		1964 22.09.		1965 29.09.		1966 20.09.		1967 14.09.		1968 23.07.		1969 24.08.		1970 23.08.
		days		345		362				88		355		372		355		359		312		395		365	
1	170	H	3724.4		3724.3		3722.2	122	3722.5		3722.5		3723.3		3723.8		3724.8		3725.7		3726.9		3725.7		3726.2
		ΔH		-0.1		-2.1				0.0		+0.8		+0.5		+1.0		+0.9		+1.2		-1.2		+0.5	
		D_h								3.5		9.4		10.1		12.7		9.0				8.1		7.2	
2	210	H	3726.9		3727.0		3724.8	160	3723.2		3722.9		3723.5		3724.0		3725.0		3725.8		3727.9		3725.8		3726.7
		ΔH		+0.1		-2.2				-0.3		+0.6		+0.5		+1.0		+0.8		+1.9		-2.1		+0.9	
		D_h								3.6		11.9		10.5		17.0		8.5				12.0		10.6	
3	248	H	3729.6		3729.9		3727.4	199	3724.9		3725.4		3725.9		3726.0		3726.8		3727.7		3728.8		3728.1		3729.1
		ΔH		+0.3		-2.5				+0.5		+0.5		+0.1		+0.8		+0.9		+1.1		-0.7		+1.0	
		D_h								4.5		12.7		11.8		15.0		12.5				14.9		12.7	
4	281	H	3731.1		3731.0		3729.3	240	3727.9		3728.4		3728.9		3729.0		3729.5		3730.5		3731.0		3730.7		3731.8
		ΔH		-0.1		-1.7				+0.5		+0.5		+0.1		+0.5		+1.0		+0.5		-0.3			
		D_h								5.0		13.2		12.8		16.0		11.8				14.6		13.5	
5	324	H	3731.5		3731.8		3730.8	279	3730.3		3730.4		3730.6		3730.8		3731.7		3732.3		3732.3		3732.2		3733.6
		ΔH		+0.3		-1.0				+0.1		+0.2		+0.2		+0.9		0.6		-0.0		-0.1			
		D_h				13.5				5.0		12.9		12.4		15.7		13.5				13.2		12.2	
6	355	H	3730.9		3732.0		3731.3	319	3731.2		3731.2		3731.7		3731.8		3732.8		3733.6		3734.1		3733.7		3734.7
		ΔH		+1.1		-0.7				0.0		+0.5		+0.1		+1.0		+0.8		+0.5		-0.4			
		D_h				15.0				5.5		11.7		13.5		14.8		13.0				15.4		14.2	
7	411	H	3731.8		3731.4		3731.1	359	3731.6		3731.4		3732.1		3732.4		3733.1		3733.6		3734.1		3734.1		3735.0
		ΔH		-0.4		-0.3				-0.2		+0.7		+0.3		+0.7		+0.5		+0.5		0.0			
		D_h				17.6				5.5		12.7		12.7		14.7		13.2				15.4		14.2	
8	464	H	3733.9		3732.8		3732.5	399	3731.5		3731.2		3731.7		3732.2		3732.9		3733.7		3734.2		3734.0		3734.8
		ΔH		-1.1		-0.3				-0.3		+0.5		+0.5		+0.7		+0.8		+0.5		-0.2			
		D_h								5.7		12.8		13.0		15.0		12.4				15.4		14.2	
9	516	H	3735.5		3735.1		3734.7	439	3730.6		3731.5		3732.3		3732.0		3733.7		3734.4		3735.0		3734.8		3735.2
		ΔH		-0.4		-0.4				+0.9		+0.8		-0.3		+1.7		+0.7		+0.6		-0.2			
		D_h								8.0		12.8		11.7		15.0		12.5				15.0		13.9	
10	564	H	3738.3		3737.8		3737.4	477	3732.8		3733.8		3734.3		3733.5		3735.0		3736.2		3736.5		3736.6		3737.0
		ΔH		-0.5		-0.4				+1.0		+0.5		-0.8		+1.5		+1.2		+0.3		+0.1			
		D_h																				15.0		13.9	
11	623	H						517	3734.8		3734.9		3735.3		3734.8		3736.7		3737.5		3738.0		3738.0		3738.3
		ΔH								+0.1		+0.4		-0.5		+1.9		+0.8		+0.5		0.0			
		D_h																				14.6		13.5	

Table 9.3.6. USSR - Changes in thickness.

Part 14 of 54 Variations in the surface elevation and horizontal component of the displacement of marked points on the glacier surface

Tien-Shan Tsentralny Tuyuksu Glacier Profile 8

point no.	distance from fixed point	date / days	1966 29.07.	418	1966 20.09.	359	1967 14.09.	312	1968 23.07.	397	1969 24.08.	364	1970 23.08.
1	121	H	3740.9		3740.1		3737.8		3736.5		3735.3		3735.5
		ΔH		-0.8		-2.3		-1.3		-1.2		+0.2	
		D_h		←		8.9				8.8		7.7	
2	183	H	3737.5		3736.5		3736.9		3736.9		3736.0		3736.3
		ΔH		-1.0		+0.4		0.0		-0.9		+0.3	
		D_h		←		9.8				9.2		8.2	
3	221	H	3737.7		3737.5		3737.9		3738.4		3737.4		3737.9
		ΔH		-0.2		+0.4		+0.5		-1.0		+0.5	
		D_h		←		9.8				10.0		8.9	
4	261	H	3738.5		3738.3		3738.5		3738.5		3737.6		3737.9
		ΔH		-0.2		+0.2		0.0		-0.9		+0.3	
		D_h		←		11.4				10.7		9.2	
5	302	H	3737.9		3737.8		3737.9		3738.2		3737.4		3737.7
		ΔH		-0.1		+0.1		+0.3		-0.8		+0.3	
		D_h		←		12.0				11.3		9.4	
6	340	H	3737.7		3738.2		3737.6		3737.8		3737.1		3737.8
		ΔH		+0.5		-0.6		+0.2		-0.7		+0.7	
		D_h								11.3		9.8	

Table 9.3.6. USSR - Changes in thickness.

Part 15 of 54 Variations in the surface elevation and horizontal component of the displacement of marked points on the glacier surface

Tien-Shan Igly Tuyuksu Glacier Profile 1

point no.	distance from fixed point	date	1959 09.10.		1961 17.09.		1962 15.09.		1963 25.09.		1964 23.09.		1965 05.10.		1966 24.09.		1967 15.09.		1968 25.08.		1969 25.08.		1970 22.08.
		days		708		363		355		363		377		354		356		345		365		362	
1	257	H	3568.7		3567.3		3565.3		3565.3		3566.5		3565.8		3567.2		3569.5		3566.0		3566.2		3566.1
		ΔH		-1.4		-2.0		0.0		+1.2		-0.7		+1.4		+2.3		-3.5		+0.2		-0.1	
		D_h				7.8														9.8		8.7	
2	304	H	3570.2		3569.6		3568.3		3568.8		3569.4		3569.1		3570.1		3571.1		3568.8		3569.7		3569.6
		ΔH		-0.6		-1.3		+0.5		+0.6		-0.3		+1.0		+1.0		-2.3		+0.9		-0.1	
		D_h				9.3																9.2	
3	353	H	3568.4		3568.8		3567.5		3568.3		3569.0		3569.3		3569.7		3570.5		3568.4		3569.2		3569.3
		ΔH		+0.4		-1.3		+0.8		+0.7		+0.1		+0.6		+0.8		-1.9		+0.8		+0.1	
		D_h		9.4																		9.6	
4	390	H	3566.6		3566.8		3565.5		3566.1		3568.4		3569.3		3567.3		3568.7		3566.2		3566.7		3567.5
		ΔH		+0.2		-1.3		+0.6		+2.3		+0.7		-1.8		+1.4		-2.5		+0.5		+0.8	
		D_h				9.1																9.4	
5	441	H	3562.8		3564.2		3561.9		3561.9		3565.1		3565.8		3566.0		3567.3		3562.8		3563.3		3564.1
		ΔH		+1.4		-2.3		0.0		+3.2		+0.7		+0.2		+1.3		-4.5		+0.5		+0.8	
		D_h				8.2														11.0			
6	481	H	3561.1		3562.2		3560.5		3562.1		3566.1		3564.5		3562.1		3565.7		3561.4		3562.0		3562.6
		ΔH		+1.1		-1.7		+1.6		+4.0		-1.6		-2.4		+3.6		-4.3		+0.6		+0.6	
		D_h				7.9																	
7	521	H	3560.6		3556.6		3554.8		3554.8		3563.7		3566.0		3559.4		3563.5		3559.8		3560.4		3561.1
		ΔH		-4.0		-1.8		0.0		+8.9		+2.3		-6.6		+4.1		-3.7		+0.6		+0.7	
		D_h				7.8																	

Table 9.3.6. USSR - Changes in thickness.

Part 16 of 54 Variations in the surface elevation and horizontal component of the displacement of marked points on the glacier surface

Tien-Shan Igly Tuyuksu Glacier Profile 2

point no.	distance from fixed point	data	1959 09.10.		1963 25.09.		1964 23.09.		1965 05.10.		1966 24.09.		1967 15.09.		1968 25.08.		1969 25.08.		1970 22.08.
		days		716		363		377		354		366		345		365		362	
		H	3618.2		3615.5		3617.3		3616.8		3617.7		3620.9		3620.9		3621.5		3623.0
1	341	ΔH		-2.7		+1.8		-0.5		+0.9		+3.2		0.0		+0.6		+1.5	
		D_h																10.6	
		H	3616.8		3614.7		3616.8		3616.8		3619.4		3618.7		3615.2		3614.2		3614.3
2	380	ΔH		-2.1		+2.1		0.0		+2.6		-0.7		-3.7		-1.0		+0.1	
		D_h																11.1	
		H	3616.7		3612.6		3614.5		3612.5		3616.7		3618.3		3615.5		3616.8		3617.7
3	423	ΔH		-4.1		+1.9		-2.0		+4.2		+1.6		-2.8		+0.3		+0.9	
		D_h																11.4	
		H	3613.7		3611.3		3612.1		3610.4		3617.1		3619.6		3618.0		3618.5		3619.2
4	467	ΔH		-2.4		+0.9		-1.7		+6.7		+2.5		-1.6		+0.5		+0.7	
		D_h																10.9	
		H	3612.3		3610.2		3611.3		3610.7		3616.7		3618.9		3616.4		3616.9		3617.2
5	510	ΔH		-2.1		+1.1		-0.6		+6.0		+2.2		-2.5		+0.5		+0.3	
		D_h																10.3	

Table 9.3.6. USSR - Changes in thickness.

Part 17 of 54 Variations in the surface elevation and horizontal component of the displacement of marked points on the glacier surface

Tien-Shan Manshuk Mametova's Glacier Profile 1

point no.	distance from fixed point	date	1963 03.10.		1964 22.09.		1965 04.10.		1966 24.09.		1967 17.09.		1968 19.08.		1969 10.08.		1970 20.08.
		days		354		377		355		358		367		356		375	
1	154	H	3628.4		3628.4		3629.2		3632.3		3630.5		3629.4		3630.0		3630.0
		ΔH		0.0		+0.8		+3.1		-1.8		-1.1		+0.6		0.0	
		Dh												7.5		8.4	
2	161	H	3629.5		3629.1		3629.4		3631.2		3630.2		3629.4		3630.2		3630.2
		ΔH		-0.4		+0.3		+1.8		-1.0		-0.8		+0.8		0.0	
		Dh												8.1		8.6	
3	172	H	3628.8		3629.4		3629.4		3629.5		3630.2		3629.3		3629.8		3630.0
		ΔH		+0.6		0.0		+0.1		+0.7		-0.8		+0.5		+0.2	
		Dh												9.2		9.0	
4	187	H	3626.4		3627.1		3626.9		3626.9		3627.1		3626.4		3626.8		3627.0
		ΔH		+0.7		-0.2		0.0		+0.2		-0.7		+0.4		+0.2	
		Dh												8.7		9.1	
5	211	H	3621.9		3623.0		3623.2		3624.1		3623.7		3622.4		3622.4		3622.5
		ΔH		+1.1		+0.2		+0.9		-0.4		-1.3		0.0		+0.1	
		Dh												7.8		9.4	
6	224	H	3623.8		3624.0		3624.4		3624.1		3624.3		3622.8		3623.4		3623.0
		ΔH		+0.2		+0.4		-0.3		+0.2		-1.5		+0.6		-0.4	
		Dh												6.8		7.7	
7	245	H	3624.0		3624.5		3624.9		3624.3		3624.0		3623.6		3624.5		3624.3
		ΔH		+0.5		+0.4		-0.6		-0.3		-0.4		+0.9		-0.2	
		Dh												5.0		5.2	

Table 9.3.6. USSR - Changes in thickness.

Part 18 of 54 Variations in the surface elevation and horizontal component of the displacement of marked points on the glacier surface

Tien-Shan Manshuk Mametova's Glacier Profile 2

point no.	distance from fixed point	date	1963 03.10.		1964 22.09.		1965 04.10.		1966 24.09.		1967 17.09.		1968 19.08.		1969 10.08.		1970 20.08.
		days		354		377		355		358		336		356		375	
		H	3648.7		3648.9		3649.0		3647.8		3646.9		3650.0		3649.5		3647.5
1	89	ΔH		+0.2		+0.1		-1.2		-0.9		+3.1		-0.5		-2.0	
		D_h												8.9		8.8	
		H	3655.3		3688.2		3662.8		3656.4		3655.4		3657.0		3657.0		3655.5
2	110	ΔH		+32.9		-25.4		-6.4		-1.0		+1.6		0.0		-1.5	
		D_h												9.3		9.0	
		H	3667.0		3683.5		3676.0		3667.4		3665.2		3667.4		3667.6		3667.1
3	130	ΔH		+16.5		-7.5		-8.6		-2.2		+2.2		+0.2		+0.5	
		D_h												9.6		9.3	
		H	3673.4		3678.3		3679.6		3674.4		3672.8		3674.2		3674.5		3675.5
4	148	ΔH		+3.9		+1.3		-5.2		-1.6		+1.4		+0.3		+1.0	
		D_h												9.7		9.4	
		H	3678.3		3673.3		3683.0		3677.8		3679.6		3679.3		3679.5		3680.0
5	163	ΔH		-5.0		+9.7		-5.2		+1.8		-0.3		+0.2		+0.5	
		D_h												10.2		10.2	
		H	3683.5		3663.0		3689.6		3685.4		3685.2		3684.7		3684.8		3685.2
6	183	ΔH		-20.5		+26.6		-4.2		-0.2		-0.5		+0.1		+0.4	
		D_h												10.8		9.3	
		H	3687.7		3652.8		3692.2		3690.0		3690.0		3690.0		3689.1		3688.9
7	202	ΔH		-34.9		+39.4		-2.2		0.0		0.0		-0.9		-0.2	
		D_h														8.7	

Manshuk Mametova's Glacier Profile 3

point no.	distance from fixed point	date	1963 03.10.		1964 22.09.		1965 04.10.		1966 24.09.		1967 17.09.		1968 19.08.		1969 10.08.		1970 20.08.
		days		354		377		355		358		336		356		375	
		H	3704.4		3705.4		3707.2		3705.9		3705.4		3705.5		3705.0		3706.0
1	101	ΔH		+1.0		+1.8		-1.3		-0.5		+0.1		-0.5		+1.0	
		D_h												8.8		9.7	
		H	3707.4		3708.4		3709.8		3708.6		3708.9		3707.2		3708.0		3707.8
2	130	ΔH		+1.0		+1.4		-1.2		+0.3		-1.7		+0.8		-0.2	
		D_h												9.0		9.7	
		H	3709.1		3709.2		3712.9		3711.0		3710.5		3709.2		3710.4		3710.1
3	144	ΔH		+0.1		+3.7		-1.9		-0.5		-1.3		+1.2		-0.3	
		D_h												8.2		9.6	
		H	3711.2		3711.1		3713.7		3713.2		3712.8		3711.7		3712.8		3712.6
4	157	ΔH		-0.1		+2.6		-0.5		-0.4		-1.1		+1.1		-0.2	
		D_h												8.9		9.5	
		H	3713.0		3712.9		3715.0		3717.4		3717.7		3716.2		3716.7		3717.5
5	182	ΔH		-0.1		+2.1		+2.4		+0.3		-1.5		+0.5		+0.8	
		D_h												10.0		9.4	
		H	3716.1		3716.9		3718.0		3718.3		3718.2		3717.6		3718.1		3718.1
6	190	ΔH		+0.8		+1.1		+0.3		-0.1		-0.6		+0.5		0.0	
		D_h												10.1		9.3	
		H	3717.5		3718.1		3718.9		3719.7		3718.7		3717.9		3718.1		3718.1
7	208	ΔH		+0.6		+0.8		+0.8		-1.0		-0.8		+0.2		0.0	
		D_h												10.6		9.2	

Table 9.3.6. USSR - Changes in thickness.

Part 19 of 54 Variations in the surface elevation and horizontal component of the displacement of marked points on the glacier surface

Tien-Shan Manshuk Mametova's Glacier Profile 4

point no.	distance from fixed point	date / days	1959 06.09.	distance from fixed point	1963 03.10.	354	1964 22.09.	377	1965 04.10.	355	1966 29.09.	358	1967 17.09.	336	1968 19.08.	356	1969 10.08.	375	1970 20.08.
1	106	H	3728.3	105	3730.0		3729.2		3729.3		3729.1		3730.7		3728.0		3729.9		3730.0
		ΔH				-0.8		+0.1		-0.2		+1.6		-2.7		+1.9		+0.1	
		D_h														8.5		8.2	
2	128	H	3728.0	127	3729.6		3728.8		3728.8		3728.7		3730.5		3728.0		3729.8		3730.3
		ΔH				-0.8		0.0		-0.1		+1.8		-2.5		+1.8		+0.5	
		D_h														7.0		7.8	
3	148	H	3727.3	147	3728.8		3727.9		3728.1		3728.1		3729.5		3727.3		3729.0		3729.4
		ΔH				-0.9		+0.2		0.0		+1.4		-2.2		+1.7		+0.4	
		D_h														7.3		7.9	
4	168	H	3726.1	168	3727.0		3727.1		3727.4		3727.2		3728.4		3726.2		3728.0		3728.2
		ΔH				+0.1		+0.3		-0.2		+1.2		-2.2		+1.8		+0.2	
		D_h														7.6		8.3	
5	181	H	3725.2	181	3725.7		3726.4		3726.2		3725.8		3726.8		3725.3		3726.9		3727.4
		ΔH				+0.7		-0.2		-0.4		+1.0		-1.5		+1.6		+0.5	
		D_h														7.7		8.3	
6	214	H	3722.8	217	3724.4		3722.7		3722.6		3722.7		3724.1		3724.3		3723.8		3726.2
		ΔH				-1.7		-0.1		+0.1		+1.4		+0.2		-0.5		+2.4	
		D_h														7.9		9.1	
7	239	H	3722.3	255	3725.7		3722.2		3722.1		3722.4		3724.3		3724.3		3723.7		3723.9
		ΔH				-1.5		-0.1		+0.3		+1.9		0.0		-0.6		+0.2	
		D_h														8.2		8.5	
8		H		264	3723.6		3722.2		3722.1										
		ΔH				-1.4		-0.1											
		D_h																	

Table 9.3.6. USSR - Changes in thickness.

Part 20 of 54 Variations in the surface elevation and horizontal component of the displacement of marked points on the glacier surface

Tien-Shan Molodezhnyy Glacier Profile 1

point no.	distance from fixed point	date / days	1959 03.09.	737	1961 19.09.	distance from fixed point	1963 04.10.	354	1964 23.09.	370	1965 28.09.	360	1966 23.09.	358	1967 16.09.	342	1968 24.08.	358	1969 17.08.	369	1970 21.08.
1	271	H	3503.1		3501.2	272	3501.4		3501.8		3500.4		3496.2		3497.8		3497.7		3494.5		3494.5
		ΔH		-1.9				+0.4		-1.4		-4.2		+1.6		-0.1		-3.2		0.0	
		D_h																1.7		1.4	
2	317	H	3506.3		3503.8	317	3503.1		3503.9		3503.1		3499.4		3501.8		3500.3		3498.4		3498.4
		ΔH		-2.5				+0.8		-0.8		-3.7		+2.4		-1.5		-1.9		0.0	
		D_h																2.2		2.0	
3	359	H	3507.3		3504.5	359	3501.3		3499.8		3502.3		3497.7		3498.7		3497.2		3496.4		3496.4
		ΔH		-2.8				-1.5		+2.5		-4.6		+1.0		-1.5		-0.8		0.0	
		D_h																2.8		2.9	
4	406	H	3506.5		3502.4	406	3494.4		3497.1		3496.0		3494.1		3495.3		3494.2		3493.6		3493.7
		ΔH		-4.1				+2.7		-1.1		-1.9		+1.2		-1.1		-0.6		+0.1	
		D_h																3.5		3.3	
5	447	H	3498.5		3498.1	447	3497.5		3494.9		3497.7		3494.5		3497.6		3495.9		3495.7		3495.7
		ΔH		-0.4				-2.6		+2.8		-3.2		+3.1		-1.7		-0.2		0.0	
		D_h																3.6		3.8	
6	486	H	3500.0		3497.9	487	3499.5		3502.2		3500.2		3496.9		3500.6		3499.1		3498.8		3498.8
		ΔH		-2.1				+2.7		-2.0		-3.3		+3.7		-1.5		-0.3		0.0	
		D_h																3.3		3.4	
7	533	H	3500.0		3499.9	534	3503.1		3507.1		3504.6		3501.2		3504.3		3502.6		3502.9		3502.9
		ΔH		-0.1				+4.0		-2.5		-3.4		+3.1		-1.7		+0.3		0.0	
		D_h																2.8		3.0	
8	590	H	3502.8		3505.0	591	3508.5		3513.1		3510.1		3506.4		3512.0		3508.0		3507.1		3507.1
		ΔH		+2.2				+4.6		-3.0		-3.7		+5.6		-4.0		-0.9		0.0	
		D_h																			
9	634	H			3508.5	644	3516.1		3518.4		3516.1		3512.7		3516.8		3514.3		3513.5		3513.5
		ΔH						+2.3		-2.3		-3.4		+4.1		-2.5		-0.8		0.0	
		D_h																			

Table 9.3.6. USSR - Changes in thickness.

Part 21 of 54 Variations in the surface elevation and horizontal component of the displacement of marked points on the glacier surface

Tien-Shan Konstitutsiya Glacier Profile 1

point no.	distance from fixed point	date / days	1961 28.07.	50	1961 16.09.	363	1962 14.09.	354	1963 03.09.	381	1964 19.09.	375	1965 29.09.	364	1966 28.09.	380	1968 12.10.	331	1969 08.09.	333	1970 07.08.
1	336	H	3427.7		3426.6		3424.6		3425.6		3425.2		3525.5		3424.7		3426.6		3422.1		3430.1
		ΔH		-1.1		-2.0		+1.0		-0.4		+0.3		-0.8		+1.9		-4.5		+8.0	
		D_h		2.4		11.2				← 30.9 →				7.5							
2	346	H	3428.6		3427.4		3425.9		3425.8		3425.6		3425.4		3424.1		3426.0		3420.9		3429.6
		ΔH		-1.2		-1.5		-0.1		-0.2		-0.2		-1.3		+1.9		-5.1		+8.7	
		D_h		2.4		11.2				← 31.9 →				8.0							
3	356	H	3428.8		3427.6		3426.0		3425.6		3425.5		3425.3		3423.7		3425.4		3419.7		3429.1
		ΔH		-1.2		-1.6		-0.4		-0.1		-0.2		-1.6		+1.7		-5.7		+9.4	
		D_h		2.2		11.4				← 33.0 →				8.5							
4	366	H	3428.5		3427.3		3425.7		3425.3		3425.8		3425.1		3423.9		3425.7		3420.1		3429.0
		ΔH		-1.2		-1.6		-0.4		+0.5		-0.7		-1.2		+1.8		-5.6		+8.9	
		D_h		2.3		11.4				← 33.8 →				10.0							
5	377	H	3428.4		3427.2		3426.1		3425.9		3425.3		3425.7		3424.1		3426.3		3420.7		3429.7
		ΔH		-1.2		-1.1		-0.2		-0.6		+0.4		-1.6		+2.2		-5.6		+9.0	
		D_h		2.5		13.4				← 33.1 →				11.5							
6	387	H	3428.4		3427.4		3426.1		3426.1		3425.6		3426.9		3424.7		3427.2		3421.3		3430.1
		ΔH		-1.0		-1.3		0.0		-0.5		+1.3		-2.2		+2.5		-5.9		+8.8	
		D_h		2.5		11.8				← 33.4 →				14.3							
7	398	H	3428.4		3427.4		3425.9		3425.8		3426.0		3426.7		3425.3		3427.6		3421.9		3430.3
		ΔH		-1.0		-1.5		-0.1		+0.2		+0.7		-1.4		+2.3		-5.7		+8.4	
		D_h		2.0		11.7				← 32.8 →				8.7							
8	408	H	3428.3		3427.1		3425.7		3425.5		3425.7		3426.4		3425.5		3427.9		3421.6		3430.8
		ΔH		-1.2		-1.4		-0.2		+0.2		+0.7		-0.9		+2.4		-6.3		+9.2	
		D_h		2.3		11.2				← 31.5 →				9.5							
9	416	H	3427.9		3426.7		3425.0		3425.8		3425.8		3426.4		3424.7		3428.3		3421.3		3431.4
		ΔH		-1.2		-1.7		+0.8		0.0		+0.6		-1.7		+3.6		-7.0		+10.1	
		D_h		2.3		10.6				← 31.1 →				9.2							

Table 9.3.6. USSR - Changes in thickness.

Part 22 of 54 Variations in the surface elevation and horizontal component of the displacement of marked points on the glacier surface

Tien-Shan Konstitutsiya Glacier Profile 2

point no.	distance from fixed point	date / days	1961 26.07.	52	1961 16.09.	363	1962 14.09.	351	1963 31.08.	385	1964 19.09.	375	1965 29.09.	374	1966 28.09.	745	1968 12.10.	330	1969 07.09.	337	1970 09.08.
		H	3467.0		3466.0		3465.4		3465.9		3466.2		3467.7		3467.9		3470.5		3471.2		3473.3
1	143	ΔH		-1.0		-0.6		+0.5		+0.3		+1.5		+0.2		+2.6		+0.7		+2.1	
		D_h		2.6		11.4		←		34.0		→		10.0							
		H	3467.2		3466.0		3465.3		3466.0		3466.6		3468.2		3468.5		3470.3		3470.9		3473.4
2	156	ΔH		-1.2		-0.7		+0.7		+0.6		+1.6		+0.3		+1.8		+0.6		+2.5	
		D_h		2.5		11.4		←		35.1		→		10.0							
		H	3467.0		3465.8		3465.5		3466.0		3466.7		3468.0		3468.6		3470.0		3470.6		3472.4
3	176	ΔH		-1.2		-0.3		+0.5		+0.7		+1.3		+0.6		+1.4		+0.6		+1.8	
		D_h		2.5		12.7		←		35.3		→		12.5							
		H	3465.6		3464.3		3463.7		3464.3		3464.9		3465.2		3465.8		3468.8		3469.9		3471.6
4	197	ΔH		-1.3		-0.6		+0.6		+0.6		+0.3		+0.6		+3.0		+1.1		+1.7	
		D_h		2.6		13.8		←		36.6		→		13.0							
		H	3463.6		3462.5		3462.3		3462.4		3463.5		3465.2		3466.2		3468.5		3470.0		3471.4
5	216	ΔH		-1.1		-0.2		+0.1		+1.1		+1.7		+1.0		+2.3		+1.5		+1.4	
		D_h		3.0		14.4		←		37.2		→		13.0							
		H	3463.1		3461.9		3461.3		3463.3		3464.0		3465.3		3465.3		3467.8		3469.3		3470.6
6	238	ΔH		-1.2		-0.6		+2.0		+0.7		+1.3		0.0		+2.5		+1.5		+1.3	
		D_h		3.1		14.2		←		40.2		→		15.0							
		H	3463.1		3461.9		3461.5		3462.2		3463.2		3464.2		3463.8		3466.6		3466.6		3468.0
7	259	ΔH		-1.2		-0.4		+0.7		+1.0		+1.0		-0.4		+2.8		0.0		+1.4	
		D_h		2.8		14.3		←		40.9		→		15.5							
		H	3462.4		3461.1		3460.9		3461.4		3461.3		3462.4		3463.4		3465.5		3464.2		3465.5
8	272	ΔH		-1.3		-0.2		+0.5		-0.1		+1.1		+1.0		+2.1		-1.3		+1.3	
		D_h		2.6		14.7		←		40.7		→		15.5							
		H	3459.8		3458.9		3458.8		3458.7		3460.5		3462.4		3462.9		3463.4		3464.0		3468.0
9	297	ΔH		-0.9		-0.1		-0.1		+1.8		+1.9		+0.5		+0.5		+0.6		+4.0	
		D_h		2.7		14.9		←		45.9		→		15.0							
		H	3462.7		3461.7		3461.2		3461.5		3464.0		3465.9		3468.6		3471.4		3475.0		3475.5
10	318	ΔH		-1.0		-0.5		+0.3		+2.5		+1.9		+2.7		+2.8		+3.6		+0.5	
		D_h		2.9		15.3		←		50.0		→		15.0							

Table 9.3.6. USSR - Changes in thickness.

Part 23 of 54 Variations in the surface elevation and horizontal component of the displacement of marked points on the glacier surface
Tien-Shan Konstitutsiya Glacier Profile 3

point no.	distance from fixed point	date	1961 24.07.		1961 15.09.		1962 11.09.		1963 30.08.		1964 20.09.		1965 29.09.		1966 28.09.		1968 11.10.		1969 08.09.		1970 09.08.
		days		53		361		353		387		374		364		744		332		335	
1	168	H	3789.3		3787.9		3787.0		3787.8		3789.2		3790.0		3789.5		3788.8		3788.2		3789.5
		ΔH		-1.4		-0.9		+0.8		+1.4		+0.8		-0.5		-0.7		-0.6		+1.3	
		D_h		4.4		24.2		←		27.0											
2	199	H			3783.7		3782.9		3783.9		3782.9		3784.9		3787.1		3785.6		3785.3		3786.0
		ΔH				-0.8		+1.0		-1.0		+2.0		+2.2		-1.5		-0.3		+0.7	
		D_h		4.8		31.2															
3	239	H			3779.3		3778.9		3779.9		3781.0		3781.0		3782.1		3780.5		3780.5		3781.0
		ΔH				-0.4		+1.0		+1.1		0.0		+1.1		-1.6		0.0		+0.5	
		D_h		5.9		36.3															
4	280	H			3773.9		3773.4		3774.1		3775.3		3775.0		3776.2		3774.7		3775.0		3775.7
		ΔH				-0.5		+0.7		+1.2		-0.3		+1.2		-1.5		+0.3		+0.7	
		D_h		6.1		42.3		←		61.0											
5	320	H			3769.2		3768.4		3769.4		3770.8		3770.8		3771.6		3770.3		3771.7		3771.0
		ΔH				-0.8		+1.0		+1.4		0.0		+0.8		-1.3		+1.4		-0.7	
		D_h		6.7		39.9															
6	363	H	3765.3		3764.4		3764.4		3765.2		3766.6		3766.4		3767.7		3766.3		3767.4		3766.8
		ΔH		-0.9		0.0		+0.8		+1.4		-0.2		+1.3		-1.4		+1.1		-0.6	
		D_h		6.6		41.1		←		70.3											
7	403	H	3761.0		3760.2		3759.6		3761.0		3762.4		3762.7		3764.7		3761.9		3764.2		3762.7
		ΔH		-0.8		-0.6		+1.4		+1.4		+0.3		+2.0		-2.8		+2.3		-1.5	
		D_h		7.1		39.2															
8	443	H	3761.0		3760.4		3761.6		3761.1		3762.0		3763.1		3764.9		3762.0		3764.7		3762.0
		ΔH		-0.6		+1.2		-0.5		+0.9		+1.1		+1.8		-2.9		+2.7		-2.7	
		D_h		7.4		38.4		←		80.7		46.3									
9	483	H	3764.1		3763.6		3764.3		3763.8		3764.3		3765.7		3767.3		3764.3		3767.2		3764.4
		ΔH		-0.5		+0.7		-0.5		+0.5		+1.4		+1.6		-3.0		+2.9		-2.8	
		D_h		7.2		38.0															
10	523	H	3768.6		3768.0		3768.2		3767.8		3768.8		3771.2		3774.0		3770.2		3773.0		3769.3
		ΔH		-0.6		+0.2		-0.4		+1.0		+2.4		+2.8		-3.8		+2.8		-3.7	
		D_h		6.8		37.7		←		78.8											
11	569	H	3773.8		3772.9		3773.2		3773.4		3774.9		3775.4		3775.6		3775.4		3776.9		3776.4
		ΔH		-0.9		+0.3		+0.2		+1.5		+0.5		+0.2		-0.2		+1.5		-0.5	
		D_h		6.7		36.9															
12	609	H	3773.6		3772.6		3774.1		3772.1		3774.0		3774.9		3774.9		3774.5		3776.0		3776.0
		ΔH		-1.0		+1.5		-2.0		+1.9		+0.9		0.0		-0.4		+1.5		0.0	
		D_h		6.0		32.8															

Table 9.3.6. USSR - Changes in thickness.

Part 24 of 54 Variations in the surface elevation and horizontal component of the displacement of marked points on the glacier surface

Tien-Shan Dimitriev's Glacier Profile 1, the left and main branch

point no.	distance from fixed point	date	1961 04.08.		1961 13.09.		1962 02.09.		1963 21.09.		1964 17.09.		1965 28.09.		1966 26.09.		1968 10.10.		1969 06.09.
		days		40		354		384		362		376		363		746		341	
1	244	H	3659.6		3658.4		3655.4		3653.3		3653.5		3649.9		3648.1		3642.7		3642.2
		ΔH		-1.2		-3.0		-2.1		+0.2		-3.6		-1.8		-5.4		-0.5	
		D_h		0.1		0.9								0.4					
2	284	H	3662.3		3661.3		3658.2		3656.5		3656.6		3652.8		3651.3		3644.4		3645.0
		ΔH		-1.0		-3.1		-1.7		+0.1		-3.8		-1.5		-6.9		+0.6	
		D_h		0.2		1.1								0.4					
3	333	H	3664.5		3663.5		3663.7		3658.6		3659.3		3655.9		3654.9		3648.7		3648.4
		ΔH		-1.0		-2.8		-2.1		+0.7		-3.4		-1.0		-6.2		-0.3	
		D_h		0.3		1.3				1.8				1.6					
4	378	H	3668.0		3666.9		3664.5		3663.2		3663.4		3660.5		3659.7		3653.6		3655.5
		ΔH		-1.1		-2.4		-1.3		+0.2		-2.9		-0.8		-6.1		+1.9	
		D_h		0.4		2.6								1.5					
5	444	H	3676.9		3676.2		3674.2		3673.0		3672.9		3670.3		3667.6		3664.0		3665.6
		ΔH		-0.7		-2.0		-1.2		-0.1		-2.6		-2.7		-3.6		+1.6	
		D_h		0.4		2.1								1.6					
6	502	H	3682.9		3682.1		3680.2		3679.3		3680.1		3677.4		3676.9		3673.1		3671.6
		ΔH		-0.8		-1.9		-0.9		+0.8		-2.7		-0.5		-3.8		-1.5	
		D_h		0.4		2.4								2.8					
7	567	H	3688.2		3687.5		3685.2		3684.6		3686.3		3683.6		3680.8		3676.7		3676.4
		ΔH		-0.7		-2.3		-0.6		+1.7		-2.7		-2.8		-4.1		-0.3	
		D_h		0.4		3.3				9.1				3.2					
8	616	H	3690.0		3689.2		3686.9		3686.2		3687.9		3685.0		3682.3		3679.0		3677.4
		ΔH		-0.8		-2.3		-0.7		+1.7		-2.9		-2.7		-3.3		-1.6	
		D_h		0.4		3.6				10.2				3.9					
9	665	H	3690.0		3689.0		3686.7		3686.2		3688.2		3685.4		3682.8		3681.3		3679.2
		ΔH		-1.0		-2.3		-0.5		+2.0		-2.8		-2.6		-1.5		-2.1	
		D_h		0.5		4.4								5.0					

Table 9.3.6. **USSR - Changes in thickness.**
Part 25 of 54 Variations in the surface elevation and horizontal component of the displacement of marked points on the glacier surface
Tien-Shan Dimitriev's Glacier Profile 2, left and main branch

point no.	distance from fixed point	date / days	1961 04.08.	40	1961 13.09.	354	1962 02.09.	365	1963 02.09.	381	1964 17.09.	376	1965 28.09.	363	1966 26.09.	745	1968 10.10.	331	1969 06.09.	339	1970 11.08.
1	133	H	3867.2		3866.9		3865.6		3866.2		3866.2		3866.2		3865.9		3863.0		3866.3		3865.1
		ΔH		-0.3		-1.3		+0.6		0.0		0.0		-0.3		-2.9		+3.3		-1.2	
		D_h		0.1		0.4		←		0.4		→		0.6							
2	234	H	3869.4		3868.9		3867.3		3868.1		3868.3		3867.9		3867.4		3865.6		3866.4		3866.1
		ΔH		-0.5		-1.6		+0.8		+0.2		-0.4		-0.5		-1.8		+0.8		-0.3	
		D_h		0.3		1.1		←		2.2		→		1.2							
3	332	H	3863.3		3862.9		3861.7		3862.7		3863.4		3863.4		3863.1		3861.9		3863.4		3862.5
		ΔH		-0.4		-1.2		+1.0		+0.7		0.0		-0.3		-1.2		+1.5		-0.9	
		D_h		0.4		1.9		←		7.3		→		1.2							
4	430	H	3862.5		3862.0		3860.0		3861.1		3861.9		3861.7		3860.0		3861.5		3860.7		3860.5
		ΔH		-0.5		-2.0		+1.1		+0.8		-0.2		-1.7		+1.5		-0.8		-0.2	
		D_h		0.5		3.0		←		10.9		→		6.3							
5	535	H	3849.1		3848.2		3848.8		3849.3		3848.7		3848.3		3847.5		3849.4		3849.0		3848.6
		ΔH		-0.9		+0.6		+0.5		-0.6		-0.4		-0.8		+1.9		-0.4		-0.4	
		D_h		0.4		3.3		←		16.1		→		9.3							
6	632	H	3840.8		3840.4		3839.6		3841.2		3844.6		3842.6		3842.0		3844.3		3843.0		3842.4
		ΔH		-0.4		-0.8		+1.6		+3.4		-2.0		-0.6		+2.3		-1.3		-0.6	
		D_h		1.0		3.2		←		21.0		→		8.4							
7	735	H	3842.6		3842.2		3840.9		3842.7		3844.1		3845.1		3845.4		3847.9		3847.4		3845.8
		ΔH		-0.4		-1.3		+1.8		+1.4		+1.0		+0.3		+2.5		-0.5		-1.6	
		D_h		1.2		3.4		←		32.3		→		10.1							
8	834	H	3843.6		3843.2		3842.4		3844.7		3845.8		3846.3		3846.8		3849.8		3849.4		3848.8
		ΔH		-0.4		-0.8		+2.3		+1.1		+0.5		+0.5		+3.0		-0.4		-0.6	
		D_h		1.8		6.0		←		42.0		→		10.8							
9	943	H	3842.3		3841.9		3841.2		3844.6		3844.2		3845.3		3846.7		3848.2		3848.9		3849.3
		ΔH		-0.4		-0.7.		+3.4		-0.4		+1.1		+1.4		+1.5		+0.7		+0.4	
		D_h		2.1		7.4		←		51.7		→		15.6							
10	1043	H			3846.7		3845.8		3847.0		3842.2		3843.5		3843.9		3851.2		3851.5		3849.1
		ΔH				-0.9		+1.2		-4.8		+1.3		+0.4		+7.3		+0.3		-2.4	
		D_h				9.5															
11	1145	H			3845.3		3843.8		3845.1		3845.2		3846.8		3846.0		3848.3		3848.6		3846.9
		ΔH				-1.5		+1.3		+0.1		+1.6		-0.8		+2.3		+0.3		-1.7	
		D_h				8.4															
12	1234	H			3849.5				3849.8		3846.2		3847.8		3849.2		3849.9		3850.3		3848.5
		ΔH								-3.6		+1.6		+1.4		+0.7		+0.4		-1.8	
		D_h																			

Table 9.3.6. USSR - Changes in thickness.

Part 26 of 54 Variations in the surface elevation and horizontal component of the displacement of marked points on the glacier surface

Tien-Shan Toguzak's Glacier Profile 1

point no.	distance from fixed point	date / days	1961 30.07.	49	1961 17.09.	362	1962 14.09.	352	1963 01.09.	385	1964 20.09.	371	1965 26.09.	366	1966 27.09.	381	1968 13.10.	359	1969 07.10.	307	1970 10.08.
1	172	H	3879.5		3878.7		3877.2		3877.0		3877.0		3875.5		3876.1		3874.0		3873.0		3873.4
		ΔH		-0.8		-1.5		-0.2		0.0		-1.5		+0.6		-2.1		-1.0		+0.4	
		D_h		1.4		7.8		←		15.0		4.0		3.0							
2	233	H	3877.9		3876.9		3875.5		3875.6		3875.8		3874.5		3876.0		3873.5		3873.5		3872.8
		ΔH		-1.0		-1.4		+0.1		+0.2		-1.3		+1.5		-2.5		0.0		-0.7	
		D_h		1.5		9.0		←		14.4		5.0		5.0							
3	287	H	3877.2		3876.5		3875.2		3875.2		3875.4		3874.0		3875.6		3873.1		3872.7		3872.8
		ΔH		-0.7		-1.3		0.0		+0.2		-1.4		+1.6		-2.5		-0.4		+0.1	
		D_h		1.7		10.1		←		18.5		5.0		5.0							
4	338	H	3878.0		3877.3		3875.9		3876.0		3876.0		3874.4		3877.8		3874.0		3872.9		3872.7
		ΔH		-0.7		-1.4		+0.1		0.0		-1.6		+3.4		-3.8		-1.1		-0.2	
		D_h		1.6		10.4		←		15.6		6.5		6.5							
5	392	H	3880.0		3879.0		3877.8		3877.7		3877.9		3876.3		3878.5		3876.0		3874.9		3875.4
		ΔH		-1.0		-1.2		-0.1		+0.2		-1.6		+2.2		-2.5		-1.1		+0.5	
		D_h		2.2		11.0		←		16.0		9.0		4.0							
6	422	H	3879.4		3878.3		3877.1		3877.1		3877.5		3875.9		3878.2		3876.1		3874.6		3874.8
		ΔH		-1.1		-1.2		0.0		+0.4		-1.6		+2.3		-2.1		-1.5		+0.2	
		D_h		2.7		9.4		←		15.1		11.0		4.5							
7	443	H	3878.8		3878.0		3876.7		3876.6		3877.2		3876.5		3878.2		3875.6		3874.1		3874.3
		ΔH		-0.8		-1.3		-0.1		+0.6		-0.7		+1.7		-2.6		-1.5		+0.2	
		D_h		2.2		10.2		←		17.1		10.0		4.0							
8	462	H	3878.5		3877.7		3875.9		3875.8		3878.0		3876.5		3876.2		3875.1		3873.9		3874.1
		ΔH		-0.8		-1.8		-0.1		+2.2		-1.5		-0.3		-1.1		-1.2		+0.2	
		D_h		1.9		10.8		←		17.7		9.0		4.5							

Table 9.3.6. USSR - Changes in thickness.

Part 27 of 54 Variations in the surface elevation and horizontal component of the displacement of marked points on the glacier surface

Tien-Shan Toguzak's Glacier Profile 2

point no.	distance from fixed point	date / days	1961 30.07.	49	1961 17.09.	714	1963 01.09.	756	1965 26.09.	366	1966 27.09.	747	1968 13.10.	332	1969 07.09.
		H	3960.5		3959.9		3959.1		3959.6		3959.8		3957.8		3958.3
1	75	ΔH		-0.6		-0.8		+0.5		+0.2		-2.0		+0.5	
		Dh		0.4											
		H	3957.1		3956.4		3956.0		3958.3		3956.1		3955.9		3956.4
2	137	ΔH		-0.7		-0.4		+2.3		-2.2		-0.2		+0.5	
		Dh		1.5											
		H	3954.6		3954.0		3953.5		3954.5		3955.8		3954.6		3954.8
3	195	ΔH		-0.6		-0.5		+1.0		+1.3		-1.2		+0.2	
		Dh		1.7											
		H	3952.5		3952.1		3851.7		3952.5		3954.5		3952.5		3953.2
4	255	ΔH		-0.4		-0.4		+0.8		+2.0		-2.0		+0.7	
		Dh		1.9		16.0		20.9		6.7					
		H	3952.4		3951.9		3851.4		3952.4		3953.6		3952.0		3952.6
5	315	ΔH		-0.5		-0.5		+1.0		+1.2		-1.6		+0.6	
		Dh		1.9		←——		——→ 35.6		7.0					
		H	3953.2		3952.6		3952.1		3953.2		3955.2		3952.3		3953.0
6	374	ΔH		-0.6		-0.5		+1.1		+2.0		-2.9		+0.7	
		Dh		1.9		←——		——→ 36.1		7.7					
		H	3953.4		3953.0		3951.8		3952.7		3954.8		3952.0		3952.9
7	438	ΔH		-0.4		-1.2		+0.9		+2.1		-2.8		+0.9	
		Dh		1.8		←——		36.7		7.0					
		H	3951.5		3951.0		3949.8		3950.7		3954.5		3950.3		3950.4
8	491	ΔH		-0.5		-1.2		+0.9		+3.8		-4.2		+0.1	
		Dh		1.8		←——		36.7		6.7					
		H	3951.9		3951.5		3950.6		3951.9		3954.9		3951.5		3952.2
9	540	ΔH		-0.4		-0.9		+1.3		+3.0		-3.4		+0.7	
		Dh		1.6		←——		37.9		7.0					
		H	3952.1		3951.6		3950.7		3952.1		3955.3		3951.8		3952.0
10	592	ΔH		-0.5		-0.9		+1.4		+3.2		-3.5		+0.2	
		Dh		1.7		15.5		13.9		6.0					

Table 9.3.6. USSR - Changes in thickness.

Part 28 of 54 Variations in the surface elevation and horizontal component of the displacement of marked points on the glacier surface

Tien-Shan Shokalskiy Glacier Profile 1, left branch

point no.	distance from fixed point	date / days	1962 25.07.		1962 26.09.		1963 11.09.		1964 07.09.		1965 16.09.		1966 25.08.		1967 27.09.		1968 05.08.		1969 02.08.		1970 08.09.
		days		63		350		362		374		343		398		314		362		402	
		H	3411.0		3408.8		3406.7		3404.4		3400.6		3400.2		3392.4		3391.5		3389.4		3385.1
1	251	ΔH		-2.2		-2.1		-2.3		-3.8		-0.4		-7.8		-0.9		-1.9		-4.3	
		D_h		1.03		1.6		←		2.3		4.4						0.2		1.0	
		H	3411.3		3408.7		3406.4		3403.5		3400.1		3399.4		3391.5		3391.3		3389.8		3383.5
2	265	ΔH		-2.6		-2.3		-2.9		-3.4		-0.7		-7.9		-0.2		-1.5		-6.3	
		D_h		0.0		1.3		←		2.7		4.5						0.2		0.7	
		H	3410.9		3407.6		3403.7		3401.8		3398.9		3398.2		3391.5		3391.5		3390.8		3387.2
3	279	ΔH		-3.3		-3.9		-1.9		-2.9		-0.7		-6.7		0.0		-0.7		-3.6	
		D_h		0.0		2.4		←		3.0		4.4						0.4		0.8	
		H	3407.3		3404.0		3401.6		3398.6		3396.2		3393.7		3387.6		3389.0		3385.7		3382.4
4	312	ΔH		-3.3		-2.4		-3.0		-2.4		-2.5		-6.1		+1.4		-3.3		-3.3	
		D_h		0.0		2.4		←		3.2		3.8						0.2		1.2	
		H	3404.5		3401.4		3398.9		3396.9		3393.9										
5	339	ΔH		-3.1		-2.5		-2.0		-3.0											
		D_h		1.18		1.8		←		2.2		3.7									
		H	3405.5		3402.1		3399.8		3396.5		3392.3										
6	333	ΔH		-3.4		-2.3		-3.3		-4.2											
		D_h		0.40		2.6		←		1.8		3.2									

Table 9.3.6. USSR - Changes in thickness.
Part 29 of 54 Variations in the surface elevation and horizontal component of the displacement of marked points on the glacier surface
Tien-Shen Shokalskiy Glacier Profile 2, left branch

point no.	distance from fixed point	date / days	1962 25.07.		1962 26.09.		1963 03.09.		1964 07.09.		1965 15.09.		1967 27.09.	distance from fixed point	1968 04.08.		1969 31.07.		1970 07.09.
		days		63		342		370		373		742				361		403	
1	162	H	3551.3		3548.9		3550.3		3547.3		3545.0		3544.0	178	3539.2		3540.6		3534.2
		ΔH		-2.4		+1.4		-3.0		-2.3		-1.0				+1.4		-6.4	
		Dh		1.9		3.3		3.4		←		12.5				7.2		8.4	
2	177	H	3551.2		3548.9		3550.3		3547.3		3544.7		3543.0	198	3538.3		3540.7		3535.5
		ΔH		-2.3		+1.4		-3.0		-2.6		-1.7				+2.4		-5.2	
		Dh		2.4		4.0		4.1		←		12.0				4.1		5.8	
3	194	H	3551.2		3549.1		3550.2		3547.3		3544.4		3542.9	219	3539.6		3541.6		3537.1
		ΔH		-2.1		+1.1		-2.9		-2.9		-1.5				+2.0		-4.5	
		Dh		2.3		3.3		←		6.5		13.1				5.5		6.6	
4	209	H	3551.2		3548.9		3550.0		3547.1		3544.3		3542.7	239	3541.0		3543.2		3538.8
		ΔH		-2.3		+1.1		-2.9		-2.8		-1.6				+2.2		-4.4	
		Dh		2.5		←						24.5				5.8		7.0	
5	224	H	3552.1		3549.8		3550.9		3548.4		3545.3		3543.9	258	3542.5		3545.5		3542.2
		ΔH		-2.3		+1.1		-2.5		-3.1		-1.4				+3.0		-3.3	
		Dh		2.4		2.6		←		7.0		15.5				6.0		7.4	
6	240	H	3553.1		3551.1		3552.3		3549.3		3546.7		3545.8	279	3546.1		3548.7		3544.9
		ΔH		-2.0		+1.2		-3.0		-2.6		-0.9				+2.6		-3.8	
		Dh		3.1		3.5		←				21.4				4.7		6.8	
7	256	H	3553.6		3551.4		3552.7		3549.8		3547.6		3547.7	301	3547.9		3552.3		3550.0
		ΔH		-2.2		+1.3		-2.9		-2.2		+0.1				+4.4		-2.3	
		Dh		2.7		3.5				6.6		15.6				4.6		8.1	
8	276	H	3555.1		3553.0		3554.2		3552.3		3550.2		3545.4	320	3551.9		3556.0		3553.5
		ΔH		-2.1		+1.2		-1.9		-2.1		-4.8				+4.1		-2.5	
		Dh		3.5		←						26.0				5.2		6.5	
9	291	H	3558.0		3555.9		3557.4		3554.6		3553.2		3551.8	341	3555.7		3559.0		3555.3
		ΔH		-2 1		+1.8		-2.8		-1.4		-1.4				+3.3		-3.7	
		Dh		3 5		2.1		←		7.4		17.3				5.0		7.3	
10	311	H	3558.8		3556.8		3557.9		3555.7		3555.0		3554.2	361	3557.3		3560.9		3557.2
		ΔH		-2.0		+1.1		-2.2		-0.7		-0.8				+3.6		-3.7	
		Dh		3.6		←						30.5				5.8		7.4	
11	336	H	3562.2		3559.9		3560.0		3558.7		3566.2		3558.1						
		ΔH		-2.3		+0.1		-1.3		-2.5		+1.9							
		Dh		4.2		2.5				6.2		26.5							
12	375	H	3568.1		3566.3		3567.0		3563.6		3561.2		3562.3						
		ΔH		-1.8		+0.7		-3.4		-2.4		+1.1							
		Dh		4.3		←						37.2							

Table 9.3.6. USSR - Changes in thickness.

Part 30 of 54 Variations in the surface elevation and horizontal component of the displacement of marked points on the glacier surface

Tien-Shan Shokalskiy Glacier Profile 3, left branch

point no.	distance from fixed point	date	1968 05.08.		1969 31.07.		1970 07.09.
		days		360		403	
1	114	H	3611.5		3615.6		3620.3
		ΔH		+4.1		+4.7	
		D_h		10.5		27.0	
2	156	H	3611.9		3616.2		3622.9
		ΔH		+4.3		+6.7	
		D_h		14.5		32.7	
3	197	H	3610.6		3613.0		3621.8
		ΔH		+2.4		+8.8	
		D_h		15.7		34.8	
4	237	H	3606.5		3611.2		3618.2
		ΔH		+4.7		+7.0	
		D_h		16.9		35.2	
5	279	H	3607.7		3610.8		3619.4
		ΔH		+3.1		+8.6	
		D_h		17.8		34.2	
6	320	H	3612.8		3616.8		3622.7
		ΔH		+4.0		+5.9	
		D_h		17.2		33.3	
7	360	H	3612.4		3615.4		3623.5
		ΔH		+3.0		+8.1	
		D_h		17.0		35.4	
8	401	H	3613.7		3618.1		3623.6
		ΔH		+4.4		+5.5	
		D_h		15.8		30.0	
9	439	H	3612.9		3615.2		3621.7
		ΔH		+2.3		+6.5	
		D_h		14.0		25.0	

Part 31 of 54 Variations in the surface elevation and horizontal component of the displacement of marked points on the glacier surface

Tien-Shan Shokalskiy Glacier Profile 4, left branch

point no.	distance from fixed point	date / days	1962 24.07.	63	1962 25.09.	354	1963 14.09.	359	1964 07.09.	373	1965 15.09.	344	1966 25.08.	398	1967 27.09.	312	1968 04.08.	361	1969 31.07.	403	1970 07.09.
1	88	H	3719.7		3718.3		3722.1		3723.8		3726.0		3728.6		3735.5		3737.4		3742.1		3744.3
		ΔH		-1.4		+3.8		+1.7		+2.2		+2.6		+6.9		+1.9		+4.7		+2.2	
		D_h		2.6										65.4		23.0		44.8		70.9	
2	113	H	3721.9		3720.4		3723.9		3725.4		3727.7		3730.6		3736.5		3738.4		3742.1		3743.9
		ΔH		-1.5		+3.5		+1.5		+2.3		+2.9		+5.9		+1.9		+3.7		+1.8	
		D_h		2.6		11.6				32.7				40.3		26.2		52.0		75.1	
3	138	H	3722.8		3721.2		3725.0		3726.8		3729.1		3731.3		3737.1		3738.8		3742.0		3743.6
		ΔH		-1.6		+3.8		+1.8		+2.3		+2.2		+5.8		+1.7		+3.2		+1.6	
		D_h		2.7										104.3		(17.3)		55.2		78.1	
4	163	H	3723.7		3722.0		3726.1		3728.0		3730.1		3738.5		3737.1		3738.6		3741.5		3743.5
		ΔH		-1.7		+4.1		+1.9		+2.1		+2.4		+4.6		+1.5		+2.9		+2.0	
		D_h		4.5		18.3				36.4				61.7		(15.5)		57.9		80.2	
5	188	H	3725.1		3723.5		3727.2		3728.9		3730.8		3732.2		3737.2		3738.5		3741.7		3743.1
		ΔH		-1.6		+3.7		+1.7		+1.9		+1.4		+5.0		+1.3		+3.2		+1.4	
		D_h		4.6										115.4		22.6		60.0		81.6	
6	213	H	3726.4		3724.9		3728.4		3730.3		3731.8		3732.3		3737.8		3738.7		3742.6		3742.4
		ΔH		-1.5		+3.5		+1.9		+1.5		+0.5		+5.5		+0.9		+3.9		-0.2	
		D_h		4.6		20.1				40.2				54.1		27.7		61.5		81.3	
7	238	H	3728.5		3726.9		3730.3		3731.7		3733.0		3733.6		3738.3		3739.6		3742.1		3742.3
		ΔH		-1.6		+3.4		+1.4		+1.3		+0.6		+4.7		+1.3		+2.5		+0.2	
		D_h		5.4										114.6		34.1		61.5		81.1	
8	264	H	3729.7		3728.4		3731.7		3733.0		3734.5		3735.8		3738.9		3739.8		3742.0		3742.6
		ΔH		-1.3		+3.3		+1.3		+1.5		+1.3		+3.1		+0.9		+2.2		+0.6	
		D_h		5.6		21.7				42.7				53.0		36.6		63.0		79.1	
9	287	H	3731.1		3729.8		3733.0		3734.1		3735.7		3737.1		3739.5		3739.7		3742.3		3743.1
		ΔH		-1.3		+3.2		+1.1		+1.6		+1.4		+2.4		+0.2		+2.6		+0.8	
		D_h		5.8										119.2		38.1		67.2		75.6	
10	312	H	3731.7		3730.4		3733.6		3733.8		3735.8		3737.2		3739.7		3739.7		3742.2		3742.9
		ΔH		-1.3		+3.2		+0.2		+2.0		+1.4		+2.5		0.0		+2.5		+0.7	
		D_h		5.8		21.8				42.7				56.2		36.3		67.6		72.8	
11	337	H	3731.5		3730.1		3793.0		3733.0		3735.0		3736.5		3739.1		3739.4		3742.0		3743.3
		ΔH		-1.4		+2.9		0.0		+2.0		+1.5		+2.6		+0.3		+2.6		+1.3	
		D_h		5.8										122.7		32.9		61.3		73.6	
12	362	H	3731.1		3729.7		3732.6		3732.4		3734.3		3735.7		3738.8		3739.3		3742.1		3744.1
		ΔH		-1.4		+2.9		-0.2		+1.9		+1.4		+3.1		+0.5		+2.8		+2.0	
		D_h		4.9		21.7				44.0				59.8		30.0		57.8		71.1	
13	387	H	3730.7		3729.4		3732.2		3732.0		3733.7		3735.9		3740.1		3740.7		3743.7		3746.3
		ΔH		-1.3		+2.8		-2.2		+1.7		+2.2		+4.2		+0.6		+3.0		+2.6	
		D_h		6.2										120.8		32.5					
14	417	H	3729.9		3728.8		3731.7		3733.2		3734.5		3737.9		3741.4		3744.0		3746.9		3748.8
		ΔH		-1.1		+2.9		+1.5		+1.3		+3.4		+3.5		+2.6		+2.9		+1.9	
		D_h		5.2		20.3				42.0				56.4		29.4					

Table 9.3.6. USSR - Changes in thickness.
Part 32 of 54 Variations in the surface elevation and horizontal component of the displacement of marked points on the glacier surface
Tien-Shan Shokalskiy Glacier Profile 1, right branch

point no.	distance from fixed point	date	1962 27.07.		1963 21.09.		1964 07.09.		1965 14.09.		1966 09.09.		1967 28.08.		1968 03.08.		1969 01.08.		1970 08.09.
		days		56		717		372		360		353		341		363		403	
1	230	H	3576.3		3574.4		3577.1				3624.9		3626.6		3621.2		3620.8		3618.2
		ΔH		-1.9		+2.7		←		+47.8		+1.7		-5.4		-0.4		-2.6	
		D_h		0.0												13.6		11.8	
2	247	H	3578.0		3576.2		3578.2				3625.6		3628.1		3621.0		3620.3		3618.0
		ΔH		-1.8		+2.0		←		+47.4		+2.5		-7.1		-0.7		-2.5	
		D_h		0.0												18.6		15.5	
3	265	H	3577.4		3575.5		3577.9				3622.5		3625.9		3620.8		3619.9		3617.7
		ΔH		-1.9		+2.4		←		+44.6		+3.4		-5.1		-7.9		-2.2	
		D_h		0.0												23.0		18.4	
4	284	H	3577.1		3575.4		3577.9		3613.7		3619.3		3625.2		3620.1		3619.0		3616.8
		ΔH		-1.7		+2.5		+35.8		+5.6		+5.9		-5.1		-1.1		-2.2	
		D_h		0.0												23.4		20.2	
5	333	H	3576.7		3575.0		3577.8		3607.0		3620.7		3625.0		3620.4		3619.8		3617.8
		ΔH		-1.7		+2.8		+29.2		+12.3		+4.3		-4.6		-0.6		-2.0	
		D_h		0.0												25.3		21.7	
6	348	H	3575.9		3573.8		3577.7				3620.7		3625.0		3620.7		3619.9		3617.5
		ΔH		-2.1		+3.9		←		+43.0		+4.3		-4.3		-0.8		-2.4	
		D_h		0.0												26.6		22.4	
7	364	H	3575.5		3573.5		3577.6				3621.2		3625.0		3621.0		3619.9		3617.2
		ΔH		-2.0		+4.1		←		+43.6		+3.8		-4.0		-1.1		-2.7	
		D_h		0.0												27.2		23.2	
8	382	H	3574.9		3573.4		3577.6				3623.0		3625.2		3621.4		3620.7		3618.5
		ΔH		-1.5		+4.2		←		+45.4		+2.2		-3.8		-0.7		-2.2	
		D_h		0.0												27.9		23.8	
9	398	H	3575.7		3573.9		3578.0				3625.5		3625.5		3622.4		3621.7		3619.7
		ΔH		-1.2		+4.1		←		+47.5		0.0		-3.1		-0.7		-2.0	
		D_h		0.0												27.7		24.2	
10	412	H	3577.1		3575.5		3578.3				3621.9		3626.7		3624.8		3623.7		3620.4
		ΔH		-1.6		+2.8		←		+43.6		+4.8		-1.9		-1.1		-3.3	
		D_h		0.0												28.0		26.4	
11	431	H	3579.9		3578.6		3579.1				3626.9		3629.8		3627.8		3626.2		3621.5
		ΔH		-1.3		+0.5		←		+47.7		+2.9		-2.0		-1.6		-4.7	
		D_h		0.0												28.5		24.8	
12	451	H	3580.2		3578.9		3579.6				3632.5		3637.3		3630.4		3628.5		3624.1
		ΔH		-1.3		+0.7		←		+52.9		+4.8		-6.9		-1.9		-4.4	
		D_h		0.0												29.5		25.7	

Table 9.3.6. USSR - Changes in thickness.

Part 33 of 54 Variations in the surface elevation and horizontal component of the displacement of marked points on the glacier surface

Tien-Shan Shokalskiy Glacier Profiel 2, right branch

point no.	distance from fixed point	date	1962 26.07.		1962 21.09.		1963 13.09.		1968 05.08.		1969 31.07.		1970 08.09.
		days		57		357		1776		360		404	
1	80	H	3761.3		3760.3		3774.2		3740.0		3734.5		3732.1
		ΔH		-1.0		+13.9		-34.0		-5.5		-2.4	
		Dh		13.7						23.0		18.8	
2	108	H	3756.8		3755.7		3766.4		3738.3		3733.3		3731.3
		ΔH		-1.1		+10.7		-28.1		-5.0		-2.0	
		Dh		16.00						22.7		18.5	
3	134	H	3752.8		3751.8		3758.4		3736.2		3732.4		3730.5
		ΔH		-1.0		+ 6.6		-22.2		-3.8		-1.9	
		Dh		16.87						22.5		18.2	
4	158	H	3747.1		3746.1		3752.6		3734.4		3731.3		3729.8
		ΔH		-1.0		+ 6.5		-18.2		-3.1		-1.5	
		Dh		16.88						22.2		17.9	
5	177	H	3745.4		3744.0		3749.2		3733.1		3730.6		3729.2
		ΔH		-1.4		+ 5.2		-16.0		-2.5		-1.4	
		Dh		16.77						22.0		17.7	
6	192	H	3744.4		3743.6		3746.1		3732.0		3730.0		3728.8
		ΔH		-0.8		+ 2.5		-14.1		-2.0		-1.2	
		Dh		16.78						21.8		17.5	
7	238	H	3744.5		3743.3		3747.7		3730.5		3726.4		3725.5
		ΔH		-1.2		+ 4.4		-17.2		-4.1		-0.9	
		Dh		17.35						21.8		17.0	
8	270	H	3745.2		3744.4		3749.4		3734.5		3728.0		3725.2
		ΔH		-0.8		+ 5.0		-14.9		-6.5		-2.8	
		Dh		17.00						22.1		17.5	
9	334	H	3741.6		3739.8		3750.9		3728.0		3721.7		3719.8
		ΔH		-1.8		+11.1		-22.9		-6.3		-1.9	
		Dh		18.40						22.2		17.2	
10	366	H1	3742.3		3740.3		3750.6		3726.5		3721.9		3720.7
		ΔH		-2.0		+10.3		-24.1		-4.6		-1.2	
		Dh		16.60						21.9		16.5	
11	397	H	3747.3		3745.4		3752.1		3726.7		3724.0		3722.0
		ΔH		-1.9		+ 6.7		-25.4		-2.7		-2.0	
		Dh		15.00						21.4		15.8	
12	416	H	3739.3		3737.1		3751.8		3726.1		3724.5		3721.0
		ΔH		-2.2		+14.7		-25.7		-1.6		-3.5	
		Dh								21.0		15.4	

Table 9.3.6. USSR - Changes in thickness.

Part 34 of 54 Variations in the surface elevation and horizontal component of the displacement of marked points on the glacier surface

Tien-Shan Shokalskiy Glacier Profile 3, right branch

point no.	distance from fixed point	date	1962 31.07.		1963 13.09.		1964 07.09.		1965 14.09.		1967 28.09.		1968 03.08.		1969 01.08.		1970 08.09.
		days		409		359		372		379		309		363		372	
		H	3873.6		3875.5		3876.8		3877.2		3878.9		3878.3		3880.5		3879.0
1	134	ΔH		+1.9		+1.3		+0.4		+1.7		-0.6		+2.2		-1.5	
		D_h												5.2			
		H	3868.3		3870.3		3871.7		3871.9		3874.2		3873.3		3874.3		3874.0
2	177	ΔH		+2.0		+1.4		+0.2		+2.3		-0.9		+1.0		-0.3	
		D_h												5.3			
		H	3861.4		3863.3		3865.3		3865.5		3867.7		3866.5		3867.1		3866.9
3	232	ΔH		+1.9		+2.0		+0.2		+2.2		-1.2		+0.6		-0.2	
		D_h												5.4			
		H	3855.0		3856.9		3859.5		3860.1		3863.1		3859.4		3859.8		3859.3
4	276	ΔH		+1.9		+2.6		+0.6		+3.0		-3.7		+0.4		-0.5	
		D_h												6.8			
		H	3844.0		3845.4		3847.7		3848.6		3851.6		3848.5		3848.0		3848.0
5	314	ΔH		+1.4		+2.3		+0.9		+3.0		-3.1		-0.5		-0.0	
		D_h												7.9			
		H	3832.4		3833.7		3836.0		3837.4		3840.8		3838.9		3836.2		3834.4
6	352	ΔH		+1.3		+2.3		+1.4		+3.4		-1.9		-2.7		-1.8	
		D_h												8.9			
		H	3823.5		3826.4		3829.9		3830.8		3833.8		3830.8		3830.3		3829.6
7	397	ΔH		+2.9		+3.5		+0.9		+3.0		-3.0		-0.5		-0.7	
		D_h												10.0			
		H	3821.8		3824.6		3828.5		3829.3		3834.3		3831.8		3831.9		3831.2
8	446	ΔH		+2.8		+3.9		+0.8		+5.0		-2.5		+0.1		-0.7	
		D_h												10.2			
		H	3819.8		3823.5		3827.5		3828.4		3835.4		3831.3		3832.2		3830.0
9	506	ΔH		+3.7		+4.0		+0.9		+7.0		-4.1		+0.9		-2.2	
		D_h												10.8			
		H	3818.6		3821.3		3825.6		3826.5								
10	561	ΔH		+2.7		+4.3		+0.9									
		D_h															
		H	3819.4		3823.2		3827.4		3828.4								
11	690	ΔH		+3.8		+4.2		+1.0									
		D_h															
		H	3824.8		3827.5		3831.7		3832.8								
12	807	ΔH		+2.7		+4.2		+1.1									
		D_h															
		H	3825.9		3829.1		3833.1		3834.2								
13	855	ΔH		+3.2		+4.0		+1.1									
		D_h															
		H	3828.2		3831.9		3835.5		3836.6								
14	941	ΔH		+3.7													
		D_h															

Table 9.3.6. USSR - Changes in thickness.

Part 35 of 54 Variations in the surface elevation and horizontal component of the displacement of marked points on the glacier surface

Tien-Shan Shokalskiy Glacier, Profile 4

point no.	distance from fixed point	date / days	1962 30.07.	51	1962 19.09.	359	1963 13.09.	360	1964 07.09.	362	1965 14.09.	361	1966 10.09.	383	1967 28.09.	310	1968 03.08.	363	1969 01.08.	403	1970 08.09.
1	80	H	3926.7		3925.7		3926.2		3927.3		3927.5		3928.3		3928.5		3928.5		3928.6		3928.4
		ΔH		-1.0		+0.5		+1.1		+0.2		+0.8		+0.2		0.0		+0.1		-0.2	
		D_h		1.2																	
2	105	H	3925.1		3924.0		3924.7		3925.7		3925.9		3926.2		3926.6		3926.5		3926.5		3926.2
		ΔH		-1.1		+0.7		+1.0		+0.2		+0.3		+0.4		-0.1		0.0		-0.3	
		D_h		0.8																	
3	132	H	3923.1		3921.9		3922.8		3923.7		3923.4		3924.0		3924.4		3924.1		3924.2		3924.0
		ΔH		-1.2		+0.9		+0.9		-0.3		+0.6		+0.4		-0.3		+0.1		-0.2	
		D_h		1.2																	
4	159	H	3920.0		3918.7		3919.6		3920.4		3919.8		3920.7		3920.8		3921.1		3921.0		3920.7
		ΔH		-1.3		+0.9		+0.8		-0.6		+0.9		+0.1		+0.3		-0.1		-0.3	
		D_h		1.2																	
5	211	H	3916.1		3915.4		3915.5		3916.6		3915.5		3916.5		3916.6		3916.3		3916.4		3917.0
		ΔH		-0.7		+0.1		+1.1		-1.1		+1.0		+0.1		-0.3		+0.1		+0.6	
		D_h		1.1																	
6	249	H	3930.3		3929.0		3919.9		3922.6		3922.1		3923.1		3922.7		3923.1		3922.4		3922.7
		ΔH		-1.3		-9.1		+2.7		-0.5		+1.0		-0.4		+0.4		-0.7		+0.3	
		D_h		1.2																	

Table 9.3.6. USSR - Changes in thickness.

Part 36 of 54 Variations in the surface elevation and horizontal component of the displacement of marked points on the glacier surface

Tien-Shan Shokalskiy Glacier, Profile 5

point no.	distance from fixed point	date	1962 30.07.		1962 19.09.		1963 13.09.		1964 06.09.		1965 13.09.		1966 10.09.		1967 28.09.		1968 03.08.		1969 01.08.		1970 08.09.
		days		51		359		360		372		362		383		310		363		403	
1	67	H	4011.5		4011.4		4011.9		4010.9		4012.6		4012.2		4012.2		4012.9		4012.1		4013.3
		ΔH		-0.1		+0.5		-1.0		+1.7		-0.4		0.0		+0.7		-0.8		+1.2	
		D_h		0.64																	
2	98	H	4015.1		4014.3		4014.9		4014.9		4015.7		4015.8		4016.2		4016.0		4015.9		4016.0
		ΔH		-0.8		+0.6		0.0		+0.8		+0.1		+0.4		-0.2		-0.1		+0.1	
		D_h		1.10																	
3	130	H	4018.0		4017.1		4017.3		4017.3		4018.4		4018.7		4019.5		4019.2		4018.8		4019.0
		ΔH		-0.9		+0.2		0.0		+1.1		+0.3		+0.8		-0.3		-0.4		+0.2	
		D_h		1.15																	
4	170	H	4019.4		4019.0		4018.8		4019.3		4020.4		4020.8		4021.0		4021.1		4020.3		4020.4
		ΔH		-0.4		-0.2		+0.5		+1.1		+0.4		+0.2		+0.1		-0.8		+0.1	
		D_h		1.38																	
5	202	H	4019.1		4018.3		4019.1		4019.1		4020.8		4020.2		4020.9		4020.3		4020.4		4020.6
		ΔH		-0.8		+0.8		0.0		+1.7		-0.6		+0.7		-0.6		+0.1		+0.2	
		D_h		1.24																	
6	273	H	4020.1				4019.0		4019.0		4021.8		4019.6		4019.6		4018.6		4017.6		4023.3
		ΔH						0.0		+2.8		-2.2		0.0		-1.0		-1.0		+5.7	
		D_h																			

Table 9.3.6. USSR - Changes in thickness.

Part 37 of 54 Variations in the surface elevation and horizontal component of the displacement of marked points on the glacier surface

Tien-Shan Teu Southern Profile 1

point no.	distance from fixed point	date	1962 01.08.		1962 02.10.		1963 12.09.		1964 08.09.		1965 17.09.		1966 12.09.
		days		62		345		362		374		360	
		H	3679.0		3677.3		3676.4		3676.3		3672.3		3672.9
1	113	ΔH		-1.7		-0.9		-0.1		-4.0		+0.6	
		Dh		0.4		1.2		1.5					
		H	3679.1		3677.7		3677.3		3677.6		3674.2		3675.1
2	136	ΔH		-1.4		-0.4		+0.3		-3.4		+0.9	
		Dh		0.4		1.3		2.2					
		H	3678.4		3676.8		3676.9		3678.2		3675.6		3677.4
3	164	ΔH		-1.6		+0.1		+1.3		-2.6		+1.8	
		Dh		0.5		1.2		2.2		2.4		2.9	
		H	3681.6		3680.2		3680.3		3681.1		3678.2		3680.1
4	182	ΔH		-1.4		+0.1		+0.8		-2.9		+1.9	
		Dh		0.2		2.8		2.2		2.2		2.4	
		H	3683.5		3682.2		3683.0		3683.4		3679.9		3681.9
5	201	ΔH		-1.3		+0.8		+0.4		-3.5		+2.0	
		Dh		0.7		3.1		2.4		2.6		2.9	
		H	3685.8		3684.4		3685.0		3685.7		3682.5		3683.9
6	226	ΔH		-1.4		+0.6		+0.7		-3.2		+1.4	
		Dh		0.7		3.3		2.8		3.5		4.0	
		H	3687.8		3686.4		3686.9		3687.5		3685.7		3685.8
7	252	ΔH		-1.4		+0.5		+0.6		-1.8		+0.1	
		Dh		0.7		3.5		3.0		3.5		4.0	
		H	3689.2		3687.9		3688.2		3689.1		3687.4		3687.6
8	273	ΔH		-1.3		+0.3		+0.9		-1.7		+0.2	
		Dh		0.8		3.6		2.8		3.7		5.0	
		H	3690.9		3689.4		3689.8		3690.8		3689.2		3690.0
9	292	ΔH		-1.5		+0.4		+1.0		-1.6		+0.8	
		Dh		0.8		3.1		4.3		3.7		3.9	
		H	3695.6				3694.5		3695.7		3692.7		3695.1
10	321	ΔH						+1.2		-3.0		+2.4	
		Dh		←				15.24				→	
		H	3700.9				3700.1		3700.1		3696.9		3699.4
11	349	ΔH						0.0		-3.2		+2.5	
		Dh											

Table 9.3.6. USSR - Changes in thickness.

Part 38 of 54 Variations in the surface elevation and horizontal component of the displacement of marked points on the glacier surface

Tien-Shan Teu Southern Profile 2

point no.	distance from fixed point	date / days	1962 31.07.	408	1963 12.09.	362	1964 08.09.	374	1965 17.09.	360	1966 12.09.
1	135	H	3834.8		3833.6		3835.1		3835.8		3833.6
		ΔH		-1.2		+1.5		+0.7		-2.2	
2	166	H	3837.2		3836.1		3837.0		3837.5		3836.1
		ΔH		-1.1		+0.9		+0.5		-1.4	
3	193	H	3838.0		3836.9		3837.5		3838.2		3836.9
		ΔH		-1.1		+0.6		+0.7		-1.3	
4	221	H	3838.7		3836.7		3837.5		3838.3		3836.7
		ΔH		-2.0		+0.8		+0.8		-1.6	
5	270	H	3837.2		3839.5		3840.4		3840.7		3839.5
		ΔH		+2.3		+0.9		+0.3		-1.2	
6	302	H	3839.3		3841.4		3842.1		3842.6		3841.4
		ΔH		+2.1		+0.7		+0.5		-1.2	
7	341	H	3840.4		3842.7		3843.5		3844.0		3842.7
		ΔH		+2.3		+0.8		+0.5		-1.3	
8	388	H	3842.4		3844.7		3845.4		3845.7		3844.7
		ΔH		+2.3		+0.7		+0.3		-1.0	

Korzhenevsky's Glacier Profile 1

point no.	distance from fixed point	date / days	1966 16.08.	363	1967 14.08.	373	1968 21.08.	331	1969 18.07.	408	1970 30.08.
1	141	H	3439.5		3438.0		3436.7		3435.8		3432.0
		ΔH		-1.5		-1.3		-0.9		-3.8	
		D_h		9.7		5.7		7.1		8.2	
2	175	H	3443.4		3441.3		3440.2		3439.8		3436.7
		ΔH		-2.1		-1.1		-0.4		-3.1	
		D_h		13.0		8.7		9.3		10.0	
3	217	H	3447.0		3444.7		3444.0		3443.3		3439.5
		ΔH		-2.3		-0.7		-0.7		-3.8	
		D_h		15.8		10.3		11.3		11.5	
4	262	H	3448.2		3446.7		3444.6		3444.3		3439.9
		ΔH		-1.5		-2.1		-0.3		-4.4	
		D_h		17.8		11.6		13.7		13.0	
5	320	H	3449.6		3449.2		3447.7		3447.3		3444.0
		ΔH		-0.4		-1.5		-0.4		-3.3	
		D_h		19.7		10.1		14.7		11.9	
6	359	H	3453.0		3450.9		3448.4		3447.8		3443.7
		ΔH		-2.1		-2.5		-0.6		-4.1	
		D_h		20.0		12.7		15.3		12.8	

Table 9.3.6. USSR - Changes in thickness.

Part 39 of 54 Variations in the surface elevation and horizontal component of the displacement of marked points on the glacier surface

Tien-Shan Korzhenevsky's Glacier Profile 2

point no.	distance from fixed point	date / days	1966 16.08.	363	1967 14.08.	369	1968 20.08.	333	1969 19.07.	405	1970 28.08.
1	189	H	3528.0		3524.2		3523.7		3525.2		3525.0
		ΔH		-3.8		-0.5		+1.5		-0.2	
		D_h		10.2		16.5		12.9		16.8	
2	239	H	3532.2		3530.2		3530.4		3531.9		3531.3
		ΔH		-2.0		+0.2		+1.5		-0.6	
		D_h		12.8		20.4		15.0		22.1	
3	285	H	3533.1		3531.4		3530.8		3531.4		3530.2
		ΔH		-1.7		-0.6		+0.6		-1.2	
		D_h		14.0		22.8		27.1		28.1	
4	352	H	3536.3		3535.1		3534.1		3534.8		3534.0
		ΔH		-1.2		-1.0		+0.7		-0.8	
		D_h		16.8		25.2		20.9		33.7	
5	438	H	3536.1		3534.1		3532.9		3532.9		3538.0
		ΔH		-2.0		-1.2		0.0		+5.1	
		D_h		18.9		27.6		23.4		31.4	
6	542	H	3541.1		3541.2		3541.1		3541.4		3543.4
		ΔH		+0.1		-0.1		+0.3		+2.0	
		D_h		20.2		27.3		27.7		34.0	
7	662	H	3540.1		3539.5		3541.1		3541.7		3543.4
		ΔH		-0.6		+1.6		+0.6		+1.7	
		D_h		20.3		31.6		28.7		34.5	
8	722	H	3541.1		3540.1		3540.6		3541.6		3543.4
		ΔH		-1.0		+0.5		+1.0		+1.8	
		D_h		19.4		33.1		31.4		33.9	
9	782	H	3542.3		3541.1		3540.6		3541.5		3543.4
		ΔH		-1.2		-0.5		+0.9		+1.9	
		D_h				32.9		29.3		33.5	

Table 9.3.6. USSR - Changes in thickness.

Part 40 of 54 Variations in the surface elevation and horizontal component of the displacement of marked points on the glacier surface

Tien-Shan Korzhenevsky's Glacier Profile 3

point no.	distance from fixed point		1964		1965		1966		1967		1968		1969		1970
		date	20.08.		25.07.		17.08.		13.08.		20.08.		25.07.		27.08.
		days		339		388		361		372		339		398	
		H	3586.9		3586.0		3584.7		3584.9		3584.6		3585.9		3585.3
1	195	ΔH		-0.9		-1.3		+0.2		-0.3		+1.3		-0.6	
		D_h		14.1		16.6				20.7		16.9		23.8	
		H	3594.4		3593.5		3591.7		3591.3		3590.3		3591.2		3591.1
2	244	ΔH		-0.9		-1.8		-0.4		-1.0		+0.9		-0.1	
		D_h		19.2		22.6				27.1		23.1		31.4	
		H	3611.4		3611.8		3611.0		3608.2		3603.9		3605.1		3605.7
3	326	ΔH		+0.4		-0.8		-2.8		-4.3		+1.2		+0.6	
		D_h		24.3		25.6				31.8		27.0		39.3	
		H	3609.8		3609.7		3606.3		3602.9		3602.6		3603.3		3604.1
4	373	ΔH		-0.1		-3.4		-3.4		-0.3		+0.7		+0.8	
		D_h		24.4		26.6				33.8		28.7		41.3	
		H	3618.0		3618.3		3616.1		3614.9		3613.8		3614.8		3612.3
5	470	ΔH		+0.3		-2.2		-1.2		-1.1		+1.0		-2.5	
		D_h		26.1		29.5				35.2		29.7		47.8	
		H	3617.0		3616.8		3616.1		3616.3		3613.6		3614.9		3614.4
6	542	ΔH		-0.2		-0.7		+0.2		-2.7		+1.3		-0.5	
		D_h		27.9		30.4				35.8		30.6		41.1	
		H	3618.6		3618.1		3617.9		3616.8		3615.9		3616.7		3615.3
7	615	ΔH		-0.5		-0.2		-1.1		-0.9		+0.8		-1.4	
		D_h		28.8		29.8				36.0		32.4		40.3	
		H	3615.7		3616.0		3615.4		3614.6		3612.7		3612.6		3615.6
8	711	ΔH		+0.3		-0.6		-0.8		-1.9		-0.1		13.0	
		D_h		27.5		29.6				35.8		30.6		38.3	
		H	3618.3		3617.6		3614.5		3615.7		3615.6		3614.7		3617.1
9	752	ΔH		-0.7		-3.1		+1.2		-0.1		-0.9		+2.4	
		D_h								34.0		26.1		37.3	

Korzhenevsky's Glacier Profile 5

point no.	distance from fixed point		1964		1965		1966		1967		1968		1969		1970
		date	18.08.		28.07.		22.08.		12.08.		15.08.		30.07.		23.08.
		days		344		390		355		367		349		389	
		H	3780.5		3782.9		3782.5		3780.7		3781.7		3780.7		3778.8
1	154	ΔH		+2.4		-0.4		-1.8		+1.0		-1.0		-1.9	
		D_h		28.0		12.4				33.4		29.2		54.2	
		H	3790.6		3796.8		3792.2		3792.5		3791.9		3790.9		3789.9
2	256	ΔH		+6.2		-4.6		+0.3		-0.6		-1.0		-1.0	
		D_h		36.2		27.5				43.3		36.4		45.7	
		H	3811.3		3811.9		3812.9		3812.0		3810.9		3810.6		3809.6
3	366	ΔH		+0.6		+1.0		-0.9		-1.1		-0.3		-1.0	
		D_h		31.7		37.8				47.0		37.3		45.5	
		H	3813.0		3812.2		3814.5		3814.6		3813.3		3813.2		3812.6
4	440	ΔH		-0.8		+2.3		+0.1		-1.3		-0.1		-0.6	
		D_h		31.5		36.1				43.4		36.0		44.1	
		H	3807.3		3807.6		3809.8		3810.5		3809.3		3809.2		3808.8
5	545	ΔH		+0.3		+2.2		+0.7		-1.2		-0.1		-0.4	
		D_h		32.4		39.7				41.2		35.1		42.1	
		H	3806.1		3805.9		3806.0		3807.4		3808.0		3808.2		3808.9
6	654	ΔH		-0.2		+0.1		+1.4		+0.6		+0.2		+0.7	
		D_h		27.8		36.4				38.4		30.2		34.8	
		H	3806.7		3806.0		3808.3		3810.1		3810.6		3812.1		3813.2
7	766	ΔH		-0.7		+2.3		+1.8		+0.5		+1.5		+1.1	
		D_h		20.8		27.2				30.6		25.0		31.0	
		H	3813.2		3812.2		3814.4		3815.6		3814.3		3815.3		3817.2
8	860	ΔH		-1.0		+2.2		+1.2		-1.3		+1.0		+1.9	
		D_h		12.0		15.6				27.5		11.1		28.0	

Table 9.3.6. USSR - Changes in thickness.

Part 41 of 54 Variations in the surface elevation and horizontal component of the displacement of marked points on the glacier surface

Tien-Shan Korzhenevsky'a Glacier Profile 6

point no.	distance from fixed point	date	1964 18.08.		1965 28.07.		1966 22.08.		1967 12.08.		1968 15.08.		1969 30.07.		1970 20.08.
		days		344		390		355		367		349		386	
		H	3780.3		3779.9		3779.8		3779.4		3778.4		3778.3		3775.5
1	254	ΔH		-0.4		-0.1		-0.4		-1.0		-0.1		-2.8	
		Dh		29.3		27.9				37.7		35.7		43.9	
		H	3783.9		3783.0		3780.6		3783.0		3780.6		3779.1		3775.9
2	297	ΔH		-0.9		-2.4		+2.4		-2.4		-1.5		-3.2	
		Dh		28.7		29.5				40.0		37.8		49.2	
		H	3793.8		3794.1		3793.2		3791.9		3789.4		3787.4		3783.2
3	380	ΔH		+0.3		-0.9		-1.3		-2.5		-2.0		-4.2	
		Dh		32.7		33.1				46.3		41.1		51.7	
		H	3779.4		3781.1		3780.2		3777.9		3776.3		3777.2		3777.8
4	464	ΔH		+1.7		-0.9		-2.3		-1.6		+0.9		+0.6	
		Dh		36.8		37.4				52.7		44.6		54.2	
		H	3775.2		3778.6		3774.9		3774.4		3773.1		3773.1		3772.0
5	543	ΔH		+3.4		-3.7		-0.5		-1.3		0.0		-1.1	
		Dh		41.2		45.0						44.0		54.6	
		H	3773.3		3776.7		3772.7		3772.9		3771.9		3772.3		3771.1
6	613	ΔH		+3.4		-4.0		+0.2		-1.0		+0.4		-1.2	
		Dh		41.3		46.6						45.0		53.7	
		H	3783.9		3787.0		3782.6		3782.3		3777.9		3781.7		3778.8
7	730	ΔH		+3.1		-4.4		-0.3		-4.4		+3.8		-2.9	
		Dh		39.4		44.8						42.5		52.3	
		H	3775.1		3774.2		3775.6		3775.0		3771.3		3772.5		3782.1
8	809	ΔH		-0.9		+1.4		-0.6		-3.7		+1.2		+9.6	
		Dh		37.5		42.7						41.8		51.5	
		H	3777.4		3776.8		3778.2		3778.3		3776.4		3777.3		3776.6
9	915	ΔH		-0.6		+1.4		+0.1		-1.9		+0.9		-0.7	
		Dh		35.1		41.5				47.1		38.6		53.5	
		H	3780.3		3779.7		3782.2		3782.3		3780.4		3781.4		3780.1
10	1013	ΔH		-0.6		+2.5		+0.1		-1.9		+1.0		-1.3	
		Dh		32.1		38.3				46.3		35.0		55.6	
		H	3787.5		3788.5		3788.5		3789.2		3787.1		3788.4		3785.4
11	1118	ΔH		+1.1		0.0		+0.7		-2.1		+1.3		-3.0	
		Dh		28.2		33.5				39.3		31.0		55.9	
		H	3787.9		3787.8		3789.1		3789.4		3788.6		3788.7		3788.5
12	1174	ΔH		-0.1		+1.3		+0.3		-0.8		+0.1		-0.2	
		Dh		29.8		35.0				35.9		29.6		59.5	

Table 9.3.6. USSR - Changes in thickness.

Part 42 of 54 Variations in the surface elevation and horizontal component of the displacement of marked points on the glacier surface

Tien-Shan Korzhenevsky's Glacier Profile 4

point no.	distance from fixed point	date	1964 21.08.		1965 26.07.		1966 17.08.		1967 12.08.		1968 20.08.		1969 27.07.		1970 26.08.
		days		339		387		360		374		341		395	
		H	3715.1		3715.7		3715.3		3715.7		3714.8		3715.3		3714.9
1	69	ΔH		+0.6		-0.4		+0.4		-0.9		+0.5		-0.4	
		D_h		23.8		29.3				31.7		27.2		29.8	
		H	3709.8		3707.8		3708.5		3709.7		3709.7		3711.1		3710.4
2	136	ΔH		-2.0		+0.7		+1.2		0.0		+1.4		-0.7	
		D_h		29.7		28.3		28.4		33.1		28.8		40.3	
		H	3704.5		3704.4		3702.8		3704.0		3703.8		3704.4		3704.4
3	229	ΔH		-0.1		-1.6		+1.2		-0.2		+0.6		0.0	
		D_h		29.2		32.1		35.2		36.8		34.2		42.5	
		H	3702.7		3703.3		3702.9		3702.7		3702.2		3702.9		3702.9
4	312	ΔH		+0.6		-0.4		-0.2		-0.5		+0.7		0.0	
		D_h		29.2		39.8		34.3		39.7		36.0		48.3	
		H	3699.7		3699.2		3699.5		3700.1		3700.2		3700.0		3698.3
5	403	ΔH		-0.5		+0.3		+0.6		+0.1		-0.2		-1.7	
		D_h		35.2		38.2				41.7		39.9		53.0	
		H	3710.1		3709.6		3709.1		3711.4		3707.2		3708.8		3706.8
6	538	ΔH		-0.5		-0.5		+2.3		-4.2		+1.6		-2.0	
		D_h		33.9		39.6				45.4		39.8		52.3	
		H	3696.7		3697.6		3698.2		3700.4		3706.2		3708.5		3711.2
7	607	ΔH		+0.9		+0.6		+2.2		-4.2		+2.3		+2.7	
		D_h		33.7		41.4						39.5		51.9	
		H	3699.6		3698.7		3697.5		3699.6		3697.7		3699.3		3697.8
8	678	ΔH		-0.9		-1.2		+2.1		-1.9		+1.6		-1.5	
		D_h		33.5		40.4						39.6		48.7	
		H	3697.7		3696.8		3694.8		3697.8		3695.4		3697.0		3694.8
9	758	ΔH		-0.9		-2.0		+3.0		-2.4		+1.6		-2.2	
		D_h		33.6		40.8		39.2				38.9		55.2	
		H	3703.5		3703.1		3702.6		3703.7		3702.0		3704.4		3701.3
10	884	ΔH		-0.4		-0.5		+1.1		-1.7		+2.4		-3.1	
		D_h		29.4		33.2		38.0				33.7		45.1	

Korzhenevsky's Glacier Profile 7

point no.	distance from fixed point	date	1964 18.08.		1965 09.08.		1966 18.08.		1967 06.08.		1968 13.08.		1969 30.07.		1970 23.08.
		days		356		374		353		372		351		389	
		H	3847.7		3847.4		3853.7		3849.9		3848.0		3849.2		3849.0
1	319	ΔH		-0.3		-6.3		-3.8		-1.9		+1.2		-0.2	
		D_h		36.5		38.3				46.1		41.2		49.2	
		H	3845.9		3846.6		3848.3		3847.8		3847.4		3849.1		3850.1
2	397	ΔH		+0.7		+1.7		-0.5		-0.4		+1.7		+1.0	
		D_h		36.8		38.8				46.3		42.3		50.6	
		H	3850.0		3852.4		3852.4		3854.1		3854.5		3855.6		3856.8
3	496	ΔH		+2.4		0.0		+1.7		+0.4		+1.1		+1.2	
		D_h		35.9		38.3				44.8		40.9		49.8	
		H	3852.5		3854.5		3854.2		3855.4		3855.0		3856.0		3856.5
4	575	ΔH		+2.0		-0.3		+1.2		-0.4		+1.0		+0.5	
		D_h		31.6		34.1				42.3		39.5		45.4	
		H	3849.0		3850.2		3852.1		3851.8		3851.2		3852.5		3853.1
	mean	ΔH		+1.2		+1.9		-0.3		-0.6		+1.3		+0.6	
		D_h													

Table 9.3.6. USSR - Changes in thickness.

Part 43 of 54 Variations in the surface elevation and horizontal component of the displacement of marked points on the glacier surface

Tien-Shan Korzhenevsky's Glacier Profile 8

point no.	distance from fixed point	date / days	1964 17.08.	357	1965 09.08.	378	1966 22.08.	347	1967 04.08.	371	1968 10.08.	355	1969 31.07.	390	1970 25.08.
1	326	H	3874.1		3873.3		3873.8		3873.2		3872.2		3872.0		3871.1
		ΔH		-0.8		+0.5		-0.6		-1.0		-0.2		-0.9	
		Dh		20.8		19.6				25.2		20.8		25.4	
2	443	H	3872.9		3871.4		3870.6		3870.0		3870.0		3869.7		3869.0
		ΔH		-1.5		-0.8		-0.6		0.0		-0.3		-0.7	
		Dh		37.8		37.9				46.0		39.2		85.3	
3	564	H	3875.5		3874.5		3874.5		3874.0		3871.0		3871.4		3871.4
		ΔH		-1.0		0.0		-0.5		-3.0		+0.4		0.0	
		Dh		47.8		49.6				56.3		49.2		107.1	
4	686	H	3876.9		3876.1		3876.4		3875.7		3874.4		3874.1		3875.0
		ΔH		-0.8		+0.3		-0.7		-1.3		-0.3		+0.9	
		Dh		58.4		55.0				71.0		59.2		91.6	
5	806	H	3882.4		3880.3		3880.4		3880.0		3872.4		3880.9		3882.1
		ΔH		-2.1		+0.1		-0.4		-7.6		+8.5		+1.2	
		Dh		65.1		64.8				82.2		65.4		80.3	
6	900	H	3896.3		3892.6		3893.8		3891.5		3884.3		3894.9		3896.4
		ΔH		-3.7		+1.2		-2.3		-7.2		+10.6		+1.5	
		Dh		62.5		63.7				83.4		68.9		74.7	
7	996	H	3902.4		3900.2		3901.5		3900.0		3893.0		3901.6		3902.3
		ΔH		-2.2		+1.3		-1.5		-7.0		+8.6		+0.7	
		Dh		60.0		62.6				77.8		68.5		69.3	
8	1088	H	3908.5		3907.6		3908.1		3906.7		3897.6		3905.8		3904.7
		ΔH		-0.9		+0.5		-1.4		-9.1		+8.2		-1.1	
		Dh										64.8		67.4	

Korzhenevsky's Glacier Profile 9

point no.	distance from fixed point	date / days	1964 17.08.	357	1965 09.08.	378	1966 22.08.	347	1967 04.08.	371	1968 10.08.	355	1969 31.07.	389	1970 24.08.
1	199	H	3863.9		3861.4		3862.5		3863.3		3861.9		3861.9		3861.8
		ΔH		-2.5		+1.1		+0.8		-1.4		0.0		-0.1	
		Dh		15.9		9.2		11.4		14.0		12.9		14.4	
2	313	H	3868.6		3867.5		3867.8		3867.3		3866.3		3865.9		3866.0
		ΔH		-1.1		+0.3		-0.5		-1.0		-0.4		+0.1	
		Dh		18.1		12.1				19.1		15.6		17.5	
3	429	H	3871.0		3869.2		3870.0		3868.0		3867.3		3867.3		3869.7
		ΔH		-1.8		+0.8		-1.1		-1.6		0.0		+2.4	
		Dh		19.7		15.7		18.3		20.3		18.3		20.9	
4	560	H	3870.5		3873.3		3873.9		3873.4		3872.0		3871.9		3871.8
		ΔH		+2.8		+0.6		-0.5		-1.4		-0.1		-0.1	
		Dh		19.9		14.9				20.7		17.5		21.0	
5	668	H	3871.9		3874.7		3874.9		3874.5		3873.0		3872.7		3873.5
		ΔH		+2.8		+0.2		-0.4		-1.5		-0.3		+0.8	
		Dh		19.8		14.0		16.6		19.9		16.9		21.0	
6	777	H	3871.7		3874.9		3874.7		3874.4		3873.1		3873.0		3874.0
		ΔH		+3.2		-0.2		-0.3		-1.3		-0.1		+1.0	
		Dh		19.7		13.9				18.7		14.5		25.1	
7	880	H	3870.5		3874.3		3874.2		3873.4		3870.7		3870.8		3869.8
		ΔH		+3.8		-0.1		-0.8		-2.7		+0.1		-1.0	
		Dh		18.4		10.0		11.8		15.5		12.4		31.2	
	mean	H	3869.7		3870.8		3871.1		3870.7		3869.2		3869.1		3869.5
		ΔH		+1.1		+0.3		-0.4		-1.5		-0.1		+0.4	
		Dh													

Table 9.3.6. USSR - Changes in thickness.

Part 44 of 54 Variations in the surface elevation and horizontal component of the displacement of marked points on the glacier surface

Tien-Shan Bogatyr Glacier Profile 1

point no.	distance from fixed point	date	1965 13.08.		1966 25.08.		1967 28.08.		1968 24.09.		1969 13.08.		1970 30.09.
		days		377		368		373		374		413	
1	158	H	- 48.8		- 53.5		- 56.8		- 61.2		- 68.1		- 69.7
		ΔH		- 4.7		- 3.3		- 4.4		- 6.9		- 1.6	
		D_h		1.3									
2	198	H	- 53.5		- 56.0		- 59.5		- 59.7		- 74.1		- 77.4
		ΔH		- 2.5		- 3.5		- 0.2		-14.4		- 3.3	
		D_h		0.4									
3	275	H	- 43.8		- 48.0		- 51.8		- 57.6		- 62.3		- 66.6
		ΔH		- 4.2		- 3.8		- 5.8		- 4.7		- 4.3	
		D_h		0.5									
4	339	H	- 44.9		- 49.3		- 53.0		- 62.1		- 63.8		- 69.6
		ΔH		- 4.4		- 3.7		- 9.1		- 1.7		- 5.8	
		D_h		0.6									
5	392	H	- 44.5		- 48.5		- 52.2		- 59.3		- 63.8		- 69.1
		ΔH		- 4.0		- 3.7		- 7.1		- 4.5		- 5.3	
		D_h		0.7									
6	445	H	- 39.5		- 43.7		- 47.2		- 54.0		- 58.2		- 64.0
		ΔH		- 4.2		- 3.5		- 6.8		- 4.2		- 5.8	
		D_h		0.05									
7	515	H	- 39.5		- 43.2		- 46.7		- 52.4		- 57.8		- 61.3
		ΔH		- 3.7		- 3.5		- 5.7		- 5.4		- 3.5	
		D_h		0.3									

Bogatyr Glacier Profile 2

point no.	distance from fixed point	date	1965 13.08.		1966 26.08.		1967 25.08.		1968 04.09.		1969 13.08.		1970 28.09.
		days		378		364		376		343		411	
1	58	H	- 29.5		- 31.0		- 33.6		- 35.2				
		ΔH		- 1.5		- 2.6		- 1.6					
		D_h		- 1.2									
2	143	H	- 18.0		- 23.2		- 27.5		- 29.9		- 38.1		- 46.8
		ΔH		- 5.2		- 4.3		- 2.4		- 8.2		- 8.7	
		D_h		- 0.9									
3	175	H	- 18.0		- 23.3		- 27.7		- 29.6		- 38.5		- 47.5
		ΔH		- 5.3		- 4.4		- 1.9		- 8.9		- 9.0	
		D_h		1.3									
4	243	H	- 20.6		- 25.7		- 30.1		- 31.8		- 39.9		- 46.4
		ΔH		- 5.1		- 4.4		- 1.7		- 8.1		- 6.5	
		D_h		1.3									
5	301	H	- 17.3		- 22.3		- 26.3		- 28.6		- 36.4		- 42.4
		ΔH		- 5.0		- 4.0		- 2.3		- 7.8		- 6.0	
		D_h		1.9									
6	567	H	- 13.2		- 17.5		- 20.7		- 22.4		- 30.2		- 35.6
		ΔH		- 4.3		- 3.2		- 1.7		- 7.8		- 5.4	
		D_h		1.7									
7	432	H	- 6.6		- 10.8		- 13.3		- 15.0		- 23.7		- 28.9
		ΔH		- 4.2		- 2.5		- 1.7		- 8.7		- 5.2	
		D_h		1.8									
8	492	H	- 3.6		- 7.2		- 10.0		- 10.2		- 18.4		- 24.5
		ΔH		- 3.6		- 2.8		- 0.2		- 8.2		- 6.1	
		D_h		2.0									
9	539	H	- 1.2		- 5.0		- 8.3		- 9.3		- 18.0		- 24.6
		ΔH		- 3.8		- 3.3		- 1.0		- 8.7		- 6.6	
		D_h		1.5									
10	591	H	- 0.3		- 3.7		- 8.9		- 10.6		- 18.5		- 24.8
		ΔH		- 3.4		- 5.2		- 1.7		- 7.9		- 6.3	
		D_h											

Table 9.3.6. USSR - Changes in thickness.

Part 45 of 54 Variations in the surface elevation and horizontal component of the displacement of marked points on the glacier surface

Tien-Shan Bogatyr Glacier Profile 3

point no.	distance from fixed point	date	1965 12.08.		1966 26.07.		1967 28.08.		1968 03.09.		1969 13.08.		1970 26.09.
		days		348		398		372		344		409	
		H	- 68.3		- 70.4		- 73.1		- 75.8		- 79.3		- 82.8
1	251	ΔH		- 2.1		- 2.7		- 2.7		- 3.5		- 3.5	
		D_h		2.0									
		H	- 63.8		- 65.9		- 69.2		- 71.8		- 75.6		- 79.6
2	311	ΔH		- 2.1		- 3.3		- 2.6		- 3.8		- 4.0	
		D_h		2.5									
		H	- 61.0		- 63.4		- 66.5		- 69.0		- 72.9		- 76.8
3	371	ΔH		- 2.4		- 3.1		- 2.5		- 3.9		- 3.9	
		D_h		2.7									
		H	- 58.1		- 60.5		- 64.8		- 67.7		- 71.4		- 72.8
4	424	ΔH		- 2.4		- 4.3		- 2.9		- 3.7		- 1.4	
		D_h		3.1									
		H	- 57.1		- 61.4		- 64.9		- 68.9		- 71.7		- 73.1
5	484	ΔH		- 4.3		- 3.5		- 4.0		- 2.8		- 1.4	
		D_h		3.3									
		H	- 60.3		- 64.6		- 68.0		- 70.5		- 73.1		- 76.7
6	540	ΔH		- 4.3		- 3.4		- 2.5		- 2.6		- 3.6	
		D_h		3.0									
		H	- 62.4		- 65.4		- 68.6		- 71.0		- 74.3		- 81.1
7	614	ΔH		- 3.0		- 3.2		- 1.4		- 3.3		- 6.8	
		D_h		3.2									
		H	- 62.5		- 65.4		- 68.4		- 69.6		- 74.0		- 78.3
8	684	ΔH		- 2.9		- 3.0		- 1.2		- 4.4		- 4.3	
		D_h		4.0									
		H	- 58.5		- 61.6		- 66.8		- 71.5		- 73.4		- 77.5
9	814	ΔH		- 3.1		- 5.2		- 4.7		- 1.9		- 4.1	
		D_h		3.0									
		H	- 70.9		- 73.9		- 79.3		- 82.2		- 83.2		- 89.5
10	880	ΔH		- 3.0		- 5.4		- 2.9		- 1.0		- 6.3	
		D_h		2.2									
		H	- 75.3		- 78.8		- 81.1		- 83.8		- 86.5		- 91.5
11	949	ΔH		- 3.5		- 2.3		- 2.7		- 2.7		- 5.0	
		D_h		1.7									
		H	- 72.8		- 75.8		- 79.4		- 82.3		- 87.7		- 90.2
12	1020	ΔH		- 3.0		- 3.6		- 2.9		- 5.4		- 2.5	
		D_h		0.4									

Table 9.3.6. USSR - Changes in thickness.

Part 46 of 54 Variations in the surface elevation and horizontal component of the displacement of marked points on the glacier surface

Tien-Shan Bogatyr Glacier Profile 4

point no.	distance from fixed point	date	1965 12.08.		1966 26.08.		1967 24.08.		1968 01.09.		1969 13.08.		1970 26.09.
		days		379		363		374		346		409	
1	171	H	- 43.2		- 46.2		- 49.2		- 49.8		- 55.2		59.3
		ΔH		- 3.0		- 3.0		- 0.6		- 5.4		- 4.1	
		D_h		5.0									
2	270	H	- 34.7		- 37.8		- 39.7		- 41.2		- 43.1		- 47.2
		ΔH		- 3.1		- 1.9		- 1.5		- 1.9		- 4.1	
		D_h		5.6									
3	330	H	- 36.5		- 40.1		- 42.4		- 43.7		- 45.8		- 50.5
		ΔH		- 3.6		- 2.3		- 1.4		- 2.1		- 4.7	
		D_h		6.4									
4	394	H	- 38.2		- 41.6		- 43.6		- 44.8		- 46.7		- 53.5
		ΔH		- 3.4		- 2.0		- 1.2		- 1.9		- 6.8	
		D_h		7.7									
5	454	H	- 40.0		- 43.8		- 45.7		- 47.1		- 48.6		- 56.2
		ΔH		- 3.8		- 1.9		- 1.4		- 1.5		- 7.6	
		D_h		8.0									
6	514	H	- 41.9		- 46.3		- 47.5		- 49.0		- 50.7		- 62.0
		ΔH		- 4.4		- 1.2		- 1.5		- 1.7		-11.3	
		D_h		7.4									
7	574	H	- 44.3		- 47.7		- 50.4		- 51.5		- 53.7		- 66.4
		ΔH		- 3.4		- 2.7		- 1.1		- 2.2		-12.7	
		D_h		7.3									
8	630	H	- 44.3		- 48.6		- 51.0		- 52.7		- 54.8		- 70.1
		ΔH		- 4.3		- 2.4		- 1.7		- 2.1		-15.3	
		D_h		7.4									
9	689	H	- 48.0		- 51.8		- 54.2		- 56.3		- 58.8		- 74.4
		ΔH		- 3.8		- 2.4		- 2.1		- 2.5		-15.6	
		D_h		6.6									
10	749	H	- 50.6		- 54.3		- 57.2		- 59.2		- 62.3		- 76.3
		ΔH		- 3.7		- 2.9		- 2.0		- 3.1		-14.0	
		D_h		6.0									
11	811	H	- 53.9		- 57.8		- 60.6		- 63.6		- 66.3		- 78.2
		ΔH		- 3.9		- 2 8		- 2.7		- 3.0		-11.9	
		D_h		4.4									

Table 9.3.6. USSR - Changes in thickness.

Part 47 of 54 Variations in the surface elevation and horizontal component of the displacement of marked points on the glacier surface

Tien-Shan Bogatyr Glacier Profile 5

point no.	distance from fixed point	date	1965 12.08.		1966 26.08.		1967 25.08.		1968 01.09.		1969 12.08.		1970 25.09.
		days		379		364		373		345		409	
1	174	H	- 23.8		- 25.3		- 25.7		- 28.3		- 29.8		- 32.5
		ΔH		- 1.5		- 0.4		- 2.6		- 1.5		- 2.7	
		D_h		2.3									
2	235	H	- 17.9		- 19.5		- 20.7		- 21.1		- 22.4		- 24.7
		ΔH		- 1.6		- 1.2		- 0.4		- 1.3		- 2.3	
		D_h		2.6									
3	296	H	- 9.1		- 10.2		- 11.9		- 12.7		- 14.1		- 16.8
		ΔH		- 1.1		- 1.7		- 0.8		- 1.4		- 2.7	
		D_h		3.1									
4	359	H	- 4.1		- 5.4		- 7.3		- 8.5		- 9.9		- 13.0
		ΔH		- 1.3		- 1.9		- 1.2		- 1.4		- 3.1	
		D_h		3.4									
5	419	H	- 2.3		- 3.8		- 5.6		- 6.7		- 8.1		- 11.0
		ΔH		- 1.5		- 1.8		- 1.1		- 1.4		- 2.9	
		D_h		3.5									
6	479	H	- 0.8		- 2.1		- 4.3		- 5.2		- 7.2		- 9.3
		ΔH		- 1.3		- 2.2		- 0.9		- 2.0		- 2.1	
		D_h		4.1									
7	538	H	- 0.1		- 1.4		- 3.7		- 4.4		- 6.2		- 8.9
		ΔH		- 1.3		- 2.3		- 0.7		- 1.8		- 2.7	
		D_h		3.7									
8	599	H	+ 0.9		- 0.1		- 2.1		- 4.2		- 6.1		- 9.2
		ΔH		- 1.0		- 2.0		- 2.1		- 1.9		- 3.1	
		D_h		3.2									
9	661	H	+ 0.2		- 0.7		- 3.7		- 5.0		- 7.3		- 9.3
		ΔH		- 0.9		- 3.0		- 1.3		- 2.3		- 2.0	
		D_h		3.4									
10	721	H	- 1.3		- 2.0		- 6.0		- 6.0		- 8.2		- 8.2
		ΔH		- 0.7		- 4.0		0.0		- 2.2		0.0	
		D_h		2.8									

Table 9.3.6. USSR - Changes in thickness.

Part 48 of 54 Variations in the surface elevation and horizontal component of the displacement of marked points on the glacier surface

Tien-Shan Zhangyryk Glacier Profile 1

point no.	distance from fixed point	date	1965 25.08.		1966 01.09.		1967 29.08.		1968 11.09.		1969 15.08.		1970 12.09.
		days		372		362		379		338		393	
		H	- 82.1		- 84.5		- 87.2		- 91.5		- 88.6		- 95.8
1	225	ΔH		- 2.4		- 2.7		- 4.3		+ 2.9		- 7.2	
		D_h		1.8									
		H	- 78.5		- 81.0		- 83.3		- 87.9		- 84.7		- 93.3
2	275	ΔH		- 2.5		- 2.3		- 4.6		+ 3.2		- 7.6	
		D_h		1.0									
		H	- 75.0		- 77.4		- 79.8		- 84.4		- 81.7		- 91.7
3	324	ΔH		- 2.4		- 2.4		- 4.6		+ 2.7		-10.0	
		D_h		3.0									
		H	- 74.4		- 76.8		- 79.1		- 83.8		- 79.8		- 92.7
4	374	ΔH		- 2.4		- 2.3		- 4.7		+ 4.0		-12.9	
		D_h		1.6									
		H	- 67.8		- 70.6		- 73.4		- 79.4		- 77.4		- 85.0
5	478	ΔH		- 2.8		- 2.8		- 6.0		+ 2.0		- 7.6	
		D_h		1.8									
		H	- 72.6		- 75.7		- 79.1		- 84.9		- 81.3		- 87.5
6	528	ΔH		- 3.1		- 3.4		- 5.8		+ 3.6		- 6.2	
		D_h		1.1									
		H	- 76.3		- 79.6		- 83.0		- 89.6		- 85.2		- 89.8
7	575	ΔH		- 3.3		- 3.4		- 6.6		+ 4.4		- 4.6	
		D_h		1.5									

Table 9.3.6. USSR - Changes in thickness.

Part 49 of 54 Variations in the surface elevation and horizontal component of the displacement of marked points on the glacier surface

Tien-Shan Zhangyryk Glacier Profile 2

point no.	distance from fixed point	date	1965 24.08.		1966 01.09.		1967 29.08.		1968 11.09.		1969 05.08.		1970 12.09.
		days		373		362		379		328		403	
1	267	H	-114.2		-114.4		-113.8		-116.0		-116.5		-117.0
		ΔH		- 0.2		+ 0.6		- 2.2		- 0.5		- 0.5	
		D_h		1.2									
2	336	H	-123.9		-124.5		-123.2		-125.8		-120.0		-126.0
		ΔH		- 0.6		+ 1.3		- 2.6		+ 5.8		- 6.0	
		D_h		1.5									
3	406	H	-125.1		-126.6		-124.1		-127.6		-122.4		-129.2
		ΔH		- 1.5		+ 2.5		- 3.5		+ 5.2		- 6.8	
		D_h		2.8									
4	477	H	-116.7		-118.5		-116.4		-119.9		-119.5		-121.6
		ΔH		- 1.8		+ 2.1		- 3.5		+ 0.4		- 2.1	
		D_h		1.8									
5	547	H	-114.4		-115.6		-114.5		-118.1		-117.0		-120.6
		ΔH		- 1.2		+ 1.1		- 3.6		+ 1.1		- 3.6	
		D_h		2.4									
6	591	H	-114.9		-116.3		-115.6		-119.5		-117.1		-120.8
		ΔH		- 1.4		+ 0.7		- 3.9		+ 2.4		- 3.7	
		D_h		2.0									
7	671	H	-121.7		-123.1		-120.8		-126.1		-123.5		-128.6
		ΔH		- 1.4		+ 2.3		- 5.3		+ 2.6		- 5.1	
		D_h		2.5									
8	731	H	-118.6		-119.5		-118.5		-122.5		-121.0		-124.5
		ΔH		- 0.9		+ 1.0		- 4.0		+ 1.5		- 3.5	
		D_h		1.0									

Zhangyryk Glacier Profile 3

point no.	distance from fixed point	date	1965 24.08.		1966 01.09.		1967 20.08.		1968 11.09.		1969 05.08.		1970 13.09.
		days		373		353		388		328		404	
1	201	H	- 61.6		- 61.3		- 62.3		- 61.1		- 62.7		- 60.0
		ΔH		+ 0.3		- 1.0		+ 1.2		- 1.6		+ 2.7	
		D_h		8.1									
2	281	H	- 66.2		- 65.9		- 67.3		- 66.8		- 67.7		- 65.4
		ΔH		+ 0.3		- 1.4		+ 0.5		- 0.9		+ 2.3	
		D_h		9.9									
3	361	H	- 65.5		- 65.0		- 67.8		- 67.1		- 69.2		- 66.4
		ΔH		- 0.4		- 1.9		+ 0.7		- 2.1		+ 2.8	
		D_h		10.7									
4	441	H	- 63.5		- 63.4		- 65.7		- 65.2		- 67.5		- 66.4
		ΔH		+ 0.1		- 2.3		+ 0.5		- 2.3		+ 1.1	
		D_h		11.4									
5	521	H	- 65.3		- 65.4		- 67.5		- 68.2		- 69.6		- 67.3
		ΔH		- 0.1		- 2.1		- 0.7		- 1.4		+ 2.3	
		D_h		10.9									
6	601	H	- 69.9		- 70.6		- 71.0		- 72.6		- 74.1		- 70.0
		ΔH		- 0.7		- 0.4		- 1.6		- 1.5		+ 4.1	
		D_h		10.2									
7	681	H	- 68.7		- 69.3		- 71.9		- 72.8		- 75.8		- 70.5
		ΔH		- 0.6		- 2.6		- 0.9		- 3.0		+ 5.3	
		D_h		8.9									
8	761	H	- 74.6		- 75.7		- 76.8		- 76.6		- 77.3		- 69.8
		ΔH		- 1.1		- 1.1		+ 0.2		- 0.7		+ 7.5	
		D_h		6.4									

Table 9.3.6. USSR - Changes in thickness.

Part 50 of 54 Variations in the surface elevation and horizontal component of the displacement of marked points on the glacier surface

Tien-Shan Zhangyryk Glacier (Southern) Profile 1

point no.	distance from fixed point		1966		1967		1968		1969		1970
		date	02.09.		29.08.		10.09.		04.08.		07.09.
		days		361		378		328		399	
		H	+ 34.0		+ 31.9		+ 27.8		+ 25.7		+ 21.1
1	168	ΔH		-2.1		-4.1		-2.1		-4.6	
		D_h									
		H	+ 40.0		+ 38.6		+ 34.6		+ 32.0		+ 32.3
2	238	ΔH		-1.4		-3.8		-2.8		+0.3	
		D_h									
		H	+ 44.0		+ 40.7		+ 37.2		+ 35.0		+ 33.0
3	273	ΔH		-3.3		-3.5		-2.2		-2.0	
		D_h		1.3							
		H	+ 42.4		+ 38.9		+ 35.9		+ 33.7		+ 29.7
4	353	ΔH		-3.5		-3.0		-2.2		-4.0	
		D_h		1.4							
		H	+ 46.9		+ 43.2		+ 40.6		+ 38.1		+ 36.9
5	433	ΔH		-3.7		-2.6		-2.5		-1.2	
		D_h		1.2							
		H	+ 51.2		+ 47.4		+ 43.9		+ 41.2		+ 38.6
6	503	ΔH		-3.8		-3.5		-2.7		-2.6	
		D_h		0.5							
		H	+ 49.8		+ 48.6		+ 46.5		+ 43.4		+ 39.5
7	539	ΔH		-1.2		-2.1		-3.1		-3.9	
		D_h		0.4							

Zhangyryk Glacier (Southern) Profile 2

point no.	distance from fixed point		1966		1967		1968		1969		1970
		date	02.09.		29.08.		10.09.		04.08.		06.09.
		days		361		378		328		398	
		H	- 16.1		- 17.0		- 18.2		- 18.4		- 20.0
1	162	ΔH		-0.9		-1.2		-0.2		-1.6	
		D_h		3.7							
		H	- 12.2		- 13.3		- 14.2		- 14.6		- 15.8
2	232	ΔH		-1.1		-0.9		-0.4		-1.2	
		D_h		4.4							
		H	- 9.6		- 10.7		- 12.0		- 12.2		- 14.1
3	288	ΔH		-1.1		-1.3		-0.2		-1.9	
		D_h									
		H	- 8.4		- 9.4		- 10.9		- 11.1		- 12.0
4	358	ΔH		-1.0		-1.5		-0.2		-0.9	
		D_h									
		H	- 8.0		- 8.5		- 10.4		- 11.5		- 11.1
5	428	ΔH		-0.5		-1.9		-0.1		-0.4	
		D_h									
		H	- 10.2		- 10.6		- 12.2		- 12.6		- 14.6
6	508	ΔH		-0.4		-1.6		-0.4		-2.0	
		D_h									
		H	- 10.9		- 11.0		- 12.6		- 13.6		- 14.5
7	588	ΔH		-0.1		-1.6		-1.0		-0.9	
		D_h									

Table 9.3.6. USSR - Changes in thickness.

Part 51 of 54 Variations in the surface elevation and horizontal component of the displacement of marked points on the glacier surface

Tien-Shan Zhangyryk Glacier (Southern) Profile 3

point no.	distance from fixed point	date	1966 02.09.		1967 29.08.		1968 10.09.		1969 04.08.		1970 05.09.
		days		361		378		328		397	
1	121	H	- 32.7		- 35.8		- 35.8		- 35.2		- 36.7
		ΔH		-3.1		0.0		+0.6		-1.5	
		Dh									
2	207	H	- 25.3		- 26.2		- 27.0		- 26.6		- 26.1
		ΔH		-0.9		-0.8		+0.4		+0.5	
		Dh									
3	277	H	- 25.3		- 25.8		- 27.1		- 26.1		- 24.2
		ΔH		-0.5		-1.3		+1.0		+1.9	
		Dh									
4	341	H	- 25.4		- 26.8		- 27.2		- 27.2		- 25.0
		ΔH		-1.4		-0.4		0.0		+2.2	
		Dh									
5	407	H	- 26.3		- 26.8		- 27.6		- 27.5		- 24.8
		ΔH		-0.5		-0.8		+0.1		+2.7	
		Dh									
6	487	H	- 29.6		- 29.6		- 31.0		- 30.2		- 28.7
		ΔH		0.0		-1.4		+0.8		+1.5	
		Dh									
7	579	H	- 24.4		- 25.1		- 26.0		- 26.0		- 24.3
		ΔH		-0.7		-0.9		0.0		+1.7	
		Dh									

Table 9.3.6. USSR - Changes in thickness.
Part 52 of 54 Variations in the surface elevation and horizontal component of the displacement of marked points on the glacier surface
Altai Bolshoy Berel Glacier

point no.	date / days	Profile A 1962 24.08.	1813	1967 13.08.	725	1969 07.08.	348	1970 21.07.	distance from fixed point
1	H	2160.2		2159.4		2158.1		2158.4	
	dH		- 0.8		- 1.3		+ 0.3		327
	D_h				46.0		23.0		
2	H	2149.0		2135.3		2133.6		2136.0	
	dH		-13.7		- 1.7		+ 2.4		400
	D_h				58.7		26.0		
3	H	2160.2		2156.8		2156.0		2158.2	
	dH		- 3.4		- 0.8		+ 2.2		456
	D_h				65.0		27.1		
4	H	2163.7		2160.0		2159.3		2161.1	
	dH		- 3.7		- 0.7		+ 1.8		497
	D_h				63.0		26.0		
5	H	2173.4		2160.7		2158.2		2158.9	
	dh		-12.7		- 2.5		+ 0.7		512
	D_h				62.6		27.2		
6	H	2170.0		2156.0		2152.8		2156.8	
	dH		- 14.0		- 3.2		+ 4.0		539
	D_h				68.1		27.4		
7	H	2165.0		2134.6		2133.4		2137.1	
	dH		-30.4		- 1.2		+ 3.7		611
	D_h				59.7		26.5		
8	H	2162.1		2134.7		2134.4		2134.8	
	dH		-27.4		- 0.3		+ 0.4		641
	D_h				54.3		24.4		
9	H	2165.6		2134.8		2134.4		2135.0	
	dH		-30.8		-0.4		+ 0.6		672
	D_h				58.0		24.8		
10	H	2171.0		2136.0		2134.6		2137.6	
	dH		-35.0		- 1.4		+ 3.0		694
	D_h				52.0		23.0		
11	H	2175.5		2151.7		2151.2		2151.8	
	dH		-23.8		- 0.5		+ 0.6		758
	D_h				47.5		19.5		
12	H								
	dH								
	D_h								

point no.	date / days	Profile A' 1962 24.08.	1812	1967 12.08.	726	1969 07.08.	346	1970 19.07.	distance from fixed point
1	H	2389.0		2388.0		2386.2		2385.5	
	dH		- 1.0		- 1.8		- 0.7		102
	D_h				30.8		10.0		
2	H	2391.2		2388.0		2386.6		2386.5	
	dH		- 3.2		- 1.4		- 0.1		132
	D_h		164.0		51.2		19.5		
3	H	2390.7		2386.3		2386.0		2384.8	
	dH		- 4.4		- 0.3		- 1.2		172
	D_h		246.2		68.7		27.5		
4	H	2389.5		2384.8		2382.8		2382.2	
	dH		- 4.7		- 2.0		- 0.6		232
	D_h		277.5		75.0		33.4		
5	H	2392.6		2389.4		2389.0		2388.0	
	dh		- 3.2		- 0.4		- 1.0		328
	D_h		303.1		96.0		39.3		
6	H	2394.0		2391.8		2390.0		2388.7	
	dH		- 2.2		- 1.8		- 1.3		387
	D_h				97.5		49.7		
7	H	2396.3		2393.6		2391.8		2390.8	
	dH		- 2.7		- 1.8		- 1.0		428
	D_h				99.8		50.1		
8	H	2398.4		2396.3		2395.6		2393.4	
	dH		- 2.1		- 0.7		- 2.2		480
	D_h				100.3		49.5		
9	H	2400.1		2397.9		2395.7		2394.5	
	dH		- 2.2		- 2.2		- 1.2		523
	D_h				97.3		47.3		
10	H	2401.8		2398.0		2396.9		2395.1	
	dH		- 3.8		- 1.1		- 1.8		570
	D_h				99.8		44.0		
11	H	2401.9		2398.2		2397.0		2394.7	
	dH		- 3.7		- 1.2		- 2.3		599
	D_h				96.7		40.7		
12	H								
	dH								
	D_h								

point no.	date / days	Profile B 1962 24.08.	1812	1967 12.08.	726	1969 07.08.	366	1970 10.07.	distance from fixed point
1	H	2356.6		2353.4		2352.9		2351.7	
	dH		- 3.2		- 0.5		- 1.2		367
	D_h				42.0		15.0		
2	H	2358.8		2354.7		2354.2		2352.4	
	dH		- 4.1		- 0.5		- 1.8		403
	D_h				64.0		25.0		
3	H	2362.8		2359.0		2358.8		2356.5	
	dH		- 3.8		- 0.2		- 2.3		487
	D_h				88.0		36.6		
4	H	2363.1		2360.1		2359.8		2357.7	
	dH		- 3.0		- 0.3		- 2.1		512
	D_h				94.0		40.0		
5	H	2363.3		2360.8		2360.3		2358.4	
	dh		- 2.5		- 0.5		- 1.9		540
	D_h		269.8		96.3		41.5		
6	H	2363.5		2360.9		2359.4		2357.5	
	dH		- 2.6		- 1.5		- 1.9		568
	D_h		286.3		96.4		43.3		
7	H	2362.4		2360.5		2358.5		2356.3	
	dH		- 1.9		- 2.0		- 2.2		603
	D_h		270.5		94.5		38.9		
8	H	2358.5		2356.4		2355.8		2354.4	
	dH		- 2.1		- 0.6		- 1.4		776
	D_h		282.5		89.2		39.4		
9	H	2357.5		2353.6		2353.2		2352.5	
	dH		- 3.9		- 0.4		- 0.7		815
	D_h				89.0		38.6		
10	H	2356.8		2351.0		2350.4		2349.3	
	dH		- 5.8		- 0.6		- 1.1		846
	D_h				84.2		38.3		
11	H	2356.8		2347.6		2347.4		2346.3	
	dH		- 9.2		- 0.2		- 1.1		880
	D_h				81.8		37.5		
12	H	2352.3		2345.4		2344.8		2343.8	
	dH		- 6.9		- 0.6		- 1.0		942
	D_h				75.5		34.0		

Table 9.3.6. USSR - Changes in thickness.

Part 53 of 54 Variations in the surface elevation and horizontal component of the displacement of marked points on the glacier surface

Altai Malyi Berel Glacier

point no.		Profile "A"							distance from fixed point	Profile "B"							distance from fixed point
	date	1962 16.08.		1967 09.08.		1969 06.08.		1970 26.06.		1962 16.08.		1967 10.08.		1969 06.08.		1970 01.08.	
	days		1817		728		324				1818		727		360		
	H	2225.0		2220.0		2218.0		2217.2		2408.6		2403.9		2401.6		2401.2	
1	dH		- 5.0		- 2.0		- 0.8		249		- 4.7		- 2.3		- 0.4		115
	D_h		94.0		17.0		4.9				124.4		35.2		17.9		
	H	2231.1		2220.9		2218.2		2216.8		2413.2		2405.1		2402.6		2402.0	
2	dH		-10.2		- 2.7		- 1.4		292		- 8.1		- 2.5		- 0.6		182
	D_h		94.0		21.2		6.4				156.8		36.0		21.8		
	H	2236.9		2224.2		2219.4		2217.6		2410.2		2406.3		2404.0		2403.6	
3	dH		-12.7		- 4.8		- 1.8		347		- 3.9		- 2.3		- 0.4		233
	D_h		101.5		25.3		5.4				171.0		40.3		23.2		
	H	2237.2		2226.5		2220.3		2218.7		2412.0		2405.1		2404.4		2404.1	
4	dH		-10.7		- 6.2		- 1.6		376		- 6.9		- 0.7		- 0.3		295
	D_h		110.5		25.7		5.5				177.0		54.1		25.0		
	H	2237.3		2226.1		2220.5		2219.4		2411.4		2406.7		2405.6		2405.4	
5	dH		-11.2		- 5.6		- 1.1		412		- 4.7		- 1.1		- 0.2		409
	D_h		110.3		24.0		7.6				187.5		55.6		24.9		
	H	2236.1		2223.9		2217.9		2217.2		2416.4		2412.5		2410.6		2410.2	
6	dH		-12.2		- 6.0		- 0.7		452		- 3.9		- 1.9		- 0.4		480
	D_h		110.7		25.3		8.1				192.0		58.0		27.0		
	H	2232.2		2219.4		2213.3		2213.3		2417.0		2414.0		2410.3		2410.0	
7	dH		-12.8		- 6.1		0.0		497		- 3.0		- 3.7		- 0.3		510
	D_h		110.2		24.3		6.6				194.1		55.8		26.1		
	H	2232.5		2223.4		2219.1		2218.8		2417.9		2411.2		2408.4		2407.5	
8	dH		- 9.1		- 4.3		- 0.3		535		- 6.7		- 2.8		- 0.9		580
	D_h		108.0		21.0		5.5				196.5		54.0		23.0		
	H	2232.9		2231.1		2226.1		2224.1		2418.8		2412.7		2409.2		2408.6	
9	dH		- 1.8		- 5.0		- 2.0		555		- 6.1		- 3.5		- 0.6		640
	D_h		107.0		16.0		5.1				189.1		54.6		25.5		
	H									2420.6		2415.1		2412.4		2412.2	
10	dH										- 5.5		- 2.7		- 0.2		720
	D_h										193.6		47.0		20.1		

Table 9.3.6. USSR - Changes in thickness.

Part 54 of 54 Variations in the surface elevation and horizontal component of the displacement of marked points on the glacier surface

Altai Małyi Berel Glacier

point no.		Profile B							distance from fixed point
	date	1962 16.08.		1967 10.08.		1969 06.08.		1970 08.07.	
	days		1818		727		336		
	H	2287.7		2286.0		2284.5		2284.4	
1	dH		- 1.7		- 1.5		- 0.1		218
	D_h		101.2		23.8		7.6		
	H	2289.0		2287.1		2284.8		2284.4	
2	dH		- 1.9		- 2.3		- 0.4		237
	D_h		122.8		29.0		10.1		
	H	2291.0		2296.3		2284.1		2283.3	
3	dH		- 4.7		- 2.2		- 0.8		284
	D_h		137.0		36.5		12.6		
	H	2293.0		2286.3		2285.6		2284.9	
4	dH		- 6.7		- 0.7		- 0.7		313
	D_h		148.5		39.2		14.8		
	H	2294.0		2289.8		2287.6		2286.7	
5	dH		- 4.2		- 2.2		- 0.9		339
	D_h		158.3		41.2		15.0		
	H	2296.4		2292.1		2290.9		2289.8	
6	dH		- 4.3		- 1.2		- 1.1		364
	D_h		165.6		42.2		15.7		
	H	2302.6		2297.6		2293.3		2293.1	
7	dH		- 5.0		- 4.3		- 0.2		398
	D_h		162.0		42.6		15.8		
	H	2310.6		2300.5		2295.6		2295.0	
8	dH		-10.1		- 4.9		- 0.6		428
	D_h		159.7		45.0		16.1		
	H	2312.6		2302.7		2298.4		2298.1	
9	dH		- 9.9		- 4.3		- 0.3		465
	D_h		175.9		47.0		16.5		
	H	2315.4		2306.0		2300.0		2299.6	
10	dH		- 9.4		- 6.0		- 0.4		525
	D_h		159.6		47.5		18.9		
	H	2320.5		2309.0		2304.8		2304.6	
11	dH		-11.5		- 4.2		- 0.2		560
	D_h		151.7		47.0		15.7		
	H	2324.0		2313.0		2308.2		2308.0	
12	dH		-11.0		- 4.8		- 0.2		600
	D_h				43.2		15.8		
	H	2324.5		2315.0		2311.0		2310.0	
13	dH		- 9.5		- 4.0		- 1.0		640
	D_h				43.8		15.0		
	H	2326.7		2316.6		2310.9		2310.6	
14	dH		-10.1		- 5.7		- 0.3		675
	D_h				37.2		12.6		

Table 9.4.1. France - Combined glaciological observations.

Glacier de Saint Sorlin 1964/65 - 1971/72

(cf. Figure 9.4.1.)

Terminus changes (- = retreat)

Period	Terminus changes
1964/65 to 1970/71	- 46 m (6.50 m $year^{-1}$)
1969/70	- 12 m
1970/71	- 11 m
1971/72	- 5 m

Mass balance between profiles

(see Figure 9.4.1. where profiles are reported)

Region between the profiles	Mass balance in 10^4 m^3 in the years				
	1965/66	1966/67	1967/68	1968/69	1969/70
Front and x = +350 m	- 1.80	- 4.16	- 1.80	- 3.20	- 3.85
+350 m > x > +250 m	- 4.05	- 10.77	- 3.50	- 7.44	- 9.18
+250 m > x > +150 m	- 3.98	- 11.65	- 3.03	- 7.07	- 9.34
+150 m > x > + 50 m	- 5.70	- 15.46	- 3.51	- 9.04	- 12.49
+ 50 m > x > - 50 m	- 5.26	- 14.69	- 2.00	- 7.32	- 10.54
- 50 m > x > -150 m	- 4.85	- 13.61	+ 0.35	- 5.88	- 8.99
-150 m > x > -250 m	- 4.47	- 11.40	+ 3.16	- 4.69	- 8.40
-250 m > x > -350 m	- 4.00	- 8.53	+ 4.90	- 3.63	- 8.10
-350 m > x > -450 m	- 3.21	- 7.06	+ 7.08	- 1.80	- 6.20
-450 m > x > -550 m	- 1.97	- 5.08	+ 9.42	+ 1.00	- 5.81

Altitude of equilibrium line and activity coefficient

Budget year	Equilibrium line m a. sl.	Activity index m of water equivalent/100m
1965/66	2781	0.66
1966/67	2838	1.35
1967/68	2738	1.74
1968/69	2797	1.31
1969/70	2831	1.08

Thickness changes 1965/66 to 1969/70

Lines of equal thickness change see figure 9.4.1.

Mean surface velocities 1965/66 - 1966/70

Iso-velocity lines see figure 9.4.1.

Fig. 9. 4. 1.

Thickness changes, surface velocities and location of profiles

GLACIER DE SAINT SORLIN

France

contour lines survey 1952

glacier terminus 1952

1964

1969

ablation stakes summer 1970

iso-velocity lines 1965–1970 m.year^{-1}

thickness changes 1965–1970 m

profiles (cf. table: mass balance values)

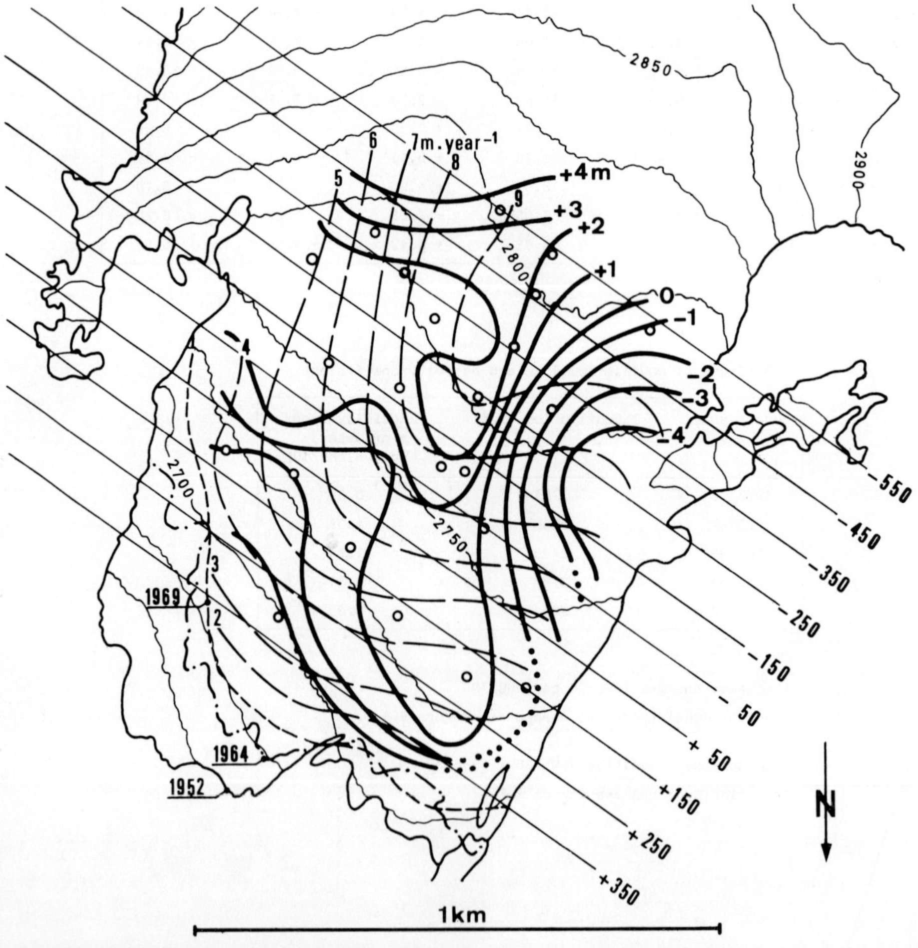

Table 9.4.2. Switzerland - Short time measurements of surface velocity.

Part 1 of 2 Aaregletscher 1969/70

The velocity was determined with the aid of photographs of an ablation stake, taken every 4 days by an automatic camera "Hasselblad". For details see P.Kasser, 1972: Variations des glaciers Suisses 1969-1970, 91e rapport, p.61.

1 Number of photograph; the photos no. 34-63 could not be used because the stake was under snow. The interpretation of the photos laking in the table was not possible because of bad weather (fog or snow fall).

2 The photos were taken at about 11 o'clock.

3 Number of days between two measurements.

4 Horizontal component of the displacement of a surface point in cm.

5 Horizontal component of the velocity of a surface point in cm/day.

6 Vertical component of the displacement of a surface point in cm.

7 Height of the top of the stake above the glacier surface, in cm.

8 Specific mass balance in cm of ice since 26.9.1969.

9 Height of snow in cm; between February 3 and June 4 1970 the height of snow was bigger than 118 cm.

10 Number of days since 26.9.1969.

11 Horizontal component of the displacement of a surface point since 26.9.1969, in cm.

12 Variation of the altitude of the top of the stake since 26.9.1969 in cm.

1	2	3	4	5	6	7	8	9	10	11	12
No.	date	days	cm	cm/day	cm	cm	cm	cm	days	cm	cm
1	26.09.69					74	0		0	0	0
		8	61	7.7	- 9						
3	4.10.69					99	- 25		8	61	- 9
		4	18	4.5	- 1						
4	8.10.69					100	- 26		12	79	- 10
		4	27	6.2	+ 2						
5	12.10.69					108	- 34		16	106	- 8
		4	23	5.8	- 1						
6	16.10.69					111	- 37		20	129	- 9
		4	23	5.8	- 2						
7	20.10.69					114	- 40		24	152	- 11
		4	19	4.8	- 2						
8	24.10.69					98			28	171	- 13
		4	22	5.5	- 2						
9	28.10.69					118	- 44	0	32	193	- 15
		4	28	7.0	- 4						
10	1.11.69					102		16	36	221	- 19
		20	100	5.0	- 11						
15	21.11.69					86		32	56	321	- 30
		16	85	5.3	- 5						
19	7.12.69					70		48	72	406	- 35
		4	20	5.0	+ 5						
20	11.12.69					84		34	76	426	- 30
		16	75	4.7	- 6						
24	27.12.69					61		57	92	501	- 36
		8	40	5.0	- 5						
26	4.01.70					65		53	100	541	- 41
		8	43	5.4	+ 3						
28	12.01.70					49		69	108	584	- 38
		8	50	6.2	- 6						
30	20.01.70					38		80	116	634	- 44
		12	71	5.9	0						
33	1.02.70					24		94	128	705	- 44
		124	851	6.9	- 7						
64	5.06.70					1		117	252	1556	- 51
		4	73	18.2	- 1						
65	9.06.70					22		96	256	1629	- 52
		4	106	26.5	- 2						
66	13.06.70					46		72	260	1735	- 54
		4	31	7.8	- 6						
67	17.06.70					74		44	264	1766	- 60
		4	52	13.0	- 8						

Table 9.4.2. Switzerland - Short time measurements of surface velocity.

Part 2 of 2 Aaregletscher 1969/70

1	2	3	4	5	6	7	8	9	10	11	12
No.	date	days	cm	cm/day	cm	cm	cm	cm	days	cm	cm
68	21.06.70.					102		16	268	1818	- 68
		4	36	9.0	- 6						
69	25.06.70					138	- 64		272	1854	- 74
		4	29	7.2	- 5						
70	29.06.70					149	- 75		276	1883	- 79
		4	36	9.0	+ 3						
71	3.07.70					159	- 85		280	1919	- 76
		4	27	6.8	- 7						
72	7.07.70					163	- 89		284	1946	- 83
		4	33	8.2	- 5						
73	11.07.70					171	- 97		288	1979	- 88
		4	33	8.2	- 3						
74	15.07.70					187	- 113		292	2012	- 91
		4	15	3.8	- 3						
75	19.07.70					192	- 118		296	2037	- 94
		4	37	9.2	- 6						
76	23.07.70					201	- 127		300	2074	- 100
		4	33	8.2	- 6						
77	27.07.70					213	- 139		304	2107	- 106
		4	34	8.5	- 10						
78	31.07.70					233	- 159		308	2141	- 116
		1	5	5.0	+ 3						
79	1.08.70					210	- 136		309	2146	- 113
		3	22	7.3	+ 5						
80	4.08.70					222	- 148		312	2168	- 108
		4	42	10.5	- 12						
81	8.08.70					258	- 184		316	2210	- 120
		4	30	7.5	- 8						
82	12.08.70					256	- 182		320	2240	- 128
		4	32	8.0	0						
83	16.08.70					274	- 200		324	2272	- 128
		3	21	7.0	- 5						
84	19.08.70					283	- 209		327	2293	- 133
		5	38	7.6	- 7						
85	24.08.70					287	- 213		332	2331	- 140
		4	28	7.0	+ 2						
86	28.08.70					299	- 225		336	2359	- 138
		4	13	3.2	- 1						
87	1.09.70					302	- 228		340	2372	- 139
		4	32	8.0	- 2						
88	5.09.70					312	- 238		344	2404	- 141
		4	39	9.8	- 7						
89	9.09.70					323	- 249		348	2443	- 148
		4	23	5.8	- 6						
90	13.09.70					337	- 263		352	2466	- 154
		4	25	6.2	+ 1						
91	17.09.70					341	- 267		356	2491	- 153
		4	25	6.2	+ 4						
92	21.09.70					340	- 266		360	2516	- 149
		4	37	9.2	- 8						
93	25.09.70					357	- 283		364	2553	- 157

Fig. 9.4.2. Aaregletscher 1969/1970

Short time measurements of surface velocity

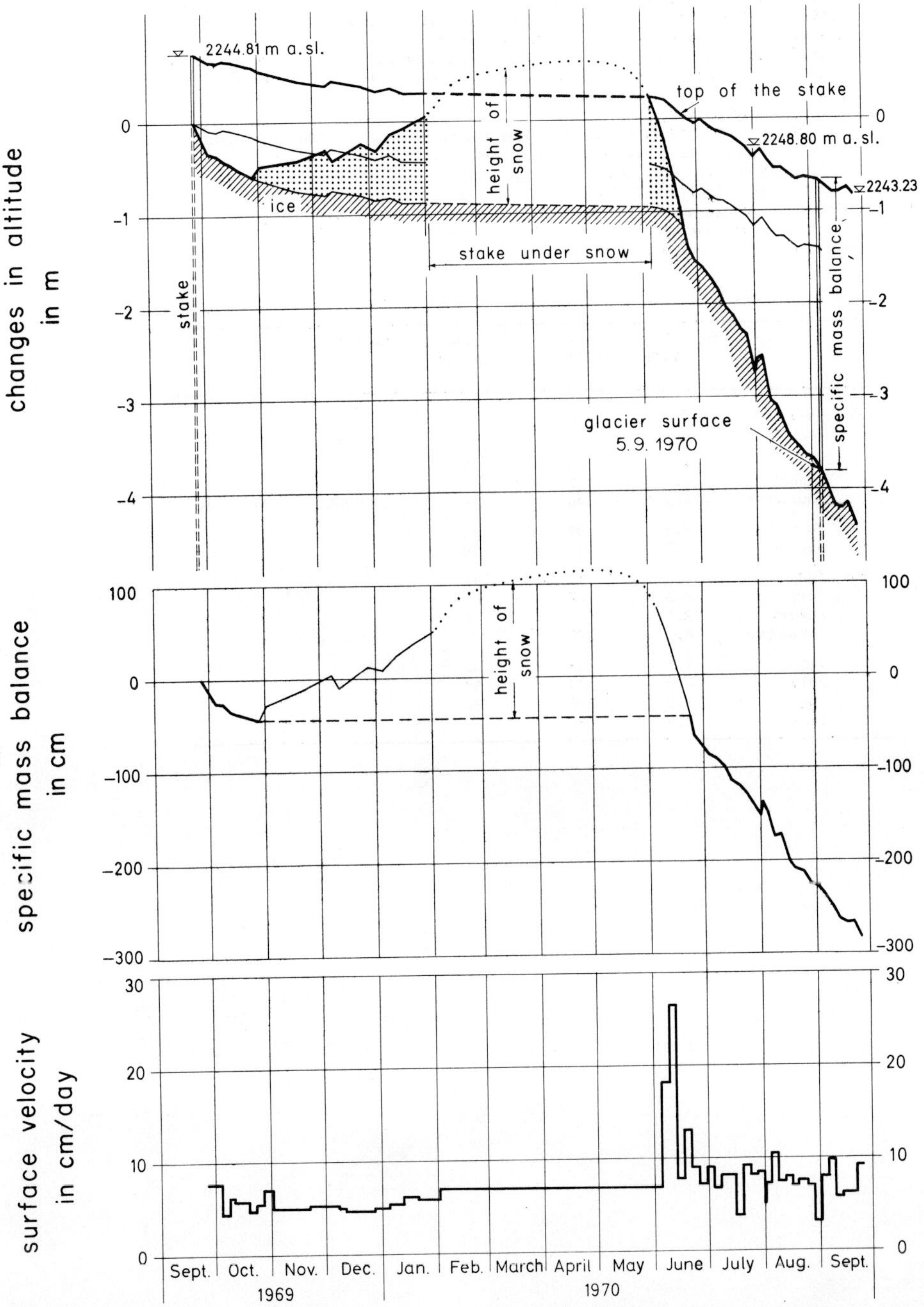

Table 9.5.1. USA - Hydrometeorological data.

Part 1 of 5 Blue Glacier 1964 - 1968

114. Blue Glacier

Washington, Olympic Mountains
Branched valley glacier
Glacier area 4.2 km^2, glacier length 4.2 km (1964)
Altitude range of glacier, 2350 m to 1275 m
No streamflow measurements near glacier
Streamflow gaging station, Hoh River near Forks, altitude 98 m, area 539 km^2
Drainage, Glacier Creek, Hoh River, Pacific Ocean
Complete meteorologic records collected at Tatoosh Island, altitude 30 m, 100 km from glacier (until July 1966), or at Quillayute, altitude 36 m, 69 km from glacier (after July 1966)
Descriptive references,
LaChapelle, E.R., 1959, Annual mass and energy exchange on the Blue Glacier: Jour. Geophys. Res., v. 64, no. 4, p. 443 - 449
LaChapelle, E.R., 1965, The mass budget of Blue Glacier, Washington: Jour. Glaciology, v. 5, no. 41, p. 609 - 623
Principal investigator, E. LaChapelle, Department of Atmospheric Sciences, University of Washington, Seattle, Washington
Sponsoring agency, University of Washington

These data were collected at the hut site, altitude 2025 m:

		Air temperature °C	Precipitation mm	Snow ablation mm	Remarks
1964	July	7.2	455	47	
	August	7.1	66	40	August 1 - 24 only
1965	July	9.8	23	54	
	August	9.1	132	40	
	September	5.9	40	16	September 1 - 16 only
1966	July	5.8	83	40	
	August	9.7	68	58	
1967	June	7.8	13	41	
	July	7.3	27		
	August	12.9	2	44	
	September	6.1	176	32	September 1 - 11 only
1968	June	4.7	169	40	June 21 - 30 only
	July	9.0	110	51	
	August	8.0	222	30	
	September	9.3	58		

Table 9.5.1. USA - Hydrometeorological data

Part 2 of 5 South Cascade Glacier 1964 - 1968

121. South Cascade Glacier

Washington, North Cascade Range
Latitude 48° 45' N, longitude 121° 03' W
Simple valley glacier
Glacier area 2.80 km^2, glacier length 3.2 km (1967)
Altitude range of glacier, 2190 m to 1610 m
Streamflow gaging station, South Fork Cascade River at South Cascade Glacier, altitude 1610 m, area 6.11 km^2
Streamflow gaging station, South Fork Cascade River near Marblemount, altitude 116 m, area 435 km^2
Drainage, South Fork Cascade River, Cascade River, Skagit River, Puget Sound, Pacific Ocean
Precipitation, temperature, snow depth, also recorded at Newhalem, Washington, altitude 160 m, 37 km from glacier
Descriptive reference,
Meier, M.F., and Tangborn, W.V., 1965, Net budget and flow of South Cascade Glacier, Washington, Jour. Glaciology, v. 5, no. 41, p. 547 - 566
Principal investigators, M.F. Meier, W.V. Tangborn, R.M. Krimmel, U.S. Geological Survey, Tacoma, Washington
Sponsoring agency, U.S. Geological Survey

	Streamflow, mm/month, averaged over drainage basin hydrologic year 1 Oct - 30 Sept.					Precipitation, * mm/month, measured in recording gage altitude 1610 m				
Month	1964	1965	1966	1967	1968	1964	1965	1966	1967	1968
Oct.	493	277	330	325	368	188	141	143	259	332
Nov.	124	110	144	76	170	302	248	244	206	152
Dec.	87	68	46	82	46	265	346	152	350	315
Jan.	87	50	38	45	103	369	373	305	392	421
Feb.	23	35	18	29	50	160	293	175	297	149
Mar.	22	24	19	20	50	284	70	181	190	198
Apr.	24	45	43	16	29	255	173	129	181	202
May	141	149	169	110	172	102	181	109	120	84
June	539	514	419	714	660	119	59	137	56	163
July	948	884	737	897	940	79	27	99	3	79
Aug.	745	895	772	826	680	131	80	36	15	99
Sep.	411	333	514	692	497	190	49	50	107	187
Total	3644	3384	3249	3832	3765	2444	2040	1760	2176	2381

	Air temperature, °C, at 1610 m (monthly means)					Hydrologic balance, meters of water averaged over drainage basin				
Month	1964	1965	1966	1967	1968	1964	1965	1966	1967	1968
Oct.	+ 4	+ 4	+ 7	+ 4	+ 3	-0.2	0	-0.2	+0.2	+0.2
Nov.	0	0	+ 1	0	0	+0.2	+0.3	+0.2	+0.4	+0.2
Dec.	+ 1	- 5	- 4	- 2	- 6	+0.6	+0.8	+0.3	+0.8	+0.7
Jan.	- 4	n	- 3	- 3	- 5	+1.2	+1.3	+0.8	+1.4	+1.2
Feb.	- 2	- 3	- 4	- 3	+ 1	+1.5	+1.8	+1.0	+1.8	+1.4
Mar.	n	- 2	- 1	- 5	- 1	+2.0	+1.9	+1.3	+2.1	+1.7
Apr.	n	+ 2	+ 1	0	n	+2.4	+2.1	+1.5	+2.3	+2.0
May	n	n	+ 3	0	n	+2.4	+2.3	+1.5	+2.4	+1.9
June	+ 5	+ 7	+ 4	+ 6	+ 4	+2.1	+1.9	+1.3	+1.8	+1.5
July	+ 8	+ 9	+ 9	+ 8	+10	+1.3	+1.0	+0.7	+0.9	+0.7
Aug.	+ 7	+10	+11	+14	+ 8	+0.8	+0.3	-0.1	+0.2	+0.2
Sept.	+ 4	+ 7	+ 9	n	+ 7	+0.7	-0.1	-0.5	-0.3	0

n = Not measured
* Average precipitation over drainage basin is approximately 1.55 times greater.

Table 9.5.1. USA - Hydrometeorological data.

Part 3 of 5 Nisqually Glacier 1964 - 1968

129. Nisqually Glacier

Washington, Middle Cascade Range
Latitude 46º 48' N, Longitude 121º 44' W
Branched valley glacier
Glacier area, 6.5 km^2, length 6.5 km (1966)
Altitude range of glacier, 4300 m to 1410 m
Streamflow gaging station, Nisqually River near Paradise, altitude 1170 m, area 16.0 km^2
Streamflow gaging station, Nisqually River near National, altitude 442 m, area 344 km^2
Drainage, Nisqually River, Puget Sound, Pacific Ocean
Precipitation, temperature, snow depth recorded at Paradise Ranger Station, altitude 1650 m
Descriptive references,

Meier, M.F., 1968, Calculations of slip of Nisqually Glacier on its bed: no simple relation of sliding velocity to shear stress: Internat. Assoc. Sci. Hyd., Bern Assembly 1967, Pub. 79, p. 49 - 57

Veatch, F.M., 1970, Analysis of a 24-year photographic record of Nisqually Glacier, Mount Rainier National Park, Washington: U.S. Geol. Survey Prof. Paper 631, 52 p.

Principal investigators, D.Richardson, A.Johnson, G.Giles, U.S. Geological Survey Tacoma, Washington
Sponsoring agency, U.S. Geological Survey

Streamflow, in mm, averaged over the drainage basin, 1968 (record started 1 March):

Mar.	82	July	697
Apr.	56	Aug.	548
May	233	Sep.	530
June	704		

Precipitation and air temperature at Paradise Ranger Station (altitude 1650 m)

	Precipitation, mm/month read daily at storage gage					Air temperature, °C (monthly means)				
Month	1964	1965	1966	1967	1968	1964	1965	1966	1967	1968
Oct.	230	159	152	253	595	5	7	8	5	3
Nov.	412	449	308	349	248	- 1	- 1	1	0	1
Dec.	302	662	295	445	499	1	- 6	- 3	- 2	- 4
Jan.	697	620	428	738	475	- 4	- 2	- 3	- 3	- 4
Feb.	188	378	263	323	420	- 2	- 2	- 3	- 2	1
Mar.	324	56	316	268	247	- 3	0	- 1	- 4	0
Apr.	173	158	110	116	183	- 1	2	1	- 1	- 1
May	88	130	n	56	52	3	3	n	4	4
June	143	27	67	54	158	6	8	16	10	7
July	100	39	34	T	13	11	13	10	12	13
Aug.	168	167	31	4	87	9	12	12	17	9
Sep.	156	70	64	83	221	7	8	11	13	9
Total	2981	2915		2688	3198					

n = Not measured
T = Trace

Table 9.5.1. USA - Hydrometeorological data.

Part 4 of 5 Maclure Glacier 1967 - 1968

133. Maclure Glacier

California, Sierra Nevada
Latitude 37° 45' N, Longitude 119° 17' W
Cirque glacier
Glacier area 0,17 km^2, glacier length 0.4 km (1966)
Altitude range of glacier, 3800 m to 3600 m
Streamflow gaging station, Maclure Creek at Maclure Glacier, altitude 3510 m, area 0.97 km^2
Streamflow gaging station, Tuolumne River near Hetch-Hetchy, altitude 1045 m, area 1184 km^2
Drainage, Tuolumne River, San Joaquin River, San Francisco Bay, Pacific Ocean
Precipitation, temperature, snow depth recorded at Yosemite National Park Headquarters, altitude 1210 m, 27 km from glacier
Principal investigator, D.R. Scully, U.S. Geological Survey, Sacramento,California
Sponsoring agency, U.S. Geological Survey

	Streamflow, mm/month averaged over drainage basin		Precipitation, mm/month, measured at storage gage, altitude 3510 m	
Month	1967	1968	1967	1968
Oct.		30		0
Nov.			319	75
Dec.			486	139
Jan.			469	124
Feb.			17	125
Mar.			276	69
Apr.			288	30
May	18[1]	22[2]	33	28
June	138	224	0	9
July	618	337	70	0
Aug.	526	168	40	13
Sep.	208	123	30	1

1 May 23 - 31
2 May 24 - 31

	Hydrologic balance meters of water, end of month, averaged over drainage basin		Air temperature, °C, altitude 3510 m (monthly means)	
Month	1967	1968	1967	1968
Oct.		0		+ 1.4
Nov.	+ 0.3	+ 0.1	- 4.4	- 3.7
Dec.	+ 0.8	+ 0.2	- 7.5	- 8.5
Jan.	+ 1.3	+ 0.3	- 7.0	- 7.5
Feb.	+ 1.3	+ 0.5	- 5.7	- 5.8
Mar.	+ 1.6	+ 0.5	- 7.5	
Apr.	+ 1.9	+ 0.6	- 8.5	
May	+ 1.9	+ 0.6	+ 1.2	
June	+ 1.7	+ 0.3	+ 5.5	+ 7.6
July	+ 1.2	0	+10.4	+ 9.6
Aug.	+ 0.7	- 0.1	+ 9.1	+ 5.5
Sep.	+ 0.5	- 0.3	+ 4.3	+ 5.5

Table 9.5.1. USA - Hydrometeorological data.

Part 5 of 5 Grinnell Glacier 1964 - 1968

134. Grinnell Glacier

Montana, North Rocky Mountains
Latitude 48° 45' N, Longitude 113° 44' W
Cirque glacier
Glacier area 1.27 km^2, glacier length 1.83 km
Altitude range of glacier: 2255 m to 1945 m
Streamflow gaging station "Grinnell Creek at Grinnell Glacier," altitude 1945 m, area 2.85 km^2
Drainage: St. Mary River, Saskatchewan River, Hudson Bay
Precipitation and temperature also recorded at Many Glacier, altitude 1480 m, 9.6 km from glacier
Descriptive references:
Alden, W.C. 1914, Glaciers of Glacier National Park, U.S. Department of the Interior;
Gibson, G.R., and Dyson, J.L., 1939, Grinnell Glacier, Glacier National Park, Montana, Bull. Geol. Soc. America, v. 50, p. 681 - 696.
Principal investigator: Arthur Johnson, University of North Dakota, Grand Forks, North Dakota
Sponsoring agencies: U.S. Geological Survey and National Park Service

	Streamflow, mm/month averaged over drainage basin, Grinnell Creek near Many Glacier					Streamflow, mm/month averaged over drainage basin, Grinnell Creek at Grinnell Glacier				
Month	1964	1965	1966	1967	1968	1964	1965	1966	1967	1968
Oct.	96	171	100	89	127					
Nov.	47	45	72	66	120					
Dec.	15	27	38	50	36					
Jan.	5	21	20	40	51					
Feb.	11	20	15	35	38					
Mar.	10	11	22	24	67					
Apr.	25	79	84	22	55					
May	287	339	502	435	460					
June	998	820	620	875	980	246	252	144	129	198
July	606	566	550	733	650	1210	1062	1080	1255	1025
Aug.	274	350	266	362	320	672	885	712	890	742
Sep.	178	136	178	200	402	260	179	547	539	816
Total	2552	2585	2467	2931	3306					

Period	Precipitation, in mm, average of 2 storage gages, altitude 1900 m, near Grinnell Glacier
19.7.63 - 30.7.64	3040
31.7.64 - 12.8.65	3880
13.8.65 - 12.8.66	2920
13.8.66 - 10.8.67	3200
11.8.67 - 25.7.68	3140

Air temperature, °C, altitude 1910 m, below Grinnell Glacier		
August	1964	9.5
August	1965	12.4
August	1966	11.7
-		
August	1968	10.0

Table 9.5.2. USSR - Hydrometeorological data.

Part 1 of 46 Arctic Ural, station Bol'shaya Khadata (240 m a.sl.)

Year		Month 9	10	11	12	1	2	3	4	5	6	7	8	Winter 9-5	Summer 6-8	Year 9-8
1958 - 1959	precipitation, mm	71.2	54.2	64.1	30.4	58.0	53.1	115.5	42.7	82.3	54.9	59.0	73.3	571.5	187.2	758.7
	air temperature, °C	- 0.6	- 3.8	-10.6	-25.9	-15.1	-17.3	-11.5	-15.9	- 2.0	4.0	10.7	9.4	-11.4	8.0	- 6.6
	degree days for temperature above 0 °C	132.4	26.0	0.9	0.0	0.0	0.0	0.0	0.7	96.0	722.9	164.06	1367.6	256.0	3731.1	3987.1
	duration of sunshine hours	29.0	8.7	0.0	0.0	0.0	7.3	63.6	202.3	114.7	265.6	256.4	176.1	425.6	698.1	1123.7
	total radiation ccal/sec	4.04	1.68	0.25	0.01	0.07	0.98	4.68	11.18	12.68	16.87	15.26	8.20	35.57	40.33	75.90
1959 - 1960	precipitation, mm	135.5	39.8	30.3	79.7	28.5	43.9	20.9	38.8	19.1	44.2	9.5	86.2	436.5	139.9	576.4
	air temperature, °C	4.9	- 6.7	- 9.4	-15.0	-22.0	-26.7	-23.6	- 8.8	- 4.1	4.8	12.1	7.5	-12.4	8.1	- 7.2
	degree days for temperature above 0 °C	593.5	2.7	2.3	0.0	0.0	0.0	0.0	19.8	77.2	590.6	1500.9	924.9	695.5	3016.4	3711.9
	duration of sunshine hours	32.4	11.3	0.0	0.0	0.0	5.0	119.4	179.0	302.7	184.6	290.4	170.4	649.8	645.4	1295.2
	total radiation ccal/sec	2.79	1.79	0.28	0.01	0.05	0.90	5.05	10.19	17.26	12.90	13.58	8.51	38.32	34.99	73.31
1960 - 1961	precipitation, mm	109.7	28.5	60.0	42.2	72.0	39.2	59.2	78.1	90.6	39.4	45.7	55.3	579.5	140.4	719.9
	air temperature, °C	3.8	- 7.2	-15.3	-11.6	-16.7	-17.0	- 8.1	-10.3	- 4.2	3.6	14.0	11.0	- 9.6	9.5	- 4.8
	degree days for temperature above 0 °C	503.2	2.6	0.6	0.0	0.0	0.0	3.6	14.4	45.8	523.2	1735.0	1366.7	570.2	3624.9	4195.1
	duration of sunshine hours	62.0	17.5	0.0	0.0	0.0	3.6	54.2	174.4	113.7	148.3	351.8	204.4	425.4	704.5	1129.9
	total radiation ccal/sec	3.22	1.64	0.28	0.00	0.14	0.90	4.08	10.17	12.40	14.22	17.12	1.02	32.83	32.36	65.19
1961 - 1962	precipitation, mm	179.9	54.9	78.2	69.5	100.5	211.8	55.2	73.8	93.4	100.7	66.1	57.7	917.2	224.5	1141.7
	air temperature, °C	3.2	- 4.2	-13.0	-18.1	-15.3	-13.3	-14.3	- 5.3	- 0.2	5.4	15.0	8.7	- 9.0	9.7	- 4.3
	degree days for temperature above 0 °C	401.6	21.6	2.3	0.0	0.0	0.0	1.3	107.5	235.6	660.4	1863.4	1072.3	769.9	3596.1	4366.0
	duration of sunshine hours	28.8	22.2	0.0	0.0	0.0	1.6	58.4	130.1	192.8	180.6	262.6	94.5	433.9	537.7	971.6
	total radiation ccal/sec	0.27	2.07	0.28	0.00	0.01	1.04	5.04	9.40	14.25	13.30	13.28	7.48	32.36	34.16	66.52

Table 9.5.2. USSR - Hydrometeorological data.

Part 2 of 46 Arctic Ural, station Bol'shaya Khadata (240 m a.sl.)

Year		Month												Winter	Summer	Year
		9	10	11	12	1	2	3	4	5	6	7	8	9-5	6-8	9-8
1962 - 1963	precipitation, mm	(66.6)														
	air temperature, ^{o}C	(6.2)	(- 3.7)	(- 8.3)	(-18.7)	(-19.3)	(-19.3)	(-24.3)	(- 9.6)	1.2	7.6	11.6	9.1	(-10.6)	(9.4)	(- 5.6)
	degree days for temperature above 0 ^{o}C	(813.2)														
	duration of sunshine hours	(49.3)														
	total radiation ccal/sec	(2.87)														
1963 - 1964	precipitation, mm															
	air temperature, ^{o}C	(2.5)	(- 1.9)	(-14.0)	(-21.9)	(-25.8)	(-17.7)	(-20.5)	(-15.7)	(- 3.4)	(4.7)	(14.2)	(8.3)	(-13.2)	(9.1)	(- 7.6)
	degree days for temperature above 0 ^{o}C															
	duration of sunshine hours															
	total radiation ccal/sec															
1964 - 1965	precipitation, mm								32.4	35.4	33.3	174.6	133.0	-	340.9	
	air temperature, ^{o}C	(3.5)	(- 3.1)	(-20.0)	(-12.0)	(-18.3)	(-19.1)	(-12.8)	- 9.6	- 5.5	3.1	11.8	10.4	(-10.8)	8.4	(- 6.0)
	degree days for temperature above 0 ^{o}C									8.9	417.0	1460.8	1290.1		3167.9	
	duration of sunshine hours								121.8	162.6	182.0	156.0	135.6		473.6	
	total radiation ccal/sec									16.27	16.19	10.32	8.02		34.53	
1965 - 1966	precipitation, mm										71.6	154.7	114.7		341.0	
	air temperature, ^{o}C	(2.4)	(- 7.9)	(-19.8)	(13.3)	(-20.8)	(-29.8)	(-18.4)	(-10.1)	(- 3.1)	3.2	11.5	9.1	(-13.4)	7.9	(- 8.1)
	degree days for temperature above 0 ^{o}C										405.4	1428.9	133.7		1968.0	
	duration of sunshine hours										143.8	195.7	133.7		473.2	
	total radiation ccal/sec										10.45	10.67	6.37		27.49	

The observations have been ceased on 27th September 1962. The air temperature has been reconstructed with the aide of a graphical regression to the temperature measured at the meteorological station Vorkut (AMSG)

Table 9.5.2. USSR - Hydrometeorological data.

Part 3 of 46 Arctic Ural, station Bol,shaya Khadata (240 m a.sl.)

Year		Month 9	10	11	12	1	2	3	4	5	6	7	8	Winter 9-5	Summer 6-8	Year 9-8
1966 - 1967	precipitation, mm							92.6	168.9	51.3	56.0	39.2	17.1		112.3	
	air temperature, °C	(2.8)	(- 9.7)	(- 9.6)	(-16.5)	(-23.4)	(-16.2)	- 6.6	- 2.6	- 3.7	4.1	13.8	10.8	(- 9.5)	9.6	(- 4.7)
	degree days for temperature above 0 °C							0.6	254.5	59.5	1157.1	3425.3	2668.1		7250.1	
	duration of sunshine hours							60.6	77.0	133.9	223.6	289.5	228.2		741.3	
	total radiation ccal/sec															
1967 - 1968	precipitation, mm	107.9	193.5	62.6	39.4	12.9	42.6	94.2	38.7	47.2	40.3	115.8	191.0	639.0	347.1	986.1
	air temperature, °C	3.7	- 1.0	- 4.6	-13.2	-19.4	-15.3	-10.2	-13.2	- 3.2	1.5	7.0	8.4	- 8.5	5.6	- 5.0
	degree days for temperature above 0 °C	912.1		69.4	0.7	0.0	0.0	0.0	5.4	141.1	474.6	1733.9	2069.0	(1128.7)	4277.5	(5406.2)
	duration of sunshine hours	75.3	22.7	0.0	0.0	0.0	6.0	22.2	127.6	130.5	148.9	135.3	114.0	384.3	398.2	782.5
	total radiation ccal/sec															
1968 - 1969	precipitation, mm	56.0	46.0	27.4	18.6	36.1	36.2	57.8	54.4	13.1	74.2	78.3	67.7	345.6	220.2	565.8
	air temperature, °C	1.2	- 5.4	-22.7	-21.9	-25.1	-21.8	-18.1	-11.8	- 7.3	2.7	12.5	6.8	-14.8	7.3	- 9.3
	degree days for temperature above 0 °C	445.6	312.1	0.0	0.0	0.0	0.0	0.0	22.8	27.5	805.9	3090.5	1671.8	808.0	5568.2	6376.2
	duration of sunshine hours	44.1	31.6	0.0	0.0	0.0	6.4	60.9	157.0	222.9	188.1	227.3	92.4	522.9	507.8	1030.7
	total radiation ccal/sec										12.36	11.78	6.81		30.95	
1969 - 1970	precipitation, mm	92.9	14.0	41.3	37.1	55.4	20.5	34.9	45.4	56.8	36.9	35.2	30.4	398.3	102.5	500.8
	air temperature, °C	2.8	- 5.9	- 6.6	-13.2	-23.3	-18.5	-11.5	-12.3	- 6.1	1.7	11.0	6.6	-10.5	6.4	- 6.3
	degree days for temperature above 0 °C	829.6	64.1	46.4	13.2	0.0	0.0	10.0	37.7	66.5	480.3	2726.4	1652.1	1067.5	4858.8	5926.3
	duration of sunshine hours	110.0	32.1	0.0	0.0	0.0	3.3	91.0	214.6	176.8	128.6	252.5	180.0	627.8	561.1	1188.9
	total radiation ccal/sec															

Since March 1967 the 8-term meteorological observations are executed by Omsk UGMS

Table 9.5.2. USSR - Hydrometeorological data.

Part 4 of 46 Caucasus

Mean values	Name of station m a.sl.	Month 10	11	12	1	2	3	4	5	6	7	8	9	Winter 10-4	Summer 5-9	Year 10-9	Observation period
Precipitation, mm	Kazbegi v/g, 3653	99	83	58	63	71	95	147	183	165	150	169	121	616	788	1404	1935-1939 1941-1965
	Koruldash, 1943	133	106	101	94	84	97	105	119	111	108	106	112	720	556	1276	1938-1965
	Mamisonskiy pass, 2854	65	59	65	73	76	81	82	110	109	86	85	77	501	467	968	1928-1930 1932-1965
	Mestia, 1441	95	76	73	68	61	70	75	85	80	78	76	81	518	400	918	1938-1965
Air temperature, °C	Kazbegi v/g, 3653	- 4.1	- 8.6	-12.3	-15.0	-15.3	-12.2	- 8.0	- 3.5	- 0.3	3.0	3.4	0.0	- 8.5	0.5	-6.1	1934-1960
	Koruldash, 1943	5.3	0.3	- 4.4	- 6.9	- 6.2	- 3.2	1.5	7.0	10.1	12.9	13.0	9.3	- 1.9	10.5	3.2	1932-1960
	Mamisonskiy pass, 2854	- 0.5	- 5.3	- 9.1	-12.0	-12.2	- 8.9	- 4.1	0.6	3.8	7.3	7.6	4.0	- 7.4	4.7	-2.4	1932-1960
	Mestia, 1441	7.1	1.6	- 4.1	- 6.0	- 4.6	- 0.5	5.2	11.0	14.0	16.4	16.3	12.0	- 0.2	13.9	5.7	1936-1960
Degree days for temperature above 0, °C	Mamisonskiy pass, 2854	41	5					9	52	134	218	228	129	55	761	816	
	Kazbegi v/g, 3653	4							4	31	106	107	39	4	287	291	
	Mestia, 1441	200	80	9	2	8	32	155	336	422	520	477	349	486	2104	2590	
	Koruldash, 1943	157	61	10	1	1	13	72	214	311	405	384	261	315	1575	1890	
Duration of sunshine, hours	Kazbegi v/g, 3653	174	157	139	147	152	179	191	205	225	239	230	194	1139	1039	2232	1935-1962
Runoff, m^3/sec	r.Chkheri-s.Kazbegi,33	0.72	0.54	0.43	0.34	0.36	0.37	0.50	0.85	1.45	2.69	2.16	1.22	0.46	1.67	0.95	1947-1962
	r.Terek-s.Kazbegi, 778	16.3	11.8	9.40	8.02	7.50	7.52	13.5	33.8	52.7	54.9	39.2	24.8	10.6	41.1	23.4	1926-1942 1953-1962
	Pirikitel'skaya Alazan' s.Dartlo, 290	5.20	3.58	3.16	2.38	2.22	2.02	3.74	8.80	14.3	20.2	12.7	8.52	3.19	12.9	7.26	1950-1962
	Pirikitel'skaya Alazan' s.Omaylo, 352	7.43	5.71	4.17	3.62	3.36	3.18	5.15	12.0	19.4	25.7	17.8	11.0	4.66	17.2	9.85	1951-1962
	r.Sel'dy-s.Kurush, 26	0.22							0.75	0.95	1.45	1.11	0.44		0.94		1958-1962
	r.Mestiachala-g.Mestia 144	11.9	7.74	3.20	2.83	3.16	3.12	7.25	16.5	30.4	40.8	40.5	25.8	5.6	30.8	16.5	1939,1940 1942,1943 1946-1962

Table 9.5.2. USSR - Hydrometeorological data.

Part 5 of 46 Caucasus

Mean values	Name of station m a.sl.	Month 10	11	12	1	2	3	4	5	6	7	8	9	Winter 10-4	Summer 5-9	Year 10-9	Observation period
Runoff, m^3/sec	r.Bol'shaya Liakhvi- -Dzhava, 646	11.0	8.53	6.95	6.00	6.22	8.78	24.2	45.2	40.2	25.7	16.0	11.5	10.2	277	17.4	1929-1962
	r.Rioni-s.Gebi, 222	7.62	4.53	2.91	2.77	2.47	3.08	12.5	22.7	25.8	23.6	15.5	9.82	5.12	19.6	11.1	1950-1955
Water equivalent of the snow cover in the third decade, mm	Kazbegi	51	38	32	52	109	172	130									
	Koruldash	30	121	216	322	476											
	Mamisonskiy pass	34	59	74	88	116											
	Mestia	10	46	99	133	144											

Table 9.5.2. USSR - Hydrometeorological data.

Part 6 of 46 Caucasus

1959 - 1960	Name of station	Month 10	11	12	1	2	3	4	5	6	7	8	9	Winter 10-4	Summer 5-9	Year 10-9
Precipitation, mm	Terekol	77	12	11	68	90	24	115	66	96	60	69	106	397	397	794
	Kazbegi v/g	167	68	26	138	223	57	236	151	230	112	115	129	915	737	1652
	Korul'dash	169	104	96	160	205	75	237	102	129	38	164	58	1046	491	1537
	Mamisonskiy pass	56	53	46	130	94	80	168	76	108	80	89	49	627	402	1029
	Mestia	78	48	69	79	155	35	90	75	102	23	69	34	554	303	857
Air temperature, °C	Terekol	5.7	0.8	- 2.2	- 4.3	- 5.3	- 5.1	1.4	7.2	6.3	12.3	10.6	8.6	- 1.3	9.6	3.2
	Kazbegi v/g	- 6.9	-10.1	- 9.8	-12.9	-13.2	-13.1	- 7.8	- 2.7	- 0.2	2.7	2.4	0.6	-10.5	0.5	- 6.0
	Korul'dash	2.0	- 0.4	- 2.3	- 4.5	- 5.3	- 4.0	1.6	7.5	10.1	13.4	11.1	10.1	- 2.1	10.4	3.1
	Mamisonskiy pass	- 4.6	- 6.6	- 6.7	- 9.6	-10.4	-10.0	- 4.4	1.3	4.1	7.4	5.8	4.7	- 7.5	4.8	- 2.4
	Mestia	3.1	1.5	- 3.6	- 2.8	- 4.4	- 1.4	5.2	11.0	13.4	16.7	14.0	12.6	- 0.2	13.5	5.5
Degree days for temperature above 0,°C	Kazbegi v/g								15	26	87	83	50		261	261
	Korul'dash	77	53	9	4	4	5	56	235	302	417	345	300	208	1699	1907
	Mamisonskiy pass	4	2						72	130	230	181	142	6	755	761
	Mestia	107	82	6	3	9	22	155	341	403	517	433	377	384	2071	2455
Discharge m^3/sec.	r.Chkheri - s.Kazbegi	0.59	0.47	0.32	0.28	0.28	0.30	0.39	0.95	1.37	2.31	2.0	1.13	0.37	1.6	0.86
	r.Terek - s.Kazbegi	18.2	13.4	9.91	8.24	7.83	7.62	14.3	48.2	63.6	56.5	38.8	23.6	11.5	46.1	25.9
	Pirikitel'skaya Alazan' s.Dartlo	6.68	4.66	3.02	2.62	2.72	2.01	4.10	10.8	16.8	2.07	14.2	8.64	3.69	1.42	8.08
	Pirikital'skaya Alazan' s.Omaylo	8.73	6.91	4.04	3.50	3.37	2.78	6.78	14.2	21.1	25.1	19.8	12.1	5.16	18.4	10.7
	r.Sel'dy - s.Kurush	0.27	0.088	0.071	0.071	0.079	0.081	0.31	0.68	0.77	1.17	0.77	0.48	0.14	0.77	0.40
	r.Mestiachala	14.9	9.56	4.19	1.82	2.03	2.46	5.97	21.1	35.6	43.0	41.0	44.2	5.85	36.9	18.83

Table 9.5.2. USSR - Hydrometeorological data.

Part 7 of 46 Caucasus

1960 - 1961	Name of station	Month 10	11	12	1	2	3	4	5	6	7	8	9	Winter 10-4	Summer 5-9	Year 10-9
Precipitation, mm	Terekol	27	112	42	68	12	66	106	170	106	171	56	76	433	579	1012
	Kazbegi v/g	85	35	55	182	40	127	149	266	203	333	68	149	673	1039	1712
	Korul'dash	103	45	28	124	29	160	187	226	92	168	19	92	676	597	1273
	Mamisonskiy pass	36	33	41	109	24	88	72	110	84	153	42	82	403	471	874
	Mestia	85	20	29	74	11	103	136	142	62	118	24	64	458	410	868
Air temperature, °C	Terekol	2.6	- 0.3	- 2.2	- 4.3	- 5.3	- 5.1	1.4	7.2	9.3	12.3	10.6	8.6	- 1.9	9.6	2.9
	Kazbegi v/g	- 2.1	- 7.0	- 8.3	-14.2	-14.1	-13.9	- 7.5	- 9.0	1.2	3.0	3.8	- 2.8	- 9.5	- 0.8	- 5.9
	Korul'dash	7.0	2.2	- 1.6	- 5.6	- 5.5	- 5.0	1.7	8.2	11.6	12.4	12.8	6.8	- 1.0	10.4	3.8
	Mamisonskiy pass	1.3	- 3.3	- 5.4	-10.4	-10.7	-11.1	- 3.7	2.5	6.1	6.8	7.4	1.0	- 6.2	4.8	- 1.6
	Mestia	8.2	3.2	- 1.5	- 5.4	- 4.0	- 2.4	5.3	11.6	15.1	15.6	16.0	10.0	0.5	13.7	6.0
Degree days for temperature above 0, °C	Kazbegi v/g	3								55	93	111	1	3	260	263
	Korul'dash	215	93	7			5	68	255	347	383	399	205	388	1589	1977
	Mamisonskiy pass	53	2						7	94	183	211	228	55	723	778
	Mestia	252	112	7	1	2	12	158	358	452	485	496	299	544	2090	2634
Discharge m^3/sec.	r.Chkheri - s.Kazbegi	0.50	0.40	0.38	0.35	0.33	0.32	0.33	0.99	2.04	6.12	1.59	1.10	0.37	2.37	1.20
	r.Terek - s.Kazbegi	16.4	12.5	11.5	8.64	7.83	7.67	14.5	47.0	57.5	59.3	39.8	18.6	11.28	44.4	25.10
	Pirikitel'skaya Alazan' s.Dartlo	5.68	2.87	2.32	2.02	2.08	1.98	3.89	10.7	17.8	25.1	10.6	5.74	2.98	14.0	7.56
	Pirikital'skaya Alazan' s.Omaylo	8.36	6.56	5.20	3.90	3.79	3.72	5.22	12.2	17.9	19.0	13.5	8.10	5.25	14.1	8.95
	r.Sel'dy - s.Kurush	0.20	0.11	0.082	0.069	0.075	0.074	0.18	1.05	0.97	1.38	1.21	0.39	0.11	1.0	0.48
	r.Mestiachala	19.4	9.18	3.61	3.25	3.19	2.61	7.51	26.1	29.2	31.0	28.4	11.3	6.96	25.2	14.89

Table 9.5.2. USSR - Hydrometeorological data

Part 8 of 46 Caucasus

1961 - 1962	Name of station	Month												Winter	Summer	Year
		10	11	12	1	2	3	4	5	6	7	8	9	10-4	5-9	10-9
Precipitation, mm	Terekol	114	16	166	17	34	50	56	155	44	162			453		
	Kazbegi v/g	50	44	152	22	93	136	71	215	75	156	133	85	568	664	1232
	Korul'dash	39	138	78	41	83	81	53	190	58	135	132	81	513	596	1109
	Mamisonskiy pass	21	77	49	27	85	42	34	75	60	69	108	55	335	367	702
	Mestia	50	147	89	52	45	47	55	106	27	56	76	50	485	315	800
Air temperature, °C	Terekol	3.9	1.0	- 2.9	- 5.9	- 5.3	0.3	2.0	7.1	9.8	13.8			- 1.0		
	Kazbegi v/g	- 5.6	- 8.6	-10.8	-12.7	-13.8	- 8.6	- 7.4	- 2.6	0.0	5.4	3.3	1.2	- 9.6	1.4	- 5.0
	Korul'dash	4.9	0.5	- 1.8	- 4.6	- 4.3	0.3	3.4	7.9	9.8	15.2	13.2	10.7	- 0.2	11.4	4.6
	Mamisonakiy pass	- 1.2	- 5.4	- 7.5	- 9.9	-10.6	- 5.9	- 3.0	1.6	4.3	10.1	7.8	5.8	- 6.2	5.9	- 1.2
	Mestia	6.6	2.0	- 1.1	- 5.0	- 3.4	2.2	6.4	11.3	14.2	18.2	16.4	12.3	1.1	14.5	6.6
Degree days for temperature above 0, °C	Kazbegi v/g								4	34	168	106	54		366	366
	Korul'dash	153	62	19	2	4	54	102	245	296	470	409	321	396	1741	2137
	Mamisonskiy pass	49	29	5				11	65	139	313	241	175	94	933	1027
	Mestia	204	84	20	1	7	89	192	350	424	564	508	368	597	2214	2811
Discharge m^3/sec	r.Chkheri - s.Kazbegi	0.64	0.39	0.34	0.34	0.35	0.35	0.34	0.41	0.39	1.43	1.53	1.28	0.39	1.08	0.68
	r.Terek - s.Kazbegi	13.7	10.9	9.74	7.51	6.90	7.89	11.8	35.1	41.8	43.6	33.2	22.7	9.60	35.3	20.40
	Pirikitel'skaya Alazan' s.Dartlo	3.96	2.89	2.49	2.26	2.25	1.79	2.80	9.42	12.0	31.2	9.31	5.58	2.63	13.5	7.16
	Pirikitel'skaya Alazan' s.Omaylo	5.97	5.60	5.12	4.31	3.97	3.76	4.24	10.6	13.8	28.3	12.9	7.75	4.71	14.7	8.86
	r.Sel'dy - s.Kurush	0.16	0.26	0,092	0.077	0.062	0.079	0.12	0.65	1.24	1.94	1.28	0.37	0.12	1.10	0.53
	r.Mestiachala	3.46	2.32	1.87	1.60	1.53	2.39	5.33	13.7	27.7	43.7	28.8	21.2	2.6	27.0	12.8

Table 9.5.2. USSR - Hydrometeorological data.

Part 9 of 46 Caucasus

1962 - 1963	Name of station	Month 10	11	12	1	2	3	4	5	6	7	8	9	Winter 10-4	Summer 5-9	Year 10-9
Precipitation, mm	Terekol	52	56	62	245	53	102	41	86	84	136	48	143	611	497	1108
	Kazbegi v/g	172	33	92	243	87	76	142	143	89	192	64	119	845	607	1452
	Korul'dash	148	39	305	403	110	191	108	84	152	166	66	124	1304	592	1896
	Mamisonskiy pass	68	19	132	163	85	41	84	121	99	189	53	87	592	549	1141
	Mestia	144	11	152	247	56	132	62	73	100	178	69	108	804	528	1332
Air temperature, °C	Terekol	3.7	- 2.0	- 5.4	- 4.3	- 4.4	- 5.2	1.9	5.1	8.8	11.8	11.0	8.4	- 2.2	9.0	2.4
	Kazbegi v/g	- 3.9	- 6.2	-10.6	-12.2	-11.6	-13.6	- 7.0	- 3.9	- 1.1	2.6	2.9	1.0	- 9.3	0.3	- 5.3
	Korul'dash	5.2	3.7	- 3.2	- 3.7	- 2.5	- 5.1	2.4	6.3	9.1	12.9	12.1	10.2	- 0.5	10.1	4.0
	Mamisonskiy pass	- 0.5	- 2.3	- 7.6	- 8.6	- 7.9	-10.9	- 3.4	- 0.6	2.7	6.9	6.4	4.9	- 5.9	4.1	- 1.7
	Mestia	6.6	4.4	- 1.6	- 3.0	- 1.8	- 2.7	4.6	10.1	12.8	16.5	15.3	12.0	0.9	13.3	5.1
Degree days for temperature above 0, °C	Kazbegi v/g	9								17	83	95	48	9	243	252
	Korul'dash	169	111	3	1	3	12	78	195	274	401	376	305	377	1551	1928
	Mamisonskiy pass	66	12						13	82	212	196	148	78	651	729
	Mestia	277	131	8	1	6	23	137	314	384	510	476	362	583	2046	2629
Discharge m^3/sec.	r.Chkheri - s.Kazbegi	0.77	0.73	0.69	0.77	0.77	0.67	0.81	1.89	2.31	3.47	3.20	2.32	0.74	2.63	1.53
	r.Terek - s.Kazbegi	14.8	9.82	8.10	7.78	7.36	6.73	12.9	38.6	63.3	86.3	49.7	33.8	9.63	54.3	13.49
	Pirikitel'skaya Alazan' s.Dartlc	3.83	2.99	2.70	2.35	2.09	2.16	4.07	8.74	13.2	28.6	17.7	9.64	2.88	15.6	8.22
	Pirikitel'skaya Alazan' s.Omaylc	5.89	4.36	3.85	3.20	3.13	3.16	5.70	11.9	18.9	38.9	30.0	15.5	4.17	23.0	12.0
	r.Sel'dy - s.Kurush	0.22	0.20	0.11	0.079	0.073	0.089	0.33	0.78	2.09	2.04	1.22	0.62	0.16	1.35	0.65
	r.Mestiachala	7.49	3.37	3.48	3.32	2.62	2.62	5.12	10.0	14.2	40.7	29.7	16.8	4.00	22.1	11.61

Table 9.5.2. USSR - Hydrometeorological data

Part 10 of 46 Caucasus

1963 - 1964	Name of station	Month 10	11	12	1	2	3	4	5	6	7	8	9	Winter 10-4	Summer 5-9	Year 10-9
Precipitation, mm	Terekol	29	76	19	42	22	45	48	125	67	88	130	102	281	512	793
	Kazbegi v/g	25	310	23	61	44	90	116	279	162	207	123	27	669	798	1467
	Korul'dash	154	77	170	27	63	77	103	177	123	90	145	132	671	667	1338
	Mamisonskiy pass	69	19	75	20	59	47	67	65	78	110	96	71	356	420	776
	Mestia	87	82	115	26	30	53	61	122	47	80	122	86	454	457	911
Air temperature, °C	Terekol	3.0	- 1.4	- 4.6	-11.7	- 6.2	- 2.2	0.7	6.1	10.6	11.2	9.2	7.7	- 3.2	9.0	3.2
	Kazbegi v/g	- 4.0	- 8.8	-12.9	-18.9	-14.4	-10.1	- 9.8	- 4.4	0.9	3.4	2.3	0.4	-11.2	0.5	- 6.3
	Korul'dash	4.9	- 0.2	- 5.2	-11.3	- 5.1	- 1.7	1.4	6.5	12.0	12.7	10.4	9.6	- 2.5	10.2	2.8
	Mamisonskiy pass	- 0.3	- 6.2	-10.4	-16.5	-11.3	- 7.6	- 5.1	- 0.1	5.7	7.1	4.4	5.4	- 8.2	4.5	- 2.9
	Mestia	6.8	0.3	- 5.2	-10.7	- 4.0	0.2	4.8	9.5	15.4	15.8	13.6	11.7	- 1.1	13.2	5.4
Degree days for temperature above 0, °C	Kazbegi v/g									32	108	78	32		250	250
	Korul'dash	151	42	14			35	63	202	362	395	324	287	305	1570	1875
	Mamisonskiy pass	46						5	40	172	219	295	134	51	860	911
	Mestia	210	55	6		4	52	145	294	461	490	422	352	472	2019	2491
Discharge m^3/sec.	r.Chkheri - s.Kazbegi	1.65	1.40	1.31	1.11	0.98	1.09	1.38	1.91	3.02	2.99	2.46	1.42	1.27	2.36	1.73
	r.Terek - s.Kazbegi	25.3	17.5	15.5	14.7	12.8	13.1	16.1	33.4	79.3	57.8	38.7	32.3	16.4	48.3	29.7
	Pirikitel'skaya Alazan' s.Dartlo	5.47	3.68	2.77	2.50	1.42	1.91	3.52	9.04	21.6	19.4	12.2	8.25	3.04	14.1	7.65
	Pirikitel'skaya Alazan' s.Omaylo	9.26	5.73	5.33	3.81	3.78	3.25	4.09	10.6	26.1	22.6	19.1	10.6	5.03	17.8	10.3
	r.Sel'dy - s. Kurush	0.59	0.29	0.54	0.11	0.065	0.19	0.19	0.77	1.11	1.34	0.86	0.41	0.28	0.90	0.54
	r.Mestiachala	7.14	1.94	1.92	2.04	2.00	2.14	3.10	7.99	25.2	26.4	20.0	15.0	2.78	18.91	9.57

Table 9.5.2. USSR - Hydrometeorological data.

Part 11 of 46 Caucasus

1964 - 1965	Name of station	Month 10	11	12	1	2	3	4	5	6	7	8	9	Winter 10-4	Summer 5-9	Year 10-9
Precipitation, mm	Terekol	100	98	68	4	56	29	94	82	100	54	86	45	449	367	816
	Kazbegi v/g	25	310	23	5	78	77	193	195	212	75	173	115	711	770	1481
	Korul'dash	58	154	29	3	151	78	118	141	131	8	89	64	591	433	1024
	Mamisonskiy pass	21	54	30	7	60	52	83	91	121	43	45	20	307	320	627
	Mestia	45	113	39	14	63	53	116	80	102	50	123	39	443	394	837
Air temperature, °C	Terekol	0.0	- 0.4	- 3.0	- 8.8	- 6.7	- 3.2	0.2	6.3	9.8	12.2	12.2	7.8	- 3.1	9.7	2.2
	Kazbegi v/g	- 4.1	-10.6	-11.8	-14.0	-15.3	-11.1	-10.4	- 3.7	0.0	3.0	4.4	0.2	-11.0	0.8	- 6.2
	Korul'dash	5.7	- 0.7	- 3.3	- 6.0	- 5.8	- 1.5	0.3	7.3	10.2	13.3	14.0	9.5	- 1.6	10.9	3.5
	Mamisonskiy pass	0.2	- 6.6	- 8.6	-11.0	-12.0	- 7.7	- 6.4	0.9	4.0	7.7	8.6	3.9	- 7.4	5.0	- 2.3
	Mestia	6.2	0.1	- 3.5	- 6.8	- 4.8	- 0.2	3.1	10.4	13.9	16.0	16.3	11.9	- 0.8	13.7	4.4
Degree days for temperature above 0, °C	Kazbegi v/g	6							1	20	94	138	33	6	286	292
	Korul'dash	184	41	11		1	28	70	229	306	411	434	286	335	1666	2001
	Mamisonskiy pass	65						7	60	120	238	267	119	72	804	876
	Mestia	193	48	3	1	3	44	129	321	418	496	506	359	421	2100	2521
Discharge m^3/sec	r.Chkheri - s.Kazbegi	0.84	0.73	0.81	0.42	0.46	0.47	0.53	1.07	1.83	2.01	4.32	2.42	0.61	2.33	1.32
	r.Terek - s.Kazbegi	26.4	11.9	9.18	7.89	7.67	8.27	11.3	32.6	58.1	48.9	40.7	27.7	11.8	41.6	24.21
	Pirikitel'skaya Alazan' s.Dartlo	5.37	3.62	2.26	2.23	1.97	1.96	3.04	8.79	15.3	16.1	11.2	6.29	2.92	11.5	6.51
	Pirikitel'skaya Alazan' s.Amaylo	7.94	6.20	7.51	4.27	3.43	2.97	5.19	13.8	30.2	30.4	19.4	14.5	5.33	21.6	12.1
	r.Sel'dy - s.Kurush	0.21	0.14	0.084	0.049	0.044	0.061	0.15	0.54	1.11	1.67	1.10	0.46	0.11	0.98	0.47
	r.Mestiachala	6.12	2.48	2.14	1.90	1.92	1.92	3.31	9.52	19.6	31.0	26.2	12.8	2.83	19.8	9.91

Table 9.5.2. USSR - Hydrometeorological data.

Part 12 of 46 Caucasus

1965 - 1966	Name of station	10	11	12	1	2	3	4	5	6	7	8	9	Winter 10-4	Summer 5-9	Year 10-9
Precipitation, mm	Terekol	11	13	48	100	10	34	64	97	58	137	169	45	280	506	786
	Kazbegi v/g	247	145	26	282	102	129	181	194	180	182	214	128	1112	898	2010
	Korul'dash	151	131	125	190	45	97	112	83	69	82	171	59	851	464	1315
	Mamisonskiy pass	53	56	58	103	23	44	55	109	103	59	110	53	392	434	826
	Mestia	100	91	81	99	41	35	87	88	97	102	148	50	534	485	1019
Air temperature, ^{o}C	Terekol	4.7	2.9	- 2.5	- 2.6	- 2.5	- 1.2	3.5	5.7	9.3	13.4	12.7	7.7	0.3	9.7	4.3
	Kazbegi v/g	- 7.2	- 8.5	- 9.4	-10.1	-10.5	-10.7	- 5.1	- 3.7	- 0.1	4.4	4.2	0.2	- 8.8	1.0	- 4.7
	Korul'dash	1.3	- 0.7	- 2.2	- 1.4	- 2.0	- 0.6	4.3	6.8	9.8	15.5	14.1	9.3	- 0.2	11.1	4.5
	Mamisonskiy pass	- 4.6	- 6.5	- 6.9	- 7.0	- 7.9	- 7.4	- 1.8	0.0	3.7	9.4	8.7	3.6	- 6.0	5.1	- 1.4
	Mestia	3.6	0.5	- 1.0	- 0.6	0.0	2.6	7.0	10.0	12.8	17.7	16.5	11.8	1.7	13.8	6.7
Degree days for temperature above 0, ^{o}C	Kazbegi v/g	1						4	1	40	142	129	42	5	355	360
	Korul'dash	75	28	13	11	13	26	130	214	295	481	438	280	296	1708	2004
	Mamisonskiy pass	19						19	41	130	292	270	110	38	843	881
	Mestia	122	41	16	21	73	85	211	312	384	832	611	352	569	2491	3060
Discharge m^3/sec.	r.Chkheri - s.Kazbegi	1.25	0.72	0.45	0.29	0.24	0.24	0.78	1.35	2.27	3.42	3.09	1.41	0.57	2.31	1.30
	r.Terek - s.Kazbegi	12.8	8.44	7.70	7.15	6.64	7.02	11.5	33.6	66.6	60.8	44.5	32.6	8.75	47.6	24.94
	Pirikitel'skaya Alazan' s.Dartlo	4.59	3.36	2.74	2.25	1.79	2.13	3.44	6.84	17.5	16.4	10.2	6.84	2.9	11.5	6.51
	Pirikitel'skaya Alazan' s.Omaylo	7.57	4.44	3.73	3.45	3.12	3.21	4.51	9.58	24.1	24.8	18.5	11.4	4.29	17.7	9.87
	r.Sel'dy - s.Kurush	0.22	0.14	.0.12	0.092	0.085	0.086	0.25	1.24	1.78	1.43	1.19	0.44	0.13	1.22	0.58
	r.Mestiachala	4.10	1.97	1.91	1.89	1.89	1.95	3.04	5.96	11.9	25.6	37.3	13.2	2.39	18.8	9.23

Table 9.5.2. USSR - Hydrometeorological data.

Part 13 of 46 Caucasus

1966 - 1967	Name of station	Month 10	11	12	1	2	3	4	5	6	7	8	9	Winter 10-4	Summer 5-9	Year 10-9
Precipitation, mm	Terekol	13	50	114	107	5	54	58	84	161	160	212	57	401	674	1075
	Kazbegi v/g	22	19	61	221	13	99	112	124	224	236	295	113	547	992	1539
	Korul'dash	35	17	106	112	23	72	41	106	163	233	194	72	406	768	1174
	Mamisonskiy pass	3	102	68	28	14	54	83	46	53	149	204	127	352	579	931
	Mestia	4	21	79	86	7	35	37	72	130	172	203	36	289	613	902
Air temperature, °C	Terekol	5.0	- 1.7	- 4.2	- 6.3	-11.0	- 3.9	0.9	7.5	7.6	9.7	10.9	6.8	- 3.0	8.5	1.8
	Kazbegi v/g	- 3.2	- 3.5	-10.6	-13.9	-18.4	-13.4	- 8.7	- 2.7	- 2.4	1.1	2.4	- 1.0	-10.2	- 0.7	- 6.2
	Korul'dash	7.6	5.7	- 1.8	- 5.6	- 9.2	- 3.0	1.6	8.2	8.3	11.4	12.6	8.5	- 0.7	9.8	3.7
	Mamisonskiy pass	1.6	0.3	- 7.4	-10.9	-15.0	- 9.6	- 4.9	1.4	2.4	5.3	7.1	2.8	- 6.6	3.8	- 2.2
	Mestia	8.7	5.9	- 0.9	- 4.9	- 7.1	- 0.1	5.0	11.7	11.7	14.8	15.2	11.4	0.9	13.0	5.9
Degree days for temperature above 0, °C	Kazbegi v/g								4	9	40	75	23		151	151
	Korul'dash	234	171	10				60	253	253	353	391	256	475	1506	1981
	Mamisonskiy pass	52	38						66	93	164	220	95	90	638	728
	Mestia	269	178	14	3		23	148	364	349	461	472	342	635	1988	2623
Duration of sunshine	Terekol	193.4	97.6	47.8	49.3	147.2	171.5	176.3	181.5	178.7	160.2	173.1	171.6	883.1	865.1	1748.2
Discharge m^3/sec.	r.Chkheri - s.Kazbegi	0.81	0.97	1.20	0.65	0.37	0.45	0.67	1.24	1.38	1.99	5.89	1.60	0.73	2.4	1.43
	r.Terek - s.Kazbegi	27.2	15.4	11.2	8.62	7.23	8.21	11.2	34.0	44.1	63.6	72.5	31.5	11.26	48.1	27.89
	Pirikitel'skaya Alazan' s.Dartlo	4.88	3.69	2.74	2.10	1.95	1.60	1.99	6.94	10.2	17.9	17.9	11.2	2.71	12.8	6.92
	Pirikitel'skaya Alazan' s.Omaylo	7.58	6.62	4.70	3.53	3.09	2.52	3.16	9.00	16.3	26.9	25.7	16.1	4.45	18.8	10.4
	r.Sel'dy - s.Kurush	0.22	0.11	0.081	0.048	0.040	0.052	0.12	1.33	1.31	1.39	0.92	0.42	0.096	1.07	0.50
	r.Mestiachala	6.80	2.76	1.95	1.97	1.97	2.00	3.09	10.2	15.0	29.4	36.8	10.0	2.93	20.3	10.2

Table 9.5.2. USSR - Hydrometeorological data.

Part 14 of 46 Caucasus

1967 - 1968	Name of station	Month 10	11	12	1	2	3	4	5	6	7	8	9	Winter 10-4	Summer 5-9	Year 10-9
Precipitation, mm	Terekol	145	23	17	138	46	126	36	52	65	112	129	102	531	460	991
	Kazbegi v/g	50	110	180	223	104	107	50	112	152	50	112	100	824	526	1350
	Korul'dash	52	153	284	152	99	80	38	33	66	53	143	115	858	410	1268
	Mamisonskiy pass	45	53	86	90	37	77	42	59	115	39	84	55	430	352	782
	Mestia	26	70	150	99	100	127	34	30	69	51	115	76	606	341	947
Air temperature, °C	Terekol	4.2	1.2	- 4.6	- 8.1	- 6.6	- 4.6	2.5	7.8	7.9	12.0	10.6	9.0	- 2.3	9.5	2.0
	Kazbegi v/g	- 3.8	- 9.8	-12.0	-16.2	-13.2	-13.2	- 6.1	- 1.6	- 1.4	3.8	2.6	0.6	-10.6	- 0.8	- 5.8
	Korul'dash	6.7	- 0.4	- 4.8	- 7.8	- 5.5	- 4.1	3.6	8.6	9.0	13.2	11.8	10.1	- 1.8	10.5	3.4
	Mamisonskiy pass	0.9	- 6.0	- 9.7	-13.4	-10.9	- 9.8	- 2.6	2.4	2.9	7.8	6.7	5.0	- 7.3	5.0	- 2.2
	Mestia	8.5	1.4	- 3.9	- 6.2	- 5.2	- 1.5	6.0	12.4	12.9	16.5	14.6	12.3	- 0.1	13.7	5.6
Degree days for temperature above 0, °C	Kazbegi v/g								3	12	120	83	40		258	258
	Korul'dash	209	74				1	109	267	271	408	368	304	393	1618	2011
	Mamisonskiy pass	51	11					1	75	93	242	208	148	63	766	829
	Mestia	264	101	2		2	8	181	385	386	512	452	367	558	2102	2660
Duration of sunshine	Terekol	135.6	94.8	84.5	53.6	117.0	178.1	182.5	207.2	205.5	250.8	204.9	172.1	846.1	1040.5	1886.6
Discharge m^3/sec.	r.Chkheri - s.Kazbegi	0.85	0.66	0.67	0.58	0.54	0.55	0.62	1.07	1.21	3.02	1.47	1.32	0.64	1.62	1.05
	r.Terek - s.Kazbegi	18.4	13.2	10.6	9.09	8.67	8.61	15.6	37.6	55.3	61.6	36.4	21.3	12.02	42.4	24.69
	Pirikitel'skaya Alazan' s.Dartlo	5.69	4.06	3.46	3.70	3.24	2.98	4.61	12.7	15.1	21.6	17.9	9.12	3.96	15.3	8.68
	Pirikitel'skaya Alazan' s.Omaylo	12.0	7.15	5.79	5.32	4.88	3.68	8.01	18.4	24.6	33.4	26.8	13.7	6.69	23.3	13.6
	r.Sel'dy - s.Kurush	0.21	0.14	0.094	0.084	0.061	0.079	0.33	0.69	0.91	1.43	0.89	0.35	0.14	0.84	0.44
	r.Mestiachala	3.15	2.05	1.97	1.97	2.00	2.09	5.05	12.1	16.2	30.2	24.8	15.7	2.61	18.9	9.78

Table 9.5.2. USSR - Hydrometeorological data

Part 15 of 46 Caucasus

1968 - 1969	Name of station	Month 10	11	12	1	2	3	4	5	6	7	8	9	Winter 10-4	Summer 5-9	Year 10-9
Precipitation, mm	Terekol	92	38	71	6	16	68	65	92	70	116	79	88	356	445	801
	Kazbegi v/g	161	39	63	240	40	79	79	105	92	79	41	30	485	347	832
	Korul'dash	241	30	140	61	156	101	63	92	112	49	73	78	792	404	1196
	Mamisonskiy pass	78	33	13	23	35	16	22	31	71	47	73	58	220	280	500
	Mestia	154	38	58	24	32	69	49	45	79	39	42	56	424	261	685
Air temperature, °C	Terekol	2.3	- 0.2	- 3.3	- 9.3	- 7.1	- 2.8	0.81	7.0	11.6	10.1	12.2	7.1	- 2.8	9.6	2.4
	Kazbegi v/g	- 3.9	- 7.8	-12.0	-15.4	-14.8	-11.8	- 8.4	- 3.0	1.7	1.8	4.6	- 1.4	-10.6	0.7	- 5.8
	Korul'dash	4.8	2.5	- 3.5	- 6.9	- 6.0	- 2.4	1.5	7.8	12.2	12.0	14.1	8.2	- 1.4	10.9	3.7
	Mamisonskiy pass	- 0.6	- 3.7	- 9.1	-12.8	-11.4	- 6.3	- 4.5	1.4	6.5	5.8	8.7	2.4	- 7.0	5.0	- 2.0
	Mestia	7.1	4.4	- 3.0	- 5.1	- 3.7	0.5	4.9	11.2	15.2	15.4	17.0	11.6	0.7	14.1	6.3
Degree days for temperature above 0, °C	Kazbegi v/g	2							6	60	68	143	23	2	443	445
	Korul'dash	149	75	6			10	70	244	366	372	437	244	310	1663	1973
	Mamisonskiy pass	31						10	69	196	179	270	102	41	816	857
	Mestia	219	133	16		1	5	151	345	457	479	524	348	525	2153	2678
Duration of sunshine	Terekol	152.5	108.0	53.8	91.6	130.4	147.7	205.4	223.0	253.9	236.9	252.4	202.7	889.4	1168.9	2058.3
Discharge m^3/sec.	r.Chkheri - s. Kazbegi	0.74	0.56	0.42	0.38	0.33	0.35	0.49	1.13	2.34	2.52	2.96	1.44	0.47	2.26	1.21
	r.Terek - s.Kaztegi	16.9	15.2	12.1	9.13	8.38	8.53	13.5	43.3	43.9	36.6	30.2	19.0	11.96	34.6	21.39
	Pirikitel'skaya Alazan' s.Dartlc	6.14	4.16	2.71	2.33	1.97	2.17	3.45	9.59	13.9	12.1	10.3	5.53	3.28	10.3	6.2
	Pirikitel'skaya Alazan' s.Omaylc	8.27	5.50	4.74	3.69	3.42	2.95	3.74	13.1	18.1	15.9	13.9	8.89	4.61	14.0	8.50
	r.Sel'dy - s.Kurush	0.16	0.13	0.061	0.059	0.053	0.062	0.17	0.88	0.91	0.98	0.87	0.33	0.099	0.79	0.39
	r.Mestiachala	6.14	4.66	5.75	5.99	6.09	6.73	8.66	23.2	31.3	30.5	34.7	18.3	6.29	27.6	15.2

Table 9.5.2. USSR - Hydrometeorological data.

Part 16 of 46 Caucasus

1969 - 1970	Name of station	Month 10	11	12	1	2	3	4	5	6	7	8	9	Winter 10-4	Summer 5-9	Year 10-9
Precipitation, mm	Terekol	86	78	56	12	140	120	74	124	66	106	136	201	566	633	1199
	Kazbegi v/g	73	60	71	29	115	146	146	197	110	154	198	152	640	811	1451
	Korul'dash	87	53	68	54	125	81	79	96	75	93	155	95	547	514	1061
	Mamisonskiy pass	50	44	74	16	101	52	57	111	67	95	89	68	395	430	824
	Mestia	77	36	78	33	123	109	79	126	99	83	169	151	535	628	1163
Air temperature, ^{o}C	Terekol	1.9	0.0	- 7.4	- 5.4	- 3.8	- 2.1	4.0	6.1	8.7	11.5	10.4	7.0	- 1.8	8.7	4.4
	Kazbegi v/g	- 5.7	- 5.6	-10.8	-12.5	-11.8	- 9.8	- 5.1	3.8	- 0.4	2.7	2.3	- 0.6	- 8.7	0.04	- 5.1
	Korul'dash	3.2	1.0	- 2.0	- 4.2	- 2.4	- 0.4	5.3	6.9	10.0	13.5	11.6	8.8	0.1	10.2	4.3
	Mamisonskiy pass	- 3.1	- 3.5	- 7.8	-10.0	- 9.6	- 7.1	- 1.0	0.3	3.5	7.3	6.3	3.6	- 6.0	4.2	- 1.8
	Mestia	5.6	1.6	- 0.7	- 3.9	- 1.9	1.0	7.8	10.0	13.3	16.3	14.6	11.4	1.3	13.1	6.3
Degree days for temperature above 0, ^{o}C	Kazbegi v/g	1							6	60	68	143	23	1	300	301
	Korul'dash	118	55	1	3	3	25	160	215	299	420	360	265	365	1559	1924
	Mamisonskiy pass	18						26	35	107	228	194	110	44	674	718
	Mestia	183	53	17	7	78	45	233	316	398	504	453	341	616	2012	2628
Duration of sunshine	Terekol	152.0	73.9	67.5	81.6	93.1	144.5	208.0	195.8	237.3	231.9	229.5	179.7	820.6	1074.2	1894.8
Discharge m^3/sec.	r.Chkheri - s.Kazbegi	0.57	0.35	0.30	0.32	0.29	0.29	0.53	0.77	1.60	2.99	2.78	1.70	0.38	1.97	1.04
	r.Terek - s.Kazbegi	14.9	12.0	10.2	8.80	7.50	8.00	22.0	34.6	52.7	48.4	45.5	25.2	11.9	41.3	24.1
	Pirikital'skaya Alazan' s.Dartlo	4.24	2.88	2.18	2.10	1.95	1.74	3.16	9.28	11.9	18.2	17.8	10.4	2.61	13.5	7.15
	Pirikitel'skaya Alazan' s.Omaylo	6.56	5.40	4.40	3.53	3.09	2.61	3.51	10.4	17.2	26.7	25.5	15.5	4.10	19.0	10.3
	r.Sel'dy - s.Kurush	0.19	0.10	0.075												
	r.Mestiachala	5.73	2.18	2.07	2.05	1.98	2.12	7.71	12.4	23.3	39.4	33.3	19.0	3.41	25.5	12.6

Table 9.5.2. USSR - Hydrometeorological data.

Part 17 of 46 Pamir-Alai and Western Tien-Shan

Mean values	Hydrological station m a.sl.	Month 10	11	12	1	2	3	4	5	6	7	8	9	Winter 10-4	Summer 5-9	Year 10-9	Observation period
Precipitation, mm	Pskem, 1256	75.2	98.5	113.7	90.4	85.5	118.4	106.0	65.3	32.2	20.3	10.0	11.4	689.7	139.2	828.9	1933-1971
	Khaydarkan, 1968	41.4	45.3	25.5	26.5	39.9	71.5	75.9	81.2	52.2	24.9	12.8	6.7	326.0	177.8	503.8	1942-1969
	Glacier Severtsova, 2780	46.6	52.9	49.1	51.9	58.0	89.7	128.9	83.4	25.8	28.2	14.1	12.9	477.1	164.4	641.5	1960-1971
	Oygaing, 2151	97.7	95.9	109.6	62.7	66.8	113.5	87.3	71.8	47.9	39.7	11.6	30.4	633.5	201.4	834.9	1963-1971
Air temperature, °C	Pskem	9.6	2.8	- 1.3	- 3.7	- 1.9	2.2	9.1	14.5	18.4	21.9	21.5	16.4	2.4	18.5	9.1	1963-1971
	Khaydarkan	6.7	0.2	- 3.5	- 5.2	- 4.2	0.6	7.0	12.0	15.9	19.0	18.1	13.2	0.2	15.6	6.6	1942-1969
	Glacier Severtsova	2.8	- 2.1	- 5.3	- 7.5	- 6.9	- 2.4	1.8	6.7	11.4	13.4	12.7	8.1	- 2.8	10.4	2.7	1960-1971
	Oygaing	2.6	- 4.3	- 7.9	-11.2	- 9.6	- 4.4	1.1	7.3	12.0	14.1	14.8	8.4	- 4.8	11.3	1.9	1963-1971
Degree days for temperature above 0, °C	Pskem	292	106	28	11	24	89	270	439	541	660	651	474	820	2765	3585	
	Khaydarkan	204	55	11	6	9	60	214	373	472	590	560	395	559	2390	2949	
	Glacier Severtsova	98	21	1	1	1	24	77	207	336	414	402	245	223	1604	1827	
	Oygaing	98	6	0	0	0	6	56	227	360	436	459	290	166	1772	1938	
Water equivalent of the snow cover, mm	Pskem		12	101	173	205	76										
	Glacier Severtsova		32	95	124	111	182	42									
	Oygaing		28	198	204	220	291	114	8								
Runoff, mm	Pskem, mouth, S=2830	40	30	29	24	21	31	69	135	192	170	111	64	244	672	916	1932-1970
	Oygaing, mouth, S=1010	38	26	21	18	15	16	34	103	188	192	126	63	168	672	840	1933-1970
	Maydantal, mouth, S=471	37	31	25	22	17	26	44	139	264	244	151	70	202	868	1070	1933-1970
	Aksu, kishlak Khazarnau, S=845	18	14	13	12	11	22	43	65	102	90	51	27	133	335	468	1927-1970
	Sokh, kishlak Sarykanda, S=2480	32	18	15	13	10	11	13	30	69	131	109	63	112	402	514	1926-1970
	Oygaing, above the mouth of r. Koksu, S=466	49	29	25	23	17	18	24	80	204	215	154	76	185	729	914	1963-1970
	Koksu, 1.3 km above the mouth, S=188	41	31	26	24	17	16	24	84	97	198	113	61	179	553	732	1963-1970

Table 9.5.2. USSR - Hydrometeorological data.

Part 18 of 46 Pamir Alai and Western Tien-Shan

1959 - 1960	Hydrological station m a.sl.	Month 10	11	12	1	2	3	4	5	6	7	8	9	Winter 10-4	Summer 5-9	Year 10-9
Precipitation, mm	Pskem	53.6	172.0	166.3	98.8	130.5	148.1	66.9	143.3	6.8	4.8	0.0	16.1	836.2	171.0	1007.2
	Khaydarkan	34.7	33.4	21.9	29.6	56.3	74.5	51.5	190.7	19.7	9.4	0.0	5.1	301.9	224.9	526.8
Air temperature, °C	Pskem	10.6	0.9	- 3.2	- 2.4	0.4	- 1.4	5.8	11.7	18.9	20.6	21.8	15.2	1.5	17.6	8.2
	Khaydarkan	7.6	- 0.6	- 4.2	- 3.1	0.0	- 2.6	4.6	9.2	17.0	18.8	19.2	12.8	0.2	15.4	6.6
Degree days for temperature above 0, °C	Pskem	329	62	4	12	37	24	193	363	567	639	676	456	661	2710	3371
	Khaydarkan	236	26	1	15	29	16	169	285	510	583	595	384	492	2357	2849
Water equivalent of the snow cover, mm	Pskem		51	170	267	312	369									
Runoff, mm	Pskem, mouth	54	42	36	31	31	38	70	198	276	272	146	77	302	969	1271
	Oygaing, mouth	50	27	20	17	15	15	23	117	267	244	131	65	167	824	991
	Maydantal, mouth	52	26	23	22	17	18	31	137	345	300	148	85	189	1015	1204
	Aksu, kishlak Khazarnau	16	12	11	10	10	12	24	73	111	92	41	22	95	339	434
	Sokh, Sarykanda	40	22	17	14	12	12	12	25	63	133	138	60	129	419	548

Table 9.5.2. USSR - Hydrometeorological data.

Part 19 of 46 Pamir Alai and Western Tien-Shan

1960 - 1961	Hydrological station m a.sl.	Month 10	11	12	1	2	3	4	5	6	7	8	9	Winter 10-4	Summer 5-9	Year 10-9
Precipitation, mm	Pskem	72.3	61.4	28.2	38.6	54.0	78.9	44.1	21.3	22.2	9.5	8.8	0.5	377.5	62.3	439.8
	Khaydarkan	23.3	71.8	5.7	23.9	19.0	52.3	39.5	58.6	14.1	19.3	18.8	1.8	235.5	112.6	348.1
	Glacier Severtsova	19.0	72.6	16.7	39.0	32.6	81.6	144.4	60.9	24.6	3.0	47.2	0.3	405.9	136.0	541.9
Air temperature, °C	Pskem	10.5	2.7	- 0.3	- 0.9	- 3.2	2.3	10.9	17.8	19.4	23.4	21.7	17.2	3.1	19.9	10.1
	Khaydarkan	7.4	0.2	- 2.7	- 4.1	- 7.7	- 0.5	7.2	14.8	17.1	20.6	17.9	14.4	- 0.1	17.0	7.0
	Glacier Severtsova	3.2	- 3.4	- 4.4	- 6.8	-10.6	- 3.8	1.1	9.6	11.1	15.0	12.1	9.6	- 3.5	11.5	2.7
Degree days for temperature above 0, °C	Pskem	326	89	27	24	1	101	327	552	582	725	673	516	895	3048	3943
	Khaydarkan	229	45	4	8	0	55	216	459	513	639	555	432	557	2598	3155
	Glacier Severtsova	99	11	0	3	0	13	59	298	333	465	375	298	185	1769	1954
Water equivalent of the snow cover, mm	Pskem			40	45	86										
	Glacier Severtsova		84	96	114	167	208	142								
Runoff, mm	Pskem, mouth	48	35	29	24	18	25	51	115	124	119	96	58	230	512	742
	Oygaing, mouth	40	26	21	18	14	16	23	87	121	134	103	58	158	503	661
	Maydantal, mouth	42	28	25	23	19	19	32	125	158	165	132	67	188	647	835
	Aksu, kishlak Khazarnau	16	13	12	10	8	15	38	58	81	68	37	24	112	268	380
	Sokh, kishlak Sarykanda	28	19	16	14	11	12	12	42	74	136	118	73	112	443	555

Table 9.5.2. USSR - Hydrometeorological data.

Part 20 of 46 Pamir Alai and Western Tien-Shan

1961 - 1962	Hydrological station m a.sl.	Month 10	11	12	1	2	3	4	5	6	7	8	9	Winter 10-4	Summer 5-9	Year 10-9
Precipitation, mm	Pskem	54.3	67.4	78.5	70.7	50.8	26.3	143.7	68.9	25.4	4.4	0.0	4.4	491.7	103.1	594.8
	Khaydarkan	11.8	35.6	13.8	25.2	56.7	26.9	73.6	31.9	19.2	9.3	7.9	13.3	243.6	81.6	325.2
	Glacier Severtsova	20.8	43.3	50.4	23.8	81.1	67.6	182.0	90.2	39.3	1.7	1.8	5.9	469.0	138.9	607.9
Air temperature, ^{o}C	Pskem	7.2	3.6	- 0.8	- 4.3	0.1	7.9	9.2	13.8	18.1	23.2	21.9	15.0	3.3	18.4	9.6
	Khaydarkan	5.2	0.9	- 2.8	- 5.5	- 1.9	4.8	6.6	11.0	15.9	21.0	18.5	12.1	1.0	15.7	7.2
	Glacier Severtsova	1.5	- 2.0	- 6.4	- 7.6	- 5.0	- 0.2	1.5	5.2	9.4	14.3	12.2	6.7	- 2.6	9.6	2.5
Degree days for temperature above 0, ^{o}C	Pskem	223	132	9	3	35	245	276	428	543	719	679	450	923	2819	3742
	Khaydarkan	162	71	1	1	19	156	198	341	477	651	574	363	608	2406	3014
	Glacier Severtsova	69	26	0	0	2	39	67	161	282	443	378	204	203	1468	1671
Water equivalent of the snow cover, mm	Pskem			92	187	161										
	Glacier Severtsova			80	97	144	77									
Runoff, mm	Pskem, mouth	37	27	25	22	20	33	57	99	168	165	104	60	221	596	817
	Oygaing, mouth	36	23	18	13	10	16	32	73	164	181	124	65	148	607	755
	Maydantal, mouth	31	25	23	18	15	29	58	115	213	238	159	90	199	815	1014
	Aksu, kishlak Khazarnau	14	11	10	10	9	13	26	48	103	98	45	24	93	318	411
	Sokh, Sarykanda	28	18	15	12	10	10	10	16	59	122	124	62	103	383	486

Table 9.5.2. USSR - Hydrometeorological data.

Part 21 of 46 Pamir Alai and Western Tien-Shan

1962 - 1963	Hydrological station m a.sl.	Month 10	11	12	1	2	3	4	5	6	7	8	9	Winter 10-4	Summer 5-9	Year 10-9
Precipitation, mm	Pskem	98.1	173.8	130.1	21.5	67.5	78.9	160.0	77.6	52.0	12.3	19.3	1.4	730.8	162.6	893.4
	Khaydarkan	14.9	63.4	16.5	2.1	45.3	54.9	142.4	79.1	57.5	16.2	0.0	14.0	339.5	166.8	506.3
	Glacier Severtsova	37.6	55.1	69.1	8.0	56.3	70.5	125.1	165.0	49.4	13.9	4.0	8.0	421.7	240.3	662.0
Air temperature, °C	Pskem	9.1	- 2.0	- 2.2	- 0.4	1.8	4.5	11.0	14.0	19.0	20.6	19.8	15.8	3.1	19.8	9.2
	Khaydarkan	6.0	- 3.6	- 3.6	- 2.0	- 0.5	1.7	8.6	11.7	17.0	18.5	17.6	13.7	0.9	15.7	7.1
	Glacier Severtsova	2.2	- 5.8	- 5.7	- 3.9	- 4.0	- 2.3	3.2	6.4	11.6	12.4	11.9	8.2	- 2.3	10.1	2.8
Degree days for temperature above 0, °C	Pskem	282	58	27	44	74	155	330	434	570	639	614	474	970	2731	3701
	Khaydarkan	186	36	8	22	41	91	258	363	510	574	546	411	642	2404	3046
	Glacier Severtsova	73	15	3	5	6	31	104	201	348	384	369	246	237	1548	1785
Water equivalent of the snow cover, mm	Pskem		39	161	162	133										
	Glacier Severtsova		50	108	101	158	122									
Runoff, mm	Pskem, mouth	38	33	29	26	25	36	90	137	236	177	106	59	277	715	992
	Oygaing, mouth	37	24	19	15	11	13	47	95	222	185	114	61	166	677	843
	Maydantal, mouth	44	31	27	19	13	17	73	144	308	255	157	72	224	936	1160
	Aksu, kishlak Khazarnau	16	13	11	10	10	18	38	80	136	91	47	25	116	379	495
	Sokh, Sarykanda	23	17	15	13	10	10	12	26	85	123	116	59	100	409	509

Table 9.5.2. USSR - Hydrometeorological data.

Part 22 of 46 Pamir Alai and Western Tien-Shan

1963 - 1964	Hydrological station m a.sl.	Month 10	11	12	1	2	3	4	5	6	7	8	9	Winter 10-4	Summer 5-9	Year 10-9
Precipitation, mm	Pskem	44.5	110.4	196.9	45.4	91.9	145.8	141.4	36.8	33.3	35.7	1.2	23.5	776.3	130.5	906.8
	Khaydarkan	20.3	41.8	17.9	21.9	68.7	82.2	122.5	72.6	31.8	46.4	2.8	2.0	375.3	155.6	530.9
	Glacier Severtsova	36.2	75.1	42.6	42.2	90.7	81.1	170.9	107.3	14.3	31.7	0.0	3.8	538.8	157.1	695.9
	Oygaing, mouth	59.8	106.4	204.2	22.5	59.2	119.2	138.4	67.4	69.0	27.6	0.7	28.2	709.7	192.9	902.6
Air temperature, ^{o}C	Pskem	11.4	3.5	- 1.1	- 7.8	- 2.2	3.1	7.9	13.7	18.3	20.2	21.0	15.2	2.1	17.7	8.6
	Khaydarkan	9.6	1.2	- 1.6	-10.0	- 4.3	1.9	5.8	11.2	16.4	18.1	18.0	13.5	0.4	15.4	6.6
	Glacier Severtsova	5.1	- 1.2	- 3.9	-12.9	- 7.0	- 0.6	0.5	5.4	10.7	12.8	12.3	7.7	- 2.9	9.8	2.4
	Oygaing	4.9	- 4.6	- 6.9	-14.1	- 8.9	- 3.5	- 0.1	6.0	11.6	13.4	14.7	9.1	- 4.8	11.0	1.8
Degree days for temperature above 0, ^{o}C	Pskem	353	114	28	0	4	110	237	425	549	626	661	456	846	2717	3563
	Khaydarkan	298	64	15	4	0	90	175	347	492	561	558	405	646	2363	3009
	Glacier Severtsova	170	12	0	0	0	39	37	168	321	397	381	231	258	1498	1756
	Oygaing	152	4	0	0	0	11	22	186	348	415	456	273	189	1678	1867
Water equivalent of the snow cover, mm	Pskem			173	188	270										
	Glacier Severtsova		32	59	118											
	Oygaing			313	330	432										
Runoff, mm	Pskem, mouth	40	39	31	26	22	40	78	155	232	222	124	63	276	796	1072
	Oygaing, mouth	34	24	21	17	14	17	29	106	221	237	140	62	156	766	922
	Maydantal, mouth	38	27	24	22	18	21	45	147	281	312	178	72	195	990	1185
	Aksu, kishlak Khazarnau	17	14	12	10	10	26	78	86	121	105	52	27	167	291	558
	Sokh, Sarykanda	28	18	15	12	11	11	13	25	70	143	131	70	108	439	547
	Oygaing, above the mouth of r. Koksu	55	33	24	31	19	22	20	79	225	259	162	80	204	805	1009
	Koksu, 1.3 km above the mouth	39	28	22	17	13	12	17	70	251	276	142	53	148	792	940

Table 9.5.2. USSR - Hydrometeorological data.

Part 23 of 46 Pamir Alai and Western Tien-Shan

1964 - 1965	Hydrological station m a.sl.	Month 10	11	12	1	2	3	4	5	6	7	8	9	Winter 10-4	Summer 5-9	Year 10-9
Precipitation, mm	Pskem	6.8	25.9	49.1	110.5	96.0	57.5	47.6	27.3	44.1	6.4	16.1	29.5	393.4	123.4	516.8
	Khaydarkan	7.6	17.4	9.6	20.4	47.2	28.3	40.0	83.8	25.0	17.9	21.5	5.2	170.5	153.4	323.9
	Glacier Severtsova	11.1	43.6	32.4	62.2	54.8	55.5	80.2	39.4	13.1	27.1	0.3	6.5	339.8	86.4	426.2
	Oygaing,	5.0	31.3	55.6	70.3	88.5	75.4	22.4	48.7	29.9	12.2	21.5	78.2	348.5	190.5	539.0
Air temperature, °C	Pskem	7.6	4.1	- 3.2	- 2.1	- 3.8	0.8	10.1	16.1	18.8	23.2	21.4	15.6	1.9	19.0	9.0
	Khaydarkan	5.2	1.4	- 5.3	- 3.5	- 5.6	- 1.2	6.9	13.2	16.4	19.6	18.0	12.6	- 0.3	16.0	6.5
	Glacier Severtsova	0.9	- 2.2	- 7.6	- 6.1	- 8.9	- 4.2	1.0	7.4	10.4	13.7	12.5	7.1	- 3.9	10.2	2.0
	Oygaing	1.3	- 3.8	- 9.2	- 9.2	-10.6	- 6.3	1.2	9.6	11.9	16.0	14.4	8.6	- 5.2	12.1	1.2
Degree days for temperature above 0, °C	Pskem	236	136	16	22	3	60	303	499	564	719	663	468	776	2913	3689
	Khaydarkan	162	80	6	16	0	24	207	409	492	608	558	378	495	2445	2940
	Glacier Severtsova	66	22	0	4	0	9	49	229	312	425	388	216	150	1570	1720
	Oygaing	75	4	0	0	0	4	53	298	357	496	446	258	136	1855	1991
Water equivalent of the snow cover, mm	Pskem			24	97	198										
	Glacier Severtsova		48	71	124		257	30								
	Oygaing		32	103	194	280	323	34								
Runoff, mm	Pskem, mouth	41	32	27	24	21	26	40	108	121	105	84	41	211	459	670
	Oygaing, mouth	38	27	21	18	15	16	24	92	111	113	100	46	159	462	621
	Maydantal, mouth	42	30	27	23	19	21	36	147	200	184	142	54	198	727	925
	Aksu kishlak Khazarnau	18	13	11	11	10	11	20	39	66	56	35	19	94	215	309
	Sokh, Sarykanda	30	20	16	14	11	11	10	29	66	139	114	53	112	401	513
	Oygaing, above the mouth of r. Koksu	45	33	33	24	17	17	20	90	129	151	133	50	189	553	742
	Koksu, 1.3 km above the mouth	36	27	23	21	18	18	19	76	146	122	94	38	162	476	638

Table 9.5.2. USSR - Hydrometeorological data.

Part 24 of 46 Pamir Alai and Western Tien-Shan

1965 - 1966	Hydrological station	Month												Winter	Summer	Year
	m a.sl.	10	11	12	1	2	3	4	5	6	7	8	9	10-4	5-9	10-9
Precipitation, mm	Pskem	179.3	177.2	27.8	138.3	70.5	174.7	41.3	71.8	4.5	15.4	5.1	18.7	809.1	115.5	924.6
	Khaydarkan	78.1	95.9	6.3	5.6	71.0	140.2	54.4	93.8	21.5	7.1	52.6	3.3	451.5	178.3	529.8
	Glacier Severtsova	55.9	63.5	19.6	51.8	78.3	163.9	41.3	118.9	10.0	47.6	65.0	27.1	474.3	268.6	742.9
	Oygaing	212.1	133.8	18.5	96.3	62.1	160.5	69.5	99.1	13.0	41.3	20.8	34.4	752.8	208.6	961.4
Air temperature, °C	Pskem	10.2	4.6	1.4	0.3	1.7	2.9	9.7	13.4	20.6	21.8	22.1	16.8	4.4	18.9	10.5
	Khaydarkan	8.3	2.8	- 1.3	0.6	- 0.1	0.1	6.2	10.7	18.0	19.0	18.1	13.9	2.4	15.9	8.0
	Glacier Severtsova	4.5	0.0	- 4.1	- 2.9	- 3.6	- 3.5	0.9	6.2	13.1	13.2	12.2	8.2	- 1.2	10.6	3.7
	Oygaing	2.8	- 3.9	- 8.5	- 7.8	- 6.6	- 4.4	0.0	6.2	13.5	14.5	14.9	10.5	- 4.1	11.9	2.6
Degree days for temperature above 0, °C	Pskem	316	141	51	32	51	108	291	415	618	676	685	504	990	2898	3888
	Khaydarkan	257	90	10	42	15	48	188	332	540	589	561	417	650	2439	3089
	Glacier Severtsova	144	30	1	2	0	12	70	193	394	408	473	246	259	1714	1973
	Oygaing	112	1	0	0	0	0	41	193	405	450	465	315	154	1828	1982
Water equivalent of the snow cover, mm	Pskem			24	130	144										
	Glacier Severtsova			44	57	133	276									
	Oygaing		142	163	262	311	450	181								
Runoff, mm	Pskem, mouth	42	49	36	29	27	41	65	138	255	186	122	68	289	769	1058
	Oygaing, mouth	38	32	25	21	18	20	42	115	242	200	126	68	196	751	947
	Maydantal, mouth	46	42	34	29	23	26	59	156	322	247	156	76	259	957	1216
	Aksu, kishlak Khazarnau	13	11	9	9	9	23	29	55	120	72	39	22	103	308	411
	Sokh, Sarykanda	24	18	14	12	9	10	12	29	126	142	140	66	99	503	602
	Oygaing, above the mouth of r. Koksu	35	29	25	21	18	22	25	73	253	247	167	89	175	829	1004
	Koksu, 1.3 km above the mouth	33	27	27	18	14	16	25	78	207	96	76	56	160	513	673

Table 9.5.2. USSR - Hydrometeorological data.

Part 25 of 46 Pamir Alai and Western Tien-Shan

1966 - 1967	Hydrological station m a.sl.	Month 10	11	12	1	2	3	4	5	6	7	8	9	Winter 10-4	Summer 5-9	Year 10-9
Precipitation, mm	Pskem	94.5	28.5	187.7	82.0	93.6	21.6	101.0	77.8	39.3	47.0	0.0	24.8	596.9	188.9	785.8
	Khaydarkan	51.6	20.9	37.7	12.8	48.4	39.3	59.8	94.8	38.0	38.1	0.0	12.7	270.5	183.6	454.1
	Glacier Severtsova	76.4	0.8	57.3	26.6	68.7	43.6	104.6	87.3	36.2	2.0	0.0	38.3	378.0	163.8	541.8
	Oygaing	63.7	12.6	159.8	75.8	78.3	29.3	42.2	39.2	77.4	60.7	0.0	47.8	461.7	225.1	686.8
Air temperature, °C	Pskem	8.7	2.4	- 1.2	- 4.0	- 3.2	4.0	9.8	13.4	17.9	20.7	21.4	16.4	2.3	19.9	8.8
	Khaydarkan	6.1	0.1	- 3.5	- 6.0	- 4.8	1.2	6.8	10.3	15.9	18.3	18.2	13.8	0.0	15.3	1.3
	Glacier Severtsova	1.4	- 1.9	- 5.5	- 8.2	- 7.3	- 3.2	1.2	5.2	10.5	13.5	13.5	9.0	- 3.4	10.3	2.3
	Oygaing	2.0	- 4.0	- 7.5	-12.0	-10.1	- 4.2	1.3	7.7	11.6	13.6	15.0	9.9	- 4.9	11.6	1.9
Degree days for temperature above 0, °C	Pskem	270	92	42	17	11	137	294	415	537	642	663	486	863	2743	3606
	Khaydarkan	189	56	13	9	4	73	212	319	477	567	564	414	556	2341	2897
	Glacier Severtsova	59	12	5	0	0	4	88	159	314	418	417	269	168	1577	1745
	Oygaing	78	0	0	0	0	6	74	240	349	422	465	296	158	1772	1930
Water equivalent of the snow cover, mm	Pskem			84	114	162										
	Glacier Severtsova			84	82	132	141									
	Oygaing			187		333										
Runoff, mm	Pskem, mouth	43	29	29	25	20	28	60	131	202	155	111	65	234	664	898
	Oygaing, mouth	41	26	22	16	13	15	30	82	164	148	122	63	163	579	742
	Maydantal, mouth	50	38	37	25	18	18	42	127	253	215	160	78	228	833	1061
	Aksu, kishlak Khazarnau	16	12	10	9	8	10	26	41	80	65	38	20	91	244	335
	Sokh, Sarykanda	26	17	14	11	9	10	11	19	60	110	135	64	98	388	486
	Oygaing, above the mouth of r. Koksu	45	28	24	21	16	15	23	67	147	140	130	78	172	562	734
	Koksu, 1.3 km above the mouth	42	33	28	25	21	21	22	64	140	135	102	53	192	494	686

Table 9.5.2. USSR - Hydrometeorological data.

Part 26 of 46 Pamir Alai and Western Tien-Shan

1967 - 1968	Hydrological station m a.sl.	Month 10	11	12	1	2	3	4	5	6	7	8	9	Winter 10-4	Summer 5-9	Year 10-9
Precipitation, mm	Pskem	128.9	67.9	169.8	85.5	28.9	196.2	165.4	42.6	50.3	22.3	27.5	0.3	842.6	143.0	985.6
	Khaydarkan	39.5	41.5	22.1	34.0	6.4	78.3	98.0	67.2	61.5	22.4	3.0	0.0	319.8	154.1	473.9
	Glacier Severtsova	54.5	90.3	42.9	79.7	12.7	135.1	128.4	95.3	26.8	5.2	12.4	0.0	543.6	139.7	683.3
	Oygaing	87.1	56.9	109.2	45.1	25.8	182.8	200.9	44.6	100.2	29.4	30.8	0.9	707.8	205.9	913.7
Air temperature, °C	Pskem	8.0	3.9	0.3	- 2.4	- 2.9	3.4	8.8	13.2	17.5	21.2	21.7	16.8	2.7	18.1	9.1
	Khaydarkan	6.0	1.8	- 1.4	- 4.2	- 5.1	1.8	6.9	10.2	15.7	18.8	19.1	14.6	0.8	15.7	7.0
	Glacier Severtsova	1.6	- 2.8	- 5.1	- 7.9	- 8.0	- 2.6	1.9	4.6	10.8	13.7	13.3	9.1	- 3.3	10.3	2.4
	Oygaing	1.6	- 4.5	- 7.1	-10.0	-10.6	- 3.8	0.6	6.9	10.7	14.1	15.2	10.2	- 4.8	11.4	1.9
Degree days for temperature above 0, °C	Pskem	240	120	46	5	23	112	266	409	525	657	673	504	812	2768	3580
	Khaydarkan	186	63	17	6	7	77	208	316	471	583	592	438	564	2400	2964
	Glacier Severtsova	64	8	0	0	0	16	81	145	324	426	412	274	169	1581	1750
	Oygaing	60	2	0	0	0	3	40	213	322	437	472	305	105	1749	1854
Water equivalent of the snow cover, mm	Pskem			113	198	176										
	Glacier Severtsova		104	130	178	225										
	Oygaing			215			405	127								
Runoff, mm	Pskem, Mulala	54	42	36	32	27	46	101	160	239	241	143	67	338	850	1188
	Oygaing, mouth	38	24	20	17	14	17	55	112	211	234	148	58	185	763	948
	Maydantal, mouth	42	32	29	28	29	26	67	150	290	290	172	67	253	969	1222
	Aksu, kishlak Khazarņau	13	11	10	8	8	17	12	58	123	111	57	24	109	373	482
	Sokh, Sarykanda	23	16	13	11	9	9	11	19	69	155	148	54	92	445	537
	Oygaing, above the mouth of r. Koksu	48	31	23	20	17	15	28	86	216	259	177	58	182	796	978
	Koksu, 1.3 km above the mouth	35	25	21	19	16	16	17	95	201	250	158	65	149	769	918

Table 9.5.2. USSR - Hydrometeorological data.

Part 27 of 46 Pamir Alai and Western Tien-Shan

1968 - 1969	Hydrological station m a.sl.	Month 10	11	12	1	2	3	4	5	6	7	8	9	Winter 10-4	Summer 5-9	Year 10-9
Precipitation, mm	Pskem	68.3	248.2	190.3	282.2	121.5	346.0	115.5	156.9	31.6	126.3	11.9	42.4	1372.0	369.1	1741.1
	Khaydarkan	16.1	15.1	74.5	56.3	52.3	120.3	109.8	151.2	66.7	62.0	5.8	39.8	444.4	325.5	769.9
	Glacier Severtsova	52.2	28.4	125.0	119.4	45.7	187.4	161.5	89.1	57.4	67.7	0.0	32.3	719.6	246.5	966.1
	Oygaing	57.3	150.4	144.6	145.5	80.2	263.4	53.8	137.6	42.6	89.0	8.5	50.4	895.2	328.1	1223.3
Air temperature, °C	Pskem	9.6	2.7	- 3.3	- 8.5	- 8.7	1.6	8.3	13.8	18.3	19.8	20.5	15.0	0.2	17.5	7.4
	Khaydarkan	12.8	2.1	- 4.0	- 9.0	- 7.9	2.2	6.8	11.4	15.8	17.4	17.9	12.5	0.4	15.0	6.5
	Glacier Severtsova	2.6	- 1.6	- 6.1	-10.1	- 9.6	- 0.9	1.2	6.2	10.8	12.6	12.7	7.2	- 3.5	9.9	2.1
	Oygaing	2.7	- 5.3	-10.1	-13.6	-17.0	- 3.2	0.1	4.8	12.0	12.9	13.7	8.3	- 6.2	10.3	0.7
Degree days for temperature above 0, °C	Pskem	288	93	7	0	0	72	249	428	549	614	636	450	709	2677	3386
	Khaydarkan	234	68	3	0	0	98	204	353	474	539	555	375	607	2296	2903
	Glacier Severtsova	89	9	0	0	0	38	73	191	325	390	394	220	208	1520	1728
	Oygaing	103	0	0	0	0	3	18	148	358	198	425	248	124	1578	1702
Water equivalent of the snow cover, mm	Pskem		43	179	424	505	469									
	Glacier Severtsova			159	261		443	252								
	Oygaing			300	527		647	455	54							
Runoff, mm	Pskem, Mulala	47	40	36	31	28	55	112	220	390	357	194	104	349	1265	1614
	Oygaing, mouth	38	25	20	16	16	18	49						182		
	Maydantal, mouth	49	42	44	45	44	39	65	181	348	361	247	129	328	1266	1594
	Aksu, kishlak Khazarnau	16	13	13	11	10	72	119	115	163	172	107	40	254	597	851
	Sokh, kishlak Sarykanda	25	16	14	12	10	11	14	31	83	144	135	60	102	453	555
	Oygaing, above the mouth of r. Koksu	30	19	17	15	12	13	19	89	274	255	153	83	125	854	979
	Koksu, 1.3 km above the mouth	42	30	23	23	17	17	38	103	215	300	94	66	180	778	958

Table 9.5.2. USSR - Hydrometeorological data.

Part 28 of 46 Pamir Alai and Western Tien-Shan

1969 - 1970	Hydrological station m a.sl.	Month 10	11	12	1	2	3	4	5	6	7	8	9	Winter 10-4	Summer 5-9	Year 10-9
Precipitation, mm	Pskem	259.0	60.6	97.8	53.8	109.1	47.8	89.0	62.0	10.0	19.6	29.1	0.0	717.1	120.7	837.8
	Glacier Severtsova	127.6	71.3	86.8	93.8	64.0	42.2	150.8	51.3	9.9	102.4	24.5	19.3	636.5	207.4	843.9
	Oygaing	245.9	50.2	68.3	19.0	46.1	36.0	61.5	91.4	29.4	54.9	7.6	0.0	527.0	183.3	710.3
Air temperature, °C	Pskem	8.8	3.9	2.3	- 3.6	0.8	2.4	11.4	15.0	17.9	20.2	22.2	16.7	3.7	18.4	9.8
	Glacier Severtsova	3.6	- 2.4	- 3.6	- 7.8	- 5.3	- 4.2	3.7	8.4	11.6	12.3	14.4	8.7	- 2.3	11.1	3.3
	Oygaing	1.3	- 4.6	- 5.6	-11.1	- 6.7	- 6.5	3.6	8.7	11.4	12.9	15.7	10.7	- 4.2	11.9	2.5
Degree days for temperature above 0, °C	Pskem	273	128	74	4	48	100	341	464	536	626	685	502	968	2813	3781
	Glacier Severtsova	112	32	1	0	0	6	114	262	348	381	445	262	265	1698	1963
	Oygaing	74	13	0	0	0	0	109	268	340	399	486	321	196	1814	2010
Water equiavalent of the snow cover, mm	Pskem			55	94	106										
	Glacier Severtsova			115	108	152	293									
	Oygaing		23	103	114	185	213									
Runoff, mm	Pskem, mouth	85	71	53	42	35	37	88	144	201	165	126	82	411	718	1129
	Oygaing, mouth				25	21	21	47	104	195	173	134	80		686	
	Maydantal, mouth	85	65	46	35	29	29	73	169	292	246	189	97	362	993	1355
	Aksu, kishlak Khazarnau	24	18	15	13	14	20	47	76	108	72	53	29	151	338	489
	Sokh, Sarykanda	26	19	15	12	10	11	15	40	95	132	152	89	108	509	616
	Oygaing, above the mouth of r. Koksu	49	33	31	21	22	22	30	76	185	199	153	92	214	705	919
	Koksu, 1.3 km above the mouth	58	45	35	28	18	16	25	99	205	201	124	206	225	725	950

Table 9.5.2. USSR - Hydrometeorological data.

Part 29 of 46 Station Kara Batkak

Year		Month												Summer	Winter	Year
		10	11	12	1	2	3	4	5	6	7	8	9	5-9	9-4	10-9
1958-1959	precipitation, mm		37	19	28	29	40	21	82	110	198	301	24	715	174	889
	air temperature °C	- 4.1	- 9.2	- 9.4	-11.9	-12.7	- 8.7	- 1.2	- 1.7	3.5	5.2	5.4	6.0	3.7	- 8.2	- 3.2
	duration of sunshine hours	52.1	91.3	68.0	89.0	102.9	117.9	182.0	170.1	172.7	174.2	148.3	195.4	860.7	803.2	1663.9
	discharge m^3/sec	0.18	0.02						0.08	0.46	1.04	0.91	0.35	0.57		
1959-1960	precipitation, mm	58	63	22	14	26	46	15	192	89	130	16	95	522	244	766
	air temperature °C	0.4	- 8.7	-12.9	-11.5	- 7.6	-10.4	- 3.8	- 0.1	3.4	6.1	7.3	3.4	4.0	- 7.8	- 2.9
	degree days for temperature above 0, °C	205.4						11.8	28.4	280.3	1841.5	4781.7	1556.7	8490.6	217.2	8707.8
	duration of sunshine hours	138.7	97.8	70.2	102.5	90.7	110.2	166.8	121.7	183.5	194.0	223.5	158.6	881.3	776.9	1658.2
	discharge m^3/sec	0.15							0.11	0.35	0.67	1.07	0.22	0.48		
1960-1961	precipitation, mm	50	34	22	10	5	24	36	18	52	135	158	60	423	181	604
	air temperature °C	- 2.0	- 9.1	-10.3	-12.7	-15.1	- 7.2	- 2.9	2.9	4.5	6.5	5.9	3.2	4.6	- 85	- 3.0
	degree days for temperature above 0, °C	357					4.5	20.0	303.5	1032.2	3707.7	3332.4	1684.7	10060.5	60.2	10120.7
	duration of sunshine hours	140.4	111.2	88.3	114.9	122.9	148.8	163.2	200.8	170.6	212.3	184.8	162.8	931.3	889.7	1821.0
	discharge m^3/sec	0.13							0.35	0.38	1.06	0.72	0.43	0.59		
1961-1962	precipitation, mm	37	43	14	13	5	20	61	91	156	84	93	74	498	193	691
	air temperature °C	- 4.8	- 8.9	-11.3	-13.9	- 9.4	- 5.3	- 3.2	1.3	3.3	6.9	5.9	1.1	3.7	- 8.1	- 3.2
	degree days for temperature above 0, °C	137					21.2	96.6	291.7	896.6	3996.9	4013.1	233.4	9431.7	131.5	9563.2
	duration of sunshine hours	138.7	130.7	106.4	115.5	139.9	156.3	149.8	124.7	171.7	164.9	176.0	136.1	773.4	937.3	1710.7
	discharge m^3/sec	0.12							0.15	0.30	0.87	0.72	0.18	0.44		

Table 9.5.2. USSR - Hydrometeorological data.

Part 30 of 46 Station Kara Batkak

Year		Month												Summer	Winter	Year
		10	11	12	1	2	3	4	5	6	7	8	9	5-9	9-4	10-9
1962-1963	precipitation, mm	32	54	12	2	15	32	82	135	109	129	90	49	512	229	741
	air temperature °C	- 2.7	-10.6	- 9.8	- 9.8	- 4.9	- 6.8	0.3	2.6	4.9	6.2	5.7	3.6	4.6	- 6.3	- 1.8
	degree days for temperature above 0, °C	14.4					5.2	3.7	205.5	502.3	3611.5	3298.5	1500	7768.2	23.3	7791.5
	duration of sunshine hours	152.1	98.4	101.5	138.7	140.5	130.3	160.7	151.4	175.2	175.6	186.9	176.3	865.4	922.2	1787.6
	discharge m^3/sec	0.11							0.13	0.33	0.83	0.71	0.23	0.45		
1963-1964	precipitation, mm	52	47	14	11	23	47	130	114	142	209	134	107	707	324	1031
	air temperature °C	- 1.7	- 6.9	- 8.0	-14.3	-13.9	- 8.9	- 3.8	- 3.2	1.9	4.8	5.1	0.3	1.8	- 8.2	- 4.0
	degree days for temperature above 0, °C	27.3						27.2	10.8	236.0	2145.8	3832.3	126.6	6351.5	54.5	6406.0
	duration of sunshine hours	133.7	117.5	90.3	104.1	97.7	142.3	113.7	147.7	143.5	145.7	186.6	152.6	776.1	799.3	1575.4
	discharge m^3/sec	0.06							0.09	0.46	0.79	0.81	0.37	0.50		
1964-1965	precipitation, mm	10	14	11	8	9	30	57	90	122	257	103	90	662	139	801
	air temperature °C	- 5.5	- 7.3	-11.5	- 9.9	-15.4	- 7.4	- 2.1	1.2	2.2	5.9	5.2	2.4	3.4	- 8.4	- 3.5
	degree days for temperature above 0, °C	19.0					16.3	20.1	225.6	276.3	3091.3	3876.5	280.6	7750.3	55.4	7805.7
	duration of sunshine hours	176.1	136.4	84.8	104.1	121.7	131.4	144.8	150.2	182.2	169.4	176.6	142.6	821.0	899.3	1720.3
	discharge m^3/sec								0.13	0.31	0.69	0.77	0.33	0.45		
1965-1966	precipitation, mm	53	44	18	15	26	41	105	100	91	195	135	82	603	302	905
	air temperature °C	- 1.5	- 6.4	-11.1	-10.5	- 9.9	- 8.3	- 3.8	0.7	5.0	4.8	5.3	2.4	3.6	- 7.3	- 2.8
	degree days for temperature above 0, °C	85.4						6.6		520.6	566.5	635.9	310.0	2033.0	92.0	2125.0
	duration of sunshine hours	146.3	111.5	103.8	96.1	123.0	93.9	149.1	168.8	183.1	179.2	165.5	139.3	835.9	823.7	1659.6
	discharge m^3/sec								0.03	0.43	0.89	1.06	0.24	0.53		

Table 9.5.2. USSR - Hydrometeorological data.

Part 31 of 46 Station Kara Batkak

Year		Month 10	11	12	1	2	3	4	5	6	7	8	9	Summer 5-9	Winter 9-4	Year 10-9
1966-1967	precipitation, mm	42	47	39	14	17	31	36	125	177	140	116	51	609	226	835
	air temperature °C	- 5.4	- 7.1	-12.9	-15.3	-13.1	- 7.4	- 3.3	1.1	3.0	3.9	4.0	2.9	3.0	- 9.2	- 4.1
	degree days for temperature above 0, °C	4.1						117.4	81.6	362.2	478.0	523.4	395.3	1840.5	121.5	1962.0
	duration of sunshine hours		127.7	72.1	124.5	124.1	155.3	163.5	155.0	146.9	198.1	190.2	152.4	842.6	767.2	1609.8
	discharge m^3/sec								0.10	0.45	0.73	0.68	0.29	0.45		
1967-1968	precipitation, mm	57	48	9	13	12	35	60	115	147	102	70	72	506	234	740
	air temperature °C	- 4.4	-11.4	- 9.2	-12.5	-14.0	- 7.5	- 4.2	0.1	3.3	6.0	4.5	1.7	3.1	- 9.0	- 4.0
	degree days for temperature above 0, °C	42.8						103.1	154.1	401.6	748.7	467.8	307.4	2079.6	145.9	2225.5
	duration of sunshine hours	148.7	121.6	105.0	118.0	135.5	111.3	125.4	134.1	167.6	204.4	194.8	173.8	874.7	865.5	1740.2
	discharge m^3/sec								0.12	0.43	1.08	0.99	0.24	0.57		
1968-1969	precipitation, mm	35	24	16	29	24	39	92	138	111	204	112	91	656	259	915
	air temperature °C	- 3.5	- 8.2	-12.1	-15.3	-15.8	- 5.3	- 3.1	0.8	2.1	5.1	5.2	0.8	2.8	- 9.0	- 4.0
	degree days for temperature above 0, °C	23.3					0.5	78.2	207.9	340.5	630.5	660.3	241.4	2080.6	102.0	2182.6
	duration of sunshine hours	151.4	118.3	73.7	56.3	90.9	96.0	133.7	192.6	155.7	184.4	212.1	150.6	895.4	720.3	1615.7
	discharge m^3/sec								0.14	0.24	0.65	0.67	0.14	0.37		
1969-1970	precipitation, mm	47	24	28	16	28	36	117	89	66	243	98	63	559	296	855
	air temperature °C	- 2.1	- 7.8	-10.2	-14.2	-13.6	-12.2	- 2.5	0.5	2.6	4.2	5.3	2.6	3.0	- 9.0	- 3.9
	degree days for temperature above 0, °C	52.7						27.6	161.2	369.8	527.9	657.4	407.0	2123.3	80.3	2203.6
	duration of sunshine hours	150.0	124.9	92.1	119.0	117.8	160.2	145.7	207.1	200.5	166.9	201.5	167.9	943.9	909.7	1853.6
	discharge m^3/sec								0.12	0.29	0.50	0.69	0.29	0.38		

Table 9.5.2. USSR - Hydrometeorological data.

Part 32 of 46 Zailiyskiy Alatau, Tien-Shan

mean values	Name of station m a.sl.	Month 10	11	12	1	2	3	4	5	6	7	8	9	Winter 10-4	Summer 5-9	Year 10-9
Precipitation, mm	1. Mynzhilki, 3013.1	46	50	34	28	33	62	103	162	154	141	95	51	356	603	959
	2. Verkhniy Gorel'nik, 2268.0	57	63	39	31	38	84	144	190	149	97	67	46	456	549	1005
	3. B.Alamatinskoye ozero, 2511.7	44	47	34	25	32	67	106	151	129	109	72	48	355	509	864
	4. Vorota, 2473.0	57	58	36	30	36	81	137	189	152	99	66	45	435	551	986
	5. Dom otd. "Medeo", 1605.0	56	57	37	29	36	77	132	183	147	95	64	45	424	534	958
Air temperature, °C	1. Mynzhilki	- 1.8	- 6.9	-10.2	-12.7	-11.4	- 7.3	- 2.6	1.5	5.0	7.4	7.5	3.1	- 7.6	4.9	- 2.3
	2. Verkhniy Gorel'nik	- 7.8	- 6.7	- 2.5	- 7.8	- 6.7	- 2.5	2.3	7.0	10.6	13.3	12.9	8.4	- 4.5	10.4	1.7
	3. B.Almatinskoye ozero	1.9	- 3.4	- 6.9	- 9.5	- 8.6	- 3.7	0.7	5.3	9.0	11.4	10.9	6.6	- 4.2	8.6	1.1
Degree days for temperature above 0, °C	1. Mynzhilki	28.2	1.0	0	0	0.8	1.9	20.4	72.6	150.7	231.1	215.5	111.8	52.3	781.7	834.0
	2. Verkhniy Gorel'nik	127.2	32.6	7.9	5.0	8.9	39.5	101.4	221.1	321.6	416.0	392.0	259.8	322.5	1610.5	1933.0
	3. B.Almatinskoye ozero	83.6	13.4	2.4	0.4	1.9	15.5	67.0	172.0	273.2	355.1	332.4	203.5	184.2	1336.2	1520.4
Duration of sunshine, hours	1. B.Almatinskoye ozero	164	114	109	112	127	150	168	190	220	251	241	197	944	1099	2043
	2. Almatinsk. selestok, 1711	107	72	68	69	85	109	127	161	209	232	197	160	637	959	1596
Water equivalent of the snow cover, mm	1. Mynzhilki	28	85	103	124	154	192	220	74							
	2. Verkhniy Gorel'nik		54	86	100	126	147	32								
	3. B.Almatinskoye ozero		57	79	90	109	141	62								
Runoff, mm	M.Almatinka - above the moth of r. Sarysay	52	40	39	31	26	24	24	47	101	171	173	88	236	580	816
	M.Almatinka - Alama-Ata	40	34	31	28	24	26	36	69	81	95	91	55	219	391	610
	B.Almatinka-above the lake	52	38	32	27	21	20	20	38	92	148	144	84	210	506	716
	Middle Talgar-alp. camp	74	48	38	36	25	24	25	51	108	182	256	120	270	717	987
	Talgar - g.Talgar	51	37	31	26	22	23	26	53	95	144	156	88	216	536	752
	Chilik - s.Malybay	16	11	9	8	7	7	8	17	32	47	48	26	66	170	236

Table 9.5.2. USSR - Hydrometeorological data.

Part 33 of 46 Zailiyskiy Alatau, Tien-Shan

1959 - 1960	Name of station m a.sl.	10	11	12	1	2	3	4	5	6	7	8	9	Winter 10-4	Summer 5-9	Year 10-9
Precipitation, mm	1. Mynzhilki	59.7	39.8	25.0	13.9	40.0	35.7	45.0	233.2	169.3	186.6	22.4	58.8	859.1	670.3	1529.4
	2. Verkhniy Gorel'nik	80.6	66.9	30.6	27.6	55.7	53.8	70.6	284.3	141.8	179.7	24.5	31.8	385.8	662.1	1047.9
	3. B.Almatinskoye ozero	71.3	54.0	29.1	22.1	69.8	44.8	52.3	220.8	145.8	142.3	17.3	49.4	343.4	575.6	919.0
	4. Vorota	85.0	60.2	35.6	28.9	66.9	48.2	61.4	275.4	164.4	183.3	27.3	45.5	386.2	695.7	1081.9
	5. Dom otdykha "Medeo"	79.8	65.5	32.9	32.4	32.0	51.3	88.4	284.0	101.5	208.5	8.0	28.0	382.3	630.0	1012.3
Air temperature, °C	1. Mynzhilki	- 0.5	- 8.5	-11.4	-10.5	- 7.1	- 9.3	- 4.2	- 1.1	4.8	6.4	7.8	2.3	- 7.4	4.0	- 2.6
	2. Verkhniy Gorel'nik	4.4	- 3.6	- 6.8	- 5.0	- 2.0	- 5.3	0.8	3.6	11.2	12.0	13.0	7.6	- 2.5	9.5	2.5
	3. B.Almatinskoye ozero	2.6	- 5.3	- 8.7	- 7.2	- 4.1	- 6.6	- 0.8	2.4	9.5	10.4	11.5	5.9	- 4.3	7.9	0.8
Degree days for temperature above 0, °C	1. Mynzhilki	46.2	0	0	0	0	0	19.7	31.2	147.9	194.3	240.0	94.7	65.9	708.1	774.0
	2. Verkhniy Gorel'nik	148.3	7.7	0	13.6	14.1	16.3	101.1	118.7	336.1	372.2	401.6	235.2	301.1	1463.8	1764.9
	3. B.Almatinskoye ozero	99.7	4.3	16.2	2.0	4.0	6.8	70.3	88.8	285.0	322.9	356.2	178.9	203.3	1231.8	1435.1
Duration of sunshine, hours	1. B.Almatinskoye ozero	142.3	99.0	85.6	92.0	107.8	105.9	222.6	126.9	235.4	232.6	282.8	155.8	855.2	1033.5	1888.7
Water equivalent of the snow cover, mm	1. Mynzhilki	54	134	162	156	216	305	229	287							
	2. Verkhniy Gorel'nik		61	86	110	158	196									
	3. B.Almatinskoye ozero		54	84	101	124	170									
Runoff, mm	M.Almatinka - above the moth of r. Sarysay	88	71	46	44	33	27	26	39	130	300	215	161	335	845	1180
	M.Almatinka - Alma-Ata	46	46	43	33	32	31	36	72	113	126	107	58	267	476	743
	B.Almatinka - above the lake	62	44	38	31	25	24	21	26	95	206	176	94	245	597	842
	Talgar - g.Talgar	66	45	38	30	25	25	27	48	103	156	147	94	256	548	804
	Chilik - s.Malybay	18	12	10	9	7	6	8	12	32	48	44	22	70	158	228

Table 9.5.2. USSR - Hydrometeorological data.

Part 34 of 46 Zailiyskiy Alatau, Tien-Shan

1960 - 1961	Name of station m a.sl.	Month 10	11	12	1	2	3	4	5	6	7	8	9	Winter 10-4	Summer 5-9	Year 10-9
Precipitation, mm	1. Mynzhilki	36.2	27.1	16.2	35.2	18.1	53.9	69.0	114.9	130.7	139.0	112.1	70.2	255.7	566.9	822.6
	2. Verkhniy Gorel'nik	84.6	44.7	25.1	39.9	36.8	91.8	103.1	120.1	124.4	84.8	84.9	76.4	426.0	490.6	916.6
	3. B.Almatinskoye ozero	64.0	29.4	28.6	33.7	23.2	73.9	77.5	129.7	137.6	81.0	84.1	69.8	330.3	502.2	832.5
	4. Vorota	73.8	37.0	23.1	46.1	27.1	79.8	84.1	129.9	111.8	117.5	91.8	92.1	371.0	543.1	914.1
	5. Dom otdykha "Medeo"	97.0	40.9	27.7	44.1	41.5	93.3	97.8	100.3	130.8	52.8	66.8	81.9	442.3	432.6	874.9
Air temperature, °C	1. Mynzhilki	- 1.5	- 8.3	- 9.0	-11.4	-13.0	- 7.5	- 6.3	- 4.0	5.4	7.8	6.3	3.9	- 8.1	3.9	- 3.1
	2. Verkhniy Gorel'nik	3.8	- 3.7	- 3.9	- 6.0	- 8.8	- 2.9	4.0	10.0	11.4	13.6	12.4	9.0	- 2.5	11.3	3.2
	3. B.Almatinskoye ozero	2.2	- 5.4	- 6.0	- 8.0	-10.4	- 4.2	2.4	8.2	9.6	11.6	10.8	7.2	- 4.2	9.5	1.5
Degree days for temperature above 0, °C	1. Mynzhilki	18.9	1.2	0	0	0	1.7	30.6	131.9	164.4	242.1	212.2	130.5	52.4	881.1	933.5
	2. Verkhniy Gorel'nik	133.3	25.1	9.1	2.7	0	35.2	128.4	313.7	340.3	420.1	385.2	270.5	333.8	1729.8	2063.6
	3. B.Almatinskoye ozero	89.6	14.4	1.0	0	0	12.5	92.3	253.9	289.2	360.3	332.9	219.9	209.8	1456.2	1666.0
Duration of sunshine, hours	1. B.Almatinskoye ozero	170.6	119.4	95.8	108.1	123.2	184.8	175.2	222.3	232.5	243.0	-	-	977.1	-	-
Water equivalent of the snow cover, mm	1. Mynzhilki	49	67	89	101	108	152	121	34							
	2. Verkhniy Gorel'nik	40	71	97	126	167	151									
	3. B.Almatinskoye ozero	18	57	80	100	100	105									
Runoff, mm	M.Almatinka - above the moth of r. Sarysay	92	51	34	30	29	30	36	66	144	186	141	92	302	629	931
	M.Almatinka - Alma-Ata	39	31	32	31	24	28	42	71	69	83	76	52	227	351	578
	B.Almatinka - above the lake	53	40	35	30	24	25	26	73	110	157	152	106	233	598	831
	Talgar - g.Talgar	63	43	36	32	25	28	30	68	113	163	144	99	257	587	844
	Chilik - s.Malybay	14	9	8	7	6	6	7	17	24	40	41	25	57	147	204

Table 9.5.2. USSR - Hydrometeorological data.

Part 35 of 46 Zailiyskiy Alatau, Tien-Shan

1961 - 1962	Name of station m a.sl.	Month 10	11	12	1	2	3	4	5	6	7	8	9	Winter 10-4	Summer 5-9	Year 10-9
Precipitation, mm	1. Mynzhilki	38.6	20.2	18.9	11.0	9.7	21.2	88.2	128.9	145.9	117.2	38.8	65.4	207.8	496.2	704.0
	2. Verkhniy Gorel'nik	59.9	30.2	29.8	23.0	14.0	34.6	113.4	150.3	151.5	69.4	27.7	60.3	304.9	459.2	764.1
	3. B.Almatinskoye ozero	46.0	21.2	24.7	14.9	9.7	23.5	88.2	149.9	147.3	82.9	34.8	57.2	228.2	472.1	700.3
	4. Vorota	30.7	24.9	28.8	22.6	6.9	19.6	104.7	141.5	145.8	89.6	28.2	52.6	238.2	457.7	695.9
	5. Dom otdykha "Medeo"	63.9	38.3	25.1	24.2	14.9	28.6	95.0	156.9	126.5	46.8	21.5	67.1	290.0	418.8	708.8
Air temperature, ^{0}C	1. Mynzhilki	- 4.8	- 6.5	-10.1	-12.3	- 8.6	- 3.6	- 3.0	2.0	5.0	7.8	7.6	2.0	- 7.0	4.9	- 2.0
	2. Verkhniy Gorel'nik	- 0.2	- 1.2	- 4.0	- 6.9	- 3.1	2.0	1.8	7.2	10.8	14.0	13.8	7.2	- 1.7	10.6	3.4
	3. B.Almatinskoye ozero	- 1.7	- 1.9	- 6.2	- 9.0	- 5.4	0.0	0.3	5.6	9.2	12.2	12.0	5.6	- 3.4	8.9	1.7
Degree days for temperature above 0, ^{0}C	1. Mynzhilki	1.9	0	0	0	0	11.7	18.0	63.9	153.5	242.0	237.9	115.3	31.6	812.6	844.2
	2. Verkhniy Gorel'nik	54.1	39.4	0	2.8	24.0	87.0	82.4	225.0	324.9	432.4	426.6	315.9	289.7	1724.9	2014.5
	3. B.Almatinskoye ozero	31.1	15.1	0	0	2.7	41.5	57.9	172.3	276.9	378.8	371.6	191.6	148.3	1291.2	1439.5
Duration of sunshine, hours	1. Mynzhilki	124.7	114.4	111.4	109.8	134.1	161.4	161.9	131.5	191.8	210.4	181.1	131.3	917.7	846.1	1763.8
Water equivalent of the snow cover, mm	1. Mynzhilki	66	78	84	106	125	92	122					27			
	2. Verkhniy Gorel'nik		32	28	54	74	57									
	3. B.Almatinskoye ozero	67	22	41	67	78	48									
Runoff, mm	M.Almatinka - above the moth of r. Sarysay	67	40	39	39	32	34	30	53	112	147	271	123	281	706	987
	M.Almatinka - Alma-Ata s=117 km^2	37	35	36	37	30	26	27	45	49	62	100	54	228	310	538
	B.Almatinka - above the lake, s=70.4 km^2	52	38	32	26	19	19	20	38	88	148	173	82	206	529	735
	Middle Talgar-alp. camp s=50.4 km^2										173	365	147			
	Talgar - g.Talgar s=431 km^2	56	33	29	24	19	21	19	40	94	137	186	94	201	551	752
	Chilik - s.Malybay s=4500 km^2	14	9	9	7	6	6	6	18	33	40	48	25	56	164	220

Table 9.5.2. USSR - Hydrometeorological data.

Part 36 of 46 Zailiyskiy Alatau, Tien-Shan

1962 - 1963	Name of station m a.sl.	Month 10	11	12	1	2	3	4	5	6	7	8	9	Winter 10-4	Summer 5-9	Year 10-9
Precipitation, mm	1. Mynzhilki	28.7	81.6	3.2	0.5	15.6	48.4	180.5	157.3	159.7	138.0	93.7	15.2	358.5	563.9	922.4
	2. Verkhniy Gorel'nik	31.6	80.2	11.6	5.0	25.3	83.9	247.5	181.6	132.0	109.0	74.6	25.6	485.1	522.8	1007.9
	3. B.Almatinskoye ozero	31.7	85.8	5.5	1.5	15.3	54.6	212.9	154.7	145.1	106.0	68.7	19.6	407.3	493.6	900.9
	4. Vorota	26.2	64.7	4.4	3.3	16.2	64.8	233.7	169.5	157.3	123.7	76.3	28.9	413.3	555.7	969.0
	5. Dom otdykha "Medeo"	44.9	69.8	25.0	13.1	34.0	92.3	225.0	175.5	107.5	59.7	72.4	19.9	504.1	435.0	939.0
Air temperature, °C	1. Mynzhilki	- 1.7	-10.9	- 8.7	- 8.2	- 7.8	- 5.2	- 2.3	2.0	5.7	6.8	6.1	3.0	- 6.4	4.7	- 1.8
	2. Verkhniy Gorel'nik	2.6	- 5.9	- 3.7	- 2.0	- 2.4	- 0.4	2.0	7.4	12.0	12.9	11.8	7.6	- 1.4	10.3	3.5
	3. B.Almatinskoye ozero	1.4	- 7.6	- 5.5	- 4.3	- 4.4	- 2.0	0.8	6.0	10.1	11.1	10.0	6.3	- 3.1	8.7	1.8
Degree days for temperature above 0, °C	1. Mynzhilki	18.6	0	0	0	0	5.6	7.9	72.9	172.4	210.8	188.3	93.0	32.1	737.4	769.5
	2. Verkhniy Gorel'nik	103.2	15.3	10.5	23.6	23.5	63.3	84.0	231.2	359.6	400.9	364.1	228.3	323.4	1584.1	1907.5
	3. B.Almatinskoye ozero	69.5	3.3	2.0	3.3	9.1	35.4	60.1	197.6	303.5	343.3	309.6	189.4	182.7	1145.8	1328.5
Duration of sunshine, hours	1. Mynzhilki	137.8	93.6	103.4	134.4	127.0	126.2	151.2	147.0	186.5	183.2	183.5	171.5	873.6	871.7	1745.3
Water equivalent of the snow cover, mm	1. Mynzhilki	10	122	127	97	108	139	313	128				9			
	2. Verkhniy Gorel'nik		10	68	69	44	35	32								
	3. B.Almatinskoye ozero		90	91	72	55	68									
Runoff, mm	M.Almatinka - above the moth of r. Sarysay	57	38	38	36	28	30	31	59	168	186	132	81	258	626	884
	M.Almatinka - Alma-Ata	43	41	37	29	23	26	51	75	119	116	93	69	250	472	722
	B.Almatinka - above the lake	46	31	20	16	13	13	11	31	139	152	94	55	150	471	621
	Middle Talgar-alp. camp	75	52	41	36	30	29	28	47	120	192	190	122	291	671	962
	Talgar - g.Talgar	42	30	27	23	19	21	26	49	125	159	125	81	188	539	727
	Chilik - s.Malybay	14	10	9	8	7	8	8	14	33	39	36	22	64	144	208

Table 9.5.2. USSR - Hydrometeorological data.

Part 37 of 46 Zailiyskiy Alatau, Tien-Shan

1963 - 1964	Name of station m a.sl.	Month 10	11	12	1	2	3	4	5	6	7	8	9	Winter 10-4	Summer 5-9	Year 10-9
Precipitation, mm	1. Mynzhilki	82.6	66.5	15.3	3.9	35.2	63.6	119.9	91.6	12.7	231.1	59.2	58.8	387.0	453.4	840.4
	2. Verkhniy Gorel'nik	94.0	90.4	21.0	21.7	57.9	112.1	169.8	115.6	113.1	119.4	54.3	41.5	566.9	443.9	1010.8
	3. B.Almatinskcye ozero	96.0	71.9	17.5	11.5	62.4	72.9	151.3	94.6	117.0	192.0	41.9	42.4	483.5	487.9	971.4
	4. Vorota	88.4	80.2	17.4	16.3	51.8	91.5	146.8	105.6	131.5	186.0	50.3	59.8	492.4	533.2	1025.6
	5. Dom otdykha "Medeo"	114.0	90.8	29.5	32.1	54.1	141.7	179.5	137.9	129.2	125.0	37.7	29.4	641.7	459.2	1100.9
Air temperature, °C	1. Mynzhilki	- 1.1	- 5.4	- 7.4	-14.4	-12.0	- 5.8	- 2.5	- 0.4	5.0	6.5	6.9	2.2	- 6.9	4.0	- 2.4
	2. Verkhniy Gorel'nik	4.2	- 0.4	- 2.7	- 9.8	- 6.8	- 1.1	2.2	5.2	10.8	12.4	12.7	7.4	- 2.1	9.7	2.8
	3. B.Almatinskcye ozero	2.4	- 2.2	- 4.7	-11.7	- 8.9	- 2.8	0.5	3.6	9.1	10.5	11.0	5.6	- 3.9	8.0	1.0
Degree days for temperature above 0, °C	3. B.Almatinskcye ozero	93.2	11.3	7.1	0	0	32.5	47.2	124.8	273.1	326.5	339.6	170.6	191.3	1234.6	1425.9
	1. Mynzhilki	14.2	0.2	0	0	0	7.4	13.2	35.5	149.0	201.1	213.9	73.3	35.0	672.8	707.8
	2. Verkhniy Gorel'nik	150.2	43.8	22.6	0	0.6	56.6	80.3	171.6	310.7	382.2	393.4	218.0	354.1	1475.9	1830.0
Duration of sunshine, hours	1. Mynzhilki	130.9	96.8	81.5	103.4	104.9	124.2	119.6	156.3	162.0	140.9	184.8	157.2	761.3	801.2	1562.5
Water equivalent of the snow cover, mm	1. Mynzhilki	73	145	165	174	252	312	380	164							
	2. Verkhniy Gorel'nik		64	78	97	136	120	40								
	3. B.Almatinskcye ozero	12	77	83	94	124	132	93								
Runoff, mm	M.Almatinka - above the moth of r. Sarysay	69	35	31	31	27	24	24	47	101	171	173	88	241	580	821
	M.Almatinka - Alama-Atà	50	39	38	33	22	34	52	65	80	109	85	58	268	397	665
	B.Almatinka - above the lake	38	26	22	20	18	16	15	24	88	135	125	74	155	446	601
	Middle Talgar-alp. camp	81	52	32	34	21	20	22	55	97	180	213	90	262	635	897
	Talgar - g.Talgar	56	39	32	26	22	23	28	47	96	160	172	90	226	565	791
	Chilik - s.Malybay	15	11	9	8	6	8	9	14	33	44	44	24	66	159	225

Table 9.5.2. USSR - Hydrometeorological data.

Part 38 of 46 Zailiyskiy Alatau, Tien-Shan

1964 - 1965	Name of stattion m a.sl.	Month 10	11	12	1	2	3	4	5	6	7	8	9	Winter 10-4	Summer 5-9	Year 10-9
Precipitation, mm	1. Mynzhilki	12.3	16.6	18.5	13.2	16.2	35.5	97.0	120.5	120.6	123.7	77.0	58.8	209.7	500.6	810.3
	2. Verkhniy Gorel'nik	26.6	42.5	34.2	16.9	31.1	59.1	123.1	143.1	101.6	56.3	69.4	55.5	333.5	425.9	759.4
	3. B.Almatinskoye ozero	13.4	18.9	22.2	13.9	19.9	42.6	92.7	124.3	103.5	88.1	88.0	54.6	223.6	458.5	682.1
	4. Vorota	18.8	30.3	32.0	13.5	28.1	52.1	105.7	132.6	104.7	93.2	76.3	65.5	280.5	472.3	752.8
	5. Dom otdykha "Medeo"	27.0	53.6	34.6	17.4	42.1	69.8	128.8	162.1	67.2	46.1	65.7	53.0	373.3	394.1	767.4
Air temperature, ^{o}C	1. Mynzhilki	- 3.6	- 5.8	- 9.8	- 2.8	-10.1	-12.1	- 8.1	- 1.8	2.0	4.8	8.2	6.6	- 7.5	4.0	- 2.7
	2. Verkhniy Gorel'nik	1.2	- 0.6	- 5.7	- 4.9	- 6.9	- 3.8	3.3	8.0	11.0	14.5	12.4	7.8	- 2.5	10.7	3.0
	3. B.Almatinskoye ozero	- 0.5	- 2.5	- 7.6	- 7.1	- 8.9	- 5.4	1.6	6.2	9.0	12.9	10.6	6.1	- 4.3	9.0	1.2
Degree days for temperature above 0, ^{o}C	1. Mynzhilki	34.4	0.2	1.2	0	0	4.8	13.0	83.3	145.1	247.7	205.2	96.5	53.6	777.8	831.4
	2. Verkhniy Gorel'nik	77.5	54.3	5.2	12.1	0.5	19.0	103.7	247.9	330.7	451.1	384.1	237.6	272.3	1651.4	1923.7
	3. B.Almatinskoye ozero	47.0	27.1	0.4	1.0	0	7.9	62.9	194.6	270.8	398.3	329.1	188.7	146.3	1381.5	1527.8
Duration of sunshine, hours	1. Mynzhilki	164.0	116.4	103.2	112.0	116.5	141.8	129.9	155.8	154.5	172.5	181.4	158.6	883.8	822.8	1707.6
Water equivalent of the snow cover, mm	1. Mynzhilki		10	29	34	53	78	80								
	2. Verkhniy Gorel'nik		31	58	72	91	88	16								
	3.Almatinskoye ozero	4	9	42	48	59	80	50								
Runoff, mm	M.Almatinka - above the moth of r. Sarysay	52	40	39	40	35	37	38	59	82	155	166	81	281	543	824
	M.Almatinka - Alma-Ata	43	35	29	33	27	27	30	57	49	64	73	29	224	272	496
	B.Almatinka - above the lake	42	29	25	23	18	17	16	44	60	126	133	56	170	419	589
	Middle Talgar-alp. camp	67	42	42	39	35	33	26	51	62	124	184	99	284	520	804
	Talgar - g.Talgar	62	47	40	30	24	25	26	45	67	96	134	67	254	409	663
	Chilik - s.Malybay	16	12	9	8	7	7	10	22	28	40	48	21	69	159	228

Table 9.5.2. USSR - Hydrometeorological data.

Part 39 of 46 Zailiyskiy Alatau, Tien-Shan

1965 - 1966	Name of station m a.sl.	Month 10	11	12	1	2	3	4	5	6	7	8	9	Winter 10-4	Summer 5-9	Year 10-9
Precipitation, mm	1. Mynzhilki	69.6	88.4	8.4	50.5	50.4	131.1	67.1	164.7	184.4	205.8	128.7	33.1	465.5	716.7	1182.2
	2. Verkhniy Gorel'nik	88.3	118.0	8.6	81.1	64.1	193.5	97.6	187.0	168.8	155.5	67.8	27.7	651.2	606.8	1258.0
	3. B.Almatinskoye ozero	67.7	106.0	12.5	55.5	59.7	160.2	81.5	174.2	158.3	133.6	102.7	33.1	543.1	601.9	1145.0
	4. Vorota	58.9	123.6	9.2	74.8	67.2	165.0	77.0	213.2	167.7	104.3	95.3	32.5	565.7	613.0	1178.7
	5. Dom otdykha "Medeo"	88.6	117.3	4.9	66.8	55.8	205.8	106.1	186.0	148.3	71.1	52.7	22.5	645.3	480.6	1125.9
Air temperature, °C	1. Mynzhilki	- 0.8	- 5.6	- 9.1	- 9.7	- 8.1	- 7.8	- 3.6	- 0.1	6.5	7.5	7.8	4.2	- 6.4	5.2	- 1.5
	2. Verkhniy Gorel'nik	3.9	0.1	- 2.8	- 3.7	- 2.6	- 3.2	0.9	5.1	12.9	13.2	13.7	9.5	- 1.1	10.9	3.9
	3. B.Almatinskoye ozero	2.3	- 2.2	- 5.3	- 6.1	- 4.8	- 4.7	- 0.6	3.6	10.4	11.5	11.8	7.8	- 3.1	9.0	1.8
Degree days for temperature above 0, °C	1. Mynzhilki	38.2	0	0	0	0	0	7.9	41.3	197.7	233.3	242.5	131.1	46.1	845.9	892.0
	2. Verkhniy Gorel'nik	136.1	42.2	8.5	13.3	11.3	20.7	75.7	167.1	385.3	409.3	427.9	284.1	307.8	1673.7	1981.5
	3. B.Almatinskoye ozero	91.8	10.6	0	2.2	0	7.0	52.9	126.2	323.5	355.0	364.9	234.2	164.5	1403.8	1568.3
Duration of sunshine, hours	1. Mynzhilki	145.3	104.7	110.6	85.9	117.4	118.6	151.3	172.7	173.3	172.5	136.9	129.9	833.8	785.3	1619.1
Water equivalent of the snow cover, mm	1. Mynzhilki	64	180	148	188	219		261	282	40						
	2. Verkhniy Gorel'nik	18	88	89	145	192	335	46								
	3. B.Almatinskoye ozero	17				178	313	128								
Runoff, mm	M.Almatinka - above the moth of r. Sarysay	58	41	40	34	30	30	33	60	225	238	224	152	266	899	1165
	M.Almatinka - Alma-Ata	24	24	22	20	18	24	35	76	135	109	108	74	167	502	669
	B.Almatinka - above the lake	41	32	25	22	16	15	14	22	166	162	169	95	165	614	779
	Middle Talgar-alp. camp	76	53	29	30	26	28	27	43	151	217	219	139	269	769	1038
	Talgar - g.Talgar	40	30	28	26	24	25	26	48	146	134	147	86	199	561	760
	Chilik - s.Malybay	14	10	8	8	6	7	9	14	54	45	52	30	62	195	257

Table 9.5.2. USSR - Hydrometeorological data.

Part 40 of 46 Zailiyskiy Alatau, Tien-Shan

1966 - 1967	Name of station m a.sl.	Month 10	11	12	1	2	3	4	5	6	7	8	9	Winter 10-4	Summer 5-9	Year 10-9
Precipitation, mm	1. Mynzhilki		37.9	100.5	23.5	15.5	18.9	90.0	124.2	195.7	149.5	114.5	35.6		619.5	
	2. Verkhniy Gorel'nik		49.4	155.4	37.9	50.7	24.3		118.5	234.0	145.4	118.2	35.5		651.6	
	3. B.Almatinskoye ozero	79.7	37.1	129.0	28.8	22.3	13.4	76.4	111.8	186.8	114.2	88.2	35.8	386.7	536.8	923.5
	4. Vorota			149.9	31.9	34.3	25.3		116.0	210.8	131.5	119.7	34.8		612.8	
	5. Dom otdykha "Medeo"		46.6	120.8	37.7	43.7	38.2		130.2	210.8	70.7	83.5	37.2		532.4	
Air temperature, °C	1. Mynzhilki		- 8.6	-11.1	-12.6	-11.0	- 6.4	- 1.8	1.5	5.3	7.0	6.3	3.2		4.7	
	2. Verkhniy Gorel'nik		- 2.9	- 5.8	- 6.5	- 6.1	- 0.8		6.4	10.6	12.7	11.3	7.9		9.8	
	3. B.Almatinskoye ozero	0.4	- 5.1	- 7.9	- 8.9	- 7.9	- 3.0	1.6	4.8	8.8	10.9	9.9	6.4	- 4.4	8.2	0.8
Degree days for temperature above 0, °C	1. Mynzhilki	5.7	0	0	0	0	1.1	50.6	71.0	159.4	217.0	194.9	106.2	57.4	748.5	805.9
	2. Verkhniy Gorel'nik		20.6	9.5	2.4	6.6	48.3		203.0	316.5	393.0	353.8	235.7		1502.0	
	3. B.Almatinskoye ozero	45.8	3.0	1.8	0	1.4	15.1	104.1	151.6	262.9	336.3	308.3	192.1	171.2	1251.2	1422.4
Duration of sunshine, hours	1. Mynzhilki		108.9	61.8	118.6	115.7	164.6	149.6	125.0	139.2	186.7	166.2	143.4		760.5	
Water equivalent of the snow cover, mm	1. Mynzhilki	32	76	224	211	238	240									
	2. Verkhniy Gorel'nik		27	172	197	211	130									
	3. B.Almatinskoye ozero	16	40	156	155	158	160									
Runoff, mm	M.Almatinka - above the moth of r. Sarysay	89	60	52	46	38	37	48	79	153	186	204	106	370	728	1098
	M.Almatinka - Alma-Ata	56	46	40	34	26	29	46	63	79	86	80	49	277	357	634
	B.Almatinka - above the lake	47	35	29	24	18	17	26	44	80	108	127	62	196	421	617
	Middle Talgar-alp. camp	75	48	40	32	27	29	39	68	118	176	138	109	290	609	899
	Talgar - g.Talgar	49	36	34	30	23	24	33	50	89	115	125	64	229	443	672
	Chilik - s.Malybay	17	12	10	9	8	9	12	17	34	34	38	22	77	145	222

Table 9.5.2. USSR - Hydrometeorological data.

Part 41 of 46 Zailiyskiy Alatau, Tien-Shan

1967 - 1968	Name of station m a.sl.	Month 10	11	12	1	2	3	4	5	6	7	8	9	Winter 10-4	Summer 5-9	Year 10-9
Precipitation, mm	1. Mynzhilki	37.2	17.6	2.7	9.1	15.3	62.6	98.5	104.3	109.3	109.2	45.2	35.9	243.0	403.9	646.9
	2. Verkhniy Gorel'nik	54.4	25.3	7.0	21.9	13.6	106.0	134.9	83.4	84.1	86.4	28.6	45.1	363.1	327.6	690.7
	3. B.Almatinskoye ozero	38.5	21.0	1.5	15.2	12.0	65.1	97.7	93.5	88.4	63.5	33.3	32.5	251.0	311.2	562.2
	4. Vorota	45.0	27.7	3.1	16.1	14.1	78.0	119.3	110.8	96.2	93.8	35.1	45.8	303.3	381.7	685.0
	5. Dom otdykha "Medeo"	50.5	16.7	5.8	26.6	8.2	93.0	135.6	53.2	50.8	72.4	31.6	29.8	336.4	237.8	574.2
Air temperature, °C	1. Mynzhilki	- 2.6	- 8.6	- 7.2	-11.0	-12.4	- 5.7	- 3.4	1.8	5.5	7.6	7.1	3.6	- 7.3	5.1	- 2.1
	2. Verkhniy Gorel'nik	2.2	- 2.9	- 2.5	- 5.3	- 7.1	- 1.2	1.4	6.8	11.0	13.3	12.6	8.6	- 2.2	10.5	3.1
	3. B.Almatinskoye ozero	0.6	- 4.9	- 4.4	- 7.3	- 9.2	- 2.9	- 0.1	5.1	9.3	11.5	11.1	7.1	- 4.0	8.8	1.3
Degree days for temperature above 0, °C	1. Mynzhilki	15.2	0	0	0	0	3.9	23.8	77.2	166.6	235.2	225.0	116.7	42.9	820.7	863.6
	2. Verkhniy Gorel'nik	78.4	13.2	7.0	5.2	0	39.9	83.1	209.5	345.1	410.7	391.0	257.0	226.8	1613.3	1840.1
	3. B.Almatinskoye ozero	51.9	3.3	1.0	0	0	21.8	57.8	163.1	280.4	356.9	345.7	218.3	135.8	1364.4	1500.2
Duration of sunshine, hours	1. Mynzhilki	141.8	121.3	95.2	90.3	123.1	113.1	135.6	113.3	142.4	186.8	181.9	179.3	820.4	803.7	1624.1
	2. Verkhniy Gorel'nik	152.8	132.4	83.3	96.3	143.7	108.7	148.4	114.5	185.4	250.6	229.4	216.8	865.6	996.7	1862.3
Water equivalent of the snow cover, mm	1. Mynzhilki		30	18	13	46	79	140								
	2. Verkhniy Gorel'nik		14		22	26	44									
	3. B.Almatinskoye ozero	13	26	11	14	25	45	18								
Runoff, mm	M.Almatinka - above the moth of r. Sarysay	61	52	40	38	32	27	41	65	71	122	209	107	291	574	865
	M.Almatinka - Alma-Ata	42	34	28	24	20	28	42	51	48	60	65	33	218	257	475
	B.Almatinka - above the lake	43	34	28	24	20	18	18	49	75	125	146	51	185	446	631
	Middle Talgar-alp. camp	75	49	37												
	Talgar - g.Talgar	52	41	36	32	26	24	27	46	73	127	162	59	238	467	705
	Chilik - s.Malybay	16	11	10	10	7	8	10	23	32	41	46	18	72	160	232

Table 9.5.2. USSR - Hydrometeorological data.

Part 42 of 46 Zailiyskiy Alatau, Tien-Shan

1968 - 1969	Name of station m a.sl.	Month 10	11	12	1	2	3	4	5	6	7	8	9	Winter 10-4	Summer 5-9	Year 10-9
Precipitation, mm	1. Mynzhilki	54.6	74.4	21.1	70.3	21.1	112.9	97.1	129.1	185.3	130.6	102.6	85.5	451.5	633.1	1084.6
	2. Verkhniy Gorel'nik	58.9	107.3	23.7	83.6	25.2	149.3	132.3	216.1	131.6	26.2	80.0	66.5	580.3	520.4	1100.7
	3. B.Almatinskoye ozero	46.5	73.1	20.9	89.7	24.6	115.9	107.6	118.8	135.4	114.0	94.2	68.0	478.3	530.4	1008.7
	4. Vorota	56.8	99.6	21.5	72.6	21.6	133.9	124.1	165.5	165.3	121.4	75.0	57.6	530.1	584.8	1114.9
	5. Dom otdykha "Medeo"	53.0	101.5	21.6	62.2	23.9	151.6	155.0	381.2	85.9	86.2	77.3	57.7	569.2	688.3	1257.5
Air temperature, ^{0}C	1. Mynzhilki	- 2.8	- 6.9	-11.3	-15.8	-15.7	- 5.2	- 1.3	2.2	5.0	7.2	6.5	2.4	- 8.4	4.7	- 3.0
	2. Verkhniy Gorel'nik	2.0	- 1.4	- 7.0	-11.8	-11.3	- 1.1	3.5	7.2	10.7	13.0	11.4	7.7	- 3.9	10.0	1.9
	3. B.Almatinskoye ozero	0.6	- 3.6	- 8.7	-13.2	-12.8	- 2.6	1.8	5.6	9.0	11.2	10.1	6.0	- 5.5	8.4	0.2
Degree days for temperature above 0, ^{0}C	1. Mynzhilki	14.2	0	0	0	0	0.7	37.6	88.4	156.2	224.7	203.5	97.0	52.5	769.8	822.3
	2. Verkhniy Gorel'nik	104.1	18.3	0.1	0	0	47.5	114.1	232.1	319.4	402.5	353.8	229.4	284.1	1537.2	1821.3
	3. B.Almatinskoye ozero	62.4	0.8	0	0	0	23.3	77.2	175.4	269.0	345.9	312.6	187.2	163.7	1290.7	1453.8
Duartion of sunshine, hours	1. Mynzhilki	142.3	98.9	74.3	59.5	36.6	88.3	127.9	154.6	145.1	162.3	200.4	142.1	627.8	804.5	1432.3
	2. Verkhniy Gorel'nik	159.5	106.9	64.1	59.2	103.4	98.0	139.9	151.2	173.2	213.3	231.2	179.3	731.0	948.2	1679.2
Water equivalent of the snow cover, mm	1. Mynzhilki	34	144	166	250	247	384	278	118							
	2. Verkhniy Gorel'nik	14	87	110	187	208	257	52								
	3. Almatinskoye ozero	12	95	107	175	208	219	148								
Runoff, mm	M.Almatinka - above the moth of r. Sarysay	65	39	35	32	26	30	30	88	167	250	195	110	257	810	1067
	M.Almatinka - Alma-Ata	27	24	20	17	15	26	34	93	110	117	77	44	163	441	604
	Talgar - g.Talgar.	39	31	29	27	20	21	28	64	120	179	158	89	195	610	805
	B.Almatinka - above the lake	40	32	28	23	18	16	16	64	116	157	148	83	173	568	741
	Chilik - s.Malybay	13	10	9	7	6	9	10	25	37	44	40	23	64	169	233

Table 9.5.2. USSR - Hydrometeorological data.

Part 43 of 46 Zailiyskiy Alatau, Tien-Shan

1969 - 1970	Name of station m a.sl.	Month 10	11	12	1	2	3	4	5	6	7	8	9	Winter 10-4	Summer 5-9	Year 10-9
Precipitation, mm	1. Mynzhilki	81.8	15.0	26.7	18.5	13.4	38.2	110.3	93.8	91.8	159.6	182.6	24.8	303.9	552.6	856.5
	2. Verkhniy Gorel'nik	126.8	24.0	22.5	39.4	33.7	62.8	134.2	108.0	65.5	153.4	147.6	29.9	443.4	504.4	947.8
	3. B.Almatinskoye ozero	78.4	14.9	23.9	18.9	17.0	45.0	105.8	90.9	79.7	135.0	145.4	21.9	303.9	776.8	1080.7
	4. Vorota	107.5	19.6	23.6	29.7	19.4	47.6	134.5	107.2	87.5	147.3	188.2	27.5	381.9	557.7	939.6
	5. Dom otdykha "Medeo"	201.1	35.5	17.1	16.1	11.6	37.9	123.3	137.4	49.0	93.2	101.0	28.7	442.6	409.3	851.9
Air temperature, °C	1. Mynzhilki	- 1.6	- 5.5	- 7.7	-12.2	-10.4	- 9.6	- 0.8	2.3	4.7	6.9	8.0	- 3.9	- 6.8	3.6	- 2.4
	2. Verkhniy Gorel'nik	3.3	- 0.4	- 2.0	- 7.0	- 4.7	- 4.7	3.9	7.7	10.5	12.7	13.3	8.7	- 1.7	10.6	3.4
	3. B.Almatinskoye ozero	1.6	- 2.0	- 4.2	- 8.9	- 6.7	- 6.5	2.5	6.1	8.4	10.8	11.8	7.2	- 3.5	8.9	1.7
Degree days for temperature above 0, °C	1. Mynzhilki	22.9	6.7	0	0	0	0	24.5	88.2	154.6	213.2	248.1	126.3	54.1	827.4	881.5
	2. Verkhniy Gorel'nik	113.7	53.2	4.7	0	6.9	19.8	119.3	248.3	314.8	387.0	412.8	270.1	317.6	1632.0	1949.6
	3. B.Almatinskoye ozero	82.8	34.2	0	0	1.0	5.4	82.6	192.1	254.9	335.2	367.3	216.5	206.0	1366.0	1572.0
Duration of sunshine, hours	1. Mynzhilki	116.5	103.9	94.0	101.1	130.1	176.5	127.6	188.6	160.2	158.5	182.7	170.4	849.7	860.4	1710.1
	2. Verkhniy Gorel'nik	128.9	110.8	83.1	107.6	136.3	191.1	137.0	218.7	204.3	197.7	217.5	206.8	894.8	1045.0	1939.8
Water equivalent of the snow cover, mm	1. Mynzhilki	44	26	106	107	126	165	96								
	2. Verkhniy Gorel'nik	14		22	59	53	70	10								
	3. B.Almatinskoye ozero	14	22	38	47	51	80									
Runoff, mm	M.Almatinka - above the moth of r. Sarysay	74	61	45	42	32	36	34	54	75	145	149	135	324	558	882
	M.Almatinka - Alma-Ata	34	31	26	26	19	27	33	49	54	70	70	49	196	292	488
	B.Almatinka - above the lake	48	31	25	21	15	14	17	49	75	105	113	82	171	424	595
	Talgar - g.Talgar	54	37	34	31	25	26	29	55	84	124	116	88	236	467	703
	Chilik - s.Malybay	16	12	10	9	8	8	11	22	32	47	45	33	74	179	253

Table 9.5.2. USSR - Hydrometeorological data.

Part 44 of 46 Station Katon-Karagay (1081 m a.sl.)

Year		Month												Winter	Summer	Year
		10	11	12	1	2	3	4	5	6	7	8	9	10-4	5-9	10-9
1881 - 1960	precipitation, mm	35	27	22	16	12	14	26	56	63	67	58	36	152	280	432
	air temperature, °C	2.8	- 8.5	-13.6	-14.8	-12.5	- 6.1	3.2	10.2	15.1	17.2	15.2	10.6	- 7.1	13.7	0.6
	water equivalent of the snow cover, mm		26	37	47	52	42									
	runoff, mm	9	6	4	3	3	3	6	33	81	62	36	17	34	229	263
1960 - 1961	precipitation, mm	36	28	40	15	6	10	8	93	89	126	115	44	143	467	610
	air temperature, °C	0.5	- 9.6	-13.0	-14.7	-11.6	- 6.1	6.8	10.1	12.0	15.1	13.5	9.2	- 6.4	12.0	0.5
	degree days for temperature above 0, °C	91	4	0	0	0	18	203	312	361	468	418	276	316	1835	2151
	duration of sunshine	184	80	69	115	159	262	282	280	254	262	245	244	1151	1285	2436
	water equivalent of the snow cover, mm	18	23	57	65	64	90									
	runoff, mm	31	23	21	11	9	11	39	136	237	166	159	72	145	770	915
1961 - 1962	precipitation, mm	45	35	23	6	4	16	16	27	60	47	14	22	145	170	315
	air temperature, °C	- 3.6	- 5.6	-11.8	-13.0	-10.4	- 4.5	3.8	13.2	15.9	17.4	17.2	11.5	- 7.0	15.0	0.7
	degree days for temperature above 0, °C	22	0	0	0	0	7	136	410	478	541	535	344	165	2308	2473
	duration of sunshine	150	88	116	135	187	240	261	319	334	345	324	216	1177	1538	2715
	water equivalent of the snow cover, mm	31	48	59	65	62	42									
	runoff, mm	31	22	17	10	8	7	14	214	177	80	73	39	109	583	692
1962 - 1963	precipitation, mm	35	17	10	6	8	16	20	35	76	53	40	19	112	223	335
	air temperature, °C	2.2	-10.1	-10.1	- 8.8	- 7.2	0.1	- 1.8	9.8	14.5	17.6	15.6	7.8	- 5.1	13.1	0.7
	degree days for temperature above 0, °C	122	0	0	0	0	40	50	305	435	544	484	233	212	2001	2213
	duration of sunshine	176	129	85	161	171	151	276	283	322	338	306	238	1149	1487	2636
	water equivalent of the snow cover, mm	18	40	22	37	44										
	runoff, mm	18	10	8												

Table 9.5.2. USSR - Hydrometeorological data.

Part 45 of 46 Station Katon-Karagay (1081 m a.sl.)

Year		Month												Winter	Summer	Year
		10	11	12	1	2	3	4	5	6	7	8	9	10-4	5-9	10-9
1963 - 1964	precipitation, mm	45	38	13	3	1	24	34	45	52	100	32	27	158	256	414
	air temperature, °C	1.7	- 4.9	-10.5	- 6.6	-11.3	- 1.2	6.7	16.3	22.2	23.1	22.3	17.8	- 3.7	20.3	1.4
	degree days for temperature above 0, °C	102	15	0	0	0	2	79	286	461	524	458	290	198	2019	2217
	duration of sunshine	163	100	96	137	230	187	204	279	331	290	274	170	1117	1344	2461
	water equivalent of the snow cover, mm		18	21	19	19	25	4								
	runoff, mm															
1964 - 1965	precipitation, mm	44	20	40	18	3	10	19	26	55	21	74	32	154	208	362
	air temperature, °C	3.3	1.0	- 8.9	-12.9	-10.9	- 5.8	3.3	12.7	15.8	19.9	14.5	9.2	- 4.4	14.4	0.8
	degree days for temperature above 0, °C	52	15	0	0	0	6	111	394	473	616	450	279	184	2012	2196
	duration of sunshine	127	80	111	193	263	251	298	251	364	236	240	159	1323	1250	2573
	water equivalent of the snow cover, mm	20	28	46	62	73	22									
	runoff, mm	24	16	11	9	6	6	13	198	175	110	72	69	85	624	709
1965 - 1966	precipitation, mm	51	33	37	49	63	26	32	70	124	63	31	10	291	298	589
	air temperature, °C	2.3	- 5.0	-13.6	-14.1	-12.9	- 8.0	- 2.4	8.3	15.0	16.0	15.7	14.0	- 7.7	14.0	0.5
	degree days for temperature above 0, °C	101	6	0	0	0	0	45	258	448	495	488	419	152	2103	2255
	duration of sunshine	96	117	85	124	165	253	267	289	322	300	260	161	1107	1332	2439
	water equivalent of the snow cover, mm		23	49	84	149	156	23								
	runoff, mm	42	29	16	8	7	7	14	92	453	278	141	51	123	1015	1138

Table 9.5.2. USSR - Hydrometeorological data.

Part 46 of 46 Station Katon-Karagay (1081 m a.sl.)

Year		Month 10	11	12	1	2	3	4	5	6	7	8	9	Winter 10-4	Summer 5-9	Year 10-9
1966 - 1967	precipitation, mm	47	57	17	9	16	18	10	23	66	70	77	28	174	264	438
	air temperature, °C	0.9	-10.8	-19.5	-14.2	-12.2	- 6.0	4.3	12.9	14.8	15.4	12.3	8.0	- 8.4	12.6	0.4
	degree days for temperature above 0, °C	87	0	0	0	0	8	168	399	429	479	381	241	263	1929	2192
	duration of sunshine	76	116	156	148	261	322	300	321	195	209	253	174	1379	1152	2531
	water equivalent of the snow cover, mm		38	62	64	65	68									
	runoff, mm	31	18	10	8	10	6	20	153	125	133	103	69	103	583	686
1967 - 1968	precipitation, mm	41	11	1	6	5	33	20	20	16	16	27	52	117	399	516
	air temperature, °C	3.0	-10.4	- 9.2	-12.6	-12.9	- 1.6	0.9	11.9	15.6	16.2	15.0	5.8	- 6.1	12.9	0.6
	degree days for temperature above 0, °C	125	0	0	0	0	44	105	366	466	502	464	419	274	2217	2491
	duration of sunshine	168	144	135	192	140	270	337	325	278	320	206	209	1386	1338	2724
	water equivalent of the snow cover, mm		20	11	13	7	38									
	runoff, mm															
1968 - 1969	precipitation, mm	22	90	36	3	6	7	16	106	122	132	44	91	180	495	675
	air temperature, °C	1.9	- 9.5	-15.0	-23.0	-22.5	- 8.4	2.8	7.7	14.6	17.3	11.8	7.8	-10.5	11.8	0.1
	degree days for temperature above 0, °C	102	0	0	0	0	7	117	238	439	533	365	234	226	1809	2035
	duration of sunshine	103	54	129	156	216	232	232	267	260	262	196	114	1122	1099	2221
	water equivalent of the snow cover, mm		71	92	87	88	78	52								
	runoff, mm															
1969 - 1970	precipitation, mm	58	14	6	8	5	8	53	68	35	42	39	29	152	213	365
	air temperature, °C	1.1	- 6.1	-11.6	-12.8	-10.7	- 9.6	3.1	8.3	13.4	15.2	14.5	9.9	- 6.7	12.3	0.5
	degree days for temperature above 0, °C	82	24	0	0	0	13	125	258	402	470	450	296	244	1876	2120
	duration of sunshine	154	140	142	158	246	251	263	281	273	292	247	182	1354	1275	2629

REMARKS ON THE ANNEXED MAPS

1 Glacier Map of Southern Norway, Scale 1 : 500 000

We are indebted for this map to Dr. Gunnar Østrem, head of the Glaciology Section of the Hydrological Division, Norwegian Water Resources and Electricity Board, P.O. Box 5091, Oslo 3 Norway. The Map is included as a location map in the first volume of the "Norwegian Glacier Atlas", which in accordance with Resolution I-12 of the International Hydrological Decade contains the detailed inventory of glaciers in Southern Norway.

Reference: G. Østrem and T. Ziegler: "Atlas over breer i Sør-Norge" (Atlas of Glaciers in Southern Norway), Norges Vassdrags- og Elektrisitetsvesen, Publication No. 20 of the Hydrological Division, 1969.

G. Østrem describes this publication as follows:

The detailed inventory comprising glaciers in Southern Norway was initiated in January and completed in December 1969. In the Atlas the glaciers are organized in relation to selected drainage basins and listed in tables comprising information about glacier location, elevation, length, area, etc. An attempt has also been made to estimate their volumes so that a rough figure for the total ice mass can be obtained.

In addition to the lists which were prepared by means of computer techniques, maps are presented of each drainage basin, showing the glacier outline and existing water gauges. Finally a description is given of the various areas together with photographs of typical glacier types within these areas. The original Norwegian text has a complete English version, all captions and legends are also printed in English. The table heads are entirely in English.

In 1973 the glacier inventory for the whole of Scandinavia was completed by the issue of a second volume which contains two further general maps to a scale of 1 : 500 000. Reference: Norges Vassdrags- og Elektrisitetsvesen and Stockholm University, G. Østrem, N. Haaken sen., O. Melander, "Atlas over breer i Nord-Scandinavia" (Atlas of Glaciers in North Scandinavia), Publication No. 22 of the Hydrological Division, Publication No. 46 of the Naturgeografiska Institutionen, 1973.

2 Map of the Mattmark Glaciers, Scale 1 : 10 000

This map was used for obtaining the data given in Table 9.3.5. Particulars of surface and volume changes of the Mattmark Glaciers have previously been published in:

D. Lütschg — Ueber Niederschlag und Abfluss im Hochgebirge, Sonderdarstellung des Mattmarkgebietes, Schweiz. Wasserwirtschaftsverband, Zurich 1926

D. Lütschg-Lötscher Zum Wasserhaushalt des Schweizer Hochgebirges, Beiträge zur Geologie der Schweiz - Geotechnische Serie - Hydrologie, 4th instalment, I Vol., Part I, 4th Chapter, Kommissionsverlag Kümmerli & Frei, Bern 1944.

P. Kasser Fluctuations of Glaciers 1959-1965, IAHS, ICSI/Unesco, 1967. Table 16, Parts 1 to 15 (Periods 1932-1946, 1946-1956, 1932-1956)

P. Kasser Die Gletscher der Schweizer Alpen 1965-1966, 87th Report, 1967. Table 3b (periods 1932-1946, 1946-1956, 1932-1956)

P. Kasser Die Gletscher der Schweizer Alpen 1968-1969, 90th Report, 1970. Table 3b (period 1956-1967).

In the evaluation of the map, discrepancies became apparent in the glaciated areas in 1956 as compared with earlier planimetering results. These discrepancies were partly caused by the scatter of errors due to the methods used in preparing the basic information for the map and in the planimetric measurements, partly by small differences in the interpretation of the glacier boundaries. No corrections are necessary in the surface and volume changes published in "Fluctuations of Glaciers 1959/1965", Vol. 1, for the periods 1932-1946 and 1946-1956. Small changes are necessary, however, as a result of the new areas determined for the 1956 survey, in the areas of the survey of 1932, 1946 and 1956 and in the mean altitude changes of the glacier surface in the periods 1932-1946 and 1946-1956. The revised values are published in "Les variations des glaciers suisses en 1969-1970", 91st Report of the Glacier Commission of the SHSN , 1972.

The mean precipitation for each year, and for winter and summer, is also entered in the map for the eleven-year period 1956-1967. Because of the large scale, it was decided not to include curves of equal precipitation. The results of stake measurements are also entered on the map. They comprise both the horizontal components of the mean annual displacement of the stakes and the mean annual net balances. Evaluation of the mean annual net balances yielded the following mean altitude for the equilibrium line for the 1956-1967 period:

Glacier	Mean altitude of equilibrium line m a. sl.	Number of stakes
Kessjen	3030	2
Hohlaub	3190	3
Allalin	3190	7
Schwarzberg	3030	5
Tälliboden	2810	2
Ofental	2810	3

3, 4 and 5 Maps of the Vernagtferner

The five maps of the Vernagtferner were supplied free of charge by the Glaciology Commission of the Bayerische Akademie der Wissenschaften for "Fluctuations of Glaciers 1965-1970", Vol. 2. The PSFG wishes to express its special thanks for this generous gift.

The changes in the glaciated areas, the glacier volumes and the mean altitudes of the glacier surfaces for the periods 1889-1912, 1912-1938 and 1938-1969 were determined from Map 4, sheets 1, 2 and 3, and set out in Table 9.3.2, Part1, for steps of 50 m equidistance for both the Vernagtferner and the neighbouring Guslarferner.

O. Reinwarth describes these maps as follows:

Five maps of the Vernagtferner [1] (Oetztal Alps) to a scale 1 : 10 000 showing the glacier topography of 1969, the variation of the glacier within distinct time intervals and the bedrock topography of the Vernagtferner and the nearby Guslarferner.

3. The map of the Vernagtferner 1969, 1 : 10 000, represents as a topographical map the state of the glacier at the end of the budget year 1968/69, based on an aerial photographic flight of 4 October 1969. Beside an exact reproduction of the glacier area, a large part of the ice-free surroundings is shown in great detail. All elevations are given in metres, the contour interval is 10 m.

The map is provided with red overprint showing position and type of scientific installations established in this area by the Commission on Glaciology of the Bavarian Academy of Sciences, including glaciological facilities such as stakes for ablation measurements, fixing the sites of accumulation measurements and serving for flow determinations, meteorological and hydrological facilities (weather hut, precipitation gages and location of a proposed runoff station) and geodetic facilities (station points for surveying the stake positions and station points for the terrestrial photogrammetric survey that is repeated almost every year).

4. Three maps with former states of the Vernagtferner demonstrate the variations of Vernagtferner and Guslarferner between 1889 and 1969, based on four complete surveyings by S. Finsterwalder (1888/89), O. v. Gruber (1912), H. Schatz (1938) and the Glaciology Commission (1969), with three map sheets for the corresponding time intervals:

Sheet 1 : period 1889 - 1912

Sheet 2 : period 1912 - 1938

Sheet 3 : period 1938 - 1969.

In these maps the boundary and contour lines (vertical distance 50 meters) of each older stage is printed in black, those of the younger stages in bluish green. The area free of ice is represented by grey contour lines.

The change in area of the glacier and in ice thickness is illustrated by coloured bands. Where diminution in area and thickness has taken place the bands have a flat tint of bluish green colour,

---

[1] 'Ferner' is the local expression for a glacier

in case of increasing area and thickness a red one.

The coloured bands showing the change in area are printed in a light shade, those showing the change in thickness in a darker shade. This special designation of increase or decrease is restricted to the 100-meter contours to avoid superposition in the tongue region.

Finally the three sheets include the position of the glacier tongue for additional dates.

5. The map of the Vernagtferner with bedrock topography and geological survey shows in black the contour lines of the glacier bed and the ice-free surrounding area (vertical interval 20 meters).

The contour lines of the glacier bed are based on seismic refraction measurements, carried out by the Institute of Applied Geophysics of Munich University.

The refraction profiles and shot points are shown in a red overprint. The contour lines of the surface topography of the glacier with the same equidistance of 20 metres are printed in the map in blue, representing the state in October 1969 and thus allowing the determination of ice thickness. The ice-free area at that time was given a flat grey tint. In this area the geological formations are also designated by special signatures.

All five maps were produced at the Institute of Photogrammetry and Cartography of Munich Technical University on behalf of and in cooperation with the Glaciology Commission of the Bavarian Academy of Sciences. The cartographic work on themapswas supported by a grant from the German Research Association, Bonn. Printing of themapswas carried out by the Institute of Applied Geodesy, Frankfurt/Main. Our appreciative thanks are due to the Society of the Friends of the Bavarian Academy of Sciences for providing the financial support necessary for printing the maps.

Corrections to "Fluctuations of Glaciers, Vol, 1, 1959-1965"

Tables 8, 10 and 18

Explanatory remark concerning the surface S of the Limmerngletscher: The value of 3.29 km^2 reported in tables 8 and 10 includes the surfaces of the Limmerngletscher (2.565 km^2) and of the neighbouring Plattalvagletscher (0.722 km^2), as mentioned in table 18.

Table 10

For the year 1964/65 the mean annual net balance values $\bar{b}$ were reported wrong for two glaciers:

Woolsey Glacier:	$\bar{b}$ = - 597 mm of water equivalent, not - 60 mm
South Cascade Glacier:	$\bar{b}$ = - 240 mm of water equivalent, not -100 mm.

Table 15

Aletsch Glaciers: the lower pages of parts 1 of 9 and 2 of 9 should be interchanged.